Geological Society Special Publications
Series Editor A. J. FLEET

GEOLOGICAL SOCIETY SPECIAL PUBLICATION NO. 104

High Resolution Sequence Stratigraphy: Innovations and Applications

EDITED BY

JOHN A. HOWELL

Department of Earth Sciences, University of Liverpool,
Liverpool, UK

and

JOHN F. AITKEN

Geology & Cartography Division, Oxford Brookes University,
Oxford, UK

1996
Published by
The Geological Society
London

THE GEOLOGICAL SOCIETY

The Society was founded in 1807 as the Geological Society of London and is the oldest geological society in the world. It received its Royal Charter in 1825 for the purpose of 'investigating the mineral structure of the Earth'. The Society is Britain's national society for geology with a membership of around 8500. It has countrywide coverage and approximately 1500 members reside overseas. The Society is responsible for all aspects of the geological sciences including professional matters. The Society has its own publishing house, which produces the Society's international journals, books and maps, and which acts as the European distributor for publications of the American Association of Petroleum Geologists, SEPM and the Geological Society of America.

Fellowship is open to those holding a recognized honours degree in geology or cognate subject and who have at least two years' relevant postgraduate experience, or who have not less than six years' relevant experience in geology or a cognate subject. A Fellow who has not less than five years' relevant postgraduate experience in the practice of geology may apply for validation and, subject to approval, may be able to use the designatory letters C Geol (Chartered Geologist).

Further information about the Society is available from the Membership Manager, The Geological Society, Burlington House, Piccadilly, London W1V 0JU, UK. The Society is a Registered Charity, No. 210161.

Published by the Geological Society from:
The Geological Society Publishing House
Unit 7
Brassmill Enterprise Centre
Brassmill Lane
Bath BA1 3JN
UK
(*Orders:* Tel 01225 445046
Fax 01225 442836)

First published 1996; reprinted 1998

The publishers make no representation, express or implied, with regard to the accuracy of the information contained in this book and cannot accept any legal responsibility for any errors or omissions that may be made.

British Library Cataloguing in Publication Data
A catalogue record for this book is available from the British Library.
ISBN 1-897799-48-9

Distributors

USA
AAPG Bookstore
PO Box 979
Tulsa
OK 74101-0979
USA
(*Orders:* Tel. (918) 584-2555
Fax (918) 548-0469)

Australia
Australian Mineral Foundation
63 Conyngham Street
Glenside
South Australia 5065
Australia
(*Orders:* Tel. (08) 379-0444
Fax (08) 379-4634)

India
Affiliated East-West Press PVT Ltd
G-1/16 Ansari Road
New Delhi 110 002
India
(*Orders:* Tel. (11) 327-9113
Fax (11) 326-0538)

Japan
Kanda Book Trading Co.
Tanikawa Building
3-2 Kanda Surugadai
Chiyoda-Ku
Tokyo 101
Japan
(*Orders:* Tel. (03) 3255-3497
Fax (03) 3255-3495)

Typeset by Bath Typesetting Ltd, Bath, UK

Printed by The Alden Press, Osney Mead, Oxford, UK

Contents

High resolution sequence stratigraphy: innovations, applications and future prospects

JOHN F. AITKEN[1] & JOHN A. HOWELL[2]

[1]Geology & Cartography Division, Oxford Brookes University, Gipsy Lane Campus, Headington, Oxford OX3 0BP, UK

[2]STRAT Group, Department of Earth Sciences, University of Liverpool, PO Box 147, Brownlow St, Liverpool L69 3BX, UK

In recent years there has been a huge expansion in studies using high resolution sequence stratigraphic techniques. Sequence stratigraphy has evolved from the original concepts of seismic stratigraphy to a scale above seismic resolution. Concepts developed at outcrop are now being successfully applied to sub-surface data sets and are providing a greatly improved understanding of facies geometries and reservoir architecture. Sequence stratigraphy has developed into a powerful, predictive facies analysis tool for both the hydrocarbon industry and academic research. The papers in this volume illustrate the depth and breadth of current sequence stratigraphic research.

Although sequence stratigraphy has a long history (Sloss 1988 and papers cited therein), the interpretation of ancient sediments has traditionally relied on comparisons with modern depositional environments and transport processes within the context of facies models and Walther's Law (e.g. Walker 1984; Reading 1986). The more widespread application of a chronostratigraphic approach was initiated by the development of seismic stratigraphy (Vail *et al.* 1977*b*). Seismic stratigraphy is a tool applicable at hydrocarbon exploration scale whereas sequence stratigraphy uses a chronostratigraphic approach at hydrocarbon production scales to cores, wireline logs and outcrops, and brings the Vail concepts to the everyday working environment of most academic geologists.

Modern, high resolution, reservoir scale sequence stratigraphy was formalized by the publications of Jervey (1988), Posamentier & Vail (1988), Posamentier *et al.* (1988) and Van Wagoner *et al.* (1990). These publications have generated much discussion about the application of sequence stratigraphic concepts (e.g. Miall 1986, 1991, 1992; Summerhayes 1986; Boyd *et al.* 1988; Walker 1990; Shanley & McCabe 1994) which has led to modifications to the original models (e.g. Schlager 1991; Hunt & Tucker 1992, 1995; Posamentier & Allen 1993*a,b*, Helland-Hansen & Gjelberg 1994; Kolla *et al.* 1995) although the basic tenets remain unchanged. Despite the shortcomings of the original generalized models, sequence stratigraphy is now widely applied both within the hydrocarbon industry and by academic researchers and there has been a rapid expansion in publications dealing with the application of sequence stratigraphy at increasingly higher resolutions (e.g. Loucks & Sarg 1993, Posamentier *et al.* 1993; Dalrymple *et al.* 1994; Weimer & Posamentier 1994*a*). This has led to the geological reappraisal of some well-studied areas, such as the Upper Carboniferous in the UK (e.g. Maynard 1992; Martinsen 1993; Church & Gawthorpe 1994; Read 1994, 1995; Hampson 1995; Hampson *et al.*; this volume). A number of reviews addressing specific sequence stratigraphic issues have been published (Posamentier & James 1993; Posamentier & Weimer 1993, Shanley & McCabe 1994; Weimer & Posamentier 1994*b*; Zaitlin *et al.* 1994).

Perspectives

Sequence stratigraphy has two main aspects, firstly the construction of and correlation with global eustatic cycle charts (e.g. Vail *et al.* 1977*a*; Haq *et al.* 1988) and, secondly, lithology prediction within time equivalent successions. The first of these has been severely criticized on the basis that many of the assumptions underpinning the curves are invalid (Miall 1986, 1991, 1992; Summerhayes 1986; Burton *et al.* 1987; Underhill 1991). In particular, the precision of the palaeontological, palaeomagnetic and radiometric dating used to constrain the curves is insufficient to delimit the duration of the published third-order curves. Consequently, diachronous curves from different basins may be erroneously superimposed to produce meaningless global eustatic curves. Nowadays it is

From Howell, J. A. & Aitken, J. F. (eds), *High Resolution Sequence Stratigraphy: Innovations and Applications*, Geological Society Special Publication No. 104, pp. 1–9.

more common to use regional rather than global sea-level curves (e.g. Partington *et al.* 1993; Armentrout this volume). In this respect it is notable that within this volume only two of the papers mention the global eustatic curves. Uličný & Špičáková (this volume) correlate their parasequences in the Bohemian Basin to the Haq *et al.* (1988) curve, whilst Armentrout (this volume) argues that only by the careful documentation of regional cycles through high resolution biostratigraphy and chronostratigraphy can local events be distinguished from globally synchronous events to construct a global eustatic chart. Despite the shortcomings of the present global cycle charts, Posamentier & James (1993) and Posamentier & Weimer (1993) argue that the global cycle charts are of value in basins where there is little or no age information available.

The validity of the eustatic cycle charts, however, does not influence the validity of sequence stratigraphy as a lithological predictor and as a tool in unravelling basin-fill history. This use of sequence stratigraphy is more widespread and more valuable than the construction of eustatic charts and is dependent upon the concept of accommodation (Jervey 1988). It is applicable in a wide variety of tectonic and depositional settings as long as it is realized that the original models are generalizations and that local factors (e.g. sediment flux, basin physiography) must be accounted for before applying sequence stratigraphic concepts (Posamentier & Allen 1993*a*; Posamentier & James 1993; Posamentier & Weimer 1993; Weimer & Posamentier 1994*b*). There has been discussion (e.g. Walker 1990, 1992; Posamentier & James 1993) concerning the relative merits of allostratigraphy (NACSN 1983), genetic stratigraphic sequences (Galloway 1989) and Exxon-style sequence stratigraphy (Jervey 1988; Posamentier & Vail 1988; Posamentier *et al.* 1988). Each of these approaches to event stratigraphy are based on cyclicity within the rock record and the determination of a relative time stratigraphic framework. The major difference between them is in the bounding surfaces between cycles. Allostratigraphy uses any disconformity to bound allostratigraphic units, genetic stratigraphy uses maximum flooding surfaces and sequence stratigraphy uses unconformities. As such, both genetic and sequence stratigraphy can be viewed as specialized, interpretative subsets of descriptive-based allostratigraphy, hence there is no conflict between genetic stratigraphy, sequence stratigraphy and allostratigraphy. This is illustrated by Plint (this volume) who develops a sequence stratigraphic interpretation within a

previously established allostratigraphic framework. More controversy, however, lies between genetic and sequence stratigraphy. Genetic stratigraphy has been widely applied within the North Sea (e.g. Partington *et al.* 1993), and the recognition of condensed sections and maximum flooding surfaces is often easier than the recognition of unconformities, especially in well logs. It is commonly the starting point for correlation. Subsequently the identification of unconformities (Exxon-style sequence boundaries) may be possible. The two techniques are not mutually exclusive, both have their advantages and disadvantages and as long as it is made clear which technique is being applied there is no reason why both techniques should not be used in the same study. However, genetic stratigraphic sequences have less potential to predict when and where sand may have been deposited than unconformity-bound (Exxon-style) sequences.

As a lithology predictor, high resolution sequence stratigraphy has important implications in terms of reservoir correlation and modelling and the definition of anisotropy (Cross *et al.* 1993; O'Byrne & Flint 1993). High resolution sequence stratigraphic approaches have been applied to well log correlations of strata at production scales (e.g. Eschard *et al.* 1993; Pulham 1994) and to object-based and stochastic modelling systems (e.g. MacDonald *et al.* 1992; Kaas *et al.* 1994; Knight *et al.* 1994). This results in geologically more realistic descriptions of reservoir units away from well control points and reduces the stochastic requirement within reservoir models. Consequently sequence stratigraphic interpretations are routinely applied within the hydrocarbon industry and yet few detailed accounts of the benefits of sequence stratigraphic analyses at the reservoir scale have been published (e.g. Eschard *et al.* 1993; Posamentier & Chamberlain 1993; Pulham 1994; Reynolds 1994). Jennette & Riley (this volume) describe cyclicity within the Lower Brent Group, North Sea, which was controlled by variations in relative sea-level and exerts a fundamental control on the geometry, stacking pattern and distribution of both reservoir and seal facies. The establishment of a sequence stratigraphic framework has resulted in more efficient reservoir management plans. In a complementary study, Wehr & Brasher (this volume) focus on permeability simulation models for the Rannoch, Etive and Ness Formations (Brent Group), using both sequence stratigraphic- and lithostratigraphic-based well log correlations. Predicted recovery efficiencies from the Rannoch lower shoreface sandstones were consistently lower in the sequence stratigraphic-based model and are more closely comparable to observed reservoir response from the Cormorant Field.

In order that sequence stratigraphy can be applied reliably in the subsurface, it is important to improve the downhole identification of chronostratigraphically significant surfaces and depositional elements. The application of sequence stratigraphic concepts in the subsurface requires skilful interpretation of well logs. Davies & Elliott (this volume) refine the possible interpretations of gamma ray logs through the use of a portable gamma ray spectrometer on exposures in the Namurian Clare Basin, Eire. They argue that spectral gamma ray profiles and their respective ratios can resolve key surfaces and systems tracts with more confidence than through the use of simple total gamma ray counts. Similarly, Mitchell *et al.* (this volume) through the application of carbon isotope (δ^{13}C) stratigraphy, illustrate that δ^{13}C excursions show a close correlation with both sequence boundaries and onlap surfaces, identified on the basis of sedimentology. It is concluded that such excursions may be of value in locating surfaces of sequence stratigraphic significance when other criteria are not readily apparent.

Within the original Exxon models (Posamentier *et al.* 1988; Van Wagoner *et al.* 1990), assuming the presence of a shelf-slope break, the lowstand systems tract comprised the lowstand fan deposited during base level fall and the lowstand wedge deposited after base level fall and during the early rise. Posamentier *et al.* (1992) similarly recognized that deposition occurs during base level fall. Recently the definition of a falling stage systems tract and the location of the sequence boundary within a forced regressive, sharp-based shoreface sandbody has become a controversial matter (e.g. Hunt & Tucker 1992, 1995; Posamentier *et al.* 1992; Ainsworth 1994; Ainsworth & Pattison 1994; Fitzsimmons 1994; Helland-Hansen & Gjellberg 1994; Kolla *et al.* 1995; Plint this volume). Hunt & Tucker (1992) defined a 'forced regressive wedge systems tract' to account for deposition during falling relative sea-level. The base of this systems tract is the basal surface of forced regression and is marked at its top by a sequence boundary, which 'represents the lowest point of relative sea-level' (Hunt & Tucker 1992, p. 6). Overlying the forced regressive wedge systems tract lies the 'lowstand prograding wedge systems tract', deposited from the time that relative sea-level lies at its lowest point to the deposition of the initial (transgressive) flooding surface. This model probably works better in carbonate depositional environments than in siliciclastic ones (cf. Hunt & Tucker 1992; Kolla *et al.* 1995). Helland-Hansen & Gjellberg (1994) provide a similar scheme to Hunt & Tucker (1992). In the original Exxon models sequence boundary formation occurs at the onset and throughout sea-level fall and not solely at the point of lowest relative sea-level (cf. Hunt & Tucker 1992; Helland-Hansen & Gjelberg 1994). It is argued that placing the sequence boundary at the time of maximum subaerial exposure means that the sequence boundary becomes a more precise chronostratigraphic surface. However, there are diversities of opinions as to where a sequence boundary should be placed within a forced regression; commonly the sequence boundary is placed at the marine erosion surface below the shoreface (e.g. Posamentier *et al*, 1992; Ainsworth & Pattison 1994; Fitzsimmons 1994; Pattison 1995), but other workers place the sequence boundary at the subaerial erosion surface above the shoreface (e.g. Ainsworth 1994; Plint this volume). Furthermore, since both type 1 and 2 sequence boundaries may occur in the same basin (e.g. Gawthorpe *et al.* 1994) it is more consistent if the same chronostratigraphic surface is correlated; this is more likely to be the subaerial erosion surface above the shoreface (D. G. Quirk pers. comm. 1995). Taking a more pragmatic approach, however, it is easier to recognize the unconformity at the base of the sharp-based shoreface in well logs, since the subaerial exposure surface above the shoreface may have little wireline expression (R. J. Fitzsimmons pers. comm. 1995). Plint (this volume) discusses the issues concerning the location of the sequence boundary in more detail.

One of the most controversial aspects of sequence stratigraphy is how relative sea-level changes effect non-marine depositional environments. Posamentier & Vail (1988) and Posamentier *et al.* (1988) applied the accommodation models of Jervey (1988) to develop conceptual models which relate fluvial deposition to changes in relative sea-level assuming a constant sediment supply. These conceptual models have been criticized for being oversimplistic on a variety of scales (e.g. Miall 1991; Shanley & McCabe 1994). In particular the assertion by Posamentier & Vail (1988) that a fall in relative sea-level will result in fluvial aggradation and coastal onlap has been shown to be improbable (e.g. Miall 1991). It is now widely understood that fluvial aggradation and coastal onlap occur on the rising limb of a relative sea-level curve, that is during the late lowstand and transgressive systems tracts (e.g. Shanley & McCabe 1993; Quirk this volume). Despite the problems outlined above sequence stratigraphic techniques have been applied to predominantly alluvial successions (e.g. Shanley & McCabe 1993, 1994; Wright & Marriott 1993; Aitken & Flint 1994, 1995; Olsen *et al.* 1995; Hampson this volume, McKie & Garden this volume, Plint this volume, Uličný & Špičáková this volume). Despite this research the application of sequence stratigraphy to fluvial systems remains controversial (Schumm 1993; Wescott 1993; Shanley & McCabe 1994; Quirk this volume) and more research is warranted. Generally parasequences are not identifiable within alluvial strata (Aitken & Flint 1994, 1995), however, Uličný & Špičáková (this volume) examine the characteristics and preservation of parasequences in fluvial to estuarine strata in the Cretaceous Bohemian Basin. Although bounded by erosional surfaces these are interpreted as parasequences rather than high frequency sequences because there is no basinward shift in facies tracts across the boundary, internally there are no notable changes in palaeobathymmetry and correlation indicates that the bounding surfaces pass down-dip into marine flooding surfaces.

As a consequence of the difficulty in identifying parasequences within alluvial strata, a variety of criteria has been developed to define systems tracts and their bounding surfaces including tidally influenced channel fills to indicate maximum flooding surfaces (Shanley & McCabe 1993), coal seams to indicate transgressive and maximum flooding surfaces (Arditto 1991; Aitken & Flint 1994, 1995; Flint *et al.* 1995; Hampson 1995) and the relative development of palaeosols to indicate systems tracts (Wright & Marriott 1993). Although coal seams have been proposed as transgressive surface equivalents this is largely on the basis that, as a consequence of relative sea-level rise, accommodation potential increases and palaeo-ground water tables rise, initiating ideal conditions for the establishment of mires (e.g. Aitken & Flint 1995; Flint *et al.* 1995; Hampson 1995). No mechanism has been proposed to exclude clastic sediment supply, other than the establishment of raised mires (e.g. Flint *et al.* 1995) and/or the trapping of sediment up dip (Aitken & Flint 1994, 1995). Similarly, Hampson *et al.* (this volume, p. 238) propose that transgressive surface equivalent mires in the UK Upper Carboniferous Rough Rock Group develop as a consequence of sediment being trapped up dip by rising base level reducing fluvial gradients and consequently reducing sediment transport efficiencies. This implies that the transgressive

surface coals are basinward of the bayline. However, many Upper Carboniferous, laterally extensive coal seams contain contemporaneous sand-filled channel deposits (e.g. Guion *et al.* 1995; J. H. Rippon pers. comm. 1995), which implies that sediment was not trapped up dip, and that the bayline lay basinward of the transgressive surface mires. In this case, it is the fact that once incised valleys are filled, a broad plain is created encompassing the top of the valley fill and the adjacent interfluves, over which the sediment supply, that was originally concentrated within the confines of the incised valley, is distributed.

Future prospects

Since the publication of Wilgus *et al.* (1988) sequence stratigraphy has developed rapidly, but it is still evolving and there are a number of aspects which are still poorly understood. Perhaps the least studied depositional environment in terms of sequence stratigraphy is the deep marine realm and turbidite systems. In deep marine settings changes in relative sea-level will be very subtle and therefore difficult to identify. Nonetheless it is possible to identify a hierarchy of cycles within turbidite systems which can be interpreted in terms of sequence stratigraphy (e.g. Mutti 1994). More recently Shanmugam *et al.* (1995) have shown that basin floor fans are far more complex than predicted by the sequence stratigraphic models. More published research is required into the application of sequence stratigraphy to turbidites and deep marine fan systems.

Although sequence stratigraphy has been applied with some success to alluvial successions (see above), it is more difficult to apply sequence stratigraphic concepts to aeolian systems, and few examples have been published. In general, studies on aeolian strata have developed entirely independently of sequence stratigraphic concepts. However, Kocurek & Havholm (1994) describe a conceptual model for aeolian systems, proposing that the major controls on stratal architecture in such depositional settings are ground water table fluctuations, sediment supply, climatic change and subsidence, the last providing the accommodation space for preservation. Yang & Nio (1994) document one of the first applications of sequence stratigraphic concepts to aeolian strata with an example from the Rotliegend Group in the North Sea. McKie & Garden (this volume) describe the stratigraphy of a mixed fluvial, aeolian and lacustrine depositional system from the Devonian Clair Group, North Sea, and argue that the cyclicity evident within these successions is controlled by changes in the equilibrium between tectonically induced accommodation and climatically controlled sediment supply. The expansion of the concept of accommodation into aeolian depositional systems is an exciting future line for high resolution sequence stratigraphy.

Another area of research attracting considerable interest is the application of sequence stratigraphic concepts to extensional basins (e.g. Posamentier & Allen 1993b; Prosser 1993; Gawthorpe et al. 1994; Howell & Flint this volume). This work illustrates how in structurally active basins the effects of relative sea-level fluctuations are recorded within the sediments. The expression of these relative sea-level changes are different from those predicted by the original sequence stratigraphic models developed for passive margins. The sequence stratigraphic signatures are controlled by complicated basin margin topography and by variability in subsidence and consequently accommodation space along strike. Consequently, a spectrum of time equivalent sequence stratigraphic components, which have previously been considered to be spatially distinct, occur within tectonically active basins. Such models illustrate that in structurally active domains the expressions of eustatic changes are still present in a predictable form, enhancing the role of sequence stratigraphy as an exploration and production tool.

Interfluvial sequence boundaries are the correlative conformities to type 1 sequence boundaries and are one of the most difficult surfaces to recognize, especially in the subsurface. It is important to be able to recognize interfluvial sequence boundaries because they are one of the key criteria to distinguish incised valley fills from major channel deposits and their presence indicates potential reservoir targets both along strike (incised valley fills) and down dip (lowstand fans). Nonetheless, these are one of the least studied aspects within modern sequence stratigraphy. Aitken & Flint (this volume) report that in fluvially dominated Pennsylvanian successions interfluve sequence boundaries are very variable in their nature. The majority of interfluve palaeosols are gleys developed under conditions of poor drainage, although there is evidence to suggest that at least some of these formed initially under freely drained conditions and subsequently became gleyed because of rising water tables in the transgressive systems tract. Furthermore, these palaeosols have low preservation potential being readily eroded by later fluvial activity (Aitken & Flint this volume). O'Byrne & Flint (this volume), from predominantly shallow marine strata, record

interfluves represented by rooted carbonaceous shales and palaeosols, which in down dip areas have been partially or wholly reworked by transgressive erosion to leave an oxidized surface overlain by a lag of siderite nodules and haematite clasts. Davies & Elliott (this volume) distinguish interfluvial palaeosols in spectral gamma ray profiles by the presence of anomalously low potassium content and an exceptionally high thorium:potassium ratio. Further studies of the sort outlined above are required of these important sequence stratigraphic surfaces.

Sequence stratigraphy has developed well beyond the global eustatic curves and has become a powerful lithostratigraphic and facies analysis tool. This volume illustrates that the integration of all lines of evidence (sedimentology, biostratigraphy, structure, seismics etc.) is essential to the development of a high resolution sequence stratigraphic interpretation. It is notable, however, that although sequence stratigraphy developed from seismic stratigraphy, few of the papers contained within this volume use seismic sections. Seismic reflections are generally thought of as chronostratigraphic surfaces, hence examination of regional stratal termination patterns can establish a chronostratigraphic framework (e.g. Armentrout this volume). Related to this well logs can be tied to seismic lines and correlations confirmed in the interwell spacing, this is vital where well spacings are wide. Furthermore, modern seismic surveys, and in particular 3-D surveys, are attaining higher resolutions than was previously possible. Since acoustic impedance can vary across sequence and parasequence boundaries, Bryant (this volume) argues that lateral variations in a reflector along a single reflection, although not revealing the vertical architecture, may enhance the lateral resolution of facies variability within systems tracts.

The majority of the papers in this volume were initially presented at a conference held in Liverpool in March 1994. The papers were compiled to fill a gap which we felt existed in the literature, although in such a rapidly expanding field, such innovations are frequently integrated and outdated as fast as they are published. We encouraged papers which were specifically orientated at new innovations in sequence stratigraphy and the application of sequence stratigraphic research to the hydrocarbon industry and other geologically-based industries where prediction is a prerequisite. Consequently, the manuscripts were generally reviewed by one industrial and one academic referee.

We are grateful to the companies whose financial assistance subsidized both the conference and the costs of colour figures in the volume: Chevron (UK) Ltd, Gaps Scott Pickford, Esso Exploration & Production

UK Ltd, BP Exploration Operating Co. Ltd, Shell UK Exploration & Production Ltd, Mobil North Sea Ltd. and Badley Ashton & Associates Ltd. The conference from which this volume is a partial outgrowth was supported by the Petroleum and British Sedimentological Research Groups of the Geological Society and the Petroleum Exploration Society of Great Britain. We would also like to thank our colleagues who have assisted in the preparation of this book, in particular Steve Flint. The scientists listed below are thanked for their reviews of the manuscripts: Ron Boyd, Pat Brenchley, Ian Bryant, Steve Cannon, Pete Ditchfield, Steve Flint, Jon Gluyas, Bob Goldstein, Malcolm Hart, Adrian Hartley, Nicky Hind, Dave Hunt, Dave Jennette, Diane Kamola, K. Kennard, Lee Krystinuk, Mike Leeder, Alister MacDonald, Jim Marshall, Tom Mckie, Pete McCabe, Rory Mortimer, David Oliver, Mark Partington, Chris Paul, Guy Plint, Sarah Prosser, Bill Read, Tony Reynolds, Ron Steel, Finn Surlyk, Maurice Tucker, Fred Wehr, Andrew Whitham. We also thank those reviewers who wish to remain anonymous.

We are also grateful to Steve Flint and Dave Quirk for their constructive comments on a previous version of the manuscript. We thank our respective employers, Oxford Brookes University and University of Liverpool, for support during the editing of the volume and to the many colleagues who have influenced our thoughts on sequence stratigraphy.

References

AINSWORTH, R. B. 1994. Marginal marine sedimentology and high resolution sequence analysis: Bearpaw–Horseshoe canyon transition, Drumheller, Alberta. *Bulletin of Canadian Petroleum Geology*, **42**, 26–54.

—— & PATTISON, S. A. J. 1994. Where have all the lowstands gone? Evidence for attached lowstand systems tracts in the Western Interior of North America. *Geology*, **22**, 415–418.

AITKEN, J. F. & FLINT, S. S. 1994. High-frequency sequences and the nature of incised-valley fills in fluvial systems of the Breathitt Group (Pennsylvanian), Appalachian Foreland Basin, eastern Kentucky. *In:* DALRYMPLE, R. W., BOYD, R. & ZAITLIN, B. A. (eds.) *Incised Valley Systems: Origin and Sedimentary Sequences.* SEPM Special Publication, **51**, 353–368.

—— & —— 1995. The application of high-resolution sequence stratigraphy to fluvial systems: a case study from the Upper Carboniferous Breathitt Group, eastern Kentucky, USA. *Sedimentology*, **42**, 3–30.

ARDITTO, P. A. 1991. A sequence stratigraphic analysis of the Late Permian succession in the Southern Coalfield, Sydney Basin, New South Wales. *Australian Journal of Earth Sciences*, **38**, 125–137.

BOYD, R., SUTER, J. & PENLAND, S. 1988. Implications of modern sedimentary environments for sequence stratigraphy. *In:* JAMES, D. P. & LECKIE, D. A. (eds.) *Sequences, stratigraphy, sedimentology: surface and subsurface.* Canadian Society of Petroleum Geologists, Memoir, **15**, 33–36.

BURTON, R., KENDALL, C. G. St. C. & LERCHE, I. 1987. Out of our depth: on the impossibility of fathoming eustasy from the stratigraphic record. *Earth Science Reviews*, **24**, 237–277.

CHURCH, K. D. & GAWTHORPE, R. L. 1994. High resolution sequence stratigraphy of the late Namurian in the Widmerpool Gulf (East Midlands, UK). *Marine and Petroleum Geology*, **11**, 528–544.

CROSS, T. A., BAKER, M. R., CHAPIN, M. A., CLARK, M. S., GARDNER, M. H. *ET AL* 1993. Applications of high-resolution sequence stratigraphy to reservoir analysis. *In:* ESCHARD, R. & DOLIGEZ, B. (eds) *Subsurface reservoir characterization from outcrop observations.* Éditions Technip, Paris, 11–34.

DALRYMPLE, R. W., BOYD, R. & ZAITLIN, B. A. (eds) 1994. *Incised-valley systems: origin and sedimentary sequences.* SEPM Special Publication, **51**.

ESCHARD, R., TVEITEN, B., DESAUBLIAUX, G., LECOMTE, J. C. & VAN BUCHEM, F. S. P. 1993. High resolution sequence stratigraphy and reservoir prediction of the Brent Group (Tampen Spur area) using an outcrop analogue (Mesaverde Group, Colorado). *In:* ESCHARD, R. & DOLIGEZ, B. (eds) *Subsurface reservoir characterization from outcrop observations.* Éditions Technip, Paris, 35–52.

FITZSIMMONS, R. J. 1994. Identification of high order sequence boundaries and the land attached forced regression. *In:* JOHNSON, S. D. (ed.) *High resolution sequence stratigraphy: innovations and applications. Abstract volume.* Department of Earth Sciences, University of Liverpool, 332–333.

FLINT, S. S., AITKEN, J. F. & HAMPSON, G. 1995. Application of sequence stratigraphy to coal-bearing coastal plain successions: implications for the UK coal measures. *In:* WHATELEY, M. K. G. & SPEARS, D. A. (eds) *European Coal Geology.* Geological Society, London, Special Publication, **82**, 1–16.

GALLOWAY, W. E. 1989. Genetic stratigraphic sequences in basin analysis I: architecture and genesis of flooding surface bounded depositional units. *American Association of Petroleum Geologists Bulletin*, **73**, 125–142.

GAWTHORPE, R. L., FRASER, A. J. & COLLIER, E. Ll. 1994. Sequence stratigraphy in active extensional basins: implications for the interpretation of ancient basin-fills. *Marine and Petroleum Geology*, **11**, 642–658.

GUION, P. D., BANKS, N. L. & RIPPON, J. H. 1995. The Silkstone Rock (Westphalian A) from the east Pennines, England: implications for sand body genesis. *Journal of the Geological Society, London*, **152**, 819–832.

HAMPSON, G. 1995. Discrimination of regionally extensive coals in the Upper Carboniferous of the Pennine Basin, UK using high resolution sequence stratigraphic concepts. *In:* WHATELEY, M. K. G. & SPEARS, D. A. (eds) *European Coal Geology.* Geological Society, London, Special Publication, **82**, 79–97.

HAQ, B. U., HARDENBOL, J. & VAIL, P. R. 1988. Mesozoic and Cenozoic chronostratigraphy and eustatic cycles. *In:* WILGUS, C. K., HASTINGS, B. S., KENDALL, C. G. St. C., POSAMENTIER, H. W., ROSS, C. A. & VAN WAGONER, J. C. (eds) *Sea-level changes: an integrated approach.* Society of Economic Paleontologists and Mineralogists, Special Publication, **42**, 71–108.

HELLAND-HANSEN, W. & GJELBERG, J. G. 1994. Conceptual bias and variability in sequence stratigraphy: a different perspective. *Sedimentary Geology,* **92**, 31–52.

HUNT, D. & TUCKER, M. E. 1992. Stranded parasequences and the forced regressive wedge systems tract: deposition during base level fall. *Sedimentary Geology,* **81**, 1–9.

—— & —— 1995. Stranded parasequences and the forced regressive wedge systems tract: deposition during base level fall – reply. *Sedimentary Geology,* **95**, 147–160.

JERVEY, M. T. 1988. Quantitative geological modeling of siliciclastic rock sequences and their seismic expression. *In:* WILGUS, C. K., HASTINGS, B. S., KENDALL, C. G. St. C., POSAMENTIER, H. W., ROSS, C. A. & VAN WAGONER, J. C. (eds) *Sea-level changes: an integrated approach.* Society of Economic Paleontologists and Mineralogists, Special Publication, **42**, 47–70.

KAAS, I., SVANES, T., VAN WAGONER, J. C., HAMAR, G., JORGENVAG, S., SKARNES, P. I. & SUNDT, O. 1994. The use of high resolution sequence stratigraphy and stochastic modelling to reservoir management of the Ness Formation of the Statfjord Field, offshore Norway. *In:* JOHNSON, S. D. (ed.) *High Resolution Sequence Stratigraphy: Innovations and Applications. Abstract Volume.* Department of Earth Sciences, University of Liverpool, 57–58.

KNIGHT, S., HEATH, A., FLINT, S. S., WALSH, J. & WATTERSON, J. 1994. Application of sequence stratigraphy to 3-D visualisation and modelling of hydrocarbon reservoirs. *In:* JOHNSON, S. D. (ed.) *High Resolution Sequence Stratigraphy: Innovations and Applications. Abstract Volume.* Department of Earth Sciences, University of Liverpool, 126–130.

KOCUREK, G. & HAVHOLM, K. G. 1994. Eolian sequence stratigraphy – a conceptual framework. *In:* WEIMER, P. & POSAMENTIER, H. W. (eds) *Siliciclastic sequence stratigraphy: recent developments and applications.* American Association of Petroleum Geologists, Memoir, **58**, 393–410.

KOLLA, V., POSAMENTIER, H. W. & EICHENSEER, H. 1995. Stranded parasequences and the forced regressive wedge systems tract: deposition during base level fall – discussion. *Sedimentary Geology,* **95**, 139–145.

LOUCKS, R. G. & SARG, J. F. 1993. *Carbonate sequence stratigraphy: recent developments and applications.* American Association of Petroleum Geologists, Memoir, **57**.

MACDONALD, A. C., HOYE, T. H., LOWRY, P., JACOBSEN, T., AASEN, J. O. & GRINDHEIM, A. O. 1992. Stochastic flow unit modelling of a North Sea coastal-deltaic reservoir. *First Break,* **10**, 124–133.

MARTINSEN, O. J. 1993. Namurian (Late Carbonifer-

ous) depositional systems of the Craven-Askrigg area, northern England: implications for sequence stratigraphic models. *In:* POSAMENTIER, H. W., SUMMERHAYES, C. P., HAQ, B. U. & ALLEN, G. P. (eds) *Sequence stratigraphy and facies associations.* International Association of Sedimentologists, Special Publication, **18**, 247–282.

MAYNARD, J. R. 1992. Sequence stratigraphy of the Upper Yeadonian of northern England. *Marine and Petroleum Geology,* **9**, 197–207.

MIALL, A. D. 1986. Eustatic sea-level changes interpreted from seismic stratigraphy: a critique of the methodology with particular reference to the North Sea Jurassic record. *American Association of Petroleum Geologists Bulletin,* **70**, 131–137.

—— 1991. Stratigraphic sequences and their chronostratigraphic correlation. *Journal of Sedimentary Petrology,* **61**, 497–505.

—— 1992. Exxon global cycle chart: an event for every occasion? *Geology,* **20**, 787–790.

MUTTI, E. 1994. Sequence stratigraphic aspects of turbidite systems. *In:* JOHNSON, S. D. (ed.) *High resolution sequence stratigraphy: innovations and applications. Abstract Volume.* Department of Earth Sciences, University of Liverpool, 323–325.

NACSN 1983. North American stratigraphic code. *American Association of Petroleum Geologists Bulletin,* **67**, 841–875.

O'BYRNE, C. J. & FLINT, S. S. 1993. High-resolution sequence stratigraphy of Cretaceous shallow marine sandstones, Book Cliffs outcrops, Utah, USA – application to reservoir modelling. *First Break,* **11**, 445–459.

OLSEN, T., STEEL, R., HØSGETH, K., SKAR, T. & RØE, S.-L. 1995. Sequential architecture in a fluvial succession: sequence stratigraphy in the Upper Cretaceous Mesaverde Group, Price Canyon, Utah. *Journal of Sedimentary Research,* **B65**, 265–280.

PARTINGTON, M. A., MITCHNER, B. C., MILTON, N. J. & FRASER, A. J. 1993. Genetic sequence stratigraphy for the North Sea Late Jurassic and Early Cretaceous: distribution and prediction of Kimmeridgian-Late Ryazanian reservoirs in the North Sea and adjacent areas. *In:* PARKER, J. R. (ed.) *Petroleum geology of North West Europe: proceedings of the 4th conference.* Geological Society, London, 347–370.

PATTISON, S. A. J. 1995. Sequence stratigraphic significance of sharp-based lowstand shoreface deposits, Kenilworth Member, Book Cliffs, Utah. *American Association of Petroleum Geologists Bulletin,* **79**, 444–462

POSAMENTIER, H. W. & ALLEN, G. P. 1993*a* Variability of the sequence stratigraphic model: effects of local basin factors. *Sedimentary Geology,* **86**, 91–109.

—— 1993*b.* Siliciclastic sequence stratigraphic patterns in foreland ramp-type basins. *Geology,* **20**, 455–458.

—— & CHAMBERLAIN, C. J. 1993. Sequence-stratigraphic analysis of Viking Formation lowstand beach deposits at Joarcam Field, Alberta, Canada. *In:* POSAMENTIER, H. W., SUMMERHAYES,

C. P., HAQ, B. U. & ALLEN, G. P. (eds) *Sequence stratigraphy and facies associations*. International Association of Sedimentologists, Special Publication, **18**, 469–486.

—— & JAMES, N. P. 1993. An overview of sequence-stratigraphic concepts: uses and abuses. *In:* POSAMENTIER, H. W., SUMMERHAYES, C. P., HAQ, B. U. & ALLEN, G. P. (eds) *Sequence stratigraphy and facies associations*. International Association of Sedimentologists, Special Publication, **18**, 3–18.

—— & VAIL, P. R. 1988. Eustatic controls on clastic deposition II – sequence and systems tract models. *In:* WILGUS, C. K., HASTINGS, B. S., KENDALL, C. G. St. C., POSAMENTIER, H. W., ROSS, C. A. & VAN WAGONER, J. C. (eds.) *Sea-level changes: an integrated approach*. Society of Economic Paleontologists and Mineralogists, Special Publication, **42**, 125–154.

—— & WEIMER, P. 1993. Siliciclastic sequence stratigraphy and petroleum geology – where to from here? *American Association of Petroleum Geologists Bulletin*, **77**, 731–742.

——, JERVEY, M. T. & VAIL, P. R. 1988. Eustatic controls on clastic deposition I – conceptual framework. *In:* WILGUS, C. K., HASTINGS, B. S., KENDALL, C. G. St. C., POSAMENTIER, H. W., ROSS, C. A. & VAN WAGONER, J. C. (eds) *Sea-level changes: an integrated approach*. Society of Economic Paleontologists and Mineralogists, Special Publication, **42**, 109–124.

——, ALLEN, G. P., JAMES, D. P. & TESSON, M. 1992. Forced regressions in a sequence stratigraphic framework: concepts, examples and exploration significance. *American Association of Petroleum Geologists Bulletin*, **76**, 1687–1709.

——, SUMMERHAYES, C. P., HAQ, B. U. & ALLEN, G. P. (eds) 1993. *Sequence stratigraphy and facies associations*. International Association of Sedimentologists, Special Publication, **18**.

PROSSER, S. 1993. Rift-related linked depositional systems and their seismic expression. *In:* WILLIAMS, G. D. & DOBBS, A. (eds) *Tectonics and seismic sequence stratigraphy*. Geological Society, London, Special Publication, **71**, 35–66.

PULHAM, A. J. 1994. The Cuisiana Field, Llanos Basin, eastern Colombia: high resolution sequence stratigraphy applied to Late Palaeocene–Early Oligocene estuarine, coastal plain and alluvial clastic reservoirs. *In:* JOHNSON, S. D. (ed.) *High Resolution Sequence Stratigraphy: innovations and applications. Abstract Volume*. Department of Earth Sciences, University of Liverpool, 63–68.

READ, W. A. 1994. High-frequency, glacial-eustatic sequences in early Namurian coal-bearing fluvio-deltaic deposits, central Scotland. *In:* DE BOER, P. L. & SMITH, D. G. (eds) *Orbital forcing and cyclic sequences*. International Association of Sedimentologists, Special Publication, **19**, 413–428.

—— 1995. Sequence stratigraphy and lithofacies geometry in an early Namurian coal-bearing succession in central Scotland. *In:* WHATELEY, M. K. G. & SPEARS, D. A. (eds) *European Coal Geology*. Geological Society, London, Special Publication, **82**, 285–297.

READING, H. G. (ed.) 1986. *Sedimentary Environments and Facies* (2nd edition). Blackwell, Oxford.

REYNOLDS, A. D. 1994. Sequence stratigraphy and the dimensions of paralic sandstone bodies. *In:* JOHNSON, S. D. (ed.) *High resolution sequence stratigraphy: innovations and applications. Abstract volume*. Department of Earth Sciences, University of Liverpool, 69–72.

SCHLAGER, W. 1991. Depositional bias and environmental change – important factors in sequence stratigraphy. *Sedimentary Geology*, **70**, 109–130.

SCHUMM, S. A. 1993. River response to baselevel change: implications for sequence stratigraphy. *Journal of Geology*, **101**, 279–294.

SHANLEY, K. W. & McCABE, P. J. 1993. Alluvial architecture in a sequence stratigraphic framework – a case history from the upper Cretaceous of southern Utah, U.S.A. *In:* FLINT, S. S. & BRYANT, I. D. (eds) *The geological modelling of hydrocarbon reservoirs and outcrop analogues*. International Association of Sedimentologists, Special Publication, **15**, 21–56.

—— & —— 1994. Perspectives on the sequence stratigraphy of continental strata. *American Association of Petroleum Geologists Bulletin*, **78**, 544–568.

SHANMUGAM, G., BLOCH, R. B., MITCHELL, S. M., BEAMISH, G. W. J., HODGKINSON, R. J. *ET AL.* 1995. Basin-floor fans in the North Sea: sequence stratigraphic models vs. sedimentary facies. *American Association of Petroleum Geologists Bulletin*, **79**, 477–512.

SLOSS, L. L. 1988. Forty years of sequence stratigraphy. *Geological Society of America Bulletin*, **100**, 1661–1665.

SUMMERHAYES, C. P. 1986. Sea-level curves based on seismic stratigraphy: their chronostratigraphic significance. *Palaeogeography, Palaeoclimatology, Palaeoecology*, **57**, 27–42.

UNDERHILL, J. R. 1991. Controls on Late Jurassic seismic sequences, Inner Moray Firth, UK North Sea: a critical test of a key segment of Exxon's original global cycle chart. *Basin Research*, **3**, 79–98.

VAIL, P. R., MITCHUM, R. M. & THOMPSON, S. 1977*a*. Seismic stratigraphy and global changes of sea-level. Part 4. Global cycles of relative changes of sea-level. *In:* PAYTON, C. E. (ed.) *Seismic stratigraphy – applications to hydrocarbon exploration*. American Association of Petroleum Geologists, Memoir, **26**, 83–98.

——, ——, TODD, R. G., WIDMIER, J. M., THOMPSON, S., SANGREE, J. B., BUBB, J. N. & HATLELID, W. G. 1977*b*. Seismic stratigraphy and global changes of sea-level. *In:* PAYTON, C. E. (ed.) *Seismic stratigraphy – applications to hydrocarbon exploration*. American Association of Petroleum Geologists, Memoir, **26**, 49–211.

VAN WAGONER, J. C., MITCHUM, R. M., CAMPION, K. M. & RAHMANIAN, V. D. 1990. *Siliciclastic sequence stratigraphy in well logs, cores and outcrops: concepts for high-resolution correlation of time and facies*. American Association of

Petroleum Geologists, Methods in Exploration Series, **7**.

WALKER, R. G. 1984. *Facies models* (2nd edition). Geological Association of Canada, Geoscience Canada Reprint Series 1.

—— 1990. Facies modeling and sequence stratigraphy. *Journal of Sedimentary Petrology*, **60**, 777–786.

—— 1992. Facies, facies models and modern stratigraphic concepts. *In:* WALKER, R. G. & JAMES, N. P. (eds) *Facies models response to sea-level change*. Geological Association of Canada, St. Johns, 1–14.

WEIMER, P. & POSAMENTIER, H. W. (eds) 1994a. *Siliciclastic sequence stratigraphy: recent developments and applications*. American Association of Petroleum Geologists, Memoir, **58**.

——, —— 1994b. Recent developments and applications in siliciclastic sequence stratigraphy. *In:* WEIMER, P. & POSAMENTIER, H. W. (eds) *Siliciclastic sequence stratigraphy: recent developments and applications*. American Association of Petroleum Geologists, Memoir, **58**, 3–12.

WESCOTT, W. A. 1993. Geomorphic thresholds and complex response of fluvial systems – some implications for sequence stratigraphy. *American*

Association of Petroleum Geologists Bulletin, **77**, 1208–1218.

WILGUS, C. K., HASTINGS, B. S., KENDALL, C. G. ST. C., POSAMENTIER, H. W., ROSS, C. A. & VAN WAGONER, J. C. (eds) 1988. *Sea-level changes: an integrated approach*. Society of Economic Paleontologists and Mineralogists, Special Publication, **42**.

WRIGHT, V. P. & MARRIOTT, S. B. 1993. The sequence stratigraphy of fluvial depositional systems: the role of floodplain sediment storage. *Sedimentary Geology*, **86**, 203–210.

YANG, C. S. & NIO, S. D. 1994. Application of high resolution sequence stratigraphy to the Upper Rotliegendes in the Netherlands offshore. *In:* WEIMER, P. & POSAMENTIER, H. W. (eds) *Siliciclastic sequence stratigraphy: recent developments and applications*. American Association of Petroleum Geologists, Memoir, **58**, 285–316.

ZAITLIN, B. A., DALRYMPLE, R. W. & BOYD, R. 1994. The stratigraphic organization of incised-valley systems associated with relative sea-level change. *In:* DALRYMPLE, R. W., BOYD, R. & ZAITLIN, B. A. (eds) *Incised-valley systems: origin and sedimentary sequences*. SEPM Special Publication, **51**, 45–60.

Carbon isotopes and sequence stratigraphy

S. F. MITCHELL,[1] C. R. C. PAUL[1] & A. S. GALE[2]

[1]*Department of Earth Sciences, University of Liverpool, Brownlow Street, Liverpool L69 3BX, UK*
[2]*Department of Earth Sciences, Imperial College, South Kensington, London SW7 4BP, UK, and Department of Palaeontology, Natural History Museum, Cromwell Road, London SW7 5BD, UK*

Abstract: The carbon isotope (δ^{13}C) stratigraphy of the late Lower Cenomanian to early Lower Turonian is presented for three sections (Folkestone in the Anglo-Paris Basin, Wünstorf in the Lower Saxony Basin and Speeton in the Cleveland Basin). The similarity between these isotope curves suggests that they were controlled by synchronous, global processes and can be used for high resolution correlation. Furthermore, sequence stratigraphic analysis of this interval reveals that each isotope excursion is associated with a sequence boundary and/or onlap surface. This is also demonstrated for the whole of the Cenomanian for the section at Speeton. We show that most (if not all) δ^{13}C excursions are synchronous within the limits of current stratigraphic resolution.

We interpret the increase in background δ^{13}C values as representing an increase in the area of ocean floor (specifically continental shelf) available for burial of marine organic carbon caused by the mid-Cretaceous rise in eustatic sea-level. Thus background δ^{13}C values may provide an independent method for estimating past eustatic sea-levels. We interpret the sharp δ^{13}C excursions as reflecting more rapid changes in the carbon cycle particularly the rate of burial of organic carbon within sediments, and/or of storage in deep and intermediate water masses, at times of rapidly changing sea-level. Carbon excursions may be useful in locating sequence boundaries when other criteria are obscure or lacking.

It has long been known that stable isotopes of carbon are related to marine primary productivity and can be used in stratigraphic and petroleum exploration studies (Scholle & Arthur 1980). Renard (1986) has presented a generalized δ^{13}C curve for the period from the late Jurassic to Recent. More detailed studies in the Cretaceous (e.g. Gale *et al.* 1993; Jenkyns *et al.* 1994; Paul *et al.* 1994*a, b*) have shown that δ^{13}C curves can be correlated between different depositional basins and that excursions are synchronous within the limits of current stratigraphic resolution.

Sequence stratigraphic analysis has also evolved rapidly in the last decade (see Van Wagoner *et al.* 1988), and considerable interest has focused on Cenomanian sections in the Anglo-Paris Basin. Haq *et al.* (1988) used the type Cenomanian of Normandy (within the Anglo-Paris Basin) to exemplify their sequence stratigraphic interpretations. Their research has been developed further by Gale (1990, 1995), Simmons *et al.* (1991), Juignet & Breton (1992), Mitchell (1993), Paul *et al.* (1994*a,b*) and Robaszynski *et al.* (1995), amongst others.

In this article we demonstrate that the interval from the late Lower Cenomanian to the early Lower Turonian can be correlated using δ^{13}C curves and that isotopic excursions used in the correlation are intimately associated with variations in sea-level recognized on sedimentological criteria for this interval and the whole Cenomanian Stage at Speeton (North Yorkshire). We suggest that background δ^{13}C values are related to, and can be used to estimate changes in, eustatic sea-level. Furthermore, in basinal hemipelagic deposits we suggest that the carbon isotopic signal may be used to recognize intervals within which critical sequence stratigraphic surfaces lie, when other criteria (e.g. facies changes, erosion surfaces) are obscure.

Methods

Sections in rhythmic, hemipelagic Cenomanian chalks were logged in detail and macrofauna collected and/or recorded. Samples for micropalaeontological and stable isotope analysis were collected and accurately located on the logs. Sampling intensity varied with locality from several samples per bed in critical intervals (e.g. Folkestone) to about one sample per metre (e.g. Wünstorf). In the laboratory, micropalaeontological samples were dissociated by

From Howell, J. A. & Aitken, J. F. (eds), *High Resolution Sequence Stratigraphy: Innovations and Applications*, Geological Society Special Publication No. 104, pp. 11–24.

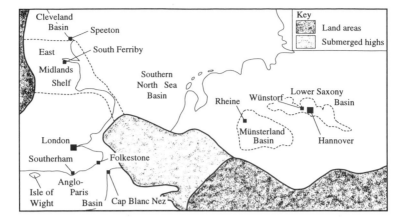

Fig. 1. Locations of sections mentioned in the text and outline of contemporary palaeogeography.

repeatedly freezing and thawing in a super-saturated solution of sodium sulphate, washed thoroughly through a 63 μm sieve, dried and size sorted into 1 or $\frac{1}{2}$ phi intervals for picking. For stable isotope analysis a small portion of each sample was ground to powder and approximately 3 mg was used to determine stable isotope ratios of carbon and oxygen in the Liverpool University Stable Isotope Laboratory. All samples were treated in a low pressure plasma oven for four hours to remove any organic matter. Gaseous CO_2 for analysis was released by reacting the powder with 2 ml anhydrous 100% orthophosphoric acid in a constant temperature bath at 25°C for at least three hours or until the reaction was complete. An acid fractionation factor of 1.01025 was used (Friedman & O'Neil 1977). Results were corrected using standard procedures (Craig 1957) and are expressed as per mil (‰) deviation from the Pee Dee Belemnite (PDB) international standard calibrated using NBS 19.

Sections studied

Here we present details of important sections (Fig. 1) from three different basins: Folkestone in the Anglo-Paris Basin (used as a standard section), Speeton in the Cleveland Basin and Wünstorf in the Lower Saxony Basin. Limited isotopic data from Rheine in the Münsterland Basin and South Ferriby on the East Midlands Shelf are used to supplement the main sections, while Jenkyns *et al*. (1994) have presented $\delta^{13}C$ curves for the entire Upper Cretaceous of Britain and Italy derived from hemipelagic, carbonate-rich sediments. To a greater or lesser extent, the sections we have studied are composed of rhythmic alternations of clay-rich and

clay-poor chalks. Bioturbated boundaries demonstrate a primary lithological contrast, more or less enhanced by subsequent diagenesis. Such rhythms are almost universally ascribed to Milankovitch processes (e.g. Hart 1987; Ditchfield & Marshall 1989; Gale 1990; Paul 1992) and at several levels individual rhythms can be correlated both within and between basins using faunal (zones and 'events' *sensu* Ernst *et al*. 1983) and lithological marker horizons (e.g. Gale 1990, 1995; Mitchell 1993). Gale (1990) introduced a numbering scheme for the Cenomanian Stage based on marl–chalk rhythms which has been subsequently refined (Gale 1995). Additional details of the sections may be found as follows: Folkestone (Gale 1989; Jenkyns *et al*. 1994; Lamolda *et al*. 1994; Paul *et al*. 1994*a*, *b*); Speeton (Jeans 1980; Mitchell 1993; Paul *et al*. 1994*a*); South Ferriby (Mitchell 1993; Paul *et al*. 1994*a*); Wünstorf (Meyer 1990).

Correlation

Standard zonal schemes for the Cenomanian stage have been erected using ammonites, planktonic foraminifera and inoceramid bivalves (see Gale (1995) for details), and the ammonite zonal nomenclature is shown for reference on Figs 2 and 3. However, preservational problems often make recognition of the ammonite zonal scheme in particular difficult and, in any case, the zones are on too coarse a timescale to test the synchroneity or otherwise of short-lived isotope excursions. In an attempt to overcome such difficulties Ernst *et al*. (1983) recognized a series of faunal and lithological 'events' within the Cenomanian and Turonian of Germany, which were further refined by Kaplan & Best (1985) and some of which have proved applic-

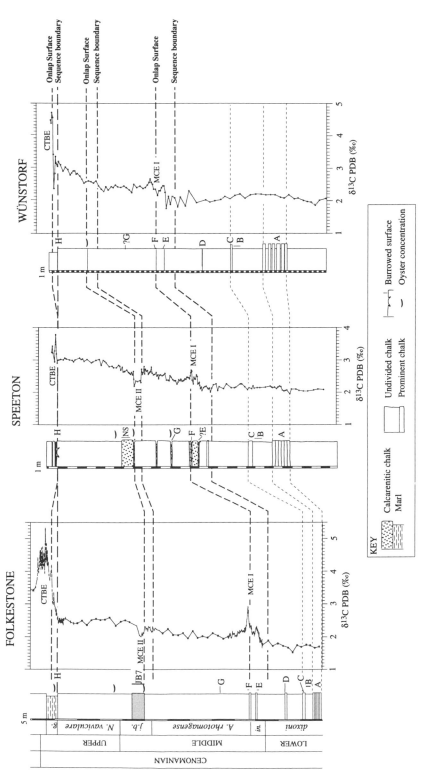

Fig. 2. Correlation of the three principal sections discussed in the text to show standard chronostratigraphy, ammonite biostratigraphy, principal lithological units, key marker horizons (A–H see text), δ^{13}C signatures, the three sequence boundaries and onlap surfaces discussed in the text, and associated carbon excursions (MCE I, MCE II and CTBE). *in.*, *Cunningtoniceras inerme* Zone; *j.b.*, *Acanthoceras jukesbrownei* Zone; *g.*, *Metoicoceras geslinianum* Zone. The Folkestone curve is a composite of our own detailed data for the excursions and the published values of Jenkyns *et al.* (1994) for the intervals between the excursions.

able to the sections we have studied. Such 'events', together with the Milankovitch timescale provided by the rhythmic sediments (Gale 1990, 1995), allow testing of synchroneity (ideally) to within ±10 000 years (10 ka). The lithological and faunal markers (A–H) we have used in our correlations (Fig. 2) are as follows:

A. Six prominent limestones, the lowermost four of which yield common *Inoceramus virgatus* Schlüter (the '*virgatus/Schloenbachia* event' of Ernst *et al.* 1983), and the top one abundant ammonites including *Mantelliceras dixoni* Spath (the '*dixoni* event' of Kaplan & Best (1985), the '*dixoni* bed' of Jenkyns *et al.* (1994)).
B. A band of the brachiopod *Orbirhynchia* cf. *mantelliana* (J. de C. Sowerby). The '*Orbirhynchia* event' of Ernst *et al.* (1983), '*Orbirhynchia* band 1' of Jeans (1980) and the lowest concentration of *Orbirhynchia* in Gale (1989) and Jenkyns *et al.* (1994, fig. 13).
C. A prominent pure or unusually thick limestone just above the *Orbirhynchia* band (B).
D. A limestone with abundant dark *Chondrites* burrows (Jenkyns *et al.* 1994).
E. A dark marl within which the small bivalve *Lyropecten* (*Aequipecten*) *arlesiensis* (Woods) occurs (the '*arlesiensis* bed' of Paul *et al.* 1994a).
F. The brief appearance of the belemnite *Actinocamax primus* Arkhangelsky at the base of a dark marl (the 'cast bed' of Price 1877, or '*primus* bed', e.g. Christensen 1990), or within the Totternhoe Stone facies (e.g. at South Ferriby, North Humberside).
G. The 'PB break' (the 'mid Cenomanian nonsequence' of Carter & Hart (1977) above which large (> 250 μm) planktonic foraminifera appear abundantly and remain a significant proportion of the microfauna of this size fraction.
H. The facies change, from pure coccolith chalk to marl, that heralds the Cenomanian–Turonian Boundary Event (CTBE).

Sequence stratigraphic interpretations

Previous sequence stratigraphic interpretations of late Lower Cenomanian to early Lower Turonian successions in NW Europe have recognized sequence boundaries in the latest Lower Cenomanian or early Middle Cenomanian (Gale 1990; Simmons *et al.* 1991; Paul *et al.* 1994a), the mid *Acanthoceras jukesbrownei* Zone (Haq *et al.* 1988; Gale 1990; Juignet & Breton 1992) and the base of the *Metoicoceras geslinianum* Zone (Gale 1990; Juignet & Breton 1992; Lamolda *et al.* 1994), while Robaszynski *et al.* (1995) have undertaken a detailed sequence stratigraphic study of the Cenomanian of the Anglo-Paris Basin.

Overall throughout the interval from the late Lower Cenomanian to the early Turonian, chalks become purer in all three basins. For example, at Folkestone the proportion of siliciclastics declines from about 20–25% in the late Lower Cenomanian to <5% in the late Upper Cenomanian and early Turonian. This is consistent with an overall rise in sea-level that was part of the mid-Cretaceous transgression (Hancock & Kauffman 1979; Hancock 1989).

Robaszynski *et al.* (1995) have demonstrated that in basinal hemipelagic facies, sequence boundaries are marked by an increase in the rate of accumulation of clay material and can be recognized by increases in insoluble residues. However, in pure chalks sequence boundaries may be poorly defined. Flooding surfaces are easier to recognize. They represent onlap surfaces, which by definition onlap the basin margins and which frequently have a coarse-grained lag at their bases. Jeans (1973, 1980) in his studies of the Lower Chalk of eastern England recognized a succession of fining upwards cycles, each represented by a thin basal unit of calcarenitic chalk resting on an erosion surface with glauconitized pebbles followed by a thick succession of coccolith-rich chalks. It is the bases of these fining upwards cycles that show onlap relationships across southwest England and these represent the onlap (flooding) surfaces of Robaszynski *et al.* (1995). In the interval from the late Lower Cenomanian to the basal Lower Turonian three sequence boundaries/onlap surfaces have been recognized (Robaszynski *et al.* 1995), as below.

1. A well defined onlap surface in the early Middle Cenomanian (basal *Turrilites costatus* Subzone of the *Acanthoceras rhotomagense* Zone) at the base of the Totternhoe Stone facies on the East Midlands Shelf (e.g. South Ferriby) or its basinal equivalent (e.g. the 'cast bed' at Folkestone or '*primus*' bed' of Wünstorf and Speeton). The underlying sequence boundary is represented in the Anglo-Paris Basin by an increase in accumulation of clay about 4 m lower. This sequence boundary is difficult to recognize in the Cleveland Basin where clay input was considerably less than in the Anglo-Paris or Saxony basins; however, the onlap surface is well defined.

2. An onlap surface high in the Middle Cenomanian (within the *Acanthoceras jukesbrownei* Zone), at the base of Jukes-Browne's Bed 7 (Jukes-Browne & Hill 1903) at Folkestone, or the base of the Nettleton Stone at Speeton. The sequence boundary is difficult to recognize in all three basinal sections studied.

3. An onlap surface at the base of bed 4 of the Plenus Marls (middle *Metoicoceras geslinianum* Zone) or its equivalent. The sequence boundary is represented by the dramatic facies change at the base of the Plenus Marls, where pure coccolith-rich chalks (acid insoluble residue values < 5%) are replaced by distinctly marly chalks (acid insoluble residues up to 20–25%). Across the East Midlands Shelf and at Speeton the onlap surface and sequence boundary coincide and lowstand deposits are absent.

Interbasinal isotope stratigraphy

Figure 2 shows the $\delta^{13}C$ signature for the three principal sections discussed here. These curves represent surface sea water $\delta^{13}C$ signatures, since bulk carbonate in chalks is predominantly coccolith material. They demonstrate first the overall increase in $\delta^{13}C$ values from the late Lower Cenomanian to the Lower Turonian, and second the close association between onlap surfaces and isotope excursions. Significant features of these signatures are as follows.

The late Lower Cenomanian, from the *virgatus/Schloenbachia* 'event' (A in Fig. 2) to a little below the *arlesiensis* bed (E in Fig. 2), is represented by a flat $\delta^{13}C$ signature in all three basins. This flat interval is terminated by the small double peaked excursion in the early Middle Cenomanian, Middle Cenomanian Event I (MCE Ia & b). Previous studies, with more widely spaced samples than ours (e.g. Jenkyns *et al.* 1994), were unable to resolve these two peaks. Both peaks are, however, clearly visible at Speeton, Folkestone and Wünstorf (Fig. 2), as well as in the sections at Cap-Blanc-Nez, France and Southerham, Sussex, UK (Paul *et al.* 1994a). The early portion of the Middle Cenomanian above MCE Ib also has a relatively flat $\delta^{13}C$ signature, although above the PB break (G in Fig. 2) $\delta^{13}C$ begins to increase. A small negative $\delta^{13}C$ excursion, Middle Cenomanian Event II (MCE II), occurs in the early portion of the *Acanthoceras jukesbrownei* Zone and can be recognized at Speeton, Folkestone and Rheine. It has not been recognized at Wünstorf where it may be cut out by an erosion surface beneath the *Pycnodonte* concentration in the Upper Cenomanian (Fig. 2). However, it is present at Büren in the Münsterland Basin (Paul unpublished data). $\delta^{13}C$ values rapidly increase above MCE II before reaching a plateau in the mid to late *Calycoceras naviculare* Zone of the Upper Cenomanian. The erosion surface at the base of the *Metoicoceras geslinianum* Zone heralds the start of the Cenomanian–Turonian Boundary Event (CTBE) with its major positive $\delta^{13}C$

excursion which has been extensively documented elsewhere, e.g. in Hart (1985); Hilbrecht & Hoefs (1986); Schlanger *et al.* (1987); Jarvis *et al.* (1988a, b); Hart *et al.* (1991); Jeans *et al.* (1991); Hilbrecht *et al.* (1992); Gale *et al.* (1993); Lamolda *et al.* (1994); Paul & Mitchell (1994) and Paul *et al.* (1994b), among others.

Thus there seems to have been a relationship between background $\delta^{13}C$ values and long-term eustatic sea-level, as well as between $\delta^{13}C$ excursions and rapid changes in eustatic sea-level. We will document the detailed relationship between $\delta^{13}C$ values and sequence boundaries/onlap surfaces for two Cenomanian carbon excursions (MCE I and II), before discussing possible mechanisms for maintaining a relationship between eustatic sea-level and marine $\delta^{13}C$ values.

Middle Cenomanian event I (MCE I)

Figure 3 shows a detailed correlation of MCE I between Wünstorf, Folkestone, Speeton and South Ferriby, showing the two small $\delta^{13}C$ excursions (MCE Ia and Ib). This composite excursion has recently been described in detail by Paul *et al.* (1994a) and only an outline of the most important details and new information will be considered here. We now place an onlap surface at the base of rhythm C1 (Fig. 3) which is equivalent to the base of the Totternhoe Stone (with its reworked basal lag) on the East Midlands Shelf and London Platform. In the Anglo-Paris Basin the sequence boundary is placed in the interval between B34 and B37 where there is an increase in the insoluble residue content. This level is less clearly recognized at Wünstorf, while at Speeton evidence of shallowing is not apparent until rhythm B40 when silt-grade bioclastic debris appears. Paul *et al.* (1994a following Jefferies 1962) argued that the belemnite *Actinocamax primus* and the bivalves *Lyropecten arlesiensis* and *Oxytoma seminudum* Dames were shallow water species that appeared in the succession at times of cooling and shallowing. New planktonic foraminiferal data (Mitchell & Carr unpublished data) from the Anglo-Paris Basin add important contributions to the mechanisms controlling the distribution of these species. Rhythm B43 at Folkestone is characterized by an influx of abundant large planktonic foraminifera (which represent more than 40% of foraminifera in the > 250 μm size fraction). This short influx of abundant large planktonic foraminifera occurs at a local temperature maximum (stable oxygen isotope data in Paul *et al.* 1994a). This implies that the two floods of apparent cold/shallow water species

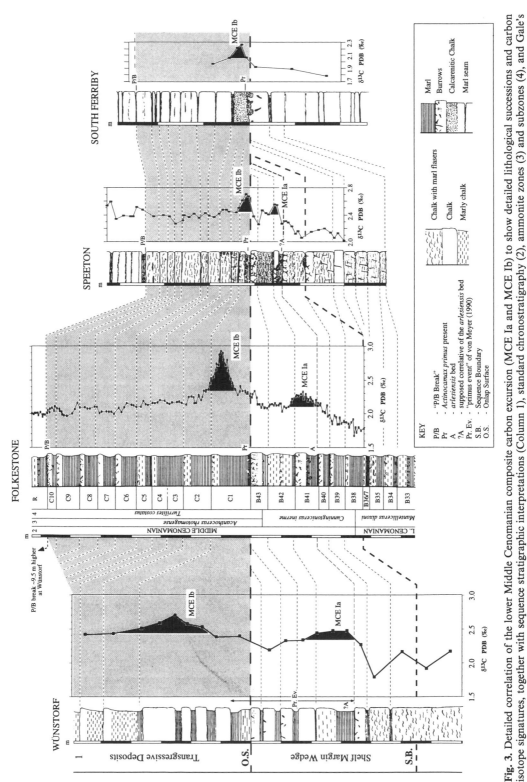

Fig. 3. Detailed correlation of the lower Middle Cenomanian composite carbon excursion (MCE Ia and MCE Ib) to show detailed lithological successions and carbon isotope signatures, together with sequence stratigraphic interpretations (Column 1), standard chronostratigraphy (2), ammonite zones (3) and subzones (4), and Gale's (1990) numbering system for rhythms (R). Transgressive deposits shaded.

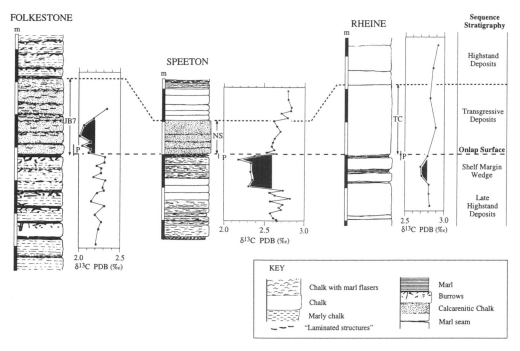

Fig. 4. Detailed correlation of the upper Middle Cenomanian (*Acanthoceras jukesbrownei* Zone) negative carbon excursion (MCE II) to show detailed lithological successions, carbon isotope signatures and sequence stratigraphic interpretations. Note that the $\delta^{13}C$ excursion is not synchronous if the *Pycnodonte* concentration and facies change are used as a datum.

were due to changes in temperature during the time of low sea-level.

The $\delta^{13}C$ signature of MCE I is represented by two small peaks, MCE Ia in the lowstand deposits (rhythms B40–41) and MCE Ib in the early transgressive deposits (C1 and C2) just above the onlap surface. The $\delta^{13}C$ signature is, therefore, clearly related to changes in sea-level. This compares well with the situation in the CTBE where the first peak of the very large positive $\delta^{13}C$ excursion occurs in the lowstand deposits and the larger second peak occurs in the early transgressive deposits (Fig. 2; fig. 10 of Robaszynski *et al.* 1995).

As to cyclostratigraphic correlation and the synchroneity of carbon excursions, the interval from the base of rhythm B40 to the top of B43 can be correlated directly between all three basinal successions (Fig. 3) and the isotope excursion is synchronous with this correlation. Similarly, the interval above the second sequence boundary that includes the upper excursion (MCE Ib) spans four chalk beds at Wünstorf, but only one rhythm (C1) at Folkestone (Fig. 3). However, under favourable light conditions, the marl of rhythm C1 at Folkestone can be seen to be composite with subtle, paler bands within it.

Hence we suggest that rhythm C1 at Folkestone is equivalent to the four rhythms at Wünstorf.

Middle Cenomanian event II (MCE II)

Figure 4 shows a detailed representation of the $\delta^{13}C$ excursion in the *Acanthoceras jukesbrownei* Zone. At this level the abundance of the oyster *Pycnodonte* sp. has been used as a biostratigraphic marker (Ernst *et al.* 1983; Gaunt *et al.* 1992). We use this marker, which is consistent with the facies change from coccolith-rich chalks to bioclastic chalks (i.e. Jukes-Browne's bed 7, the Nettleton Stone, and its equivalent in Germany), to correlate the $\delta^{13}C$ signatures between the sections at Folkestone, Speeton and Rheine (Fig. 4). This demonstrates that the isotope excursion appears synchronous between Speeton and Rheine, but occurs four rhythms later at Folkestone. Two possible conclusions can be drawn: either the isotope event is diachronous or the concentration of *Pycnodonte* and the facies change are diachronous. Whatever the case, the small negative $\delta^{13}C$ excursion is clearly associated with a sea-level event (either with the lowstand deposits (Speeton and Rheine) or the transgressive deposits (Folkestone).

The isotopic response of MCE II is different to those of MCE I and the CTBE. MCE II has a negative $\delta^{13}C$ excursion within either the lowstand or transgressive deposits. MCE I and the CTBE excursions are positive, with $\delta^{13}C$ peaks in both the lowstand and transgressive deposits. For MCE I and II and the CTBE we have evidence of a correlation between carbon excursions and sequence boundaries/onlap surfaces from more than one sedimentary basin, suggesting that these carbon excursions reflect widespread (eustatic) sea-level changes.

Other mid-Cretaceous carbon excursions

Figure 5 shows a $\delta^{13}C$ curve from the late Albian to early Turonian from the section at Speeton. It includes four additional examples of $\delta^{13}C$ excursions associated with sequence boundaries and/or onlap surfaces. The condensed nature of the early Lower Cenomanian successions in the Anglo-Paris Basin means that two of the sequences provisionally recognized here were not recognized by Robaszynski *et al.* (1995).

The Albian–Cenomanian boundary excursion (ACBE, Fig. 5) crosses the Albian–Cenomanian boundary (based on brachiopod and benthic foraminiferal data) at Speeton. A sequence boundary (SB 1) is placed at a significant increase in insoluble residue and an onlap surface (OS 1) is tentatively placed in the upper part of the marly unit above an omission surface. Jenkyns *et al.* (1994) show a small $\delta^{13}C$ peak across their inferred Albian–Cenomanian boundary in Italy, while Leckie *et al.* (1992) have reported lithological and faunal changes across the Albian–Cenomanian boundary in Canada, similar to those found at the Cenomanian–Turonian boundary in Europe which we interpret as a major sea-level fall. The isotope excursion has a characteristic three-pronged form (with a few minor spikes beneath) and is also present across the expanded Albian–Cenomanian boundary in sections in the Vocontian Trough, Southern France (Gale, unpublished data).

LCE I is also associated with a marly unit (Fig. 5). A sequence boundary (SB 2) is placed at the base of a rhythm where there is a significant increase in insoluble residues. The onlap surface (OS 2) is placed at a marked omission surface which has a thin bioclastic lag. The isotope excursion is represented by a single well-defined spike. This excursion may be equivalent to the poorly defined $\delta^{13}C$ excursion in the basal Glauconitic Marl (Jenkyns *et al.* 1994), which rests on an unconformity cutting down into the

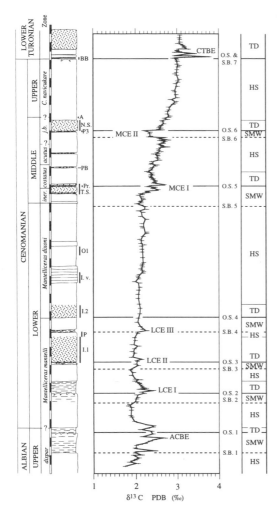

Fig. 5. A $\delta^{13}C$ curve for the whole Cenomanian Stage from the section at Speeton, showing background $\delta^{13}C$ values (from < 2.0‰ in the late Albian to > 3.0‰ in the early Turonian). $\delta13C$ excursions are associated with sequence boundaries and/or onlap surfaces. *iner., Cunningtoniceras inerme* Zone; *j.b., Acanthoceras jukesbrownei* Zone; I.1 and I.2, first and second *Inoceramus* beds (Jeans 1973); P, *Pycnodonte* concentration; A, *Amphidonte* concentration; I.v., *Inoceramus virgatus* concentration; O1, *Orbirhynchia* band; PB, PB break; BB, Black Band; T.S., Totternhoe Stone facies; N.S., Nettleton Stone; ACBE, Albian–Cenomanian boundary excursion; LCE I–III, Lower Cenomanian excursions 1–3; MCE I–II, Middle Cenomanian excursions 1–2; CTBE, Cenomanian–Turonian boundary excursion; S.B., sequence boundary; O.S., Onlap surface; TD, transgressive deposits; HS, highstand deposits; SMW, shelf margin wedge. See Fig. 2 for key to lithologies.

late Albian (*Stoliczkaia dispar* Zone) at Folke-stone.

LCE II is associated with the First *Inoceramus* Bed (of Jeans 1973). We place an onlap surface (OS 3) at the base of the First *Inoceramus* Bed and a sequence boundary (SB 3) a little lower where there is a slight increase in insoluble residue values. The $\delta^{13}C$ excursion is represented by a small positive peak in the lowstand and early transgressive deposits.

LCE III appears to be associated with the Second *Inoceramus* Bed (of Jeans 1973). An onlap surface (OS 4) is placed at the base of the Second *Inoceramus* Bed and a sequence bound-ary (SB 4) at the prominent marl below. The small positive $\delta^{13}C$ excursion is associated with the lowstand deposits. This is the third sequence recognized by Robaszynski *et al.* (1995).

We have, therefore, very clear evidence of a link between short-term fluctuations in sea-level and significant $\delta^{13}C$ excursions.

Interpretation of $\delta^{13}C$ values and the link with sea-level

The widespread distribution of each isotope excursion, together with its similar form, means that it is highly unlikely they are the result of diagenesis. Each section has undergone its own diagenetic history. For instance the chalks at Speeton show abundant calcite overgrowths on coccolith plates while the coccoliths from Folke-stone show relatively little overgrowth, yet both sections show similar $\delta^{13}C$ signatures. Carbon isotopic values are only weakly susceptible to diagenetic alteration associated with shallow burial diagenesis, because the volume of carbon in the carbonate reservoir is much greater than the volume of dissolved carbon species in pore waters (Marshall 1992), and the temperature-controlled fractionation effect is much less for carbon isotopes than it is for oxygen isotopes at near-surface burial temperatures (Emrich *et al.* 1970). The most likely causes of diagenetic alteration of the carbon signature are the presence of cements derived from bicarbonate formed from the oxidation of organic matter (producing a lighter carbon isotope signal) or from bacterial methanogenesis (giving a heavier signal). These processes occur in organic-rich sediments within, and beneath, the sulphate reduction zone, respectively. Given the extre-mely low proportion of organic carbon (less than 0.5%, Paul *et al.* 1994*a*) in these sediments (except those associated with the CTBE), it is unlikely that either process was important in the generation of these isotopic signals. They may

have been important, however, in modifying the $\delta^{13}C$ signal in CTBE sections rich in organic carbon.

Primary $\delta^{13}C$ values of marine (coccolith) chalks represent a measure of the $\delta^{13}C$ of dissolved inorganic carbon (ΣCO_2) in surface water, since there is little fractionation asso-ciated with the precipitation of skeletal calcite (see below). Changes in $\delta^{13}C$ of chalk reflect changes in carbon cycling in the ocean, particu-larly between reservoirs of significantly different isotopic composition. The Cenomanian was a time of long-term sea-level rise (Hancock & Kauffman 1979; Haq *et al.* 1988) which may be linked to increasing volumes of mid-ocean ridges (Mackenzie & Pigott 1981). Although mid-ocean ridges are an important source of CO_2, this CO_2 has a $\delta^{13}C$ value of between −3 and −8 ‰ PDB (Hoefs 1987) and cannot be responsible for the observed rise in $\delta^{13}C$ values. The most important process affecting $\delta^{13}C$ of ΣCO_2 is phytoplankton photosynthesis which strongly discriminates against ^{13}C, preferentially incorporating ^{12}C in organic matter ($\delta^{13}C = -20$). Thus changes in phytoplankton productivity, or bacterial oxida-tion of organic matter in the water column, will significantly affect surface water $\delta^{13}C$ values. Surface water phytoplankton productivity is largely controlled by changes in the input and output fluxes of nutrients to and from surface water. Nutrients are added from two main sources: riverine input from terrestrial weath-ering processes and the recycling of nutrients in deep ocean water at upwelling zones and through ocean mixing processes. Nutrients are also removed from the system by burial of organic matter. Over relatively long periods (> 250 ka) this system will reach equilibrium. We believe that on this long time scale two factors control $\delta^{13}C$ values: (1) changes in burial of organic carbon related to increases in the area of ocean floor (i.e. proportional to sea-level), and (2) increasing lock-up of particulate organic matter (POM) in low-oxygen deep and inter-mediate water masses. It is perhaps particularly significant that the rise in background $\delta^{13}C$ values during the Middle Cenomanian corre-sponds with global continental onlap (e.g. Wood *et al.* 1982; Haq *et al.* 1988; Simmons *et al.* 1991; Pascal *et al.* 1993; Robaszynski *et al.* 1995; Gale 1995).

In short term systems (of up to 1 Ma) exchanges at mid ocean ridges are too slow to affect the carbon system and changes in $\delta^{13}C$ must reflect changes in surface water produc-tivity and transport of POM to deep and inter-mediate water masses and sediments.

Sea-level falls lead to the exposure of continental shelves which are then subject to pedogenesis. Biological soils develop within a period of several tens of thousands of years (Ugolini & Saltenstein 1992) and lock up nutrients, particularly silica. With subsequent sea-level rise, low lying soils will be rapidly reworked by transgression promoting an increased flux of nutrients to marine surface waters. This change in the composition of surface sea water during transgressive intervals will directly affect the composition of the primary (and secondary) producing communities in surface water. Evidence of these changes can be obtained from studies of changes in biodetrital sedimentary style (i.e. accumulation of organic carbon, biogenic silica and biogenic carbonate).

Organic carbon accumulation. The CTBE is recognized as an interval of globally enhanced organic carbon-rich deposition (Schlanger *et al.* 1987). More generally, the association of organic carbon-rich sediments with transgressive deposits (or onlap surfaces) is widely recognized, and such deposits have been recorded in Canada (Leckie *et al.* 1992) for the ACBE, in the Vocontian Trough (Gale 1995) for MCE I, as well as at many other intervals in the geological column. This has led to the development of the concept of the 'Oceanic Anoxic Event' (Schlanger & Jenkyns 1976).

Biogenic silica accumulation. Silica is an important biolimiting element and free silica is rapidly incorporated into the shells of phytoplankton and zooplankton. The increased flux into the ocean of silica from soils during transgressive intervals would, therefore, have a significant effect on the composition of primary producers in surface water. Sea water is undersaturated with respect to silica and the development of biogenic silica-rich sediments (diatomites and radiolarites) is largely controlled by silica rain-rates (i.e. primary production rates) and carbonate deposition rates. Silica dissolution is enhanced by the presence of carbonate (i.e. higher pH values). Globally significant increases in the accumulation of biogenic silica must, therefore, reflect changes in the primary production of silica versus carbonate. The mid Cretaceous was characterized by intervals of enhanced biosilica deposition, i.e. biosiliceous events (Thurow *et al.* 1992).

Biogenic carbonate accumulation. The accumulation of calcium carbonate also varies significantly during the Cenomanian. Calculations for the Cenomanian–Turonian boundary (Paul & Mitchell 1994) demonstrated a marked reduction in carbonate (i.e. coccolith) accumulation rates. This implies that there is a reduction in the burial of carbonate from calcareous nannoplankton related to either a decrease in production or increased dissolution. The solubility of carbonate is increased by the presence of organic carbon (Archer & Maier-Reimer 1994). However, decreases in carbonate accumulation occur in sites where no organic carbon is preserved, suggesting that the decrease in carbonate accumulation is not solely due to increases in organic carbon rain rate.

The implication from these studies is that there was a change in the primary production from a high carbonate to a high silica–organic carbon productivity system. This change in productivity style will have had a significant effect on $\delta^{13}C$ since $\delta^{13}C$ of organic matter is dependent on the carbon source used (dissolved CO_2 versus bicarbonate). The majority of primary producers use dissolved CO_2 which will lead to a kinetic fractionation of about $-20‰$ PDB during photosynthesis (Farquhar *et al.* 1989). Coccolithophores use both dissolved CO_2 and bicarbonate as carbon sources (Chave 1984) which implies a lower level of fractionation (Hollander & Mckenzie 1991). Consequently, a shift from a coccolith-rich productivity system to a coccolith-poor productivity system will result in a significant change in the $\delta^{13}C$ flux of POM from surface to deep waters and lead to a positive shift in surface water $\delta^{13}C$. This idea needs to be tested.

Transgressive intervals can, therefore, be recognized as intervals during which there is an increased flux of nutrients (particularly silica) to surface sea water. This flux caused both an increase in primary production and also a change in the composition of primary producers leading to a change in the nature of the sediments accumulating during transgressive intervals. Increases in primary production will generate an increased rain rate of organic carbon. In the saline stratified Cretaceous ocean (Brass *et al.* 1992) this would have promoted oxygen depletion in deep slow circulating water masses. Locally it would have led to increased storage of organic carbon in these water masses and, where rain rates exhausted oxygen, to an increase in the burial of organic carbon. This removal of organic carbon from surface sea water would have resulted in a positive shift in surface water $\delta^{13}C$, the difference between background $\delta^{13}C$ before and after the event providing a measure of the amount of organic carbon permanently removed from surface waters.

Other important processes also occur during transgressive intervals. In deltaic-river environments, coal (peat) accumulation occurs in raised mires associated with high water tables (McCabe 1984). High water tables are promoted by transgressions where water levels are rising. Associations of coal bearing strata with transgressive intervals are well known (Arditto 1987). The widespread burial of coal during transgressive systems tracts will lead to an increase in $\delta^{13}C$ values of atmospheric CO_2 which will feed through to the oceanic system via ocean–atmosphere exchange of CO_2. Subsequent, transgressive reworking of peat deposits formed early in transgressive systems might account for the high proportion of terrestrial organic matter identified in Cretaceous black shales (Arthur *et al.* 1985).

The majority of the $\delta^{13}C$ excursions in the Cenomanian have a roughly symmetrical form. This cannot be explained solely by burial of organic material since this would permanently remove organic carbon from surface water and the $\delta^{13}C$ response would show a sharp positive $\delta^{13}C$ shift, followed by a gradual negative drift towards a new steady state value with the cessation of burial of organic carbon. Instead, $\delta^{13}C$ excursions rapidly fall to new steady state values. This implies that light carbon was temporarily removed from the surface water reservoir. We propose that two processes are involved: (i) post-event oxidation and methanogenesis of buried organic matter (supplying isotopically light carbon to the oceanic system) and (ii) decreased storage of light carbon in deep, slowly circulating water masses. The former process means that in a large number of settings the only organic material remaining (buried organic carbon) will be highly inert and this agrees with observations (Summerhayes 1987).

MCE II is characterized by a negative $\delta^{13}C$ shift. We believe that the response of the carbon system during lowstands is more variable than during transgressive systems tracts. Two effects are important: decreased burial rates of organic carbon in marine settings linked to increases in basinward migration of coarse grained clastics (within which organic carbon is less easily buried) and, increased incorporation of organic carbon in soils and biota associated with soil formation on exposed continental shelves. We believe that the dominance of the former leads to short-lived negative $\delta^{13}C$ excursions, while the dominance of the latter leads to the positive $\delta^{13}C$ excursion beginning during the forced regression. The different responses of $\delta^{13}C$ to different sea-level falls may be due to the rate at which sea-level fell or possibly to differences in climatic conditions prevailing at the time.

Problems and consequences of the burial of organic carbon

An increased burial of organic carbon should deplete dissolved carbon (CO_2) in surface water. This will drive a flux of CO_2 from the atmosphere to the ocean leading to a lowering of pCO_2. Since CO_2 is an important greenhouse gas this mechanism has been suggested to have led to glacial conditions in the late Ordovician (Brenchley *et al.* 1994) and mid Miocene (Vincent & Berger 1985) events. Glacial lock-up of water as ice on land will lead to a fall in sea-level. This mechanism, however, cannot be employed for the Cenomanian $\delta^{13}C$ excursions since the $\delta^{13}C$ excursions usually occur after the sea-level falls (sequence boundaries) and are *not* coincident with them! This implies that insufficient CO_2 was extracted from the atmosphere, by changing burial rates of organic carbon, to cause significantly lowered temperatures and to generate land-locked ice caps. Other mechanisms to control sea-level variation must be considered (e.g. tectonic uplift, changing ridge volumes, recharging aquifers, etc.).

Conclusions

1. Detailed isotope curves for the late Lower Cenomanian to early Lower Turonian have been given for three sections in different basins, and a curve for whole Cenomanian stage for the section at Speeton.

2. Where we have independent methods of correlating sections on the same scale as the $\delta^{13}C$ excursions, the excursions appear to be synchronous within stratigraphic resolution (except for MCE II where either the isotope excursion or the biostratigraphic markers are diachronous by about 80 ka at most between some sections).

3. The close correlation between background $\delta^{13}C$ values and long-term increase in eustatic sea-level during the mid-Cretaceous suggests that $\delta^{13}C$ might be considered as an independent method for estimating eustatic sea-level.

4. $\delta^{13}C$ excursions are caused by changes in carbon cycling associated with rapid regressive–transgressive cycles, specifically due to increased burial of organic carbon in transgressive deposits (exemplified by the widespread association of organic-rich black shales with transgressions),

and storage of organic carbon in deep and intermediate water masses.

Part of this work was undertaken during the tenure of an NERC Research Studentship (to SFM) and fieldwork was supported by NERC small grant GR9/1220. Stable isotope analyses were undertaken in the Liverpool University Stable Isotope Laboratory which is funded by grants to J. D. Marshall from the University and NERC. We thank the Rugby Portland Cement plc for allowing us free access to their quarry at South Ferriby. The British Council and DAAD supplemented fieldwork in Germany by CRCP.

References

ARCHER, D. & MAIER-REIMER, E. 1994. Effects of deep sea sedimentary calcite preservation on atmospheric CO_2 concentration. *Nature*, **367**, 260–263.

ARDITTO, P. A. 1987. Eustacy, sequence stratigraphic analysis and peat formation: a model for widespread late Permian coal deposition in the Sydney Basin, NSW. *Advanced studies of the Sydney Basin: 21st Newcastle symposium proceedings.* 11–17.

ARTHUR, M. A., DEAN, W. E. & CLAYPOOL, G. E. 1985. Anomalous ^{13}C enrichment in modern marine organic carbon. *Nature*, **315**, 216–218.

BRASS, G. W., SOUTHAM, J. R. & PETERSON, W. H. 1982. Warm saline bottom water in the ancient ocean. *Nature*, **296**, 620–623.

BRENCHLEY, P. J., MARSHALL, J. D., CARDEN, G. A. F., ROBERTSON, D. B. R., LONG, D. G. F. *ET AL.* 1994. Bathymetric and isotopic evidence for a short-lived late Ordovician glaciation in a greenhouse period. *Geology*, **22**, 295–298.

CARTER, D. J. & HART, M. B. 1977. Aspects of mid-Cretaceous stratigraphical micropalaeontology. *Bulletin of the British Museum, Natural History (Geology)*, **29**, 1–135.

CHAVE, A. 1984. Physics and chemistry of biomineralization. *Annual Review of Earth and Planetary Science*, **12**, 293–305.

CHRISTENSEN, W. 1990. *Actinocamax primus* Arkhangelsky (Belemnitellidae; Upper Cretaceous), biometry, comparison and biostratigraphy. *Paläontologische Zeitschrift*, **64**, 75–90.

CRAIG, H. 1957. Isotopic standards for carbon and oxygen fractionation factors for mass-spectrometric analysis of carbon dioxide. *Geochimica et Cosmochimica Acta*, **12**, 133–149.

DITCHFIELD, P. W. & MARSHALL, J. D. 1989. Isotopic variation in rhythmically bedded chalks: paleotemperature variation in the Upper Cretaceous. *Geology*, **17**, 842–845.

EMRICH, K., EHALT, D. H. & VOGEL, J. C. 1970. Carbon isotope fractionation during the precipitation of calcium carbonate. *Earth and Planetary Science Letters*, **8**, 363–371.

ERNST, G., SCHMID, F. & SEIBERTZ, E. 1983. Eventstratigraphie im Cenoman und Turon von NW-Deutschland. *Zitteliana*, **10**, 531–554.

FARQUHAR, G. D., EHLERINGER, J. R. & HUBICK, K. T. 1989. Carbon isotope discrimination and photosynthesis. *Annual Review of Plant Physiology and Plant Molecular Biology*, **40**, 503–537.

FRIEDMAN, I. & O'NEIL, J. R. 1977. Compilation of stable isotope fractionation factors of geochemical interest. *In:* FLEISCHER, M. (ed.) *Data of Geochemistry.* United States Geological Survey Professional Paper, **440**, KK1–KK12.

GALE, A. S. 1989. Field meeting at Folkestone Warren, 29th Nov., 1987. *Proceedings of the Geologists' Association*, **100**, 73–82.

——— 1990. A Milankovitch scale for Cenomanian time. *Terra Nova*, **1**, 420–425.

——— 1996. Cyclostratigraphy and correlation of the Cenomanian stage in Western Europe. *In:* HOUSE, M. R. & GALE, A. S. (eds) *Orbital forcing timescales and cyclostratigraphy.* Geological Society, London, Special Publication, **85**, 177–197.

———, JENKYNS, H. C., KENNEDY, W. J. & CORFIELD, R. 1993. Chemostratigraphy versus biostratigraphy: data from around the Cenomanian–Turonian boundary. *Journal of the Geological Society, London*, **150**, 29–32.

GAUNT, G. D., FLETCHER, T. P. & WOOD, C. J. 1992. *Geology of the country around Kingston upon Hull and Brigg.* British Geological Survey Sheet Memoirs.

HANCOCK, J. M. 1989. Sea-level changes in the British region during the late Cretaceous. *Proceedings of the Geologists' Association*, **100**, 565–594.

——— & KAUFFMAN, E. G. 1979 The great transgressions of the Late Cretaceous. *Journal of the Geological Society, London*, **136**, 175–186.

HAQ, B. U., HARDENBOL, J. & VAIL, P. R. 1988. Mesozoic and Cenozoic chronostratigraphy and cycles of sea-level change. *In:* WILGUS, C. K., HASTINGS, B. S., KENDALL, C. G. ST. C., POSAMENTIER, H. W., ROSS, C. A. & VAN WAGONER, J. C. (eds) *Sea-level changes – an integrated approach.* SEPM Special Publication, **42**, 71–108.

HART, M. B. 1985. Oceanic anoxic event 2 on-shore and off-shore S.W. England. *Proceedings of the Ussher Society*, **6**, 183–190.

——— 1987. Orbitally induced cycles in the Chalk facies of the United Kingdom. *Cretaceous Research*, **8**, 335–348.

———, DODSWORTH, P., DITCHFIELD, P. W., DUANE, A. M. & ORTH, C. J. 1991. The late Cenomanian event in eastern England. *Historical Biology*, **5**, 339–354.

HILBRECHT, H. & HOEFS, J. 1986. Geochemical and palaeontological studies of the $\delta^{13}C$ anomaly in Boreal and north Tethyan Cenomanian–Turonian sediments in Germany and adjacent areas. *Palaeogeography, Palaeoclimatology, Palaeoecology*, **53**, 169–189.

———, HUBBERTON, H. W. & OBERHANSLI, H. 1992. Biogeography of planktonic foraminifera and regional carbon isotope variations: productivity and water masses in Late Cretaceous Europe. *Palaeogeography, Palaeoclimatology, Palaeoecology*, **93**, 407–421.

HOEFS, J. 1987. *Stable isotope geochemistry*, 3rd edn, Springer-Verlag, Berlin.

HOLLANDER, D. J. & MCKENZIE, J. A. 1991. CO_2 controls on carbon-isotope fractionation during aqueous photosynthesis: a paleo-pCO_2 barometer. *Geology*, **19**, 929–932.

JARVIS, I., CARSON, G. A., HART, M. B., LEARY, P. N. & TOCHER, B. A. 1988*a*. The Cenomanian-Turonian (late Cretaceous) anoxic event in SW England: evidence from Hooken Cliffs near Beer, SE Devon. *Newsletters on stratigraphy*, **18**, 147–164.

——, ——, COOPER, M. K. E., *ET AL.* 1988*b*. Microfossil assemblages and the Cenomanian-Turonian (late Cretaceous) oceanic anoxic event. *Cretaceous Research*, **9**, 3–103.

JEANS, C. V. 1973. The Market Weighton structure: tectonics, sedimentation and diagenesis during the Cretaceous. *Proceedings of the Yorkshire Geological Society*, **39**, 409–444.

—— 1980. Early submarine lithification in the Red Chalk and Lower Chalk of eastern England: a bacterial control model and its implications. *Proceedings of the Yorkshire Geological Society*, **43**, 81–157.

——, LONG, D., HALL, M. A., BLAND, D. J. & CORNFORD, C. 1991. The geochemistry of the Plenus Marls at Dover, England: evidence for fluctuating oceanographic conditions and of glacial control during the development of the Cenomanian–Turonian $\delta^{13}C$ anomaly. *Geological Magazine*, **128**, 603–632.

JEFFERIES, R. P. S. 1962. The palaeoecology of the *Actinocamax plenus* Subzone (lowest Turonian) in the Anglo-Paris Basin. *Palaeontology*, **4**, 609–647.

JENKYNS, H. C., GALE, A. S. & CORFIELD, R. 1994. Carbon- and oxygen-isotope stratigraphy of the English Chalk and Italian Scaglia and its palaeoclimatic significance. *Geological Magazine*, **131**, 1–34.

JUIGNET, P. & BRETON, G. 1992. Mid-Cretaceous sequence stratigraphy and sedimentary cyclicity in the western Paris Basin. *Palaeogeography, Palaeoclimatology, Palaeoecology*, **91**, 197–218.

JUKES-BROWNE, A. J. & HILL, W. 1903. *The Cretaceous rocks of Britain. Volume 2, The Lower and Middle Chalk.* Memoirs of the Geological Survey of the United Kingdom.

KAPLAN, U. & BEST, M. 1985. Zur Stratigraphie der tiefen Oberkreide im Teutoburger Wald (NE-Deutschland). Teil 1. Cenoman. *Bericht des Naturwissenschaftlichen Veriens Bielefeld*, **27**, 81–103.

LAMOLDA, M. A., GOROSTIDI, A. & PAUL, C. R. C. 1994. Quantitative estimates of calcareous nannofossil changes across the Plenus Marls (latest Cenomanian), Dover, England: implications for the generation of the Cenomanian/Turonian boundary event. *Cretaceous Research*, **14**, 143–164.

LECKIE, D. A., SINGH, C., BLOCH, J., WILSON, M. & WALL, J. 1992. An anoxic event at the Albian-Cenomanian boundary. The fish-scale marker bed, northern Alberta, Canada. *Palaeogeography, Palaeoclimatology, Palaeoecology*, **92**, 139–166.

MCCABE, P. J. 1984. Depositional environments of coal and coal-bearing strata. *In:* RAHMANI, R. A. & FLORES, R. M. (eds) *Sedimentology of coal and coal-bearing sequences.* International Association of Sedimentologists, Special Publication, **7**, Blackwell Scientific Publications, Oxford, 13–42.

MACKENZIE, F. T. & PIGOTT, J. D. 1981. Tectonic controls of Phanerozoic rock cycling. *Journal of the Geological Society, London*, **138**, 183–196.

MARSHALL, J. D. 1992. Climatic and oceanographic isotopic signals from the carbonate rock record and their preservation. *Geological Magazine*, **129**, 143–160.

MEYER, T. 1990. Biostratigraphische und sedimentologische Untersuchungen in der Plänerfazies des Cenoman von Nordwestdeutschland. *Mitteilungen aus dem Geologischen Institut der Universität Hannover*, **30**, 1–114.

MITCHELL, S. F. 1993. *The mid Cretaceous of north-east England: macrofauna, microfauna, sedimentology, stable isotope geochemistry and correlation with sections in southern England and north-west Germany.* PhD Thesis, University of Liverpool.

PASCAL, A. F., MATHEY, B. J., ALZOUMA, K., LANG, J. & MEISTER, C. 1993. Late Cenomanian–Early Turonian shelf ramp, Niger, West Africa. *In:* TONI SIMO, J. A., SCOTT, R. W. & MASSE J.-P. (eds) *Cretaceous carbonate platforms*, American Association of Petroleum Geologists, Memoir, **56**, 145–154.

PAUL, C. R. C. 1992. Milankovitch and microfossils: principles and practice of palaeoecological analysis illustrated by Cenomanian chalk-marl rhythms. *Journal of Micropalaeontology*, **11**, 95–105.

—— & MITCHELL, S. F. 1994. Is famine a common factor in marine mass extinctions? *Geology*, **22**, 679–682.

——, ——, MARSHALL, J. D., LEARY, P. N., GALE, A. S., DUANE, A. M. & DITCHFIELD, P. W. 1994*a*. Palaeoceanographic events in the Middle Cenomanian of northwest Europe. *Cretaceous Research*, **15**, 707–738.

——, ——, LAMOLDA, M. A. & GOROSTIDI, A. 1994*b*. The Cenomanian–Turonian boundary event in northern Spain. *Geological Magazine*, **131**, 801–817.

PRICE, F. G. H. 1877. On the beds between the Gault and Upper Chalk near Folkestone. *Quarterly Journal of the Geological Society, London*, **33**, 431–448.

RENARD, M. 1986. Pelagic carbonate chemostratigraphy (Sr, Mg, ^{18}O, ^{13}C). *Marine Micro-palaeontology*, **10**, 117–164.

ROBASZYNSKI, F., JUIGNET, P., GALE, A. S., AMEDRO, F. & HARDENBOL, J. 1995. *Sequence stratigraphy in the Upper Cretaceous of the Anglo-Paris Basin: exemplified by the Cenomanian Stage.* SEPM Special Publication: Progress in sequence stratigraphy conference, Dijon, in press.

SCHLANGER, S. O. & JENKYNS, H. C. 1976. Cretaceous oceanic anoxic events: causes and consequences. *Geologie et Mijnbouw*, **55**, 179–184.

——, ARTHUR, M. A., JENKYNS, H. C. & SCHOLLE, P. A. 1987. The Cenomanian–Turonian oceanic anoxic event, I. Stratigraphy and distribution of organic carbon-rich beds and the marine $\delta^{13}C$ excursion. *In:* BROOKS, J. & FLEET, A. J. (eds) *Marine Petroleum Source Rocks*. Geological Society, London, Special Publication, **26**, 371–399.

SCHOLLE, P. A. & ARTHUR, M. A. 1980. Carbon isotope fluctuations in Cretaceous pelagic limestones: potential stratigraphic and petroleum exploration tool. *American Association of Petroleum Geologists Bulletin*, **64**, 67–87.

SIMMONS, M. D., WILLIAMS, C. L. & HART, M. B. 1991. Sea-level changes across the Albian–Cenomanian boundary in south-west England. *Proceedings of the Ussher Society*, **7**, 408–412.

SUMMERHAYES, C. P. 1987. Organic-rich Cretaceous sediments from the North Atlantic. *In:* BROOKS, J. & FLEET, A. J. (eds) *Marine Petroleum Source Rocks*, Geological Society, London, Special Publication, **26**, 301–316.

THUROW, J., BRUMSACK, H.-J., RULLKÖTTER, J., LITTKE, R. & MEYERS, P. 1992. The Cenomanian/Turonian boundary event in the Indian Ocean – a key to understanding the global picture. *In: Synthesis of results from scientific drilling in the Indian Ocean*. American Geophysicist's Union, Geophysical Monograph, **70**, 253–273.

UGOLINI, F. C. & SALTENSTEIN, H. 1992. 7. Pedosphere. *In:* BUTCHER, S. S., CHARLSON, R. J., ORIANS, G. H. & WOLFE, G. V. (eds) *Global biogeochemical cycles*. Academic Press, London, 123–153.

VAN WAGONER, J. C., POSAMENTIER, H. W., MITCHUM, R. M., VAIL, P. R., SARG, J. F., LOUTET, T. S. & HARDENBOL, J. 1988. An overview of the fundamentals of sequence stratigraphy and key definitions. *In:* WILGUS, C. K., HASTINGS, B. S., KENDALL, C. G. St. C., POSAMENTIER, H. W., ROSS, C. A. & VAN WAGONER, J. C. (eds) *Sea-level changes – an integrated approach*. SEPM Special Publication, **42**, 39–45.

VINCENT, E. & BERGER, W. H. 1985. Carbon dioxide and polar cooling in the Miocene: the Monterey hypothesis. *In:* SUNDQUIST, E. T. & BROECKER, W. S. (eds) *The carbon cycle and atmospheric CO_2: nature variations Archean to present*. American Geophysicist's Union, Geophysical Monograph, **32**, 455–468.

WOOD, C. J., BIGG, P. J. & MEDD, A. W. 1982. The biostratigraphy of the Upper Cretaceous (Chalk) of the Winterbourne Kingston borehole, Dorset. *Institute of Geological Sciences Report*, **81/3**, 19–27.

Spectral gamma ray characterization of high resolution sequence stratigraphy: examples from Upper Carboniferous fluvio-deltaic systems, County Clare, Ireland

S. J. DAVIES & T. ELLIOTT

Department of Earth Sciences, University of Liverpool, PO Box 147, Liverpool L69 3BX, UK

Abstract: The application of high resolution sequence stratigraphy requires the ability to recognize key surfaces which record fluctuations in relative sea-level. In sub-surface studies, gamma ray logs have been used to identify maximum flooding surfaces, but their full potential has not been realized. Gamma ray profiles produced using a portable spectrometer on exposed Upper Carboniferous fluvio-deltaic deposits in western Ireland reveal that key surfaces and systems tracts can be characterized more comprehensively and recognized with greater confidence if spectral gamma ray data (K, U, Th and their respective ratios) are used in conjunction with traditional total count data. Maximum flooding surfaces can be distinguished from lesser flooding surfaces by a distinctive U peak (> 5 ppm) and low Th/U ratio (< 2.5). Erosional unconformities and their associated incised valley fills are characterized by consistently low total counts (40–50 cps) and high Th/K ratios (> 6). Laterally correlative interfluves are represented by distinctive palaeosols that can be clearly identified in spectral gamma ray data by their anomalously low K content (< 0.4%) and exceptionally high Th/K ratio (> 17). Finally, the stacking pattern of parasequence sets can be identified using the trends of Th/K ratios from sandstones in successive parasequences. These results have widespread implications for the recognition of high resolution sequence stratigraphic signatures in the stratigraphic record, with particular reference to the sub-surface analysis of fluvio-deltaic deposits.

The application of sequence stratigraphy to basin-fill successions is based on the recognition of key stratigraphic surfaces that are interpreted to reflect fluctuations in relative sea-level. The realization that these surfaces can often be recognized at relatively close vertical spacings, from metres to tens of metres, has led to the development of high resolution sequence stratigraphy which may be applied at finer scales than seismic stratigraphy (e.g. Van Wagoner *et al.* 1990). Whilst recognition of these key surfaces in the sub-surface is possible using core, data availability is limited and surfaces may more commonly be interpreted from gamma ray well log data, a vital source of information which should be fully utilized. To date, maximum flooding surfaces have been recognized as gamma ray 'spikes' associated with uranium concentrations in condensed horizons (Myers & Wignall, 1987; Leeder *et al.* 1990; Herron, 1991). In this study we argue that a more rigorous use of gamma ray data, particularly spectral gamma ray data, permits the recognition of a wider range of key surfaces that can facilitate the application of high resolution sequence stratigraphy in the sub-surface. This objective will be achieved via the reporting of a suite of gamma ray profiles measured through an exposed Upper Carboniferous deltaic succession produced using a portable, hand-held gamma ray spectrometer.

Methodology

Gamma ray logs are a measure of the natural radioactivity of rocks and total count gamma ray logs are used extensively in the hydrocarbon industry to interpret lithology and effect correlation in the sub-surface. The logs record changes in mineralogy as opposed to simply changes in grain-size. Potassium (K) is present in K-feldspars, micas and illitic clays. Thorium (Th) is concentrated in sand- and silt-sized heavy minerals, such as the monazite and zircon groups, or in the fine-grained fraction, in association with selected clay minerals (Herron & Matteson 1993) and as authigenic phosphates (Hurst 1990; Hurst & Midlowski 1994). Uranium (U) also occurs within the heavy minerals suite but can, additionally, be concentrated in anoxic sediments (Anderson *et al.* 1989; Lovley *et al.* 1991). Correct interpretations of gamma ray data therefore require an appreciation of the mineralogy of the strata. Misconceptions and misinterpretations of gamma ray data as lithological indicators are discussed by Myers & Bristow (1989) and Rider (1990).

From Howell, J. A. & Aitken, J. F. (eds), *High Resolution Sequence Stratigraphy: Innovations and Applications*, Geological Society Special Publication No. 104, pp. 25–35.

Natural gamma ray spectrometry (NGS) logs report the three main naturally occurring radio-elements, potassium (^{40}K), uranium (^{238}U) and thorium (^{232}Th). NGS logs are commonly run during hydrocarbon exploration and develop-ment programmes, but the spectral data tend not to be used despite recent advances in interpre-tation. Significant new interpretations include using Th/K ratios and Th:K cross-plots to distinguish deltaic facies (Myers & Bristow 1989), integrating biostratigraphy, U concentra-tions and geochemistry to locate condensed horizons (Leeder *et al.* 1990), and characterizing basin-fill using spatial radioelement variations related to lithology and source chemistry (Davies 1993). Here we explore the links between spectral gamma ray data and high resolution sequence stratigraphic signatures (key surfaces and systems tracts).

There are, naturally, differences between sub-surface and outcrop natural gamma ray spectro-metry, but these differences do not invalidate application of the principles derived from the outcrop analogues to the sub-surface. In the sub-surface, downhole measurements are usually recorded continuously every *c.* 0.15 m to produce a moving average (Rider 1986). As a general rule, the main contribution to the radioactivity detected in downhole measurements, horizon-tally and vertically, derives from within 0.3 m of the detector (Rider 1986) although this will be a larger interval at faster logging speeds. In contrast to static outcrop measurements, resolu-tion of downhole measurements is a function of activity and logging speed and is averaged in a vertical sense. High logging speeds, possible contamination by K-bearing additives in drilling mud and adsorption of gamma rays in barite within the drilling mud may all contribute to errors in subsurface NGS data. Recent develop-ment and improvement of tools has, however, counteracted some of the precision errors associated with the logging speeds (S. Herron pers. comm.).

At outcrop a scintillation detector and spec-trometer unit (the GPS-21 and GR-256 respec-tively) were used to collect the results. The spectrometer displays K, U and Th in counts per time period and converts these results to K%, U-ppm and Th-ppm. The detector is placed in contact with a planar rock surface and remains stationary whilst measuring. The sample dia-meter is 2 m with the main contribution being from a 0.5 m diameter around the detector (Løvborg & Mose 1987). The surface should be planar because both minor irregularities (i.e. surface undulations) and major irregularities (cliff edges and overhangs) influence the amount of rock the detector samples (*c.* 50 kg). If sampling areas vary significantly, changes in radioelement abundances will be a function of surface effects as well as mineralogical changes. Where surface irregularities cannot be avoided, the radioelement ratios (K/U, Th/U and Th/K) are still valid.

For this outcrop study, sample selection was based on the lithological variability, the spacing of lithological contacts and logistical constraints related to the geometry or accessibility of exposure. A spacing of between 0.5–1 m was commonly used for vertical outcrop profiles. Where a more detailed profile was sought across intervals of particular interest (e.g. flooding surfaces, erosional surfaces, palaeosols) the spacing was reduced to 0.15 m equivalent to the sub-surface. Long count times of 3 minutes for fine-grained lithologies and 4–6 minutes for sandstones were used to achieve a precision of < 10% for each radioelement in an individual reading.

Despite the differences between outcrop and sub-surface data collection, gamma ray data generated using a hand-held spectrometer on exposed strata provide an invaluable source of analogue data for improved understanding and interpretation of well log information. The particular advantage of outcrop studies lies with the wealth of available sedimentological data that can be calibrated directly to the NGS data.

The Clare Basin, Ireland

The Namurian Clare Basin is exposed in Atlantic-facing cliffs in the west of Ireland that provide a regional scale, depositional strike section through the basin fill (Fig. 1). The upper part of the basin fill is a deltaic succession 915 m in thickness which overlies deeper water slope and turbidite deposits (Rider 1974; Martinsen 1989; Pulham 1989; Collinson *et al.* 1991). The delta systems are predominantly fine-grained (mud/silt:sand ratio; 80:20), fluvial-dominated and unstable in that they include abundant examples of slumps, slides, growth faults and diapirs. Both shallow water, shelf-deltas and deeper water, shelf-edge delta systems occur, with the latter being dominant in the exposed sections (Pulham 1989).

The deltaic succession has traditionally been divided into 5 cyclothems defined by faunal concentrate, condensed horizons referred to as 'marine bands' (Rider 1974; Pulham 1989; Fig. 1). Each defined cyclothem demonstrates an overall coarsening-upward succession tradition-ally interpreted as a prograding delta system.

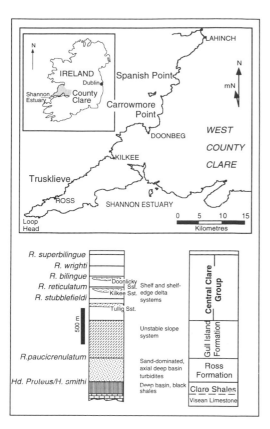

Fig. 1. Location and stratigraphy of the Upper Carboniferous Clare Basin, western Ireland. Names in italic are important marine bands which can be used on a basin and a regional scale.

Key surfaces, systems tracts and their gamma ray expression

The three lowermost cyclothems (the Tullig, Kilkee and Doonlicky cyclothems) have been the subject of a detailed reappraisal and are now interpreted as a series of sequences bound by erosional unconformities and equivalent surfaces with the marine bands being interpreted as maximum flooding surfaces (Elliott & Pulham 1990, 1993; Elliott & Davies 1994).

The mineralogical differences between the mature, quartz-rich sandstones and the mudrocks permit lithologies to be readily distinguished using gamma ray data. The sandstone mineralogy of the basin fill is dominated by strained quartz (70–90%) derived from a mature, second-cycle sedimentary source from the west. This results in a low total gamma ray response corresponding to low concentrations of K, U and Th. Subsidiary detrital components include: (i) feldspar (1–5%), split evenly between plagioclase and K-feldspar; (ii) mica (5%); (iii) heavy minerals (1–2%); and locally derived mudstone intraclasts (< 15%). The mudrocks are composed of illite, chlorite with variable amounts of kaolinite and minor detrital quartz.

In this synthesis of gamma ray responses the following key surfaces and systems tracts are discussed: (i) sequence boundaries and associated incised valley fills; (ii) initial and maximum flooding surfaces and the related transgressive systems tract; and (iii) parasequence flooding surfaces and parasequence sets. Examples are taken from a range of stratigraphic levels within the delta systems of the Central Clare Group (Fig. 1).

Characteristics of sequence boundaries

In the deltaic succession sequence boundaries comprise two elements: firstly, regional scale erosional unconformities overlain by multistorey fluvial deposits interpreted as incised valley fills; and secondly, highly distinctive palaeosols that are interpreted as laterally correlative interfluves to the erosional unconformities and incised valley fills.

Erosional unconformities and incised valley fills

The lower three cyclothems of the Central Clare Group are extensively exposed and each includes a major erosive-based sandstone body in the upper part of the cyclothem, referred to as the Tullig, Kilkee and Doonlicky Sandstones respectively (Fig. 1). These units are regional scale, sandstone-dominated fluvial complexes composed of vertically and laterally stacked channel-fill members. The thickness of these bodies averages 35 m (range 15–50 m) and one example has a mappable width, normal to palaeoflow of 20 km (Kilkee Sandstone, Fig. 1).

These multistorey fluvial complexes are interpreted as incised valley fills for the following reasons. Firstly, the scale and character of the fluvial complexes are dramatically different from those of interpreted distributary channels in the delta front of these systems. The delta front deposits underlying the fluvial complexes comprise isolated, 15 m-thick sandstone bodies of 2–3 km lateral extent that are interpreted as mouth-bars developed at the mouths of widely spaced, narrow distributary channels. The contrast between narrow, widely spaced distributary channels and regional scale, multistorey fluvial

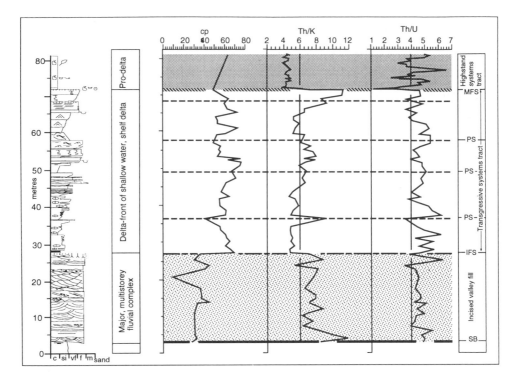

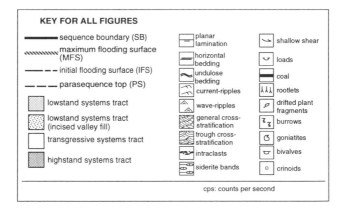

Fig. 2. The Tullig Sandstone and overlying systems tracts, Carrowmore Point (Fig. 1) This section com-prises an erosional unconformity (sequence boundary) with the associated incised valley fill (IVF) and a hierarchy of flooding surfaces: parasequence, initial and maximum. The base of the IVF (3 m) is characterized by a high Th/K ratio (12) as there is no basal intraclast conglomerate (cf. Fig. 3). The initial flooding surface (27 m) is represented by a large shift towards higher total counts and a drop in the Th/K ratio (to < 6) in comparison to the incised valley fill. The maximum flooding surface (72 m) has the lowest Th/U ratio (1.2) and successive faunal concentrations are characterized by Th/U < 3.8.

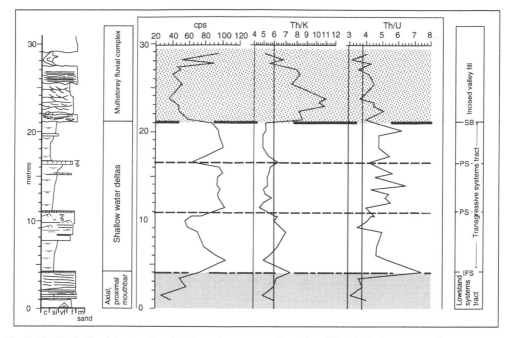

Fig. 3. The Tullig Sandstone and underlying systems tract, Trusklieve (Fig. 1). In the transgressive systems tract (4.5–21 m), the retrogradational stacking pattern, with decreasing sand : silt ratio upwards, is characterized by a trend towards lower Th/K ratios (compare the successive parasequences). Low total counts (< 50 cps) characterize the incised valley fill. A basal intraclast conglomerate overlying the erosional unconformity subdues the Th/K ratio at the base of the incised valley fill (Th/K ratio = 12, cf. Fig. 2) and higher ratios are associated with the overlying storeys (Th/K 12–14).

complexes records a radical change in the nature of the fluvial system that is felt to be a response to a basinward shift and net erosional phase during falling sea-level, followed by infilling of the erosional valley during early rise of relative sea-level. Secondly, erosional relief on the base of the fluvial complexes is greater than the estimated depth of the individual channels in the complex, suggesting that the channels were incised relative to large-scale relief on the erosional unconformity. This also provides further evidence of an initial erosional phase and later depositional phase in the evolution of the fluvial complex. Thirdly, in areas adjacent to the fluvial complexes the correlative interfluves of the complex can be identified by distinctive palaeosol horizons, suggesting that the complex was incised below a perched terrace and that there was no coeval finer-grained deposition beyond its lateral limits (see below). The erosive bases of the fluvial complexes are, therefore, interpreted as regional scale erosional unconformities and the deposits of the fluvial complex as incised valley fills.

Incised valleys fills are represented by a pronounced decrease in total counts and in-crease in Th/K values (Figs 2 and 3). Total counts of 20–30 cps and Th/K ratios greater than 6 prevail, with the latter having an average of 8.38±1.43 (1 standard deviation from the mean value). The erosional unconformity surface is generally characterized by the highest Th/K ratio for the incised valley fill, but the response may be subdued by the presence of intraclast conglomerates which increases the K values and slightly lowers the Th/K ratio (Fig. 3; 21 m). The Th/K ratio of the incised valley fill is determined by low amounts of K-bearing minerals (less than 2%) and the presence of heavy minerals which contain a significant percentage of Th.

The variable character of the Th/K ratio of the incised valley fill is produced by intervals of lower Th/K ratios associated with either an increase in quartz abundance reflected by low K (0.8–1%) and low Th (3–5 ppm), or localized fine-grained intervals interpreted either as channel plug facies (Fig. 2) or slumped overbank deposits (Fig. 3). Fine-grained intervals within the incised valleys are characterized by high total counts compared to other parts of the incised valley fill (Fig. 3, 28 m).

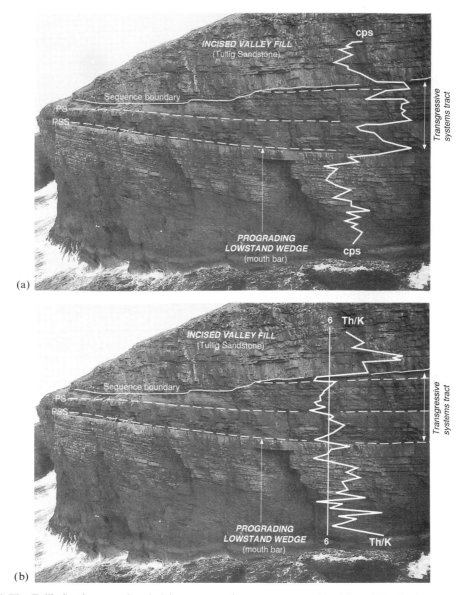

Fig. 4. The Tullig Sandstone and underlying transgressive systems tracts, Trusklieve (Fig. 1). Gamma ray logs superimposed onto the exposure at Trusklieve demonstrate the response at the parasequence flooding surfaces in a retrogradational stacking pattern and at the erosional unconformity. The erosion on the base of the incised valley is 6–7 m over a distance of approximately 55 m. Approximate cliff height is 45 m. (a) Total counts (cps), readings at same scale as Fig. 3. (b) Th/K ratio, reference line at Th/K = 6; readings at same scale as Fig. 3.

Interfluves

Interfluves are identified by palaeosol horizons that are distinctive by virtue of being exceptionally mature and well-drained. The Kilkee and Doonlicky incised valley fills are aligned approximately normal to the depositional strike section of the basin fill and examples of interfluves have been identified at the valley margins (Fig. 5a). The palaeosols are light grey in colour, exhibit abundant rootlets, *in situ* *Stigmaria* and *Calamites* and well developed profiles of ironstone rhizoconcretions (Fig. 5b). Bedding plane exposures of the upper, Doonlicky example also exhibit a well developed polygonal crack pattern that testifies to well

(a)

(b)

(c)

Fig. 5. Interfluve equivalent to the Doonlicky Incised Valley Fill, Spanish Point (Fig. 1). (**a**) Stratigraphic context of an interfluve palaeosol: the palaeosol is located in the centre of the photograph below the present day boulders on the left side; the initial flooding surface directly overlies the palaeosol and the maximum flooding surface with subsequent highstand deposits are present in the upper part of the photograph. The photograph can be directly compared with the log and gamma ray profile of Fig. 6 which was produced at this location. (**b**) Detail of the interfluve palaeosol profile illustrating the rhizoconcretions (dark, cylindrical to nodular bodies) and rootlets (fine dark streaks) in the upper part of the profile. Photograph is equivalent to Fig. 4, 3.5–4.25 m. Lens cap is 0.05 m in diameter. (**c**) Bedding plane view of the upper surface of the interfluve palaeosol; this upper surface is extremely irregular and exhibits polygonal patterned ground (to left and below lens cap). Individual polygons average 0.1–0.15 m in diameter. Associated with the patterned ground are cracks a few millimetres across and infilled with clean, well sorted sandstone.

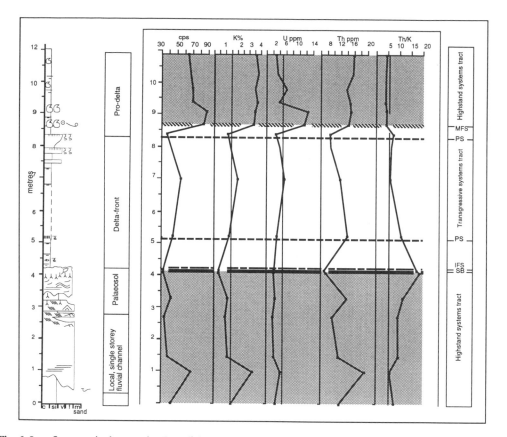

Fig. 6. Interfluve equivalent to the Doonlicky Incised Valley Fill, Spanish Point (Fig. 1). Spectral gamma ray signature of an interfluve palaeosol: the interfluve (3.5–4.25 m) is characterized by a marked decrease in total counts (31 cps), a concomitant decrease in K (0.4%) and Th (7 ppm) and consequently a high Th/K ratio of 17 (cw. Th/K$_{mean}$ = 8.38±1.4 for incised valley fill sandstones). The initial flooding surface is represented by a decrease in the Th/K ratio and is distinguished from the maximum flooding surface by low U. The maximum flooding surface is characterized by a marked increase in the total counts, of which the major contributor is U.

drained conditions that pertained in the soil profile despite the humid-tropical climate that prevailed in this region during the Upper Carboniferous (Fig. 5c). This well drained character suggests that the horizon was perched above the regional water table for a prolonged period during the cutting and filling of the incised valley. In contrast, palaeosols associated with bay filling are thin, immature, poorly drained palaeosols that lack evidence of prolonged exposure. Interfluve palaeosols are directly overlain by mudrocks that contain a condensed faunal concentrate horizon interpreted as the initial flooding surface (Fig. 5a, Fig. 6 at 4.25 m).

The interfluve palaeosol surface cannot be identified using the total count gamma ray data in isolation (Fig. 6). However, using the spectral data, interfluves can be recognized by a distinctive gamma ray signature involving exceptionally low K (less than 0.4%) and resulting high Th/K ratios, that are in excess of 17 (Fig. 6). The low K concentrations are considered to reflect leaching of K during palaeosol formation. Above the interfluve the initial flooding surface is characterized by a return to higher total counts and low Th/K ratios. The combination of a high Th/K ratio followed immediately by an abrupt decrease is a distinctive and easily recognized signature. The immature, wetted, bay-fill palaeosols lack this distinctive gamma ray signature.

Characteristics of maximum and initial flooding surfaces

In these Upper Carboniferous strata maximum flooding surfaces are represented by regionally widespread, goniatite-bearing marine bands that have historically been used to correlate successions. The horizons record reduced clastic input, condensed sedimentation and a pronounced deepening that are typically associated with the maximum rate of sea-level rise. The maximum flooding surfaces are consistently represented by high U, as noted by other workers (Leeder *et al.* 1990; Herron, 1991) and by a concomitantly low Th/U ratio 2.28±0.72 (Figs 2 & 4). Most terrestrial igneous rocks, from which sediments are derived, have Th/U ratios of approximately 3.8 (Taylor & McLennan 1985). Sediments deposited in an oxidizing environment will be characterized by higher Th/U ratios, through loss of uranium in its soluble 6+ form, whereas the low Th/U ratios encountered in marine bands result from the concentration of insoluble U^{4+}. These low Th/U ratios of less than 3.8 are diagnostic of maximum flooding surfaces in this succession.

The initial flooding surface overlies the incised valley and adjacent interfluves and is the first significant flooding surface above a sequence boundary (Figs 2 & 4). This marine transgressive surface formed during the early stages of sea-level rise as the infilling of the valley was completed and the adjacent interfluves flooded. In the Central Clare Group, initial flooding surfaces are represented by intensely bioturbated horizons, dominated by *Zoophycos*, which are overlain by marine mudstones containing a sparsely fossiliferous horizon (Pulham 1989). Initial flooding surfaces are consistently represented by a change to higher total counts, via an increase in all 3 radioelements, and lower Th/K ratios. The increase in U is of a lower magnitude than that associated with the maximum flooding surface. The initial flooding surface therefore lacks the identifiable U peak that is typical of maximum flooding surfaces.

Characteristics of parasequences and parasequence sets

In the Clare Basin, coarsening-upwards parasequences are well developed in the lowstand systems tract (shelf-edge deltas of the prograding lowstand wedge), the transgressive systems tract (shelf–delta complexes) and to a lesser extent the highstand systems tract (mud-dominated, distal parasequences). Typically, the parasequences are bound by intensely bioturbated surfaces that record an abrupt deepening of facies but lack any concentration of preserved body fauna.

The parasequences are characterized by cycles of decreasing upward total counts (Figs 2 & 3), a feature also documented by Van Wagoner *et al.* (1990). However, using the spectral gamma ray data, in this case the Th/K ratio, improves the resolution of parasequences and can also be used to identify parasequence stacking patterns. In the transgressive systems tract preserved beneath the sequence boundary associated with the Tullig Sandstone (Figs 3 & 4) the parasequences are 6–7 m thick and display an upward-decreasing sand/silt ratio that suggests a retrogradational stacking pattern. This stacking pattern is revealed by the total counts and, more emphatically, by the Th/K ratio (Figs 3 & 4). In successive parasequences, lower Th/K ratios (< 6) become predominant, with the third and incomplete parasequence having a maximum Th/K ratio of c. 5. This gamma ray pattern is a consistent feature at the equivalent interval across the basin. In common with both maximum and initial flooding surfaces, parasequence flooding surfaces are characterized by a change from low total counts (low K, U and Th) and high Th/K ratios at the bounding surface to high total counts and low Th/K ratios above.

Discussion

The results of this study suggest that the use of spectral gamma ray data, in addition to total count data, permits greater precision and confidence in the identification of key surfaces and systems tract. For example from the outcrop examples, (i) a wider range of flooding surfaces can be identified accurately, with maximum flooding surfaces being distinguished from initial, parasequence set and parasequence flooding surfaces by their associated U peak (> 6 ppm) and low Th/U ratios (< 2.5); (ii) erosional unconformities and incised valley fills are characterized by low total counts and high, but variable, Th/K ratios (> 6), with the variability reflecting the multistorey nature of the fill; (iii) interfluve sequence boundaries can be identified using their spectral characteristics of low K (< 0.4%) and exceptionally high Th/K ratios (> 17); and (iv) stacking patterns of parasequences can be resolved using Th/K ratios as a means of more accurately identifying the proportion of sandstone facies in successive parasequences.

Perhaps most significantly, spectral gamma ray data permit the identification of significant surfaces which cannot be recognized in total count data. For example, peaks in total gamma ray response do not always coincide with the maximum flooding surface identified in core (e.g. Bhattacharya 1993). By using spectral data it should be possible to decide whether the increase in total counts results from an increase in all radioelements or whether it results from a pronounced increase in U. Only in the latter case will the gamma ray data identify a maximum flooding surface. A further example concerns the search for evidence of sequence boundaries in sub-surface data in which the ability to recognize interfluves is critical. These surfaces provide a means of extending correlations from incised valleys interpreted elsewhere in a dataset, or predicting the presence of incised valleys at a particular stratigraphic level. Commonly, interfluves cannot be recognized due to an absence of core (e.g. Bhattacharya 1993), or because the interfluve palaeosol has been partially reworked by shoreface erosion associated with the overlying initial flooding surface (e.g. Van Wagoner et al. 1990). The radiochemical signature reported in this study may provide a means of identifying these important surfaces.

The results of this study have implications for the sub-surface analysis of hydrocarbon-bearing fluvio-deltaic successions in many parts of the world (e.g. North Slope, Alaska, the Mahakam (Indonesia) and Barram (Borneo) deltas, the Gulf Coast, U.S.A.). Providing that the mineralogy of the sediments is conducive to producing sensitive gamma ray responses, the integration of spectral gamma ray data into the overall analysis can yield significant insights into the high resolution sequence stratigraphy of these successions.

Conclusions

The analysis of the different spectral components of subsurface NGS logs using the ratio method illustrated in this paper could substantially improve geological interpretation and correlation. Subsurface NGS logs recorded at sufficiently slow logging speeds could provide critical information for a high resolution sequence stratigraphic interpretation that would be difficult or even impossible to obtain from other wireline logs. Where NGS data has been calibrated to core in one well, interpretations could be extended to additional wells within a field with greater confidence if NGS, rather than standard gamma ray, data are available.

Conoco UK Ltd are gratefully acknowledged for a research grant that enabled the high resolution sequence stratigraphy and gamma ray spectrometry to be investigated. The authors thank the referees Dr Ian Bryant and Dr Lee Kristinik for their comments on improving this manuscript.

References

ANDERSON, R. F., LEHURAY, A. P., FLEISHER, M. Q. & MURRAY, J. W. 1989. Uranium deposition in the Saanich Inlet sediments, Vancouver Island. *Geochimica et Cosmochimica Acta*, **53**, 2205.

BHATTACHARYA, J. P. 1993. The expression and interpretation of marine flooding surfaces and erosional surfaces in core; examples from Upper Cretaceous Dunvegan Formation, Alberta foreland basin, Canada. *In:* POSAMENTIER, H. W., SUMMERHAYES, C. P., HAQ, B. U. & ALLEN, G. P. (eds) *Sequence stratigraphy and facies associations.* International Association of Sedimentologists, Special Publication, **18**, 125–160.

COLLINSON, J. D., MARTINSEN, O. J., BAKKEN, B. & KLOSTER, A. 1991. Early fill of the western Irish Namurian Basin: a complex relationship between turbidites and deltas. *Basin Research*, **3**, 223–242.

DAVIES, S. J. 1993. *The radiochemical evolution of the Devonian Orcadian Basin, NE Scotland and comparison with coeval clastic systems from Wales, Norway and the Clair Field.* PhD Thesis, University of Leicester.

ELLIOTT, T. & DAVIES, S. J. 1994. High resolution sequence stratigraphy of an Upper Carboniferous basin-fill succession, Co. Clare, western Ireland: Fieldtrip Guide B2. *High resolution sequence stratigraphy: Innovations and applications.* University of Liverpool.

—— & PULHAM, A. J. 1990. Incised valley fills and the sequence stratigraphy of Upper Carboniferous shelf-margin deltaic cyclothems. *American Association of Petroleum Geologists Bulletin*, **75**, 874.

—— & —— 1993. High resolution sequence stratigraphy of Upper Carboniferous shelf-margin delta systems: analogues for Gulf Coast delta systems. *In:* American Association of Petroleum Geologists, abstract, 1993 Annual Convention Programme, New Orleans, 97.

HERRON, M. M. & MATTESON, A. 1993. Elemental composition and nuclear parameters of some common sedimentary minerals. *Nuclear Geophysics*, **7**, 383–406.

HERRON, S. L. 1991. *In situ* evaluation of potential source rocks by wireline logs. *In:* MERRIL, R. (ed.) *Source and migration processes and techniques for evaluation.* American Association of Petroleum Geologists, 127–134.

HURST, A. 1990. Natural gamma ray spectrometry in hydrocarbon-bearing sandstones from the Norwegian Continental Shelf. *In:* HURST, A., LOVELL, M. A. & MORTON, A. C. (eds) *Geological Applications of Wireline Logs.* Geological Society, London, Special Publication, **48**, 211–222.

—— & MIDLOWSKI, A. E. 1994. *Characterisation of clays in sandstones: Thorium content and spectral log data.* Transactions of the European Formation Evaluation Symposium, Paper S.

LEEDER, M. R., RAISWELL, R., AL-BIATTY, H., MCMAHON, A. & HARDMAN, M. 1990. Carboniferous stratigraphy, sedimentation and correlation of well 48/3-3 in the southern North Sea Basin: integrated use of palynology, natural gamma/ sonic logs and carbon/sulphur geochemistry. *Journal of the Geological Society, London,* **147,** 287–300.

LØVBORG, L. & MOSE, E. 1987. Counting statistics in radioelement assaying with portable spectrometer. *Geophysics,* **52,** 555–563.

LOVLEY, D. R., PHILIPS, E. J. P., GORBY, Y. A. & LANDA, E. R. 1991. Microbial reduction of uranium. *Nature,* **350,** 413–416.

MARTINSEN, O. J. 1989. Styles of soft-sediment deformation on a Namurian (Carboniferous) delta slope Western Irish Namurian Basin, Ireland. *In:* WHATELEY, M. & PICKERING, K. T. (eds) *Deltas: Sites and Traps for Fossil Fuels.* Geological Society, London, Special Publication, **41,** 167–177.

MYERS, K. J. & BRISTOW, C. S. 1989. Detailed sedimentology and gamma ray characteristics of a Namurian deltaic succession. Part II: Gamma ray logging. *In:* WHATELEY, M. & PICKERING, K. T. (eds) *Deltas: Sites and Traps for Fossil Fuels.* Geological Society, London, Special Publication, **41,** 81–88.

—— & WIGNALL, P. B. 1987. Understanding Jurassic organic-rich mudrocks – new concepts using gamma ray spectrometry and palaeoecology: Examples from the Kimmeridge Clay, Dorset and the Jet Rock of Yorkshire. *In:* LEGGETT, J. K. & ZUFFA, G. G. (eds) *Marine Clastic Sedimentology.* Graham & Trotman, London, 172–189.

PULHAM, A. J. 1989. Controls on internal structure and architecture of sandstone bodies within Upper Carboniferous fluvial-dominated deltas, County Clare Western Ireland. *In:* WHATELEY, M. & PICKERING, K. T. (eds) *Deltas: Sites and Traps for Fossil Fuels.* Geological Society, London, Special Publication, **41,** 179–203.

RIDER, M. H. 1974. The Namurian of west County Clare. *Proceedings of the Royal Irish Academy,* **74B,** 125–142.

—— 1986. *The geological interpretation of well logs.* Whittles Publishing, Caithness.

—— 1990. Gamma ray log shape used as a facies indicator; critical analysis of an oversimplified methodology. *In:* HURST, A., LOVELL, M. A. & MORTON, A. C. (eds) *Geological Applications of Wireline Logs.* Geological Society, London, Special Publication, **48,** 27–37.

TAYLOR, S. R. & MCLENNAN, S. M. 1985. *The Continental Crust: its Composition and Evolution,* Blackwell Scientific Publications, Oxford.

VAN WAGONER, J. C., MITCHUM, R. M., CAMPION, K. M. & RAHMANIAN, V. D. 1990. *Siliciclastic sequence stratigraphy in well logs core and outcrops.* AAPG, Methods in Exploration Series, **7.**

'Base profile': a unifying concept in alluvial sequence stratigraphy

DAVID G. QUIRK

Oxford Brookes University, Geology & Cartography Division, Gipsy Lane Campus, Headington, Oxford OX3 0BP, UK

Abstract: The sequence stratigraphy of sedimentary strata is governed by the creation and removal of accommodation space. However, current sequence stratigraphic models do not properly account for changes in accommodation space in the alluvial environment. In order to rectify this, a concept is proposed here called base profile. Base profile can be used to describe and explain the deposition and removal of fluvial sediments analogous to the way the term base level (relative sea-level) is used in the sequence stratigraphic analysis of coastal and marine strata. Base profile is the surface measured relative to a chronostratigraphic datum within a drainage basin to which rivers would regrade were conditions to remain constant. It represents the continental extension of base level. Changes in accommodation space occur if base profile rises or falls as a result of (a) subsidence or uplift, (b) variations in sediment supply, (c) variations in river discharge, (d) eustatic fluctuations, and (e) progradation or retrogradation of the coast. During a rise in base profile, coarse-grained siliciclastic sediment is trapped on the alluvial plain due to fluvial aggradation and mostly only fine-grained sediment will reach the marine environment. A fall in base profile is associated with fluvial degradation and maximum sediment input to the marine environment. Base profile forms a simple basis for describing, correlating and interpreting alluvial strata. Simple models are developed here using base profile which show how different continental and marine processes can lead to the deposition of laterally extensive, coarse-grained fluvial strata.

The correlation and interpretation of marine and coastal sediments have been revolutionized in recent years by the development and widespread application of sequence stratigraphy (e.g. Vail *et al.* 1977; Van Wagoner *et al.* 1988; Posamentier *et al.* 1988; Posamentier & Vail 1988; Galloway 1989; Shanley & McCabe 1991). However, many authors have indicated that there are limitations in using current sequence stratigraphic models to explain the deposition and erosion of sediment by rivers (e.g. Leopold & Bull 1979; Galloway 1989; Miall 1991; Schumm 1993; Wescott 1993; Wright & Marriott 1993; Koss *et al.* 1994; Shanley & McCabe 1994). One of the main problems is that, at present, it is difficult to describe alluvial sediments in sequence stratigraphic terms without first interpreting whether they may, for example, have developed during a rise or a fall in relative sea-level. Even where marine influence is unequivocal, workers often have problems in agreeing whether fluvial deposits should be assigned to lowstand, transgressive or highstand systems tracts (e.g. compare Maynard 1992; Bristow & Maynard 1994; Church & Gawthorpe 1994; Hampson *et al.* 1996).

There are three important reasons why it is essential that sequence stratigraphy can be applied successfully to alluvial sediments. Firstly, fluvial sandstones form excellent hydrocarbon reservoirs and any improvement in their prediction and correlation is of great economic value. Secondly, the deposition and removal of sediment on the alluvial plain directly affects the sediment flux to, and hence the sequence stratigraphy of, coastal and marine deposits. Thirdly, it is preferable if geologists and geomorphologists can describe and interpret alluvial sediments in a consistent way.

The purpose of the first part of this paper is to introduce a new concept that describes the dynamic nature of the drainage basin similar to the way that the concept of base level describes the dynamic nature of the marine basin. The second part of the paper consists of a discussion of some simple models and implications which arise from the new concept.

Terminology

Some confusion has been caused in recent years by different usage of terms such as 'base level' and attempts to introduce new terminology such as 'stream equilibrium profile' (e.g. Posamentier *et al.* 1988). In this paper base level is used to mean relative sea-level. The term stream equilibrium profile is dropped in favour of 'graded profile'. In addition, the term 'drainage basin' refers to all parts of the continent that are drained or traversed by rivers that ultimately deliver water to a similar part of the coast; the network of rivers and streams within a drainage

From Howell, J. A. & Aitken, J. F. (eds), *High Resolution Sequence Stratigraphy: Innovations and Applications*, Geological Society Special Publication No. 104, pp. 37–49.

basin is known as the 'river system'; the term 'hinterland' refers to the highland or proximal area of the drainage basin from which coarse-grained sediment is supplied; the term 'alluvial plain' indicates all parts of the drainage basin between the hinterland and the coast, be they valleys, floodplains or interfluvial areas; the word 'coast' is used in preference to bayline as the lower limit of alluvial deposits and as the upper limit of paralic and estuarine deposits; and the term 'marine' includes paralic and estuarine environments.

The graded river

An easy way of understanding sedimentary processes in the alluvial environment is by considering the concept of the graded river. A graded river is defined as one in which no significant erosion or deposition occurs along its length that is of permanent effect on the overall profile of the river (Davis 1902; Mackin 1948).

A river reaches the graded state when it is capable over the year or a period of years to transport all parts of the bed-load along all portions of the river at rates equivalent to sediment supply; i.e. the power of a graded river, as determined by the discharge, velocity and the form of the channel, exactly balances the energy required to carry the bed-load. A river is not graded if the power of the river over a period of years is either (a) insufficient to carry all of the bed-load, or (b) greater than is required to carry all of the bed-load. In the case of (a), the floor of the alluvial plain will build upwards (aggrade) as part of the bed-load is left behind; in the case of (b), the river will cut downwards (degrade) because some of the excess power of the river is spent picking up new bed-load material.

The amount of bed-load which a river can carry is determined by the amount and proportion of different grain-sizes present (Mackin 1948). Generally, the greater the amount and/or the proportion of coarser grain-sized material, the greater the power of the river that is required to transport it; i.e. a graded river carrying coarse bed-load will be steeper (i.e. faster flowing), and/or will have greater discharge, and/or will have a more efficient channel form than a similar river carrying finer grained bed-load (see Lane 1955).

Generally the profile of a graded river flattens towards the coast which partially reflects greater efficiencies in channel form and relatively higher discharge rates (Mackin 1948). Thus, a large, slow-moving river can typically transport all the bed-load provided by its faster flowing tributaries. A further reason for this flattening may be that the proportion of coarse grain-sizes in the

bed-load decreases with distance from the source of the sediment due to abrasion during transport.

Controls on the graded profile

The overall shape of the graded profile of a river reflects: (i) the length of the river from the watershed to the coast; (ii) the change in elevation along the length of the river; (iii) the amount and variation in discharge; and (iv) the amount and grain-sizes of the sediment.

A river may go out of grade if changes occur in the height and position of the watershed; the height and position of the coast; the slope of the alluvial plain; the amount of water entering the river; and weathering, erosion, vegetation and bed-rock in the sediment source area. These variations in turn simply reflect the effects of subsidence and uplift within the drainage basin, changes in climate and the relative height of sea-level, and sedimentological and geomorphological processes in the hinterland and at the coast.

In order to get back in grade, rivers will tend either to degrade or aggrade. Much of the erosion that results from fluvial degradation will occur during periods when the river is at peak discharge; much of the deposition associated with aggradation will occur when flow begins to wane after periods of peak discharge. Provided that there is sufficient time, most changes in grade along any part of a river will also affect the grade of adjoining tributaries and distributaries by altering the height of the junctions, the velocity of the river and (temporarily) the amount of sediment transported downstream.

Significant changes in the graded profile of a river due to variations in factors such as climate or the height of sea-level, are likely to affect large parts of the drainage basin. Therefore, such changes, which represent periods of aggradation or degradation, may serve as useful correlatable events within alluvial strata. In order to describe and explain these events over geological time, a concept is proposed known as 'base profile'.

Base profile

The sequence stratigraphy of marine strata is greatly influenced by the effects of subsidence and changes in eustatic sea-level. In order to simplify observations and interpretations the concept of base level is used. Base level is the relative height of sea-level above a chronostratigraphic datum (Fig. 1). An increase in the relative height of sea-level (a rise in base level) may be due to subsidence or a eustatic increase in sea-level or a combination of both. Base level

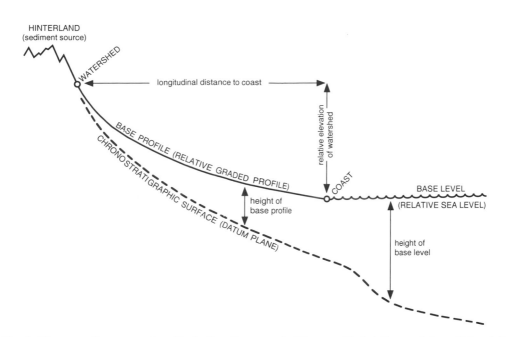

Fig. 1. Diagrammatic cross-section illustrating the concept of base profile (relative graded profile) and its relationship to base level (relative sea-level).

controls accommodation space for coastal and marine sediment as well as the position of the coast. In contrast, accommodation space for alluvial sediment is related to the graded profile of a river. The easiest way of explaining changes in accommodation space in the alluvial environment is by describing them relative to a chronostratigraphic surface or datum similar to the way base level is used in the marine environment (Fig. 1). If the height of the graded profile increases relative to the datum then accommodation space for alluvial sediment is created. If the height of the graded profile decreases relative to the datum then accommodation space is removed. Therefore, a concept of relative graded profile is proposed. Relative graded profile is given the name base profile in this paper in keeping with the name base level. Base profile is defined as the ideal graded profile of a drainage basin at a specific moment in time relative to a chronostratigraphic datum. Rivers will tend to aggrade up to or degrade down to the base profile.

The base profile of a drainage basin intersects sea-level at the coast and it is thus the terrestrial equivalent of base level (Fig. 1). Unlike base level, changes in the height of base profile are related, not only to eustatic sea-level and subsidence, but also to river discharge and hinterland tectonics. Over time, compaction will

tend to cause strata to lower in height relative to the original position of any particular base profile. In theory, the effect of compaction should be corrected for; in practice, it can usually be ignored (Quirk 1994).

Shanley & McCabe (1994), in recognizing differences in the definition of base level between different workers and disciplines, coined the term stratigraphic base level to mean a dynamic equilibrium surface that separates erosion from deposition of continental as well as marine strata. Although they do not clearly indicate what the distinction is between stratigraphic base level and terms such as graded profile and stream equilibrium profile, it is different from the concept of base profile for the following reasons.

(i) Base profile constitutes the position of the graded profile relative to a chronostratigraphic datum, stratigraphic base level does not.

(ii) The effect of stratigraphic base level on fluvial architecture is described as different to those of climate, sediment supply and hinterland tectonics (pp 556, 558 and 564 in Shanley & McCabe 1994), whereas base profile includes these effects.

(iii) Stratigraphic base level is synonymous with the groundwater table in aeolian strata (page 553 in Shanley & McCabe 1994), base

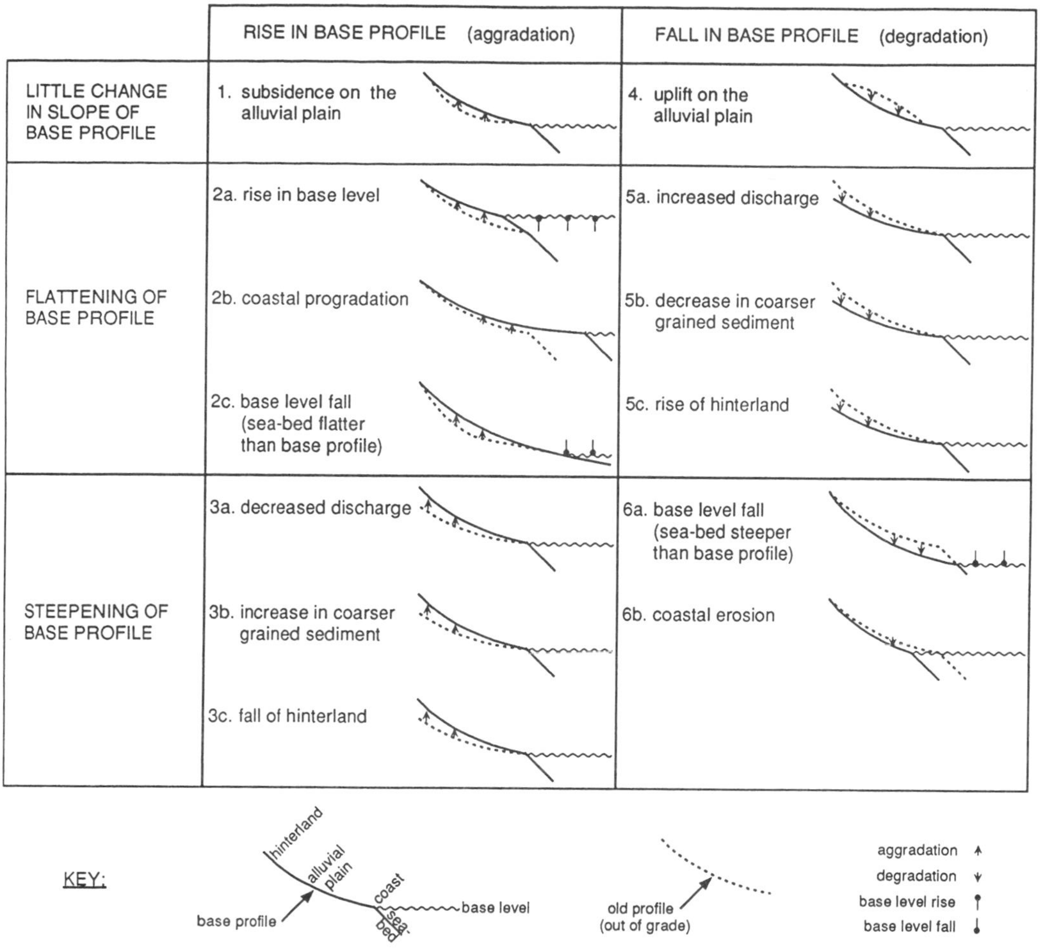

	RISE IN BASE PROFILE (aggradation)	FALL IN BASE PROFILE (degradation)
LITTLE CHANGE IN SLOPE OF BASE PROFILE	1. subsidence on the alluvial plain	4. uplift on the alluvial plain
FLATTENING OF BASE PROFILE	2a. rise in base level	5a. increased discharge
	2b. coastal progradation	5b. decrease in coarser grained sediment
	2c. base level fall (sea-bed flatter than base profile)	5c. rise of hinterland
STEEPENING OF BASE PROFILE	3a. decreased discharge	6a. base level fall (sea-bed steeper than base profile)
	3b. increase in coarser grained sediment	6b. coastal erosion
	3c. fall of hinterland	

KEY:

hinterland · alluvial plain · coast · sea · delta · base profile · base level · old profile (out of grade)

aggradation ⋏
degradation ⋎
base level rise ↑
base level fall ↓

Fig. 2. The causes and effects of changes in base profile within a drainage basin.

profile is meant only to apply to alluvial strata.

To avoid confusion with other definitions, it is thought better to limit the use of base level to meaning relative sea-level.

Controls on base profile

Each portion of a graded river may accommodate a certain range of possible slopes by variations in channel form but it must be in equilibrium with each adjacent upstream and downstream portion of the river system. Therefore, although flexibility exists in the system, at any one time a drainage basin will have one unique base profile which will depend on the following parameters: the elevation of the watershed relative to sea-level; the longitudinal distance from the watershed to the coast; the

river discharge; the amount and type of bed-load (coarse-grained) sediment being carried by the rivers; the forms of the river channels.

Base profile may change as a result of these parameters being affected by uplift or subsidence on the continent, a rise or fall in base level, a change in climate, a change in sediment supply, or an increase or decrease in the longitudinal distance to the coast. After a change in base profile, the drainage basin will be out of grade until the rivers have adjusted to the position of the new base profile. However, there is unlikely to be enough time available for the river system to become completely regraded before another change in base profile occurs. In fact, factors such as subsidence will cause base profile to be continually moving; i.e. base profile, similar to base level, is dynamic.

The position of any new base profile may lie

above or below the old base profile. The change is described as either a rise in base profile or a fall in base profile (Fig. 2). A rise in base profile means that there is an increase in the height of the ideal graded profile relative to the datum plane. A fall in base profile means that there is a decrease in the height of the ideal graded profile relative to the datum plane. A rise in base profile will tend to lead to the deposition of alluvial sediment (fluvial aggradation) when new accommodation space is created. A fall in base profile will tend to lead to erosion on the alluvial plain (fluvial degradation) when accommodation space is removed.

Rivers can compensate for certain changes in the conditions affecting a drainage basin by changing their channel form (Schumm 1993; Wescott 1993) rather than base profile. However, these changes will not be preserved in the sedimentary record unless a rise in base profile occurs.

Changes in base profile will cause variations in sediment flux to the marine environment as sediment is trapped on, or removed from, the continent by rivers. Not until the alluvial plain has readjusted to the new base profile will the sediment flux fully stabilize; i.e. when the amount of sediment reaching the coast approximately matches the supply of sediment from the hinterland.

Changes in base profile

Three types of changes in base profile can occur:

- where the river aggrades or degrades in order to maintain the original slope of the river (1 and 4 in Fig. 2);
- where the river aggrades or degrades leading to a reduction in the slope of the river (2a–c and 5a–c in Fig. 2);
- where the river aggrades or degrades leading to an increase in the slope of the river (3a–c and 6a–b in Fig. 2).

Rises in base profile which cause a change in the slope of a river will normally be associated with an increase in efficiency in channel form (particularly in the cases of 2a, 2b, 3a, 3b and 3c in Fig. 2). Falls in base profile which cause a change in the slope of a river will normally be associated with a decrease in efficiency in channel form (particularly in the cases of 5a, 5b, 5c, 6a and 6b in Fig. 2).

The main controls on changes in base profile are considered separately in the discussion below although they may well occur in combination with one another. The discussion only considers those changes in base profile that are of sufficient magnitude to affect the sedimentary record over fairly large distances, i.e. those that can be correlated.

Effect of subsidence or uplift within the drainage basin

The simplest change in base profile is that caused by subsidence and uplift on the alluvial plain. In the case of subsidence, a river will build up to the level of the alluvial plain by aggradation in order to maintain grade, i.e. a rise in base profile occurs (1 in Fig. 2). The rate of rise in base profile will correspond to the rate of subsidence. Uplift will cause the opposite effect, namely that rivers will degrade due to a fall in base profile (4 in Fig. 2).

A similar change in base profile occurs towards the hinterland of a drainage basin if it is affected by tilting, mountain building, mountain collapse or isostatic readjustments. A rise in the height of the hinterland will cause an increase in the slope and hence an increase in the velocity of a river. Therefore, rivers will often tend to erode and degrade. This change manifests itself as a fall in base profile (5c in Fig. 2). The new base profile has a slope similar to, or greater than, that of the old graded profile prior to uplift, depending on adjustments in channel form. During the process of regrading, the downstream portions of a river will be affected due to a change in the amount of bed-load being carried by the river. However, these effects will only be temporary, provided that the river has time to adjust to the new base profile. A fall in the height of the hinterland will cause rivers to slow down and coarse-grained material will tend to accumulate (aggrade) until grade is once more re-established. This equates with a rise in base profile (3c in Fig. 2).

Effect of changes in river discharge and sediment supply

A permanent change in discharge relative to the amount of coarse-grained sediment entering a river will tend to cause a change, not only in the position of the base profile, but also ultimately in the slope of the drainage basin. For example, if the amount of coarse-grained sediment decreases, a river has surplus power which will lead to down-cutting and flattening of the profile. The river will continue to flatten and slow down until it is back in grade. This corresponds to a fall in base profile and the effects will be most pronounced towards the hinterland (5b in

Fig. 2). The same effects will be seen where the amount of river discharge increases (5a in Fig. 2).

Where river discharge decreases and/or the amount of bed-load increases, the river will be incapable of carrying the coarser-grained sediment. This will cause material to build up or aggrade in the hinterland until the slope and hence the velocity of the river is again in balance. This represents a rise in base profile, the effect of which decreases to zero at the coast (3a and 3b in Fig. 2). The amount of fine-grained sediment that a river can carry in suspension is almost unlimited (Mackin 1948) and will therefore have little effect on base profile.

Changes in climate will often lead to both a change in river discharge and sediment supply. For instance, a change from humid to semi-arid conditions will tend to cause a decrease in river discharge and an increase in sediment supply (Schumm 1977). In this particular example, both effects will probably combine to produce a rise in base profile (3a + 3b in Fig. 2). However, further discussion of this is beyond the scope of this paper.

Effect of changes in the position of the coast

Coastal progradation will result in a rise in base profile (2b in Fig. 2). This occurs in order that the river or rivers build up sufficient slope to carry the bed-load across the extended part of the alluvial plain. The new slope is generally small because the extended rivers are generally more efficient than higher up on the alluvial plain. Therefore, the amount of rise in base profile on the alluvial plain caused by coastal progradation alone is fairly limited. However, the rate of subsidence within a basin may increase in the direction of progradation, in which case an additional component of rising base profile will become more important as the alluvial plain extends. This effect is equivalent to rising base level which is discussed below.

A fall in base profile will occur if coastal erosion cuts back into the lower end of the drainage basin (6b in Fig. 2). This fall is necessary in order to compensate for the decrease in the length of the river.

Effect of changes in base level

The greatest control on base profile close to the coast will come from changes in base level. A rise in base level will have an analogous effect to that of the tide coming in to an estuary. The river will become sluggish in its lower reaches causing the deposition of the coarsest part of the

bed-load. This has the effect of decreasing the slope of the adjacent upstream portion of the river which in turn causes it too to become more sluggish and deposit coarse-grained fluvial material. Thus a chain of fluvial aggradational events, equivalent to a rise in base profile, will work back up the river until grade is re-established (2a in Fig. 2). Koss et al. (1994) have elegantly shown in modelling experiments how deposition of this fluvial sediment will occur as a series of backstepping lobes. The new graded portion of the river will have a slightly shallower slope and a more efficient channel form than the old profile but is otherwise similar. The amount of transgression that occurs during a rise in base level depends on the balance between the rate of fluvial aggradation and the rate of change of base level. The exact shape of the new base profile is also dependent on this balance. If the coast retreats past any point on the continent, base level will take over from base profile as the main control on accommodation space and marine strata will overstep alluvial deposits (e.g. (1) in Fig. 3).

The distance inland to which fluvial deposition is affected by base level rise is strongly influenced by the type and relative amount of bed-load sediment transported by the river, the amount of change in slope that variations in channel efficiency can accommodate and the time available between each increment of base level rise. For example, a river which is only carrying fine-grained sediment in suspension will tend to show little change in base profile in response to a rise in base level. Also, if a rise in base level proceeds very rapidly, the rate of fluvial aggradation may be insufficient to prevent the continent from quickly flooding.

A fall in base level may cause either a fall or a rise in base profile, depending on whether the slope of the sea-bed in front of the old river mouth is steeper or flatter than the slope of the new portion of graded river that will flow down it (Miall 1991; Posamentier et al. 1992; Wescott 1993; Shanley & McCabe 1994). Usually the sea-bed in front of a major river is steeper than the graded profile (e.g. Nummedal et al. 1993), in which case a fall in base level will mean that base profile will also fall and the river will tend to degrade (6a in Fig. 2). In rare cases where the sea-bed is flatter than the graded profile, a rise in base profile will occur due to a fall in base level (2c in Fig. 2). Such a situation may arise in an internal drainage basin where rivers feed into a shallow-bottomed lake or where new rivers develop across a recently exposed continental shelf.

Schumm (1993) and Shanley and McCabe

(1994) have concluded that the effects of a change in base level are unlikely to extend more than a few hundred kilometres inland. The maximum inland position to which the effect of a change in base level extends is where the new base profile intersects the old base profile. The location of this intersection point depends on the change in the height of base level and the slope of the new base profile relative to the old base profile (see (1) in Fig. 3). The slope of the new base profile in turn is related to the changes in channel form that the rivers can accommodate.

Dominance and order of changes in base profile

The causes and effects that lead to changes in base profile have been treated here separately. However, in reality, more than one of these processes is likely to be working at the same time, but at different magnitudes and at different rates. Over the long term, subsidence is likely to be the dominant process that leads to the preservation of thick packages of alluvial sediment, i.e. in an overall sense, base profile will continue to rise at a rate roughly equivalent to the rate of subsidence. Variations in climate, hinterland tectonics and base level may cause higher order rises and falls in base profile which are superimposed on this overall rate. Many of the causes of these high order variations are likely to be cyclical. Thus, in areas of high subsidence, repetitious alluvial sequences might develop, separated by unconformities or periods of slowing down or stillstand in the rate of rise of base profile.

Sequence stratigraphic context

Lateral and vertical extent of alluvial sequences

Changes in base profile may not affect all parts of a single drainage basin. Many changes in base profile will decrease to almost zero before either the hinterland or the coast is reached. This is mainly because variations in channel form can allow a river to accommodate certain changes in slope (Schumm 1993). Also there may not be enough time for distant parts of the drainage basin to react to a change in base profile. Nonetheless, in areas where thick packages of alluvial sediment have been preserved, it should be possible to correlate important changes in base profile for distances of several tens to hundreds of kilometres.

Where a fluvial sand is overlain by a similar fluvial sand, the preserved thickness of the underlying sand (corrected for compaction) gives an approximate measurement of the amount of rise in base profile prior to the deposition of the overlying sand. In contrast, it will usually be difficult to estimate the amount of fall in base profile associated with any particular continental unconformity. In fact, it may be difficult to distinguish unconformities from channel scour features.

Aggradation

A rise in base profile is associated with the accumulation of coarse-grained and associated floodplain sediment on the continent by fluvial aggradation. Fluvial aggradation occurs along portions of the river that have slopes less than the base profile (although the final profile may be steeper than the old base profile). Therefore, coarse-grained sediment cannot easily be transported to the coast until the rivers have been regraded and/or the channel form has become more efficient, i.e. mostly only fine-grained sediment will reach the marine environment via rivers during a rise in base profile. Such fine-grained marine deposits are often equated with transgression (rapid rise in base level). However, Fig. 2 shows that there are other causes of rises in base profile that can trap coarse-grained sediment on the continent (1–3c in Fig. 2), such as a relative decrease in river discharge. Away from a river channel, aggradation of floodplain sediment may accompany a rise in base profile. Some of this material may be removed later due to channel avulsion depending on the rate of aggradation (Bridge & Leeder 1979).

In cases where a rise in base profile is due to a rise in base level (2a in Fig. 2), coarse-grained fluvial strata will tend to backstep as the continent becomes progressively flooded (see Koss et al. 1994). This backstepping pattern occurs because successive new base profiles intersect the old profile progressively nearer to the hinterland as the coast retrogrades ((1) in Fig. 3). The youngest fluvial sediments prior to transgression are expected to show a fining- and thinning-upwards trend (e.g. Shanley & McCabe 1991). A condensed sequence and a maximum flooding surface will usually occur above such a fining-upwards package. Inland from the coast, the time of maximum flooding will tend to correspond with the upper part of a coarse-grained fluvial interval (Quirk 1994) and possible tidal influence (Shanley et al. 1992).

Fluvial sandstones underlying a flooding surface are often interpreted as representing incised

valley fills (e.g. Van Wagoner *et al.* 1990).
However, incision (due to a fall in base profile)
need not always precede a rise in base profile.
Indeed, incision may be absent in areas of high
subsidence. Where incision has occurred, aggradational fluvial sandstones will tend to be
constrained within the valley unless and until
base profile rises above the valley sides. Where
incision has not occurred, for example above a
type 2 sequence boundary (e.g. Posamentier *et
al.* 1988), fluvial sandstones may form amalgamated, sheet-like bodies if the rate of aggradation is low to moderate (see Bridge & Leeder
1979) or the amount of coarse-grained bed-load
is high (Quirk 1994). However, transgression
will generally proceed at a faster rate over an
alluvial plain that has not been incised than
along an incised valley with a similar gradient.
This is because the products of fluvial aggradation that accompany the rise in base profile are
spread over a larger area on a flat plain than in
the confines of an incised valley. Also, changes
in the slope of the drainage basin are accommodated more easily on a flat plain where lateral
changes in channel form are not constrained
(Schumm 1993). It is therefore probable that
alluvial sequences which do not occur within
incised valleys are likely to be thinner but more
extensive than their incised counterparts.

Where rises in base profile have occurred due
to purely continental processes, such as climate
change or tectonics (1, 3a–3c in Fig. 2), then
aggradation of coarse-grained sediments will
tend to begin upstream and advance downstream. Subtle downlap may occur on the
underlying alluvial surface ((3) in Fig. 3) and
coarsening- and thickening-upwards alluvial
strata are predicted to develop.

Over the long term, degradation will tend to
outweigh aggradational processes towards the
hinterland and for this reason many effects on
base profile due to changes in climate and
orogenesis will not be preserved, except in areas
of active continental subsidence.

Channel form

It is often possible to distinguish at outcrop or
in the subsurface between fluvial sediments that
were deposited by braided rivers and those that
were deposited by meandering or low sinuosity
rivers. Braided rivers are more common in areas
with high gradients; meandering (high-medium
sinuosity) rivers are more common in areas with
medium–low gradients; and straight (low sinuosity) rivers are more typical of areas with very
low gradients (Allen 1970; Ouchi 1985). Hence,
braided rivers are generally found closer to the

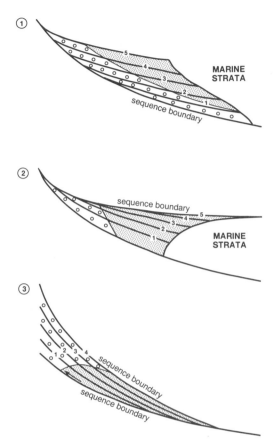

Fig. 3. Idealized cross-sections through alluvial intervals deposited as a result of rising base profile: (1) due
to base level rise (transgression); (2) due to highstand
progradation; (3) due to continental processes such as
decreased discharge, increased coarse-grained sediment supply or subsidence within the hinterland. The
hinterland is on the left of each cross-section. Circles
represent alluvial deposits associated with braided
rivers; dots represent alluvial deposits associated with
meandering and low sinuosity rivers; numbered lines
indicate the position of successive alluvial surfaces that
may or may not have caught up with rising base
profile.

hinterland, where the slope of the drainage basin
is high, and low sinuosity rivers (which may or
may not anastomose) on flat coastal plains.
Meandering rivers usually occupy the intervening part of the alluvial plain. Channel form is
also influenced by discharge and the amount of
coarse-grained bed-load. For example, for the
same slope braided rivers tend to have higher
discharges than meandering rivers (Allen 1970).
Therefore, where it is possible to log changes in

channel form within fluvial strata, inferences can be made about changes in discharge relative to sediment supply or changes in the slope of the river due to tectonics or base level effects. These inferences may in turn be useful in interpreting what caused the rise in base profile that led to the deposition of fluvial strata in the first place. Thus, continental processes that cause a rise in base profile, such as a decrease in relative discharge or a fall in the height of the hinterland (3a–c in Fig. 2), will lead initially to a decrease in sinuosity or a change from braided to meandering channels as the rivers are forced to become more efficient. However, as fluvial aggradation proceeds, the profile of the drainage basin will gradually steepen ((3) in Fig. 3). Therefore, a change from low sinuosity to high sinuosity channel deposits or from meandering to braided channel deposits is predicted upwards in the sedimentary succession. In contrast, a rise in base profile due to a rise in base level is likely to lead to a different effect, namely a change from high sinuosity to low sinuosity channel deposits or a change from braided to meandering deposits as the profile of the drainage basin gradually flattens upwards with time and the rivers become more efficient ((1) in Fig. 3).

A more general observation can also be made that eustatic changes in sea-level will cause cyclical variations in channel patterns, i.e. a eustatic fall will usually lead to an increase in the graded slope of the lower part of the drainage basin (6a in Fig. 2) and a eustatic rise will usually lead to a decrease in the graded slope (2a in Fig. 2). It is therefore possible to explain why braided fluvial channels are more typical of lowstand and early transgressive systems tracts and meandering and low sinuosity channels are more typical of late transgressive and highstand systems tracts.

Variations in channel form essentially represent an increase or decrease in the efficiency of the river. A river that is confined within an incised valley or flows through resistant material cannot easily change its channel form (Schumm 1993) and the river will consequently show greater susceptibility to changes in base profile than those that are unconfined. It is also difficult for highly braided rivers to become less efficient or very straight rivers to become more efficient.

Other changes in the efficiency of river channels may occur during changes in base profile, such as a change to a smoother or rougher bedform. The reader is referred to Schumm (1977) who discusses in great detail how channel form reacts to changes in water discharge, bed-load and gradient.

Degradation

During a fall in base profile the alluvial plain will be subject to erosion and sediment will by-pass the continent. Fluvial degradation occurs along portions of the river system which, during the process of regrading, have slopes steeper than the base profile. The point where the new profile intersects the old graded profile is called the knick point. The rate that the knick point migrates upstream from the mouth of the river is inversely proportional to the square root of the time since base level was lowered (see Begin 1988). Provided that there is enough time, the products of degradation, in addition to sediment from the hinterland, will be transported to the coast. Hence, unconformities in the alluvial environment will correlate with periods of maximum sediment influx in the marine environment and consequently most marine sands will have no alluvial counterpart of the same age.

Subaerial unconformities are often interpreted as type 1 sequence boundaries indicative of a fall in base level (e.g. Aigner & Bachmann 1992). However, any fall in base profile can produce an erosional surface by fluvial degradation (4–6b in Fig. 2). Whether this unconformity is due to a fall in base level or some other process, such as increased discharge, can, in theory, be assessed by whether the amount of erosion increases or decreases towards the coast.

Much of the erosion associated with a fall in base level tends to be confined to incised valleys, indicating that the rate of fall of base profile was relatively rapid. Outside of such valleys, changes in base profile may be marked by subtle changes within palaeosols rather than fluvial aggradation or degradation (e.g. Wright & Marriot 1993). For example, a wetter palaeosol might develop as the water table rises in association with fluvial aggradation within an adjacent valley. In contrast, a drier palaeosol may form as the water table falls due to valley incision. Where a fall in base profile occurs more slowly, then peneplanation rather than incision may occur.

Any surface that has developed within alluvial strata as a result of a fall in base profile may be regarded as a continental sequence boundary (e.g. Fig. 3).

Edge effects

Where a rise or fall in base profile occurs relatively quickly, the change in base profile towards the edges of the area affected by the change may be in a sense opposite to the change in the main part of the area, i.e. the change in base profile will smooth out the effect rather than wholly counteract it. For example, Nummedal *et al.* (1993) have shown examples of rivers that are aggrading along most of their length due to coastal progradation, the effect of which is apparently compensated by fluvial degradation in the upper reaches of the rivers. Ouchi (1985) has demonstrated that a small amount of aggradation may occur downstream and upstream of an area of significant uplift; similarly, a small amount of degradation may occur at the edges of an area of subsidence. However, these effects are likely to be ephemeral because the area affected by a change in base profile will not remain fixed.

Susceptibility of drainage basin to changes in base profile

Processes of aggradation and degradation in a drainage basin will tend to occur as a series of chain reactions downstream or upstream from the place of maximum incremental change in base profile (Fig. 3). It is therefore unlikely that there will be enough time available between every change in base profile for the drainage basin to completely adjust, particularly in cases where the change in base profile is relatively small. Also in cases such as subsidence or uplift or a rise or a fall in base level (1, 2a, 2c, 3c, 4, 5c and 6a in Fig. 2), base profile is dynamic rather than fixed, i.e. while rivers react the position of base profile continues to change. Although a drainage basin will always attempt to catch up with base profile, it may not be able to before it is overwhelmed by some other event such as marine transgression. Also certain geomorphic thresholds may not be overcome. For example, a fall in base profile may not be sufficient to cause erosion of a resistive bed.

Schumm (1993) has also pointed out that changes in the grade of a drainage basin are partially accommodated by variations in channel form, i.e. a new base profile may contain slopes that are significantly steeper or flatter than the old base profile. Therefore, the difference in the height between the new and the old base profile will decrease almost to zero before the coast or the hinterland is reached. In such cases the new base profile will intersect the old base profile on the alluvial plain. If this intersection point migrates with time, onlap, downlap and toplap patterns may develop within alluvial strata (see Fig. 3).

It is clear, therefore, that a rise in base level will lead to very little, if any, aggradation in the higher reaches of the river system. Similarly, very little fluvial aggradation will occur close to the coast as a result of a relative decrease in discharge. Mature palaeosols may develop in areas of a drainage basin which have not been affected by a change in base profile (see Wright & Marriott 1993).

Local disturbances

Ephemeral features in a river system such as small lakes or waterfalls may lead to localized events of aggradation or degradation unrelated to base profile. If these temporary adjustments are recorded in the sedimentary column then they are unlikely to correlate for distances of more than a few kilometres. The effects of major autocyclical processes, such as stream capture, may be of regional extent but such events will be of random occurrence within any particular sedimentary package.

Intra-continental sequences

Some drainage basins feed large internal lakes rather than oceans. In such cases, base level may be regarded as relative lake level and base profile is the theoretical graded profile between the hinterland and the lake edge. It is therefore possible to correlate and interpret intra-continental sequences using the same criteria as for river systems that supply the sea.

Discussion

Alluvial sequences typically consist of various thicknesses of aggradational fluvial and floodplain deposits separated by unconformities or soils (e.g. Boyd & Diessel 1994). Such sediments record changes in base profile. On the basis of the previous discussion, predictions can be made concerning the nature of intervals associated with different types of alluvial sequences. These are summarized in Table 1 and illustrated in Fig. 3 for three intervals associated with higher order rises in base profile. These intervals can be defined as system tracts on the basis that they consist of a depositional system linked in time and space. Intervals 1 and 2 depicted in Fig. 3 and Table 1 are equivalent to the alluvial parts of transgressive systems tracts and highstand systems tracts, respectively.

Changes in base profile are thought to be cyclical similar to base level. Therefore, a rise in base profile is usually preceded by and/or succeeded by a fall in base profile, except in areas of very high subsidence. Such a fall in base profile will lead to the formation of a continental sequence boundary. Hence, in areas of low—

Table 1. *List of attributes that characterize alluvial intervals (systems tracts) depicted in Fig. 3*

1 Alluvial interval associated with marine transgression	**2** Alluvial interval associated with coastal progradation	**3** Alluvial interval associated with changes in hinterland
base profile backsteps and flattens with time	base profile builds out and flattens with time	base profile steepens with time
interval pinches out towards the hinterland	interval is thin and pinches out towards the hinterland	interval pinches out towards coast
strata at base of interval show onlapping relationship towards the hinterland	strata at top of interval display toplap or are truncated	strata at base of interval show downlapping relationship towards the coast
strata fine and thin upwards	strata thin upwards	strata coarsen and thicken upwards
braided channel deposits tend to occur at base of interval	meandering-low sinuosity channel deposits are most common	braided fluvial deposits tend to occur at top of interval
evidence for increasing marine influence upwards	evidence for decreasing marine influence upwards	little evidence of marine influence
continental sequence boundary at base of interval	continental sequence boundary at top of interval	continental sequence boundary at base and top of interval

moderate subsidence, continental sequence boundaries are predicted to occur at the bases of intervals 1 and 3 in Fig. 3 (Table 1) and to truncate the tops of intervals 2 and 3 in Fig. 3 (Table 1). On this basis, interval 3 represents a complete alluvial sequence which consists of only one systems tract. Intervals 1 and 2 (Fig. 3, Table 1) are likely to occur at the bases and tops of sequences which also contain marine strata; they represent improvements in the treatment of alluvial strata compared to the traditional sequence stratigraphic models of, for example, Posamentier & Vail (1988).

Conclusions

Base profile is defined as the ideal graded profile of a drainage basin at a specific moment in time relative to a chronostratigraphic datum. It can be used to describe changes in accommodation space in the alluvial environment. Rivers will tend to aggrade up to or degrade down to base profile. New accommodation space is created in a drainage basin by a rise in base profile which may be due to one or more of the following reasons: subsidence on the alluvial plain; a decrease in river discharge and/or an increase in the amount of coarse-grained sediment; a decrease in the height of the hinterland; a rise in base level; coastal progradation; a fall in base level across a shallow-bottomed lake or a flat shelf.

Thick packages of alluvial sediment are only likely to be preserved over the long term in areas

of active subsidence. However, higher order rises and falls in base profile within such packages may be due to the effects of climate, hinterland tectonics and eustasy. The explanation for any higher order rise in base profile can be assessed by whether the effect decreases or increases towards the hinterland, by whether the sedimentary interval coarsens or fines upwards, by whether there is a change in the form of the fluvial channels, by whether there is marine influence and by the lap-out geometries of the strata.

Continental unconformities, which may be interpreted as sequence boundaries, develop as a result of falls in base profile. Falls in base profile occur not only because of falls in base level but also due to uplift on the continent, due to increased discharge relative to coarse-grained sediment supply and due to the effects of coastal erosion. However, fluvial incision need not necessarily precede fluvial aggradation.

The sediment flux in the marine environment will vary according to changes in the grade of a drainage basin. For example, a rise in base profile caused by changes in the hinterland will cause starvation of coarse-grained sediment in the marine environment, i.e. coarse-grained fluvial sediments will usually correlate with fine-grained marine beds.

I am indebted to Peter Vail for the original inspiration and the discussions that followed. I also thank colleagues at Shell and Oxford Brookes University who critically read early versions of the manuscript, in particular Wim Moeshart, Nigel Banks, Dan den Hartog-Jager, Peter de Boer and Gerhard Bloch. I am also grateful to Mike Leeder, Ron Boyd and Susan Marriott for helpful comments directed at earlier versions of the manuscript. Lisa Hill and Erwin Vrieling helped to draft the diagrams.

References

AIGNER, T. & BACHMANN, G. H. 1992. Sequence-stratigraphic framework of the German Triassic. *Sedimentary Geology*, **80**, 115–135.

ALLEN, J. R. L. 1970. *Physical processes of sedimentation*. George Allen and Unwin, London.

BEGIN, Z. B. 1988. Application of a diffusion–erosion model to alluvial channels which degrade due to base-level lowering. *Earth Surface Processes and Landforms*, **13**, 487–500.

BOYD, R. & DIESSEL, C. 1994. The application of sequence stratigraphy to non-marine clastics and coal. *In:* POSAMENTIER, H. W. & MUTTI, E. (eds) *Proceedings of 2nd High Resolution Sequence Stratigraphy Conference, 20–27 June 1994, Tremp, Spain.*

BRIDGE, J. S. & LEEDER, M. R. 1979. A simulation model for alluvial stratigraphy. *Sedimentology*, **26**, 617–644.

BRISTOW, C. & MAYNARD, J. 1994. Alternative sequence stratigraphic models for the Rough Rock Group: a Carboniferous delta in the Pennine Basin, England. *In:* JOHNSON, S. D. (ed.) *Abstract Volume: High Resolution Sequence Stratigraphy: Innovations and Applications.* University of Liverpool, 353–357.

CHURCH, K. D. & GAWTHORPE, R. L. 1994. High resolution sequence stratigraphy of the late Namurian in the Widmerpool Gulf (East Midlands, UK). *Marine and Petroleum Geology*, **11**, 528–544.

DAVIS, W. M. 1902. Base-level, grade and peneplain. *Journal of Geology*, **10**, 77–111.

GALLOWAY, W. E. 1989. Genetic stratigraphic sequences in basin analysis I: architecture and genesis of flooding-surface bounded depositional units. *American Association of Petroleum Geologists Bulletin*, **73**, 125–142.

HAMPSON, G. J., ELLIOT, T. & FLINT, S. S. 1996. Critical applications of high resolution sequence stratigraphic concepts to the Rough Rock Group (Upper Carboniferous) of Northern England. This volume.

KOSS, J. E., ETHRIDGE, F. G. & SCHUMM, S. A. 1994. An experimental study of the effects of base-level change on fluvial, coastal plain and shelf systems. *Journal of Sedimentary Geology*, **B64**, 90–98.

LANE, E. W. 1955. The importance of fluvial morphology in hydraulic engineering. *American Society of Civil Engineers Proceedings*, **81**, 745.1–745.17.

LEOPOLD, L. B. & BULL, W. B. 1979. Base level, aggradation and grade. *Proceedings of the American Philosophical Society*, **123**, 168–202.

MACKIN, J. H. 1948. Concept of the graded river. *Bulletin of the Geological Society of America*, **59**, 463–512.

MAYNARD, J. R. 1992. Sequence stratigraphy of the Upper Yeadonian of northern England. *Marine and Petroleum Geology*, **9**, 197–207.

MIALL, A. D. 1991. Stratigraphic sequences and their chronostratigraphic correlation. *Journal of Sedimentary Petrology*, **61**, 497–505.

NUMMEDAL, D., RILEY, G. W. & TEMPLET, P. L. 1993. High-resolution sequence architecture: a chronostratigraphic model based on equilibrium studies. *In:* POSAMENTIER, H. W., SUMMERHAYES, C. P. HAQ, B. U. & ALLEN, G. P. (eds) *Sequence Stratigraphy and Facies Associations.* International Association of Sedimentologists, Special Publication, **18**, 55–68.

OUCHI, S. 1985. Response of alluvial rivers to slow active tectonic movement. *Geological Society of America Bulletin*, **96**, 504–515.

POSAMENTIER, H. W. & VAIL, P. R. 1988. Eustatic controls on clastic deposition II – conceptual framework. *In:* WILGUS, C. K., HASTINGS, B. S., KENDALL, C. G. St. C., POSAMENTIER, H. W., ROSS, C. A. & VAN WAGONER, J. C. (eds) *Sea Level Changes: An Integrated Approach.* Society of Economic Paleontologists and Mineralogists, Special Publication, **42**, 125–154.

——, JERVEY, M. T. & VAIL, P. R. 1988. Eustatic controls on clastic deposition I – conceptual framework. *In:* WILGUS, C. K., HASTINGS, B. S., KENDALL, C. G. St. C., POSAMENTIER, H. W., ROSS, C. A. & VAN WAGONER, J. C. (eds) *Sea Level Changes: An Integrated Approach.* Society of Economic Paleontologists and Mineralogists, Special Publication, **42**, 109–124.

——, ALLEN, G. P., JAMES, D. P. & TESSON, M. 1992. Forced regressions in a sequence stratigraphic framework: concepts, examples and exploration significance. *American Association of Petroleum Geologists Bulletin*, **76**, 1687–1709.

QUIRK, D. G. 1994. The Upper Carboniferous of the Southern North Sea – implications for basin analysis. *In: Extended Abstracts of Papers, European Association of Petroleum Geoscientists and Engineers (EAPG), 6th Conference, Vienna.* EAPG, Zeist, The Netherlands.

SCHUMM, S. A. 1977. *The fluvial system.* John Wiley and Sons, New York.

—— 1993. River response to baselevel change: implications for sequence stratigraphy. *Journal of Geology*, **101**, 279–294.

SHANLEY, K. W. & McCABE, P. J. 1991. Predicting facies architecture through sequence stratigraphy – an example from the Kaiparowits Plateau, Utah. *Geology*, **19**, 742–745.

—— & —— 1994. Perspectives on the sequence stratigraphy of continental strata. *American Association of Petroleum Geologists Bulletin*, **78**, 544–568.

——, —— & HETTINGER, R. D. 1992. Tidal influence in Cretaceous fluvial strata from Utah, USA: a key to sequence stratigraphic interpretation. *Sedimentology*, **39**, 905–930.

VAIL, P. R., MITCHUM, R. M., TODD, R. G., WIDMIER, J. M., THOMPSON, S. III, *ET AL.* 1977. Seismic stratigraphy and global changes of sea level. *In:* PAYTON, C. E. (ed.) *Seismic Stratigraphy Applications to Hydrocarbon Exploration.* American Association of Petroleum Geologists, Memoir, **26**, 49–212.

VAN WAGONER, J. C., MITCHUM, R. M., CAMPION, K. M. & RAHMANIAN, V. D. 1990. *Siliciclastic sequence stratigraphy in well logs, cores and outcrops: concepts for high-resolution correlation of time and facies.* American Association of Petroleum Geologists, Methods in Exploration Series, **7.**

——, POSAMENTIER, H. W., MITCHUM, R. M., VAIL, P. R., SARG, J. F., LOUTIT, T. S. & HARDENBOL, J. 1988. An overview of sequence stratigraphy and key definitions. *In:* WILGUS, C. K., HASTINGS, B. S., KENDALL, C. G. St. C., POSAMENTIER, H. W., ROSS, C. A. & VAN WAGONER, J. C. (eds) *Sea Level Changes: An Integrated Approach.* Society of Economic Paleontologists and Mineralogists Special Publication, **42**, 39–45.

WESCOTT, W. A. 1993. Geomorphic thresholds and complex response of fluvial systems – some implications for sequence stratigraphy. *American Association of Petroleum Geologists Bulletin,* **77,** 1208–1218.

WRIGHT, V. P. & MARRIOTT, S. B. 1993. The sequence stratigraphy of fluvial depositional systems: the role of floodplain sediment storage. *Sedimentary Geology,* **86**, 203–210.

The application of physical measurements to constrain reservoir-scale sequence stratigraphic models

IAN D. BRYANT

Schlumberger-Doll Research, Old Quarry Road, Ridgefield, CT 06877-4108, USA

Abstract: Sequence and parasequence boundaries are characterized by lateral extents that are typically greater than development well spacings and frequently more extensive than individual oil and gas fields. These surfaces result from shifts of facies belts that generate laterally extensive changes over widespread areas. Commonly only facies interpretations of lithological signatures (e.g. hardgrounds) and biostratigraphic signatures (e.g. faunal changes) are used to identify these boundaries in both surface outcrops and subsurface reservoirs. However, a wide variety of physical measurements may also be made, especially in the subsurface, that may be used as additional criteria to recognize surfaces with sequence stratigraphic significance and hence to infer reservoir architectures.

Palaeomagnetic reversal stratigraphies may assist in establishing a chronostratigraphic framework within which to identify significant hiatuses and to constrain sequence stratigraphic interpretations, of both outcrops and subsurface reservoirs. In reservoirs, sequence and parasequence boundaries may be detected by a variety of other methodologies, all of which rely upon the spatial coherence of changes across these significant stratal surfaces. The geochemistry of both formation waters, and hydrocarbons, may show significant changes across these surfaces as a consequence of original variations in pore-water chemistry and/or the inability of diffusive mixing processes to equilibrate chemical compositions across such laterally extensive barriers to vertical communication. Similarly, formation pressure profiles, measured after depletion from initial reservoir pressures, frequently show significant discontinuities across these surfaces.

Seismic data record the varying strengths of reflections that emanate from geological interfaces as a consequence of varying contrasts in acoustic impedance across geological surfaces. These surfaces (seismic reflectors) are commonly assumed to be of chronostratigraphic significance. It follows that spatially coherent patterns of variation in the physical properties of sediments between these surfaces may serve to delineate the location of facies belts (systems tracts) during the period of geological time represented by the sediments between reflectors. In favourable circumstances these patterns are imaged with a high degree of spatial precision by 3D reflection seismic data. The integration of the geometry of reflections with variations in reflection strength or other seismic attributes, may therefore serve to define both the chrono- and lithostratigraphic framework of reservoir sequences.

It follows that qualitative inferences regarding reservoir architecture that are derived from sequence stratigraphic principles may be tested, refined and given improved spatial definition by the careful integration of diverse measurements with conceptual sedimentological models.

Sequence stratigraphic interpretation of depositional sequences hinges on the recognition of significant surfaces. In the Exxon school of sequence stratigraphy the most significant surfaces are deemed to be sequence boundaries that are defined as 'unconformities or their correlative conformities' (Mitchum 1977), with subordinate importance attached to parasequence boundaries, that are defined as 'marine-flooding surfaces or their correlative surfaces' (Van Wagoner *et al.* 1988). In an alternative scheme, proposed by Galloway (1989), maximum flooding surfaces are designated as the most significant boundaries for sequence analysis. Both of these methodologies differ from traditional sedimentary facies analysis by focusing on the recognition of surfaces, rather than lithofacies sequences. Indeed, it is acknowledged that facies sequences will show lateral variation within the envelope defined by these geological surfaces (Fig. 1). Sequence stratigraphical interpretations, therefore, typically attempt to define geological models that are predictive over greater length scales than traditional facies-based sedimentological models.

It has been argued that the lateral extent of sequence boundaries is commonly less than is inferred from seismic data, as a consequence of the limited vertical resolution of the data, and that maximum flooding surfaces (*sensu* Galloway 1989) are likely to be more extensive markers (Cartwright *et al.* 1993). Whilst this is certainly true in many instances, most sequence boundaries are nonetheless extensive relative to the scale of individual hydrocarbon accumulations. Maximum flooding surfaces are often recognized as condensed horizons that are often distinguished on the basis of biostratigraphic

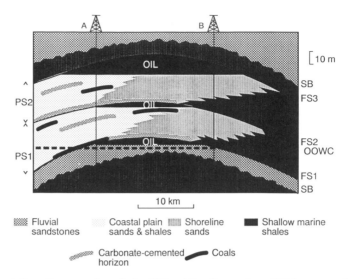

Fig. 1. Schematic representation of sequence boundaries (SB) and flooding surfaces (FS) in a reservoir sequence to illustrate their practical significance to field development. Note that flooding surfaces (FS) bound two parasequences (PS1 and PS2). Note that the surfaces are of fieldwide extent but have different lithological expression in wells A and B. The surfaces are associated with barriers to vertical communication and have trapped oil after rise of the field oil–water contact from the original oil–water contact (OOWC).

and/or geochemical data and may be traceable at a basin-scale (Loutit *et al.* 1988). Flooding surfaces, inferred to be of less widespread extent, bound parasequences and are frequently distinguished on lithological rather than biostratigraphical criteria.

Recognition of these significant surfaces is most commonly achieved by somewhat subjective facies interpretation of sedimentary deposits that over- and underlie a given surface. The subjectivity of this approach is immediately apparent when we consider the problem of identifying the 'correlative conformity' associated with an unconformity on the basis of wireline logs alone. What criteria are available to select the 'correlatable conformity' surface from the plethora of surfaces identified by changes in log response?

The sequence stratigraphic methodology as espoused by Van Wagoner *et al.* (1990) emphasizes the integration of time and facies to provide this constraint (Posamentier 1991) but this often proves to be an elusive goal when working at the reservoir-scale, because the resolution of biostratigraphic techniques is frequently inadequate to provide an effective time constraint.

The two principal attributes that distinguish sequence and parasequence boundaries from other bedding surfaces, are that they are laterally extensive and represent significant time discordances. It is these properties that endow such surfaces with significance in analysing sedimentary facies architecture. These properties, particularly the widespread geographical extent of the surfaces, also enable detection of these surfaces by a variety of physical measurements that are complementary to the lithological and biostratigraphical criteria that traditionally form the basis for sequence stratigraphic interpretations.

This paper illustrates the contribution that a variety of measurements may make to constraining sequence stratigraphic interpretation, by reviewing and re-interpreting selected published examples, in addition to considering the physical basis for interpretation of seismic reflection data in terms of reservoir-scale sequence architecture. These measurements may aid in the detection of stratigraphically significant surfaces by either recognizing the time significance of the surface and/or by inferring the widespread extent of the surface.

Significance of laterally extensive surfaces

Sequence and parasequence boundaries are of significance to reservoir-scale geological models because they are laterally extensive and consequently may act as fieldwide markers (the latter are often used as datum horizons) and because

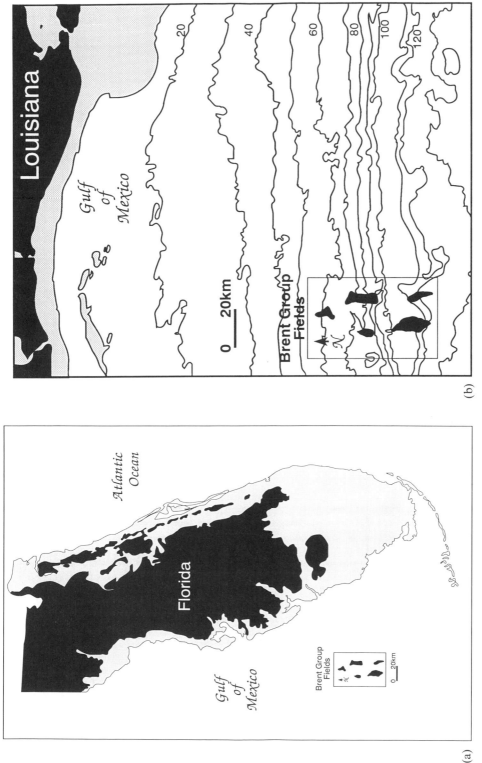

Fig. 2. Comparison of the effects of a 10 m rise or fall of sea-level in the Gulf of Mexico. (**a**) The area of Florida that would be drowned by a 10 m rise of sea-level, shown by stipple (modified from Houghton & Woodwell 1989). (**b**) The area of the Louisiana shelf that would be exposed by a 10 m fall of sea-level, shown by stipple (modified from Berryhill *et al.* 1986). Both figures include an inset showing a number of Brent Group oil fields from the UK, North Sea at the same scale. These examples indicate that flooding surfaces should be expected to be more correlatable in coastal plain sequences than sequence boundaries.

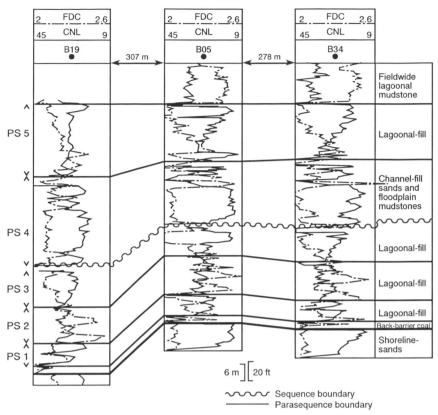

Fig. 3. Illustration of parasequences (P51 to P55) in the Ness Formation of the Brent field that are correlatable over distances of more than 44 km² (modified from Bryant & Livera 1991).

they may be associated with barriers to reservoir communication (Cross *et al.* 1993).

Let us consider the effects of a simple rise or fall of sea-level in the order of 10 m on the modern passive margin of the western coast of the USA. Matthews & Frolich (1991) suggest that high frequency fluctuations of this magnitude were common throughout the Pleistocene and Late Tertiary as a result of orbital forcing of glacioeustacy. It is apparent that much larger areas would be inundated by a sea-level rise (Fig. 2a) than by a sea-level fall of the same magnitude (Fig. 2b). This is a natural consequence of the lower gradient of the coastal plain with respect to the gradient of the inner continental shelf. This asymmetry in response to small-scale sea-level changes suggests that we might expect to recognize more frequent and widespread evidence for small-scale rises of relative sea-level than evidence for small-scale falls of relative sea-level, in coastal and deltaic reservoirs, except perhaps in shelf edge deltas. Thus in coastal plain sequences flooding surfaces should be easier to recognize and correlate than

sequence boundaries. Furthermore, the examples illustrated in Fig. 2 demonstrate that the extents of both flooding and exposure surfaces are more widespread than the lateral extent of most oil fields.

The importance of recognizing such surfaces is illustrated schematically in Fig. 1. Flooding surfaces (FS1, FS2 and FS3) serve to subdivide the reservoir sequence into stacked, parasequences (PS1 and PS2) that comprise a variety of laterally restricted facies belts, of much less widespread extent. Note that the flooding surfaces are associated with different facies at different locations, e.g. FS1 is associated with a carbonate-cemented tight streak, separating fluvial and coastal sands in well A but with a shallow marine shale within coastal sands at well B. The marine shale overlying FS3 has sealed the hydrocarbon accumulation. The shales, carbonate-cemented horizons and coal associated with the other two flooding surfaces have not affected the accumulation of hydrocarbons in the reservoir and, consequently a uniform, fieldwide oil–water contact (OOWC) existed at initial condi-

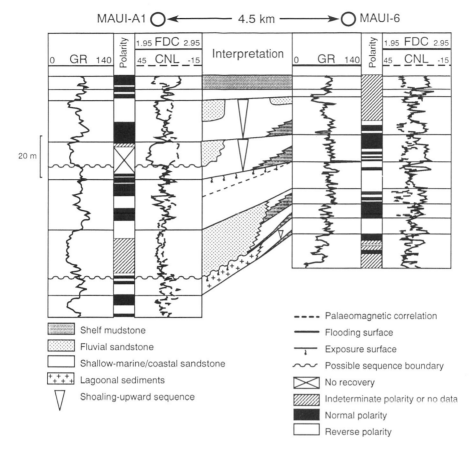

Fig. 4. Correlation of two wells from different areas of the Maui field, New Zealand to show palaeomagnetic zonation and sequence stratigraphic interpretation of the C1 Sands reservoir interval. Note that several of the flooding surfaces correspond to changes in polarity of the sediments. (From Turner & Bryant 1995.)

tions. However, after production of hydrocarbons from wells A and B, the oil–water contact has risen but PS1 and PS2 have functioned as separate drainage units, such that bypassed hydrocarbons remain trapped beneath the flooding surfaces. Clearly optimal development of this reservoir would require recognition of the significance of the flooding surfaces prior to the commencement of production.

Field examples indicate that the situation illustrated by Fig. 1 is not uncommon, e.g. Gibbons *et al.* (1993) indicate that 70–75% of the maximum flooding surfaces and 60–65% of the sequence boundaries that they recognized in the Sognefjord Formation of the Troll field are associated with carbonate-cemented horizons that may affect intra-reservoir communication. Similarly, lagoonal mudstones associated with parasequence boundaries recognized in the Ness Formation of the Brent field are laterally

continuous over an area of more than 44 km² and compartmentalize the reservoir into separate drainage units (Bryant & Livera 1991; Bryant *et al.* 1991) (Fig. 3). Once the genetic and engineering significance of these surfaces is recognized in a developed field, their likely impact on reservoir drainage may be inferred in nearby fields. It follows that the timely recognition of laterally extensive surfaces enables increased confidence in reservoir geological models and optimization of field development planning.

Age measurements

In classical, seismic sequence stratigraphic analysis, sequence boundaries are indicated by reflection patterns and calibrated by palaeontological dating of samples from wells. These techniques are appropriate for dating sequence boundaries where the sequences typically span 1–5 Ma (Vail *et al.* 1977) and are, therefore, applicable to exploration-scale analysis of basin architecture. However, analysis of reservoir-scale architecture most frequently requires resolution of reservoir units that accumulated over time scales of thousands to tens of thousands of years. Biostratigraphic dating often lacks the precision required to provide an effective constraint on correlation at this scale for most reservoir rocks (Taylor & Gawthorpe 1993). Quantitative biostratigraphic techniques have shown promise for identifying parasequence and sequence boundaries (e.g. Poumot 1989; Pasley & Hazel 1990) but these techniques are frequently incapable of providing the time constraint that rigorous reservoir-scale, sequence stratigraphic analysis requires.

The global palaeomagnetic reversal timescale provides a resolution of 0.1–0.01 Ma for much of the Tertiary (Cande & Kent 1992*a*). The possibility that variations in the Earth's magnetic field with a duration of *c.* 10 ka are also detectable (Cande & Kent 1992*b*) offers the potential for palaeomagnetic measurements to provide the resolution required to constrain reservoir-scale interpretations. For example Plint (1988) was able to revise significantly his sedimentologically-based interpretation of the Eocene succession of the Hampshire basin by incorporation of a magnetostratigraphic chronology, whilst Lerbekmo & Demchuk (1992) were able to date a sequence boundary, at outcrop, by combining magnetostratigraphic and biostratigraphic data. Palaeomagnetic reversal stratigraphies have recently seen increasing application to constrain correlation of Jurassic and Triassic reservoir intervals in the North Sea (Rey *et al.* 1993; Hauger *et al.* 1994; Turner & Turner 1995).

An example of the application of this approach is illustrated by Fig. 4, correlation of the C1 Sands reservoir from the developed eastern area to the undeveloped western area of the Maui field was accomplished by integration of sedimentological interpretation of cores with chronostratigraphical constraints provided by a palaeomagnetic reversal stratigraphy (Turner & Bryant 1995). It is suggested that the hiatus in deposition associated with ravinement surfaces and sequence boundaries increases the likelihood that a depositional break will correspond to a reversal event. The same pause in sedimentation may help to explain the association of carbonate-cements with such surfaces (Gibbons *et al.* 1993). This hypothesis was used to interpret the location of the significant breaks in three of the cored Maui

wells and hence to provide chronostratigraphic constraints on sequence stratigraphic interpretation (Fig. 4).

Fluid and rock property measurements

The physical and chemical properties of sediments and fluids above and below sequence boundaries and flooding surfaces may be markedly different. Consequently, a variety of geophysical and geochemical methods may be used to identify sequence and parsequence boundaries.

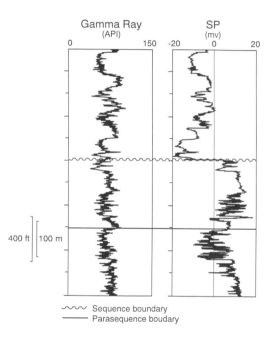

Fig. 5. Wireline logs from a Gulf of Mexico well to show significant variation of formation water resistivity across a sequence boundary. The spontaneous potential (SP) log provides an indication of formation water salinity and shows marked shifts across both the flooding surface and sequence boundary. (Modified from Herron *et al.* 1992.)

Both sequence and parasequence boundaries may be associated with barriers to vertical communication; however, this is most frequently the case with parasequence boundaries, where the flooding surface may be either overlain by shales (Galloway 1989) or associated with carbonate-cemented horizons (Gibbons *et al.* 1993; O'Byrne & Flint 1993). Barriers to vertical

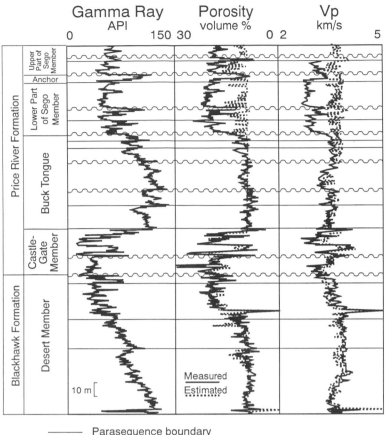

Fig. 6. Wireline logs from Sego Canyon, Utah to show changes in physical rock properties across a sequence boundary between the Buck Tongue and Sego Canyon Members of the Price River Formation. This is indicated by the divergence of estimates of porosity and compressional velocity (Vp), derived from a relationship established from the lower part of the sequence, from actual measurements in the upper part of the sequence. Note that the sequence and parasequence boundaries were identified from the adjacent outcrop and that many would be difficult to recognize on the basis of wireline logs alone. (Modified from Herron *et al.* 1992 and Van Wagoner *et al.* 1990.)

communication are probably less frequently associated with sequence boundaries that are more commonly represented by erosion surfaces overlain by sands. These barriers to vertical permeability may serve to isolate reservoir fluids above the laterally extensive surface from those below. Figure 5 illustrates an example of a spontaneous potential (SP) log from the Gulf of Mexico. This log reflects formation water salinity and indicates how formation waters above and below a sequence boundary are markedly different (Fig. 5). Similarly, hydrocarbons may have different properties above and

below such surfaces. Kaufman *et al.* (1990) provide an illustration of how comparison of gas chromatography of oil samples recovered by a repeat formation tester from above and below a shale indicated that the oils were in separate hydrocarbon pools. These data on the lateral extent of the shale barrier were particulary important since they provided information, after the drilling of only the discovery well, that could guide further field appraisal and development of the field.

Analysis of the properties of sediments from the Sego Canyon #2 well, drilled behind an

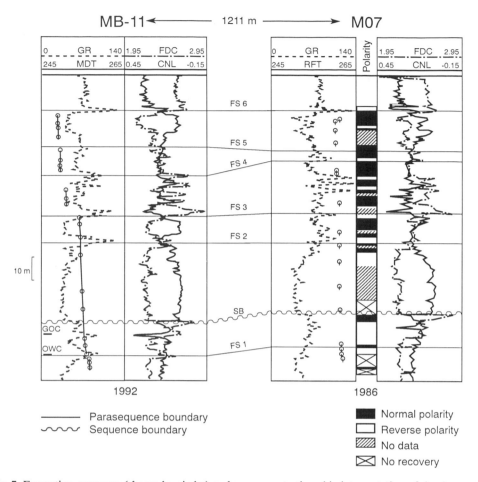

Fig. 7. Formation pressures (shown by circles) and sequence stratigraphic interpretation of development well MB-11 from the Maui field, New Zealand. Correlation with the nearby appraisal well M-07 shows the relationship of the pressure zones to the palaeomagnetic zonation and high resolution sequence stratigraphy of the C1 Sands reservoir. Pressures collected earlier in field life from M-07, show less pronounced pressure discontinuities. Note that neither the sequence boundary, nor FS2, gives rise to pressure discontinuities. GOC, gas–oil contact; OWC, oil–water contact. (Modified from Fett *et al.* 1994 and Turner & Bryant 1995.)

outcrop in Utah illustrates how the mechanical properties of rocks may vary across significant geological surfaces. A sequence boundary is interpreted to separate the lower part of the Sego Member from the Buck Tongue Member of the Price River Formation (Van Wagoner *et al.* 1990). Compressional velocities measured by wireline logging are significantly different above and below this surface (Fig. 6). Empirical relationships were established to estimate porosity and compressional velocity from wireline log measurements made below this sequence boundary (Herron *et al.* 1992). These relationships are not valid for similar lithologies above this surface, as indicated by the divergence of the

estimated values from the observed values (Fig. 6). This information suggests that the stress histories of the rocks above and below this surface are different and would add confidence to selecting this surface as a significant boundary in the absence of the supplementary (outcrop) information available at this location.

Mineralogical data may also be used to provide constraints on sequence architecture. Hart *et al.* (1992) were able to show that vertical trends in the composition of early diagenetic siderite were able to corroborate the sequence stratigraphic interpretations of Cretaceous rocks at outcrop. Carbon/sulphur ratios from sidewall cores, complemented by measurements of

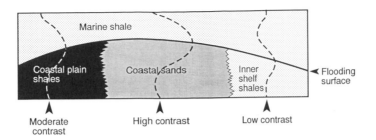

Fig. 8. Schematic representation of reservoir rocks to show the relationship of seismic attributes and geometries to sequence architecture as a consequence of lateral variation of acoustic impedance contrasts.

uranium abundance from spectral gamma ray logging have been used to identify flooding surfaces in Carboniferous strata of the North Sea (Leeder *et al.* 1990). $^{87}Sr/^{86}Sr$ ratios from residual salts extracted from cores have also been successfully used to identify reservoir compartments (Smalley & England 1994).

Reservoir pressure measurements

As discussed in the preceding section, sequence and parasequence boundaries are often associated with barriers to vertical permeability. Whilst these surfaces may not act as barriers to communication during charging of the reservoir over geological timescales (millions of years), they frequently function as barriers to pressure equilibration and fluid movement over production timescales (years to tens of years).

Following the inference that tight-streaks and thin shales associated with flooding surfaces would act as barriers to vertical communication in the western area of the Maui field (Fig. 4), subsequent drilling of the first development well in this area offered the opportunity to test this contention by measuring formation pressures within the depleted C1 Sands reservoir (Fett *et al.* 1994). These measurements confirm that several of the parasequence boundaries form barriers to vertical communication as indicated by discontinuities in the vertical pressure profile (Fig. 7). Note that not all of the parasequence boundaries identified in the cores are associated with significant pressure steps, such that the pressure information is complementary to other criteria in defining the sequence architecture. In this case many different types of independent data (sedimentological, palaeomagnetic, formation pressure) combine to give a consistent indication of the lateral extent to the ravinement

and flooding surfaces (Figs 4 and 7). Note also that the sequence boundary is associated with erosion of underlying sediments and is not associated with a barrier to vertical communication (Fig. 7).

Reflection seismic measurements

The foregoing discussion has illustrated that discontinuities of borehole measurements of reservoir rock and fluid properties may correspond with sequence and parasequence boundaries. In the case that they do, it is expected that this abrupt change in physical properties will occur over a wide area. Changes in density, velocity and/or fluid content of sediments over widespread areas will also give rise to contrasts in acoustic impedance. Under favourable circumstances, these contrasts will give rise to coherent seismic reflections. Indeed, an underlying assumption of seismic sequence stratigraphy is that seismic reflections emanate from chronostratigraphic surfaces rather than facies boundaries. Classical seismic sequence stratigraphy places emphasis on reflection geometries between seismic events that are assumed to represent isochronous surfaces (e.g. Brown & Fisher 1977). In theory the same approach could be used to carry out reservoir-scale analysis but, in practice, the frequency content of seismic waves at the depth of most hydrocarbon reservoirs is too low to enable resolution of the various reservoir elements from vertical seismic sections. Indeed, the low vertical resolution of the seismic data may give a false impression of the lateral extent of stratal discontinuities (Cartwright *et al.* 1993). The perceived limitations of the vertical resolution of seismic in high resolution sequence stratigraphic analysis, probably explains why it is not considered in standard

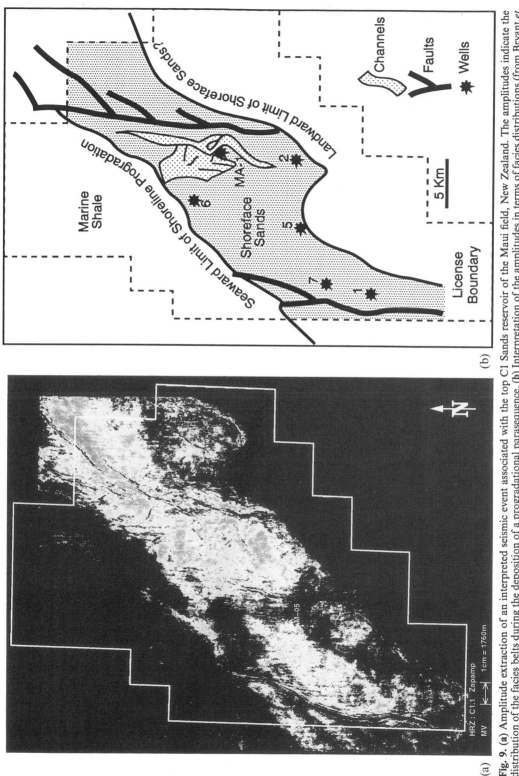

Fig. 9. (a) Amplitude extraction of an interpreted seismic event associated with the top C1 Sands reservoir of the Maui field, New Zealand. The amplitudes indicate the distribution of the facies belts during the deposition of a progradational parasequence. (b) Interpretation of the amplitudes in terms of facies distributions (from Bryant *et al.* 1994).

references (e.g. Galloway 1989; Van Wagoner *et al.* 1990). However, where the thickness of a sequence or parasequence approaches half of the thickness represented by the seismic wavelet, information contained in the character of the reflection may provide important information on the lateral distribution of reservoir facies in such an interval (Meckel & Nath 1977).

The effect of lateral variations in acoustic impedance, associated with facies variation is shown schematically in Fig. 8. The shale unit that forms the top seal to the reservoir has relatively uniform properties across the field but variation in the underlying reservoir facies gives rise to spatially coherent changes in the acoustic impedance of the rocks. This, in turn, generates spatial variation in the strength of the seismic reflection. A field example of this phenomenon is provided by an amplitude display of the seismic event associated with the top of the C1 Sands reservoir of the Maui field (Fig. 9a). The vertical resolution of the seismic is such that it is unable to resolve the individual parasequences that comprise the C1 Sands reservoir interval (Bryant *et al.* 1994), nonetheless the distribution of facies tracts in the uppermost parasequence is readily interpreted from the seismic response (Fig. 9b). By carrying out similar analyses, sequentially through the 3D seismic data, it has been possible to identify the lateral extent of many of the parasequences that comprise the C1 reservoir interval and thereby enhance the spatial resolution of the reservoir description (Bryant *et al.* 1994, 1995).

Conclusions

Sequence and parasequence boundaries are of significance at the reservoir-scale because they are typically of greater lateral extent than field dimensions. Flooding surfaces, that often constitute parasequence boundaries, are frequently associated with barriers to vertical permeability. Consideration of modern analogues suggests that such flooding surfaces are most likely to provide surfaces that will allow high resolution correlation of intra-reservoir markers.

A variety of measurements of oilfield sediments and fluids may provide evidence for the recognition of flooding surfaces. This evidence is independent of the sedimentological and biostratigraphic criteria that are routinely used in sequence stratigraphic analyses. Rarely are these data sufficient to identify unambiguously such surfaces when considered in isolation but, rather, they provide complementary evidence to constrain sequence stratigraphic interpretations. In favourable circumstances, analysis of

the attributes of 3D seismic data may provide information concerning the distribution of facies belts within individual parasequences. Thus, even where the seismic data have insufficient vertical resolution to reveal the details of the vertical organization of the sequence architecture, the data may greatly enhance the lateral resolution of the reservoir architecture. By integrating the evidence provided by the various physical measurements with that provided by traditional methods, confidence in the predictive value of the resulting reservoir geological models should be enhanced.

I gratefully acknowledge Susan Herron and Carl Greenstreet for provision of illustrations and comments on earlier versions of this manuscript. Referees Alister MacDonald and Jon Gluyas are thanked for their constructive reviews.

References

BERRYHILL, H. L., SUTER, J. R. & HARDIN, N. S. 1986. *Late Quaternary facies and structure, northern Gulf of Mexico: interpretations from seismic data.* American Association of Petroleum Geologists, Studies in Geology, **23**.

BROWN JR, L. F. & FISHER, W. L. 1977. Seismic-stratigraphic interpretation of depositional systems: examples from Brazilian rift and pull-apart basins. *In:* PAYTON, C. E. (ed.) *Seismic stratigraphy appplications to hydrocarbon exploration.* American Association of Petroleum Geologists, Memoir, **26**, 213–248.

BRYANT, I. D. & LIVERA, S. E. 1991. Identification of unswept oil volumes in a mature field by using integrated data analysis: Ness Formation, Brent field, UK North Sea. *In:* SPENCER, A. M. (ed.) *Generation, accumulation and production of Europe's hydrocarbons.* European Association of Geoscientists, Special Publication, **1**, 75–88.

——, GREENSTREET, C. W. & VOGGENREITER, W. V. 1995. Integrated 3-D geological modeling of the C1 Sands reservior, Maui field, offshore New Zealand. *American Association of Petroleum Geologists Bulletin,* **79**, 351–374

——, PAARDEKAM, A. H. M., DAVIES, P. & BUDDING, M. C. 1991. Integrated reservoir charcterisation of Cycle III, Brent Group, Brent field, UK North Sea for reservoir management. *In:* SNEIDER, R., MASSELL, W., MATHIS, W., LOREN, D. & WICHMAN, P. (eds) *The Integration of Geology, Geophysics, Petrophysics and Petroleum Engineering in Reservoir Delineation, Description and Management.* American Association of Petroleum Geologists, 405–422.

——, MARSHALL, M. G., GREENSTREET, C. W., VOGGENREITER, W. V., COHEN, J. M. & STROEMMEN, J. F. 1994. Integrated geological reservoir modelling of the Maui Field, Taranaki Basin, New Zealand. *In: 1994 New Zealand Petroleum*

Conference Proceedings. New Zealand Ministry of Commerce, 256–280.

CANDE, S. C. & KENT, D. V. 1992a. A new geomagnetic polarity time scale for the Late Cretaceous and Cenozoic. *Journal of Geophysical Research,* **97,** 13917–13951.

—— & —— 1992b. Ultrahigh resolution marine magnetic anomaly profiles: a record of continuous paleointensity variations? *Journal of Geophysical Research,* **97,** 15075–15083.

CARTWRIGHT, J. A., HADDOCK, R. C. & PINHEIRO, L. M. 1993. The lateral extent of sequence boundaries. *In:* WILLIAMS, G. D. & DOBB, A. (eds) *Tectonics and Seismic Sequence Stratigraphy.* Geological Society, London, Special Publication, **71,** 15–34.

CROSS, T. A., BAKER, M. R., CHAPIN, M. A., CLARK, M. S., GARDNER, M. H. *ET AL.* 1993. Applications of high-resolution sequence stratigraphy to reservoir analysis. *In:* ESCHARD, R. & DOLIGEZ, B. (eds) *Subsurface Reservoir Characterization from Outcrop Observations.* Editions Technip, Paris, 11–33.

FETT, T., BRYANT, I. D. & GREENSTREET, C. W. 1994. Integration of state-of-the-art wireline reservoir delineation devices with core data from an offset well, to optimise reservoir management: Maui field, New Zealand. *In: 1994 New Zealand Petroleum Conference Proceedings.* New Zealand Ministry of Commerce, 231–239.

GALLOWAY, W. E. 1989. Genetic stratigraphic sequences in basin analysis I: architecture and genesis of flooding-surface bounded depositional units. *American Association of Petroleum Geologists Bulletin,* **73,** 125–142.

GIBBONS, K., HELLEM, T., KJEMPERUD, A., NIO, S. D. & VEBENSTAD, K. 1993. Sequence architecture, facies development and carbonate-cemented horizons in the Troll Field reservoir, offshore Norway. *In:* ASHTON, M. (ed.) *Advances in Reservoir Geology.* Geological Society, London, Special Publication, **69,** 1–31.

HART, B. S., LONGSTAFFE, F. J. & PLINT, A. G. 1992. Evidence for relative sea level change from isotopic and elemental composition of siderite in the Cardium Formation, Rocky Mountain Foothills. *Bulletin of Canadian Petroleum Geology,* **40,** 52–59.

HAUGER, E., LØVLIE, R. & VAN VEEN, P. 1994. Magnetostratigraphy of the Middle Jurassic Brent group in the Oseberg oil field, northern North Sea. *Marine and Petroleum Geology,* **11,** 375–388.

HERRON, S. L., HERRON, M. M. & PLUMB, R. A. 1992. Identification of clay-supported and framework-supported domains from geochemical and geophysical well log data. SPE, Paper **24726.**

HOUGHTON, R. A. & WOODWELL, G. M. 1989. Global climatic change. *Scientific American,* **260**(4), 36–44.

KAUFMAN, R. L., AHMED, A. S. & ELSINGER, R. J. 1990. Gas chromatography as a development and production tool for fingerprinting oils from individual reservoirs–applications in the Gulf of Mexico. *In:* SCHUMACHER, D. & PERKINS, B. (eds) *Gulf coast oils and gases.* Proceedings of 9th Annual Research Conference, GCSSEPM, 263–282.

LEEDER, M., RAISWELL, R., AL-BIATTY, H., McMAHON, A. & HARDMAN, M. 1990. Carboniferous stratigraphy, sedimentation and correlation of well 48/3-3 in the southern North Sea Basin: integrated use of palynology, natural gamma/sonic logs and carbon/sulphur geochemistry. *Journal of the Geological Society, London,* **147,** 287–300.

LERBEKMO, J. F. & DEMCHUK, T. D. 1992. Magnetostratigraphy and biostratigraphy of the continental Paleocene of the Red Deer Valley, Alberta, Canada. *Bulletin of Canadian Petroleum Geology,* **40,** 24–35.

LOUTIT, T. S., HARDENBOL, J., VAIL, P. R. & BAUM, G. R. 1988. Condensed sections: the key to age determination and correlation of continental margin sequences. *In:* WILGUS, C. K., HASTINGS, B. S., KENDALL, C. G. St. C., POSAMENTIER, H. W., ROSS, C. A. & VAN WAGONER, J. C. (eds) *Sea-level changes – an integrated approach.* SEPM, Special Publication, **42,** 183–212.

MATTHEWS, R. K. & FROLICH, C. 1991. Orbital forcing of low-frequency glacioeustacy. *Journal of Geophysical Research,* **96,** 6797–6803.

MECKEL JR, L. D. & NATH, A. K. 1977. Geologic considerations for stratigraphic modeling and interpretation. *In:* PAYTON, C. E. (ed.) *Seismic stratigraphy applications to hydrocarbon exploration.* American Association of Petroleum Geologists, Memoir, **26,** 417–438.

MITCHUM, R. M., 1977. Seismic stratigraphy and global changes of sea-level, Part 1: Glossary of terms used in sequence stratigraphy. *In:* PAYTON, C. E. (ed.) *Seismic stratigraphy applications to hydrocarbon exploration.* American Association of Petroleum Geologists, Memoir, **26,** 53–62.

O'BYRNE, C. J. & FLINT, S. 1993. High-resolution sequence stratigraphy of Cretaceous shallow marine sandstones, Book Cliffs outcrops, Utah, USA – application to reservoir modelling. *First Break,* **11,** 445–459.

PASLEY, M. A. & HAZEL, J. E. 1990. Use of organic petrology and graphic correlation of biostratigraphic data in sequence stratigraphic interpretations: example from the Eocene-Oligocene boundary section, St. Stephens Quarry, Alabama. *Transactions of the Gulf Coast Association of Geological Societies,* **XL,** 661–684.

PLINT, A. G. 1988. Global eustacy and the Eocene sequence in the Hampshire Basin, England. *Basin Research,* **1,** 11–22.

POSAMENTIER, H. W. 1991. An overview of sequence stratigraphic concepts. *In: Proceedings of 1991 NUNA Conference on High-Resolution Sequence Stratigraphy.* Geological Association of Canada, 62–74.

POUMOT, C. 1989. Palynological evidence for eustatic events in the tropical Neogene. *Bulletin du Centre de la Recherche Exploration-Production Elf-Aquitaine,* **13,** 437–453.

REY, D., TURNER, P. & YALIZ, A. 1993. Palaeomagnetic study and magnetostratigraphy of the Triassic Skagerrak Formation, Crawford Field,

UK North Sea. *In:* NORTH, C. P. & PROSSER, D. J. (eds) *Characterization of Fluvial and Aeolian Reservoirs*. Geological Society, London, Special Publication, **73**, 339–420.

SMALLEY, P. S. & ENGLAND, W. A. 1994. Reservoir compartmentalization assessed with fluid compositional data. *SPE Reservoir Engineering*, **9**, 175–180.

TAYLOR, A. M. & GAWTHORPE, R. L. 1993. Application of sequence stratigraphy and trace fossil analysis to reservoir description: examples from the Jurassic of the North Sea. *In:* PARKER, J. R. (ed.) *Petroleum Geology of Northwest Europe: Proceedings of the 4th Conference*. Geological Society, London, 317–335.

TURNER, G. M. & BRYANT, I. D. 1995. Application of a palaeomagnetic reversal stratigraphy to constrain well correlation and sequence stratigraphic interpretation of the Eocene C1 Sands, Maui Field, New Zealand. *In:* TURNER, P. & TURNER, A. (eds) *Palaeomagnetic Applications in Hydrocarbon Exploration and Production*. Geological Society, London, Special Publication, **98**, 205–221.

TURNER, P. & TURNER, A. 1995. *Palaeomagnetic Applications in Hydrocarbon Exploration and Production*. Geological Society, London, Special Publication, **98**.

VAIL, P. R., MITCHUM, R. M. & THOMPSON III, S. 1977. Seismic stratigraphy and global changes of sea level, part 3: relative changes of sea level from coastal onlap. *In:* PAYTON, C. E. (ed.) *Seismic stratigraphy applications to hydrocarbon exploration*. American Association of Petroleum Geologists, Memoir, **26**, 63–97.

VAN WAGONER, J. C., MITCHUM, R. M., CAMPION, K. M. & RAHMANIAN, V. D. 1990. *Siliciclastic sequence stratigraphy in well logs, cores and outcrops: concepts for high-resolution correlation of time and facies*. American Association of Petroleum Geologists, Methods in Exploration Series, **7**.

——, POSAMENTIER, H. W., MITCHUM, R. M., VAIL, P. R., SARG J. F., LOUTIT, T. S. & HARDENBOL, J. 1988. An overview of sequence stratigraphy and key definitions, *In:* WILGUS. C. K., HASTINGS, B. S., KENDALL, C. G. St. C., POSAMENTIER, H. W., ROSS, C. A. & VAN WAGONER, J. C. (eds) *Sea-level changes – an integrated approach*. SEPM Special Publication, **42**, 39–45.

High resolution sequence biostratigraphy: examples from the Gulf of Mexico Plio-Pleistocene

JOHN M. ARMENTROUT

Mobil Exploration and Production Technical Center, PO Box 650232, Dallas, Texas 75265-0232, USA

Abstract: High resolution sequence biostratigraphy provides a framework for the construction of palaeogeographic maps for each phase of relative sea-level change. The late Pliocene and Pleistocene of the Gulf of Mexico provides a case history of the integration of this methodology. The strata are abundantly fossiliferous and stratigraphically thick so that exploration well samples provide a record of glacial/interglacial cycles with a 200–300 ka frequency. The methodology developed is suitable for other stratigraphic intervals where microfossils are relatively abundant.

Depositional cyclicity is a fundamental part of the stratigraphic record (Einsele 1982). Recognition of depositional cycles from multifold seismic reflection profiles has become a major research topic since the introduction of sequence stratigraphic methodology by Vail *et al.* (1977) and Mitchum *et al.* (1977). Biostratigraphy has been used in sequence stratigraphic studies principally to correlate local depositional cycles into a regional framework (Haq *et al.* 1988). This paper focuses on the broader contribution of biostratigraphy to cycle stratigraphy and palaeogeographic map construction.

When integrated with wireline log and palaeobathymetric data, microfossil abundance patterns are particularly useful for locating condensed sections and sequence boundaries, and for constraining the interpretation of depositional systems tracts. Armentrout (1987, 1991), Armentrout *et al.* (1990), Shaffer (1987, 1990), Pacht *et al.* (1990a, 1990b) and Vail & Wornardt (1990) use microfossil abundance peaks to locate condensed intervals in sequence stratigraphic analysis of Gulf of Mexico Plio-Pleistocene strata. These same authors suggested that fossil abundance minima are often associated with sequence boundaries. Additionally, variations in both abundance patterns and specific fossil content can be used to characterize depositional systems tracts (Armentrout *et al.* 1990).

The Gulf of Mexico Plio-Pleistocene provides an excellent study area for defining fossil patterns useful for stratigraphic analysis (Fig. 1). The stratigraphic section of this area is highly fossiliferous with foraminifera and calcareous nannoplankton being the most widely used for correlation and biofacies analysis. The 12 000–16 000 ft (3660–4880 m) of Plio-Pleistocene strata

represents rapid sediment accumulation and enables stratigraphic separation of short-term events which are recognized from biostratigraphic, lithostratigraphic and seismic-stratigraphic data. Multifold seismic reflection profiles provide a clear definition of the stratigraphic framework of the Plio-Pleistocene section, independent of the biostratigraphic correlations. The numerous exploration wells in the area make it possible to obtain samples from wells very close to regional, seismic reflection profiles which parallel depositional dip, facilitating the direct comparison of seismic-stratigraphic and biostratigraphic correlations (Armentrout & Clement 1990).

Sequence stratigraphic biostratigraphy

The primary horizons used in sequence stratigraphic analysis are the sequence boundary, a transgressive surface and the maximum flooding surface which occurs within the condensed section in areas of low sediment accumulation rates. These regionally correlative surfaces can be identified on seismic reflection profiles, on wireline logs, in stratigraphic sections and from checklists of fossil abundance and diversity. Although each surface is slightly time-transgressive and varies in its lithologic expression across regional study areas, each represents a nearly regionally synchronous surface and can be used as a correlation horizon for partitioning depositional cycles into discrete phases of relative sea-level rise, highstand, fall and lowstand. Sediments, deposited during each phase of relative sea-level change, form depositional systems tracts that consist of all correlative deposits of

From Howell, J. A. & Aitken, J. F. (eds), *High Resolution Sequence Stratigraphy: Innovations and Applications,* Geological Society Special Publication No. 104, pp. 65–86.

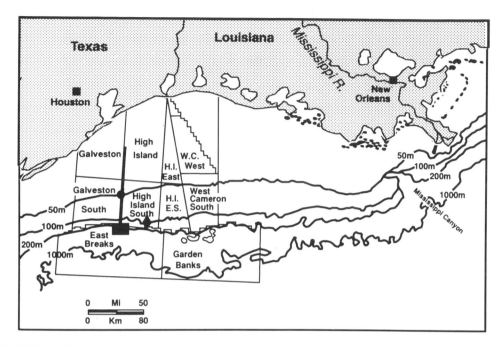

Fig. 1. Map of the study area showing offshore Texas exploration areas, bathymetry and location of the transect (thick line) along which the Galveston South Mobil A-158 #1 well (dot) and East Breaks 160 #3 well (black rectangle) are located. The black diamond in South High Island area is the location of the seismic reflection profile of Fig. 8.

lowstand, of transgressive, of highstand or of falling phases of relative sea-level.

The sequence boundary is an unconformity formed during relative lowering of and during lowstand of sea-level (Van Wagoner *et al.* 1988). The unconformity correlates into basinal areas of continuous sedimentation where the age of the unconformity is determined using age-diagnostic fossils. This unconformity surface is usually recognized by the erosional truncation of underlying strata and the onlap of overlying strata of the next sequence. The sequence boundary may be characterized by a marked shift in biofacies assemblages from deeper below the boundary to shallower above the boundary (Loutit *et al.* 1988). It may also be marked by an increase in reworked or displaced fossils, and by a decrease in both fossil abundance and diversity due to rapid accumulation of sediment or destruction of shell material in shallow water, high-energy environments associated with the erosional unconformity (Armentrout 1991).

The transgressive surface is another important correlation surface within each depositional cycle. This surface represents the landward stepping onlap of deposition related to relative rise in sea-level (Van Wagoner *et al.* 1988). Along basin margins, the transgressive surface merges with the sequence boundary unconformity. Within the basin depocentre, the transgressive surface will separate the lowstand and transgressive systems tracts. This surface is recognized by regionally correlative flooding surfaces above prograding coarse-grained sediments. Biofacies tend to be shallow below and deepening above the transgressive systems tract.

The maximum flooding surface is defined as the surface correlative to the most landward onlap of transgressive strata immediately below the subsequent downlapping progradational unit (Van Wagoner *et al.* 1988). The precise age of this surface will vary along any basin margin because of the interplay of sediment supply and accommodation space. Away from the locus of major input of sediment this surface is a clay-rich condensed section formed by slow accumulation of sediment. This interval is often represented by significant increase in fossil abundance and diversity, by the deepest-water biofacies assemblage of the transgressive–regressive cycle, and by the abundance of authigenic minerals such as phosphate and glauconite (Armentrout 1987, 1991; Loutit *et al.* 1988).

Despite their lateral variability, the seismic, lithologic and biostratigraphic signatures of the condensed section and of the associated maximum flooding surface are generally the most easily recognized and precisely dated regionally correlative surfaces (Armentrout 1987, 1991; Coleman & Roberts 1988; Galloway 1989). Similar to the condensed section are local condensed intervals which form during relative sediment starvation on the abandonment surface of basin floor fans, slope fans and deltaic lobes. Detailed faunal analysis and high resolution correlation grids help differentiate local condensed intervals from regional condensed sections.

Abundance patterns

In the Gulf of Mexico Pliocene and Pleistocene, microfossil abundance patterns provide high resolution observations for identifying sequence stratigraphic surfaces. Total microfossil abundance patterns reflect changes in sediment accumulation rates, provided that changes in these rates are much larger than these rates of biogenic productivity and preservation of the microfossils. If this condition holds during the reduced rate of sediment accumulation associated with transgression, the middle shelf and deeper transgressive phase deposits will be characterized by an increase in fossil abundance, due to relative terrigenous sediment starvation and consequent concentration of fossil material. If the same conditions of biotic productivity hold during the increased rate of sediment accumulation associated with a prograding system, the accumulated sediments will be characterized by a decrease in fossil abundance due to dilution and environmental stress.

Sediment accumulation rates calculated for the offshore Texas late Pliocene study area range from 2–4 ft (0.6–1.2 m) per thousand years (Armentrout *et al.* 1980). Short-term variations in biologic productivity would be averaged in the 30 ft (9 m) interval well-cutting samples, each representing between 15 and 8 ka. Systematic changes in microfossil abundance, therefore, may reflect significant, long-term dynamic changes in sediment accumulation rates characteristic of different settings within a depositional cycle.

Patterns of foraminiferal and calcareous nannoplankton abundance are shown in Figs 2

and 3 for the Mobil A-158#1 well. Figure 2 is a checklist of the foraminifera and nannoplankton recovered from every other 30 ft cutting sample. The observed occurrence of fossils is noted by a black symbol which on the full scale chart (Armentrout 1989) represents any one of four abundance categories. This photo-reduced display of the checklist serves to illustrate the pattern of alternating stratigraphic intervals with few fossils (arrows) versus intervals with abundant and diverse assemblages. The abundance data are redisplayed in Fig. 3 as a histogram which facilitates analysis of the patterns.

The abundance patterns of both foraminifera and calcareous nannoplankton in general parallel each other. The checklists for these two fossil groups were compiled by two different palaeontologists. The similarity in abundance patterns is interpreted to represent similarity in environmental conditions, principally those of deposition, that affected the concentration or dilution of both microfossil groups.

In addition to patterns of abundance, variations in electric log response and biofacies distribution are analysed. The gamma ray log display provides a measure of sediment type, with curve deflections to the left suggesting increased sand content while high values to the right indicate increases in clay content. Biofacies are interpreted using benthonic foraminiferal assemblages indicative of water mass conditions (Tipsword *et al.* 1966; Skinner 1966; Culver 1988; Armentrout 1991).

Stratigraphic intervals rich in calcareous nannoplankton and foraminiferal fossils and having maximum gamma ray values are interpreted to correlate with condensed depositional intervals. Some of these condensed intervals correlate with the condensed section associated with the maximum flooding surface of highstands in sea-level (Loutit *et al.* 1988). Typical candidates for condensed section intervals are shown in Fig. 3 by the two maximum fossil abundance peaks between 2457 and 3114 ft.

Above the upper condensed section candidate is a marked decrease in fossil abundance, with a low value immediately above 2457 ft (Fig. 3). The gamma ray pattern correlative with the faunal abundance low shows a forestepping, upward decrease in values to a minimum at about 2350 ft and then a steady, upward backstepping with increased values. The nearly mirror image character of these abundance and gamma ray patterns suggests that the decrease in abundance is associated with decrease in clay, and the increase in abundance is associated with the increase in clay. These patterns are asso-

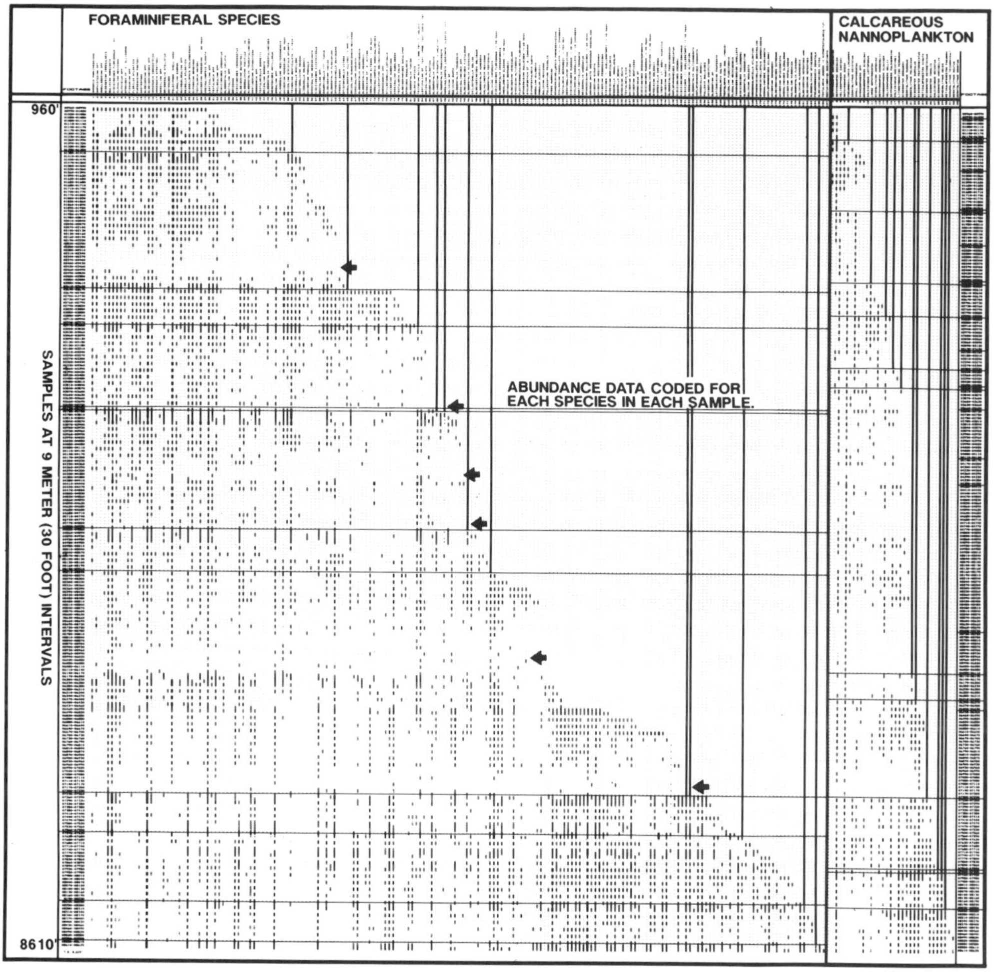

Fig. 2. Biostratigraphic checklist for the Galveston South Mobil A-158 #1 well showing species present and their abundance in each 30 ft (9 m) sample. The main purpose of this figure is to show the pattern of sample intervals with few or no fossils (arrows) versus sample intervals with many fossils. Species names are listed across the top for 191 foraminifera and 33 selected calcareous nannoplankton. Thirty-foot sample intervals from 960–8610 ft (293–2625 m) are listed along the left side. The occurrence of fossils is indicated by a black symbol for each species present in the samples examined at every 60 ft (18 m) interval, with infill samples for critical intervals. The statistical data are redisplayed in histogram format in Fig. 3. Foraminifera checklisted by J. F. Clement; calcareous nannoplankton checklisted by T. C. Huang (from Armentrout 1989).

ciated with a cycle of water depth decrease and then increase as shown by the associated biofacies pattern.

The faunal minimum just above 2457 ft can be interpreted as an interval within which a sequence boundary candidate may occur. This sequence boundary candidate is above the interpreted prograding depositional facies, asso- ciated with upward decreasing fossil abundances

and associated shallowing, and below the inter- preted backstepping, transgressive depositional facies with upward increasing fossil abundance and water depth. The underlying faunal max- imum between 2457 and 3114 ft, interpreted as a candidate for a condensed section, is in the appropriate position for a condensed section if the prograding depositional facies are inter- preted as a highstand systems tract (Vail 1987;

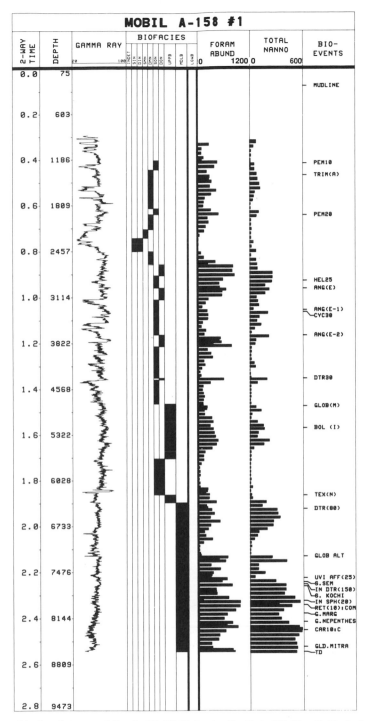

Fig. 3. Biostratigraphic data interpreted for the Mobil Galveston South A-158 #1 well. Vertical scale is in two-way-time for integration with seismic reflection profiles and depth in feet below Kelly-bushing. Patterns of shallow versus deep biofacies, and fossil abundance concentration versus dilution can be correlated with progradation of sandstone versus shale interpreted from wireline log patterns. Bioevents (abbreviated acronyms such as PEM10 and Trim (A)) and faunal discontinuity events (abbreviated FI(P10) and FI(TA)) provide correlation datum. These patterns can be correlated to other wells and evaluated against correlations constructed from seismic reflection profiles. Foraminiferal abundance (Foram Abund) scale is 0–1200 specimens; Calcareous nannofossil abundance (Total Nanno) scale is 0–600 specimens. Biofacies include middle bathyal (MDLB, 500–1000 m), upper bathyal (UPPB, 200–500 m), deep outer neritic (DON, 150–200 m), shallow outer neritic (SON, 100–150 m), deep middle neritic (DMN, 75–100 m), shallow middle neritic (SMN, 50–75 m), deep inner neritic (DIN, 25–50 m) and shallow inner neritic (SIN, 0–25 m).

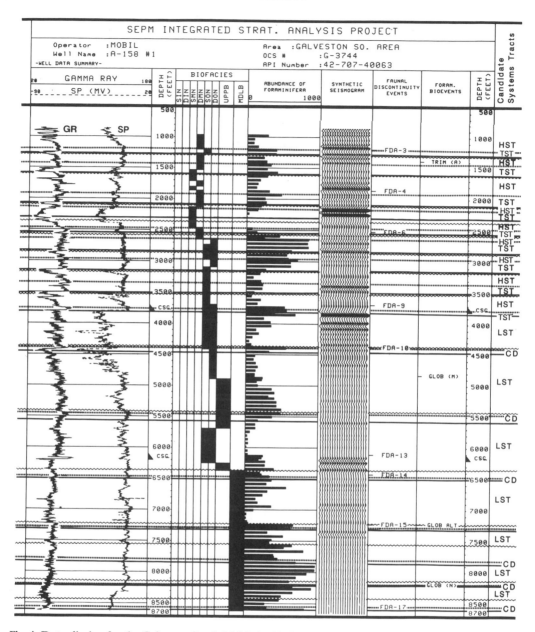

Fig. 4. Data display for the Galveston South Mobil A-158 #1 well with interpretation of sequence boundaries (wavy line), transgressive surfaces (heavy solid line), and maximum flooding surfaces (broken line) for sixteen depositional sequences. Systems tracts and condensed sections (CD) are noted to the right: LST, lowstand systems tract; TST, transgressive systems tract; HST, highstand systems tract.

Vail & Wornardt, 1990).

Numerous candidates for sequence boundaries and condensed sections can be selected from the patterns of fossil abundance, biofacies shifts, and gamma ray log measurements (Fig. 4). Correlation with similar patterns from other wells, within a chronostratigraphic framework, is necessary to test whether or not the candidate surfaces have a regional distribution and are truly sequence boundaries and condensed sections in the context of sequence stratigraphy (Armentrout 1991).

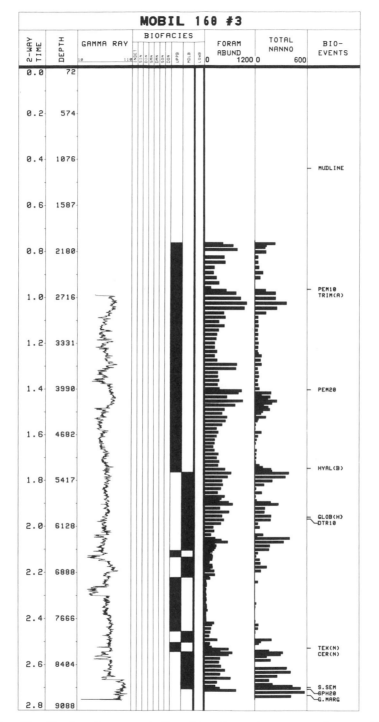

Fig. 5. Biostratigraphic calibration of sequence stratigraphic surfaces and systems tracts interpreted for the Mobil East Breaks 160 #3 well. See Fig. 3 for format explanation. Note that the biofacies interpretation for this well is entirely within the bathyal zone varying between upper bathyal and middle bathyal. This contrasts with the biofacies analysis of the Mobil A-158 #1 well where the neritic biofacies occurs in several shallowing upward cycles.

Data from the Mobil 160 #3 well, illustrated in Fig. 5, which occurs about 25 miles (15.5 km) south of the A158 #1 well, show patterns of foraminiferal and calcareous nannoplankton abundance that occur entirely within the bathyal zone, which is deeper than 600 ft (200 m), and below the effect of most sea-level changes. Changes in relative sea-level that have a more profound impact in the shallow neritic environments of less than 150 ft (50 m) water depth are recognized by the coarsening-upward sandstone cycles in the upper part of the A158 #1 well. The gamma ray pattern indicates much less sandstone in the 160 #3 well which is in keeping with regional observations of most bathyal depositional environments in the Gulf of Mexico.

The persistence of the variations of concentration and dilution events within the fossil abundance patterns of the 160 #3 well are interpreted to reflect changes in relative rates of sediment accumulation, even in this relatively sandstone poor, deep water environment. These abundance patterns may provide potential for correlation if they can be demonstrated to be age equivalent to similar patterns in other wells.

Pattern correlation

Correlation between wells are constrained primarily by the first-downhole-occurrence of chronostratigraphically significant species, and by faunal discontinuities recognized by rapid changes in biofacies assemblages, fossil abundance, and diversity (Stude 1984; Armentrout & Clement 1990). Such bioevents are listed in the right-hand column of Figs 3 and 5 where first-downhole-occurrences are listed as acronyms such as PEM10 and TRIM(A) and faunal discontinuities as FI(P10) and FI(P20).

Correlation of four wells along a depositionally-dip orientated transect are shown in Fig. 6. Both the South Galveston Mobil well A158 #1 (Fig. 3) and the East Breaks Mobil well 160 #3 (Fig. 5) are included in this transect (Fig. 1). The correlation horizons are based on the bioevents listed at the right of the figure. The observation criteria for each line of correlation varies along the section but has been tested using a fully integrated biostratigraphic and seismic stratigraphic analysis (Armentrout & Clement 1990; Armentrout 1991).

The correlation horizons occur at or in close association with fossil abundance peaks, deposited during the most effective transgression. This is the phase of a depositional cycle when the oceanic waters make their most landward incursion and fossil-rich, shale-prone depositional environments are most widely distributed.

In the Gulf of Mexico, these maximum flooding surface/condensed sections provide the most useful correlation horizons. They are identified by microfossil abundance peaks, intervals of high gamma ray values, and on seismic by regional downlap surfaces (Armentrout 1987, 1991; Shaffer 1987, 1990; Galloway 1989).

Highstand of relative sea-level is characterized by deposition of prograding sands. These are low in fossil abundance due to dilution by high sediment accumulation rates, low diversity populations resulting from stressed environments, and destruction of shell material either during deposition or post-depositionally through dissolution by fluids migrating within the sediments. As relative sea-level begins to fall, an erosional surface develops and extends basinward following the regressing shoreline. This erosional unconformity is the sequence boundary (Vail et al. 1977) and occurs within the interval of lowest fossil abundance.

The general pattern of abundances described above is well expressed for most of the correlation horizons of Fig. 6 but lateral variations result in exceptions. In Well A70-1 (Fig. 6), the patterns of fossil concentration and dilution are poorly developed due to the shallow water depositional environments represented in this well. Additionally, extensive by-pass and erosion associated with the sequence boundary result in modified abundance patterns. Only through careful assembly of the data and integration of multiple datasets can consistently correlatable patterns be correctly interpreted.

Systems tract analysis

Based on foraminiferal and wireline log patterns, such as those discussed above, Armentrout et al. (1990) constructed a preliminary model relating patterns of foraminiferal abundance to systems tracts. The preliminary model presents two cases: (1) middle to outer shelf patterns, and (2) intraslope basin floor patterns (Fig. 7).

On the shelf, where rates of subsidence, sediment supply and sea-level change are favourable, both transgressive and regressive sediment packages may be preserved. Sand-rich sediments deposited during the initial transgressions, suggested by gamma ray and spontaneous potential minima, typically have very low concentration of microfossils (Fig. 7). Foraminiferal biofacies in these sediments indicate palaeobathymetry minima and biofacies excursions. In cuttings these shallow water biofacies may sometimes be interpreted as being of a deeper water origin than their living depth. This is due

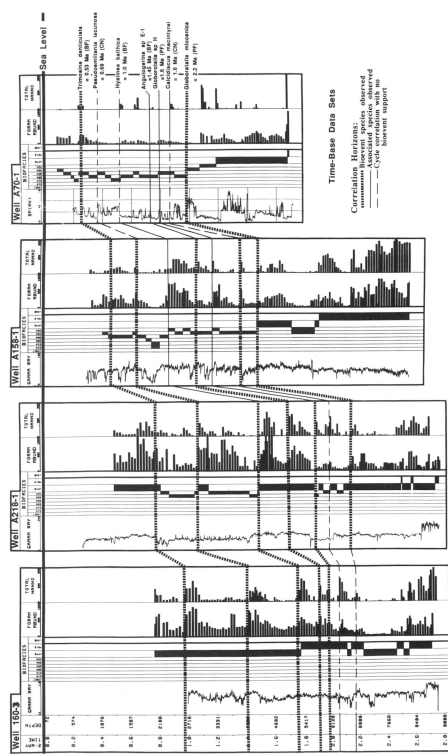
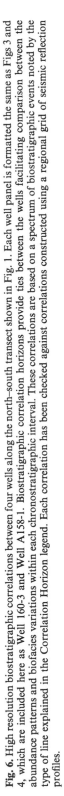

Fig. 6. High resolution biostratigraphic correlations between four wells along the north–south transect shown in Fig. 1. Each well panel is formatted the same as Figs 3 and 4, which are included here as Well 160-3 and Well A158-1. Biostratigraphic correlation horizons provide ties between the wells facilitating comparison between the abundance patterns and biofacies variations within each chronostratigraphic interval. These correlations are based on a spectrum of biostratigraphic events noted by the type of line explained in the Correlation Horizon legend. Each correlation has been checked against correlations constructed using a regional grid of seismic reflection profiles.

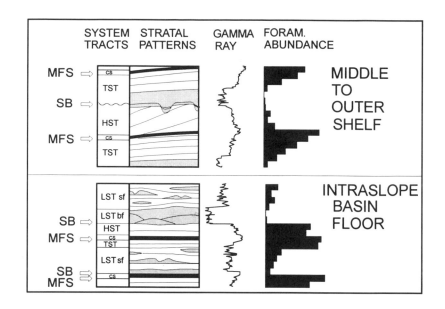

Fig. 7. Schematic illustration of possible systems tracts and stratal pattern interpretations based on wireline log gamma ray and well-cutting foraminiferal abundance for the Mobil well A158-1. Most abundance histograms will have additional abundance peaks (usually single sample) not associated with the maximum flooding surface which probably represent local condensed intervals. MFS, maximum flooding surface; SB, sequence boundary. Systems tract abbreviations are given in Fig. 4. This figure by R. J. Echols is modified after fig. 3 in Armentrout *et al.* (1990).

to the few *in situ* shallower facies microfossils being swamped by contaminants from the overlying intervals, deposited at greater water depth, and the tendency to average palaeoecologic interpretations over several samples to avoid interpretation patterns suggesting rapid oscillation of water depth. Careful examination of each sample and the regional pattern of biofacies from coeval stata in many wells will help define the correct biofacies interpretation (Armentrout *et al.* 1990).

Sediments of the transgressive systems tracts, beginning with transgressive sands and terminating at a condensed interval, are identified by upward increasing microfossil abundance, and diversity, and deeper biofacies, associated with gamma ray and spontaneous potential patterns that suggest increasing clay content. Within the transgressive system tracts of shelf facies of the A158-1, increasing foraminiferal abundance may be a more regular feature than back stepping in the wireline log patterns and, thus, a very useful diagnostic feature. The robustness of the mircofossil pattern is even more evident in the deeper water environments represented by

the A218-1 and 160-3 wells (Fig. 6).

On the shelf, relatively regular changes in the gamma ray log response from maxima at the condensed section to minima at the next overlying sand package, are interpreted as representing the fore-stepping, prograding beds of the highstand systems tract. The most profound change in properties of the foraminiferal populations in the A158-1 seems to occur just above the condensed section and appears to reflect abruptly increasing sedimentation rates with the arrival of the prograding highstand deltas (Fig. 7).

Markedly asymmetrical foraminiferal abundance patterns may be associated with relatively symmetrical gamma ray patterns around a shelf condensed section (Schaffer 1987, 1990). This phenomena is well illustrated in A158 #1 at 1900 and 3000 ft (Fig. 4). Relationships in the prolonged cycle from 2900–2350 ft in A-158 #1 are complicated by the 150 ft offset between the equally pronounced gamma ray maxima at 2500 ft and the foraminiferal abundance peak at 2650 ft. This difference seems greater than that expected from calibration errors and may actually reflect two distinct condensed sections.

The earlier condensed section at 2650 ft is clearly represented in the foraminiferal abundance data, but not the gamma ray, and the reverse is the case at this later case located at 2500 ft (Fig. 4). R. J. Echols in Armentrout *et al.* (1990) suggested this is an example of a higher frequency, lower amplitude, sea-level event (4th order) superimposed on a lower frequency, higher amplitude, event (3rd order) (Mitchum & Van Wagoner 1990).

Sedimentary sequences of the intraslope basin floor tend to be dominated by the lowstand systems tracts (Vail 1987). At one extreme, sediments of the highstand systems tract may be reduced to a thin condensed interval interbedded between lowstand sediments. Alternatively, highstand sediments may be well developed on the proximal intraslope basin floor, provided they bypass the shelf into the basin, as for example, where a highstand delta progrades to the shelf edge. The examples of candidates for condensed sections and sequence boundaries of the intraslope basin floor chosen for illustration from the A158 #1 are of the latter type. The sequence boundary is placed at the base of a rapid biofacies shallowing at 6375 ft, and a condensed interval is suggested at 6450–6500 ft where a foraminiferal abundance peak approximates a gamma ray high. The interval between the two is interpreted as a candidate highstand systems tract.

The above interpretation of a highstand systems tract is based partly on evidence from regional correlation (Fig. 6). The sequence boundary at 6375 ft and the associated events are important because they mark the onset of rapid filling of the basin to shelf depths, primarily by lowstand slope front-fill deposits. This is best expressed in the A158 #1 well by the shallowing biofacies and the increase in sand content in the upper half of the well. Sedimentation rates calculated for the interval above the sequence boundary at 6375 ft are 2.5–7.5 times higher than below, with higher values within this range being more likely than lower. The range of uncertainty is caused by the probability that the last occurrence of *Globoratalia miocenica* at 4830 ft is somewhat depressed in this well.

Overall, the contributions of systematic foraminiferal abundance changes to diagnosing basin systems tracts is not as straightforward as appears to be the case for shelf environments in the A158-1 well. Abundance peaks occur at several condensed intervals within lowstand systems tracts, in addition to the maximum flooding surfaces between the transgressive and highstand systems tracts (Armentrout 1987;

Shaffer 1987, 1990); these include fossil-rich shale intervals at the top of basin floor fans and the top of slope fan systems.

Calibration of depositional cycles

The sequence stratigraphic framework constructed using the correlation horizons, and the systems tract analysis discussed above, provide definition of depositional cycles. Each cycle may consist of a lowstand systems tract deposited seaward of its age equivalent shelf edge, a transgressive systems tract onlapping the basin margin, and highstand systems tract consisting of a shelf prograding complex (Vail 1987; Van Wagoner *et al.* 1988). Data from the area of the East Breaks 160 field (black rectangle of Fig. 1) provide an example of cycle identification and calibration.

Seismic reflection profile grids in the East Breaks area image sigmoidal clinoforms (Fig. 8) interpreted as a late Pleistocene shelf-margin delta (Sutter & Berryhill 1985). Each set of shingled clinoforms is bounded by a regionally extensive, continuous, parallel and uniformly high amplitude reflection. Abundance histograms of planktonic foraminifera from corehole MSC 160-1 are correlated to these two seismic facies (Fig. 9).

The regional continuous high amplitude reflections correlate with intervals of abundant *Globorotalia menardii* characteristic of warm waters (Kennett & Srinivasan 1983; Martin & Fletcher 1993) and interpreted as interglacial regional clays. The shingled clinoforms contain abundant *Globorotalia inflata*, characteristic of cooler water (Kennett & Srinivasan 1983; Martin & Fletcher 1993) and interpreted as glacial shelf-edge deltas. These biofacies and seismic facies patterns suggest two cycles of glacially forced deposition. In sequence stratigraphic nomenclature, the regional clays are interpreted as the condensed section of the distal highstand and transgressive systems tracts, and the shelf-edge deltas are interpreted as shelf margin lowstand prograding complexes (Van Wagoner *et al.* 1988).

In the High Island South Addition area (black diamond in Fig. 1), 20 miles east of the East Breaks area discussed above, four distinct and three less well defined prograding clinoforms occur within a salt-withdrawal basin (Fig. 10). Each clinoform package is separated by thick intervals of either parallel or chaotic reflections. The clinoforms have been mapped into a nearby well and correlated to both a depositional cycle chart and an oxygen isotope curve using time-significant bioevents (Fig. 11). The clinoforms,

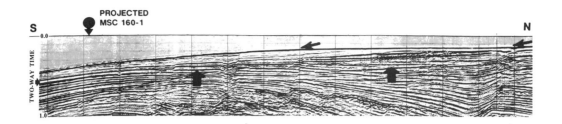

Fig. 8. North–South seismic-reflection profile from the East Breaks area, offshore Texas (black rectangle of Fig. 1), illustrating the seismic facies of a shelf-edge prograding complex correlated to core-hole MSC 160-1. The arrow at the left marks the trough above the parallel, continuous reflection that underlies the clinoforms (upward directed arrows point to the correlative reflection across the two faults). The clinoforms toplap to the right (north) against the sea-floor reflection (left directed arrows). Seismic published with permission of GECO Geophysical Company, Inc.

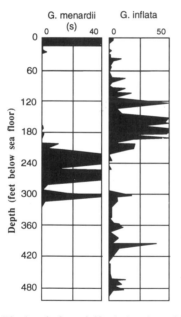

Fig. 9. Planktonic foraminiferal abundance histogram showing the warm/cool climate intervals interpreted from *G. menardii* (s = sinistral = warm) and *G. inflata* (cool) from the MSC 160-1 core-hole. This core-hole is located within the black rectangle of Fig. 1. The *G. menardii* abundance interval between 190 and 310 ft correlates with the seismic interval below the clinoforms (the trough noted by arrow on Fig. 8). Data provided by G. M. Ragan, Mobil Oil.

in general, correlate with the positive increases in isotope values, suggesting progradation during onset of glacial climates as a consequence of lowering sea-level. Six of the observed cycles occur between bioevent Trim A (0.6 Ma) and the sea floor, suggesting a cycle duration of approximately 100 ka. These six cycles correlate within the Tejas B 3.1 (0.8–0.0 Ma) third-order cycle of Haq *et al.* (1988), suggesting they are fourth-order cycles (Mitchum & Van Wagoner 1990).

Regional cycle chart

The construction of depositional cycle charts for the Gulf of Mexico extends back to at least Kolb & Van Lopik (1958) and Frasier (1976). Figure 12 is a composite of nine such studies encompassing Beard *et al.* (1982) through Martin & Fletcher (1993). Each of these studies has been calibrated to the same time scale using the same bioevent 'Marker Taxa' and is in turn correlated with the global foraminiferal zones and the magnetostratigraphic polarity scale as defined by Berggren *et al.* (1985), and the oxygen isotope chronology of Joyce *et al.* (1990). The composite of each local study appears under the column 'Sum of Sequences' and shows twenty depositional cycles, three of which occur in only one or two studies and are considered to be local autocyclic events.

The youngest six cycles of the cycle chart occur between the *Pseudoemiliania lacunosa* bioevent (0.8 Ma) and the sea floor (0.0 Ma), and average approximately 130 ka duration. The ten older cycles were deposited between *Globigerinoides mitra* (4.15 Ma) and *P. lacunosa* (0.8 Ma) bioevents and average *c.* 330 ka duration.

This cycle chart synthesizes Pliocene and Pleistocene studies ranging from offshore east

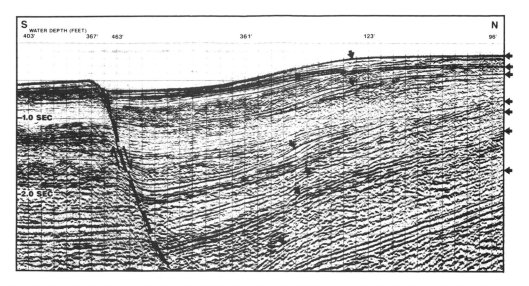

Fig. 10. Seismic reflection profile from the High Island South Addition area, illustrating the vertical stacking of depositional cycles 1–7 within a fault-bounded salt-withdrawal basin. This seismic profile is located at the black diamond in Fig. 1. Arrows point at the top of each prograding clinoform and the equivalent topset reflection to the right. Seismic published with permission of GECO Geophysical Company, Inc.

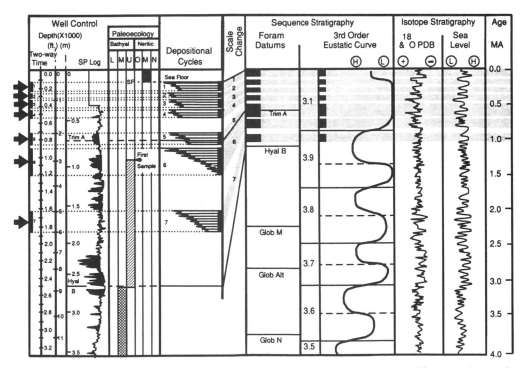

Fig. 11. The seven prograding clinoforms observed in Fig. 10 (arrows at left) are correlated into a local well using a grid of seismic reflection profiles. Using the time significant events in that well, the extinction events of *Hyalinea balthica* (Hyal B) and *Trimosina denticulata* (Trim A), and the present-day sea floor, the clinoforms are in turn correlated with the third-order eustatic curve of Haq *et al.* (1988), and the oxygen isotope curve of Williams & Trainor (1987). H, high sea-level; L, low; +, cold; —, warm.

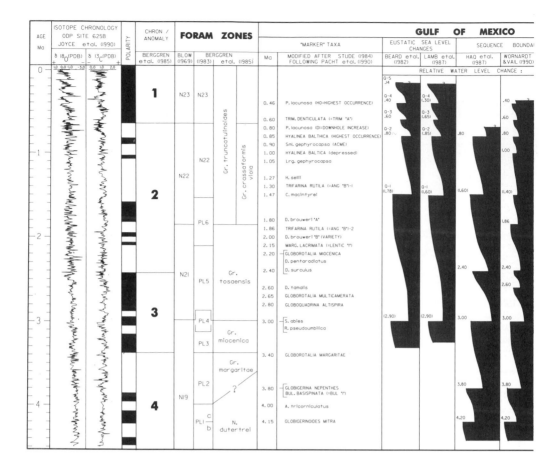

Fig. 12. Chronostratigraphic chart of Gulf of Mexico Pliocene and Pleistocene depositional and biostratigraphic events. Local cycle charts from nine studies are correlated using the same bioevent 'marker Taxa' and are in turn correlated with global foraminiferal zones, the magnetostratigraphic polarity scale as defined by Berggren *et al.* (1985), and the isotope chronology of Joyce *et al.* (1990). The resulting 'sum' of the depositional cycles or sequences and their associated condensed sections (Shaffer 1987, 1990) are illustrated at the right. Of the twenty events recognized, three are considered to be local (0.90 Ma, 1.60 Ma and 2.85 Ma). The sawtooth pattern of each cycle chart reflects relative rise in sea-level to the left and a fall to the right. The magnitude of coastal onlap is shown schematically and is not to scale.

Texas in the Galveston South area eastward to the Mississippi Delta, an area *c.* 400 miles in length, and 200–300 miles north–south. The cycle chart is of regional significance, and if correlated to similar studies in other areas could be included in a globally significant cycle chart. To construct such a globally significant chart would require documentation of each regional cycle pattern using high resolution biostratigraphy and chronostratigraphy to assure that local events are distinguished from truly globally synchronous events.

Palaeogeographic maps

With the construction of a regional cycle chart calibrated with high resolution biostratigraphy, it is possible to construct palaeogeographic maps for each phase of a depositional cycle. One such map type is the biofacies map, defined by the distribution of benthic foraminiferal biofacies (Fig. 13). The map set illustrated here (Figs 14 and 15) is of the *Globoquadrina altispira* depositional cycle (= *Glob Alt*) deposited between *c.* 2.8 and 3.4 Ma (Fig. 12).

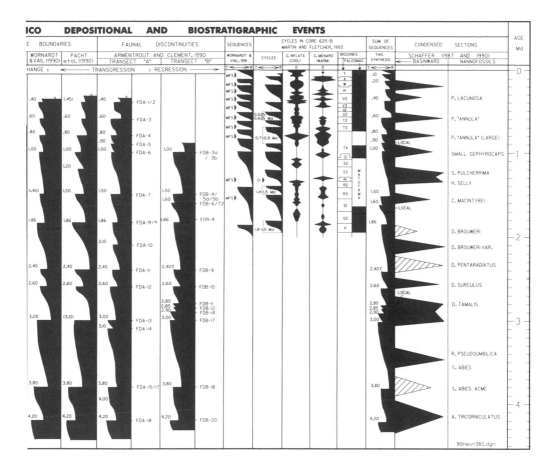

The *Globoquadrina altispira* (*Glob Alt*) biofacies maps are based on palaeoecologic analysis of benthic foraminifera in the 48 wells that penetrate the *Glob Alt* depositional cycle in the High Island–East Breaks study area (Fig. 1). The *Glob Alt* depositional cycle is bounded above by the *Glob Alt* condensed section/maximum flooding surface datum and below by the top of the underlying depositional cycle, also recognized by a condensed section/maximum flooding surface (Fig. 14). The condensed sections represent the distal facies of the transgressive and highstand systems tracts, placing the three mapped intervals within the lowstand systems tract.

The map patterns (Fig. 15a, b and c) for each lowstand interval are defined by distribution of benthic foraminiferal biofacies (Armentrout 1991). These three maps show the biofacies distribution below, within and above the *Glob Alt* sandstone interval. In upward stratigraphic order, these three intervals are interpreted as the sediment accumulated during lowering, low, and

rising phases of sea-level, respectively. This interpretation is based on the following analysis of both biofacies and seismic facies observations.

The maps show the relationship of the seismic stratigraphically-defined shelf/slope break to the biofacies distribution and the southwestward downslope excursion of the outer neritic and upper bathyal biofacies (Fig. 15b). The *Glob Alt* sandstone interval shelf/slope break is shown by seismic data to occur within the inner neritic biofacies. This suggests that the palaeoecologic map of the *Glob Alt* sandstone interval represents an interval of deposition at a sea-level lowstand (Fig. 13b).

During the lowering of sea-level the biofacies distributions and sites of maximum sediment accumulation move seaward where they are deposited on top of the maximum flooding surface. Within the initial phase of sea-level lowering, the rate of slope and intraslope basin sediment accumulation increases with fine-grained deposits above the lower maximum

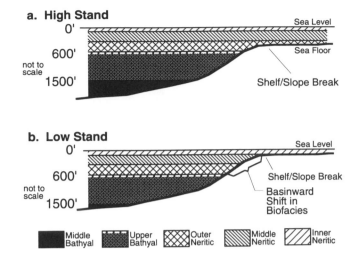

Fig. 13. Schematic diagram showing the impact of sea-level drop on the distribution of water mass biofacies. (**a**) Highstand. (**b**) Lowstand. As sea-level drops the biofacies boundaries move basinward, down the physiographic profile of the basin margin. Biofacies maps constructed for each phase of sea-level (Fig. 15), combined with other types of facies maps, provide useful information about the distribution of depositional environments.

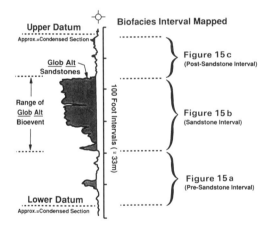

Fig. 14. Schematic spontaneous potential electric-log diagram showing the mapping intervals for slope and basinal facies of the *Globoquadrina altispira (Glob Alt)* depositional cycle. The upper and lower datums are correlative with locally significant condensed sections at 2.8 Ma and 3.1 Ma which are encompassed within the *Discoaster tamalis* condensed section of Shaffer (1990) of Fig. 12. The strata between these two condensed section datums are divided into pre-sandstone, sandstone and post-sandstone intervals based on the dominant rock type. The interval marked Range of *Glob Alt* Bioevent notes the stratigraphic interval within which *Glob Alt* occurrences are first observed. These occurrences vary relative to the abundance of shale containing foram fossils, and are therefore locally controlled by lithofacies within sand-prone intervals.

flooding surface (Fig. 14). The biofacies pattern of this pre-sandstone interval shows an irregular, basinward excursion of the outer neritic and upper bathyal assemblages (Fig. 15a).

As sea-level lowering progresses, the river systems incise the shelf or occupy submarine canyons and transport sand seaward where it is deposited directly on the upper slope. Remobilized sand, and sand supplied directly from the rivers during floods, may be transported down the slope by gravity-flow processes (Prior *et al.* 1987). These sands accumulate at changes in the depositional gradient as slope fans and basin floor fans (Bouma 1982). The biofacies associated with this depositional phase, the sandstone interval (Fig. 15b), show the same basinward excursion in biofacies as deposited during the pre-sandstone phase (Fig. 15a). The shallow water biofacies of the sandstone interval show a seaward shift, relative to the seismic stratigraphically-defined shelf/slope break, interpreted as an indication of progradation associated with lowered sea-level.

Once sea-level begins to rise, the sandy-sediment supply is cut off and mudstones accumulate during the post-sandstone interval, culminating with the overlying upper maximum flooding surface datum (Fig. 14). The biofacies pattern for this post-sandstone interval shows a northward shift toward the basin margin as the coastline regresses across the shelf during transgression (Fig. 15c).

The occurrences of neritic biofacies of the

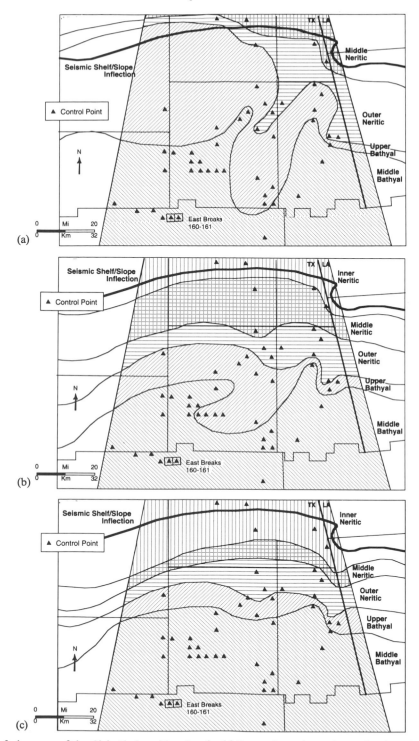

Fig. 15. Biofacies maps of the *Glob Alt* depositional cycle. (**a**) Pre-sandstone interval map showing a basinward excursion of outer neritic and upper bathyal biofacies. (**b**) Sandstone interval map in which this biofacies excursion is less pronounced. (**c**) Post-sandstone interval map in which the biofacies excursion is absent. Interpretations of these maps are given in the text.

post-sandstone interval (Fig. 15c) do not shift northward as far as their position during the pre-sandstone interval (Fig. 15a). This is interpreted as sediment accumulation rate exceeding the rate of development of accumulation space by sea-level rise plus basin subsidence. The consequence of this is coastal progradation, as suggested by comparing the mapped patterns of *Glob Alt* depositional cycle pre-sandstone and post-sandstone biofacies in the High Island–East Addition area of Fig. 15a and 15c (see Fig. 1 for areas).

The southward excursion of the *Glob Alt* interval upper bathyal biofacies is based on the presence of only upper bathyal but no middle bathyal taxa in intervals of *Glob Alt* samples from wells on depositional strike with wells having typical middle bathyal assemblages. The southward biofacies excursion could have been caused by: (1) a local topographic high, elevating the sea floor into upper bathyal water depths; (2) the downslope resedimentation of upper bathyal sediments masking the presence of *in situ* middle bathyal assemblages; (3) an ecologic shift in environment excluding the middle bathyal species from this area; or (4) inconsistent interpretation of biofacies by using data from several different biostratigraphers.

Inspection of seismic profiles gives no indication of a palaeotopographic high during this time in the area of the biofacies excursion. The study area was a major late Pliocene and early Pleistocene deep water re-entrant into the continental margin, as delineated by the biofacies map and the seismic stratigraphically-defined shelf/slope break, which is best observed on seismic profiles parallel with depositional slope.

Repeated examination of the cutting samples and sidewall core samples by a single biostratigrapher did not yield any typical middle bathyal taxa in those wells defining the upper bathyal biofacies excursion, suggesting that there is no down slope mixing of assemblages (G. M. Ragan 1987, pers. comm.). If downslope resedimentation were the cause of the biofacies excursion, the foraminiferal assemblage should contain abundant neritic taxa, as do some gravity-flow sands (Woodbury *et al.* 1978).

The most likely cause of the biofacies excursion is a change in the optimal environment for the benthic foraminiferal biofacies assemblages, perhaps similar to the 'delta-depressed fauna' described by Pflum & Frerichs (1976). These authors found that typical modern inner and middle neritic benthic foraminiferal assemblages shift basinward onto the outer shelf when the Mississippi River alters the inner shelf environment due to high sedimentation rates, coarser-grained sediments, abundant terrigenous organic matter and fresh water. This pattern of biofacies shift occurs on the inner shelf during the highstand of Holocene sea-level (Poag 1981). Modification of living environments caused by major sediment input at times of lower sea-level would probably impact facies distribution far down onto the slope where gravity-flow processes would extend down slope the environmental modifications initiated by point source sediment input (Fig. 15a and b).

The *Glob Alt* biofacies maps suggest a major sediment input area coincident with the biofacies excursion. Basinward, the *Glob Alt* interval contains gravity-flow sandstones with a hummocky to mounded seismic facies (Armentrout *et al.* 1991). Both the pattern of shelf-edge progradation and basinal gravity-flow sandstones, plus the absence of mixed biofacies assemblages and no indication of a local topographic high, argue for water mass and substrate control of the biofacies excursion pattern. This type of biofacies excursion occurs during lowstand depositional intervals of several of the cycles mapped in the study area and is not an artifact of special conditions occurring during the *Globoquadrina altispira* depositional cycle.

The mapped pattern of Fig. 15b suggests a maximum shift of inner neritic biofacies in a seaward direction to the seismic stratigraphically-defined upper slope, resulting in a basinward shift of about 300 ft (100 m) water depth from the current biofacies, bathymetric and physiographic relationships in the Gulf of Mexico (Fig. 13) (Armentrout 1991). The Global Cycle Chart, constructed from seismic sequence analysis (Haq *et al.* 1988) also shows a 300 ft (100 m) lowering of relative water depth for cycle Tejas B 3.7, correlative with the *Globoquadrina altispira* depositional cycle.

Current research

The patterns of biofacies distribution and microfossil abundance described above are indicative of the high resolution stratigraphic data available from palaeontology. High resolution foraminiferal and calcareous nannoplankton biostratigraphy have been used for reservoir correlation (Ragan & Abbott 1989), and palynomorph cycles have been used to correlate wells in non-marine facies (Poumot 1989; Armentrout *et al.* 1993*b*). These studies and the data presented in this paper are from Neogene strata, but the methodology applies to most if not all of the Phanerozoic. High resolution correlations of 100–200 ka duration have been made in the Pennsylvanian using goniatites (Elliott & Davies

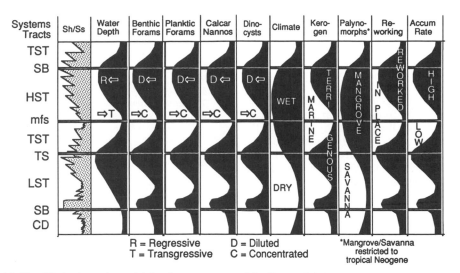

Fig. 16. Simplified schematic model for the occurrence of fossil material characteristic of depositional systems tracts. The patterns of fossil distribution are related to the interplay of water depth, climate and sediment accumulation rate. Fossil distributions along any depositional dip or strike transect will vary from this generalized model. Additionally, fossil occurrences will have a serrated to spiky pattern rather than the schematically smoothed distribution shown (compare with Figs 3 and 5).

1994). Nannoplankton abundance patterns have been used to define sequence surfaces and systems tracts within upper Triassic and earliest Cretaceous strata by Gardin & Manivit (1993). Gregory & Hart (1992) used palynofacies to characterize depositional systems tracts in Cretaceous and Palaeogene strata. Baum & Vail (1987) defined microfossil occurrence and isotopic value patterns for Palaeogene systems tracts in the North American Atlantic and Gulf Coast coastal plain. Mudge & Copestake (1992) and Armentrout *et al.* (1993*a*) integrated biostratigraphic data with log-motif analysis and sequence stratigraphy to define North Sea Palaeogene depositional cycles. These studies are only a few from the rapidly increasing literature on the sequence stratigraphic signature of biostratigraphy.

Future research in sequence biostratigraphy will continue to define the pattern of fossil occurrence within depositional sequences. Figure 16 is a compilation of patterns reported here or identified in the references cited above. Studies of such patterns from different geological ages, and from a spectrum of locations along transects, will provide further definition of fossil occurrence in a sequence stratigraphic context. Additional studies are needed to contrast the occurrence of fossils of similar taxonomic groups within global hothouse versus icehouse climates, along latitudinal gradients, and between carbonate and siliciclastic environments. The isotopic signal of fossil shells needs to be defined at the same scale of resolution and within a sequence stratigraphic framework.

All of these datasets will provide additional information for constructing high resolution correlation frameworks and for mapping palaeogeography at a significantly improved scale of precision compared with most current maps.

Summary

High resolution biostratigraphy within a sequence stratigraphic context is rapidly emerging as a valuable interpretation tool for reconstructing depositional patterns for each phase of a depositional cycle. Identification of each candidate sequence boundary and its correlative conformity, and of each candidate transgressive and maximum flooding surface, is achieved by careful recognition of patterns of fossil abundance, stratal termination and log-patterns, and by correlation of the regionally correlative surfaces between measured sections or wireline logs, using chronostratigraphically significant bioevents. If the candidate surfaces correlate throughout the depositional basin or sub-basin, they should be considered regionally significant sequence boundaries, transgressive surfaces, and maximum flooding surfaces. If the surfaces are limited to local areas, they are probably local discontinuities associated with local structural events or with autocyclic shifting of sediment accumulation such as delta or fan lobes.

Within the chronostratigraphic framework, depositional systems tracts indicative of each phase of the depositional cycle can be characterized by fossil species assemblages, abundances, biofacies, lithofacies and electrofacies, and can be mapped away from well control using seismic facies analysis. Biostratigraphy provides the most readily available and highest resolution chronostratigraphy currently available. The sea-level lowstand, rising, highstand, and falling phase biofacies maps constructed within the high resolution chronostratigraphic framework, combined with similarly high resolution lithofacies, seismic facies and electrofacies maps, will provide the geosciences with the necessary temporal and spatial resolution to unravel a new generation of detailed Earth history.

Many co-workers at Mobil have contributed to the author's understanding of biostratigraphy, most especially G. M. Ragan, R. J. Echols, R. C. Becker, T. D. Lee and L. B. Fearn. The original manuscript benefited from reviews by R. J. Echols, G. M. Ragan, N. Hine and M. A. Partington. Mobil Exploration and Producing Technology Center has given permission to publish this paper.

References

ARMENTROUT, J. M. 1987. Integration of biostratigraphy and seismic stratigraphy: Pliocene–Pleistocene, Gulf of Mexico. *In: Gulf Coast Section Society of Economic Paleontologists and Mineralogists, Eighth Annual Research Conference.* 6–14.

—— 1989. Integrated stratigraphic analysis: Plio-Pleistocene, High Island/Galveston/East Breaks areas, Gulf of Mexico. Unpublished research project for the Society of Economic Paleontologists and Mineralogists, Tulsa, Oklahoma, 7 volumes. (copy on file at SEPM Headquarters).

—— 1991. Paleontologic constraints on depositional modeling: Examples of integration of biostratigraphy and seismic stratigraphy, Gulf of Mexico. *In:* WEIMER, P. & LINK, M. H. (eds) *Seismic Facies and Sedimentary Processes of Submarine Fans and Turbidite Systems.* Springer-Verlag, Frontiers in Sedimentary Geology Series, 137–170.

—— & CLEMENT, J. F. 1990. Biostratigraphic calibration of depositional cycles: A case study in High Island–Galveston–East Breaks areas, offshore Texas. *In: Gulf Coast Section Society of Economic Paleontologists and Mineralogists, Eleventh Annual Research Conference,* 21–51.

——, ECHOLS, R. C. & LEE, T. D. 1990. Patterns of foraminiferal abundance and diversity: Implications for sequence stratigraphic analysis. *In: Gulf Coast Section Society of Economic Paleontologists and Mineralogists, Eleventh Annual Research Conference,* 53–58.

——, MALECEK, S. J., BRAITHWAITE, P. & BEEMAN, C. R. 1991. Seismic facies of slope basin turbidite reservoirs, East Breaks 160–161 Field: Pliocene–Pleistocene, Northwestern Gulf of Mexico. *In:* WEIMER, P. & LINK, M. H. (eds) *Seismic Facies and Sedimentary Processes of Submarine Fans and Turbidite Systems.* Springer-Verlag, Frontiers in Sedimentary Geology Series, 223–239.

——, ——, FEARN, L. B., SHEPPARD, C. E., NAYLOR, P. H. ET AL. 1993a. Log-motif analysis of Paleogene depositional systems tracts, Central and Northern North Sea: defined by sequence stratigraphic analysis. *In:* PARKER, J. R. (ed.) *Petroleum Geology of Northwest Europe, Proceedings of the 4th Conference.* Geological Society, London, 45–57.

——, BECKER, R. C., FEARN, L. B., ALI, W. M., HUDSON, D. G. ET AL. 1993b. Exploration applications of integrated biostratigraphic analysis, Plio-Pleistocene, offshore southern Trinidad. *American Association of Petroleum Geologists Bulletin,* **77**, 303.

BAUM, G. R. & VAIL, P. R. 1987. Sequence stratigraphy, allostratigraphy, isotope stratigraphy and biostratigraphy: Putting it all together in the Atlantic and Gulf Paleogene. *In: Gulf Coast Section Society of Economic Paleontologists and Mineralogists, Eighth Annual Research Conference,* 15–23.

BEARD, J. H., SANGREE, J. B. & SMITH, L. A. 1982. Quaternary chronology, paleoclimate, depositional sequences, and eustatic cycles. *American Association of Petroleum Geologists Bulletin,* **66**, 158–169.

BERGGREN, W. A., KENT, D. V. & VAN COUVERING, J. A. 1985. Jurassic to Paleogene: Part 2. Paleogene geochronology and chronostratigraphy. *In:* SNELLING, N. J. (ed.) *The Chronology of the Geological Record.* Geological Society, London, Memoir, **10**, Blackwell Scientific Publishing, Oxford, 211–260.

BLOW, W. H. 1969. The late Middle Eocene to Recent planktonic foraminiferal biostratigraphy. *In: First Planktonic Conference, Proceedings, Geneva, 1967.* 199–422.

BOUMA, A. H. 1982. Intraslope basins in northwest Gulf of Mexico: A key to ancient submarine canyons and fans. *In:* WATKINS, J. S. & DRAKE, C. L. (eds) *Studies in Continental Marine Geology.* American Association of Petroleum Geology, Memoir, **34**, 567–582.

COLEMAN, J. M. & ROBERTS, H. H., 1988. Sedimentary development of the Louisiana continental shelf related to sea level cycles. *Geo-Marine Letters,* **8**, 63–119.

CULVER, S. J. 1988. New foraminiferal depth zonation of the northwestern Gulf of Mexico. *Palaios,* **3**, 69–85.

EINSELE, G. 1982. General remarks about nature, occurrence, and recognition of cyclic sequences. *In:* EINSELE, G. & SEILACHER, A. (eds) *Cyclic and event stratification.* Springer-Verlag, Berlin, 3–7.

ELLIOTT, T. & DAVIES, S. 1994. High resolution sequence stratigraphy of an upper Carboniferous basin-fill succession, Co. Clare, Western Ireland. *In: High Resolution Sequence Stratigraphy: Innovations and Applications.* University of Liverpool, Conference Fieldtrip B2.

FRASIER, D. E. 1974. *Depositional episodes: Their relationship to the stratigraphic framework in the Gulf Basin.* Texas University Bureau of Economic Geology, Circular, **74-1**, 28.

GALLOWAY, W. E. 1989. Genetic stratigraphic sequences in basin analysis I: Architecture and genesis of flooding-surface bounded depositional units. *American Association of Petroleum Geologists Bulletin*, **73**, 145–154.

GARDIN, S. & MANIVIT, H. 1993. *Upper Tithonian and Berriasian calcareous nannofossils from the Vocontian Trough (SE France): Biostratigraphy and sequence stratigraphy.* Elf Aquitaine Production, Boussens. 277–290.

GREGORY, W. A. & HART, G. F. 1992. Towards a predictive model for palynologic response to sea-level changes. *Palaios*, **7**, 3–33.

HAQ, B. U., HARDENBOL, J. & VAIL, P. R. 1988. Mesozoic and Cenozoic chronostratigraphy and cycles of sea-level change. *In:* WILGUS, C. K., POSAMENTIER, H., ROSS, C. A. & KENDALL, C. G. St. C. (eds) *Sea Level Change: An integrated approach.* Society of Economic Paleontologists and Mineralogists, Special Publication, **42**, 71–108.

JOYCE, J. E., TJALSMA, L. R. C. & PRUTZMAN, J. M. 1990. High-resolution planktic stable isotope record and spectral analysis for the last 5.35 myr: ODP Site 625 northeast Gulf of Mexico. *Paleoceanography*, **5** 507–529.

KENNETT, J. P. & SRINIVASAN, S. 1983. *Neogene Planktonic Foraminifera.* Hutchinson Ross Publishing Company, Stroudsburg, Pennsylvania.

KOLB, C. R. & VAN LOPIK, J. R. 1958. *Geology of the Mississippi River deltaic plain, southeastern Louisiana.* U.S. Army Engineer Waterways Experiment Station, Corps of Engineers, Vicksburg, Mississippi, Technical Report, **3-483**.

LAMB, J. L., WORNARDT, W. W., HUANG, T. C. & DUBE, T. E. 1987. Practical application of Pleistocene eustacy in offshore Gulf of Mexico. *In:* ROSS, R. A. & HAMAN, D. (eds) *Timing and depositional history of eustatic sequences: Constraints on seismic stratigraphy.* Cushman Foundation for Foraminiferal Research, Special Publication, **24**, 33–39.

LOUTIT, T. S., HARDENBOL, J., VAIL, P. R. & BAUM, G. R. 1988. Condensed sections: the key to age determination and correlation of continental margin sequences. *In:* WILGUS, C. K., POSAMENTIER, H., ROSS, C. A. & KENDALL, C. G. St. C. (eds) *Sea Level Changes: An integrated approach.* Society of Economic Paleontologists and Mineralogists, Special Publication, **42**, 183–213.

MARTIN, R. E. & FLETCHER, R. R. 1993. Biostratigraphic expression of Plio-Pleistocene sequence boundaries, Gulf of Mexico. *In: Gulf Coast Section Society of Economic Paleontologists and Mineralogists, Fourteenth Annual Research Conference*, 119–126.

MITCHUM, R. M. JR. & VAN WAGONER, J. C. 1990. High-Frequency sequences and eustatic cycles in the Gulf of Mexico basin. *Gulf Coast Section Society of Economic Paleontologists and Mineralogists, Eleventh Annual Research Conference*, 257–267.

——, VAIL, P. R. & THOMPSON, S. III. 1977. Seismic stratigraphy and global sea level, Part 2: The depositional sequence as a basic unit for stratigraphic analysis. *In:* PAYTON, C. E. (ed.) *Seismic Stratigraphy – Applications to hydrocarbon exploration.* American Association of Petroleum Geologists, Memoir, **26**, 53–62.

MUDGE, D. C. & COPESTAKE, P. 1992. Lower Paleogene stratigraphy of the northern North Sea. *Marine and Petroleum Geology*, **9**, 287–301.

PACHT, J. A., BOWEN, B. E., BEARD, J. H. & SCHAFFER, B. L. 1990a. Sequence-stratigraphy of Plio-Pleistocene depositional facies in the offshore Louisiana south additions. *Gulf Coast Association of Geological Societies Transactions*, **XL**, 1–18.

——, ——, SCHAFFER, B. L. & POTTORF, B. R. 1990b. Sequence stratigraphy of Plio-Pleistocene strata in the offshore Louisiana Gulf Coast – Applications to hydrocarbon exploration. *Gulf Coast Section Society of Economic Paleontologists and Mineralogists, Twelfth Annual Research Conference*, 269–285.

PFLUM, C. E. & FRERICHS, W. E. 1976. *Gulf of Mexico deep water foraminifers.* Cushman Foundation for Foraminiferal Research, Special Publication, **14**.

POAG, C. W. 1981. *Ecologic atlas of benthic foraminifera of the Gulf of Mexico.* Marine Science International, Woods Hole, Massachusetts.

POUMOT, C. 1989. *Palynological evidence for eustatic events in the tropical Neogene.* Society National Elf Aquitaine, Production, Boussens.

PRIOR, D. B., BORNHOLD, B. D., WISEMAN, W. J. JR. & LOWE, D. R. 1987. Turbidity current activity in a British Columbia fjord. *Science*, **237**, 1330–1333.

RAGAN, G. M. & ABBOTT, W. H. 1989. High-resolution correlation of Gulf of Mexico Pliocene–Pleistocene sands. *American Association of Petroleum Geologists Bulletin*, **73**, 402.

SCHAFFER, B. L. 1987. The potential of calcareous nannofossils for recognizing Plio-Pleistocene climatic cycles and sequence boundaries on the shelf. *Gulf Coast Section Society of Economic Paleontologists and Mineralogists, Eighth Annual Research Conference.* 142–145.

—— 1990. The nature and significance of condensed sections in Gulf Coast late Neogene sequence stratigraphy. *Gulf Coast Association of Geological Societies Transactions*, **XL**, 186–195.

SKINNER, H. C. 1966. Modern paleoecological techniques: An evaluation of the role of paleoecology in Gulf Coast Exploration. *In:* ELLISON, S. P. (ed.) *Biostratigraphy and paleoecology of Gulf Coast Cenozoic foraminifera.* Gulf Coast Association of Geological Societies, Readings in Gulf Coast Geology, **3**, 57–77.

STUDE, G. R. 1984. Neogene and Pleistocene biostratigraphic zonation of the Gulf Coast basin. *Gulf Coast Section Society of Economic Paleontologists and Mineralogists, Fifth Annual Research Conference*, 92–101.

SUTTER, J. R. & BERRYHILL, H. L. JR. 1985. Late Quaternary shelf-margin deltas, northwest Gulf of Mexico. *American Association of Petroleum Geologists Bulletin*, **69**, 77–91.

TIPSWORD, H. L. J., SETZER, F. M. & SMITH, F. L. JR. 1966. Interpretation of depositional environment in Gulf Coast exploration from paleoecology and related stratigraphy. *Gulf Coast Association of Geological Societies Transaction*, **XVI**, 119–130.

VAIL, P. R. 1987. Seismic stratigraphy interpretation procedure. *In:* BALLY, A. W. (ed.) *Atlas of Seismic Stratigraphy Volume 1*. American Association of Petroleum Geologists, Studies in Geology, **27**, 1–10.

—— & WORNARDT, W. W. 1990. Well log – seismic stratigraphy: A new tool for exploration in the 90's. *Gulf Coast Section Society of Economic Paleontologists and Mineralogists, Eleventh Annual Research Conference*. 379–388.

——, MITCHUM, R. M. JR. & THOMPSON, S. III. 1977. Seismic stratigraphy and global changes of sea level, Part 4: Global cycles of relative changes of sea level. *In:* PAYTON, C. E. (ed.) *Seismic Stratigraphy – Applications to hydrocarbon exploration*. American Association of Petroleum Geologists, Memoir, **26** 83–97.

VAN WAGONER, J. C., POSAMENTIER, H. W., MITCHUM, R. M. JR., VAIL, P. R., SARG, J. F., LOUTIT, T. S. & HARDENBOL, J. 1988. An overview of the fundamentals of sequence stratigraphy and key definitions. *In:* WILGUS, C. K., POSAMENTIER, H., ROSS, C. A. & KENDALL, C. G. St. C. (eds) *Sea Level Changes: An integrated approach*. Society of Economic Paleontologists and Mineralogists, Special Publication, **42**, 39–45.

WILLIAMS, D. F. & TRAINOR, D. M. 1987. Integrated chemical stratigraphy of deep-water frontier areas of the northern Gulf of Mexico. *Gulf Coast Section Society of Economic Paleontologists and Mineralogists Eighth Annual Research Conference*. 151–158.

WOODBURY, H. O., SPOTTS, J. H. & ATKERS, W. H. 1978. Gulf of Mexico continental-slope sediments and sedimentation. *In:* BOUMA, A. H. (ed.) *Framework facies, and oil-trapping characteristic of the upper continental margin*. American Association of Petroleum Geologists, Studies in Geology, **7**, 117–137.

WORNARDT, W. W. & VAIL, P. R. 1990. Revision of the Plio-Pleistocene cycles and their application to sequence stratigraphy of shelf and slope sediments in the Gulf of Mexico. *Gulf Coast Section Society of Economic Paleontologists and Mineralogists, Twelfth Annual Research Conference*. 391–397.

Influence of relative sea-level on facies and reservoir geometry of the Middle Jurassic lower Brent Group, UK North Viking Graben

DAVID C. JENNETTE[1,2] & CHEYENNE O. RILEY[1,3]

[1]*Esso EXPRO UK Ltd. Esso House, Victoria St, London SW1E 5JW, UK*

[2]*Present address: Exxon Production Research Co., Box 2189, Houston TX 77252-2189, USA*

[3]*Present address: Exxon Company USA, Box 2180, Houston TX 77252-2180,USA*

Abstract: Detailed sequence stratigraphy of several neighbouring fields in the East Shetland basin indicates that a higher order of cyclicity is superimposed onto the lower Brent Group. Both relative falls and rises in sea-level affected the strongly progradational nearshore/barrier-island system and induced important variations in the geometry, distribution and stacking patterns of the reservoir facies. The majority of the shallow marine strata belong to the transgressive and highstand systems tract and consist of the lower-shoreface Rannoch, the upper shoreface and strand-plain Etive and the back-barrier and lagoonal Lower Ness. High frequency increments of shoreline progradation are preserved in the Rannoch as gently seaward-dipping bedsets. The small-scale units mark asymmetric cycles in grain size and mica content and together, with stratiform cement, impart a strong permeability anisotropy on the Rannoch. The nearshore Etive Formation represents a sediment-laden, barred nearshore complex and is subdivided into (1) a uniformly thick upper shoreface and (2) an overlying emergent strand plain. The two subfacies are commonly separated by a thin interval of foreshore which is locally enriched in heavy minerals. The shoreface deposits show systematic variations in stacking geometries ranging from strongly progradational to aggradational.

Two relative falls in sea-level led to the subaerial erosion of nearshore Etive deposits and the juxtaposition of coastal-plain mudstones, coals and fluvial deposits with shale-clast lags on lower shoreface Rannoch strata. Although lithostratigraphically part of the Etive, these deposits bear no resemblance to the well ordered nearshore-Etive deposits. The channel complexes are vertically and laterally heterogeneous and are confined to incised valleys separated by non-depositional interfluves. In Cormorant and Tern fields, well-performance anomalies such as early water breakthrough and unpredictable water injection and oil production rates are associated with these lowstand incised valley deposits. An enigmatic coarse-grained unit occurs in the lower Rannoch in Eider field and is provisionally interpreted as the distal expression of the lowstand shoreface deposits associated with the older incised valley. Following the lowstand deposition nearshore-marine sedimentation re-established during the transgressive/early highstand systems tract. Under the influence of rising relative sea-level the barrier complex aggraded in place with minimal landward translation of the shoreline position. The volume of sediment supplied to the Etive shoreface was sufficiently high to keep pace with phases of increased accommodation. The shoreline system subsequently stacked with a strongly progradational geometry as the high sediment supply filled the available accommodation space during the latter part of the highstand systems tract.

Most previous authors viewed the Middle Jurassic Brent Group in the East Shetland basin to represent the long-term progradation and subsequent retrogradation of a wave dominated delta and associated back-barrier and coastal-plain system (Budding & Inglin 1981; Johnson & Stewart 1985; Graue *et al.* 1987; Richards 1992 for review). However, a more complicated picture of the higher frequency architecture of the system emerges at the field-scale where a greater density of subsurface data allows more confident chrono- and lithostratigraphic correlation. This paper focuses on the stratigraphic relationships of the sandstone-rich lower part of the Brent Group (Rannoch, Etive and Lower Ness formations) and suggests that variations in sea-level governed the geometry, distribution and overall stacking pattern of these reservoir facies. The six fields examined in this study are located in UKCS Quads 210 and 211 on the western shoulder of the North Viking Graben in the East Shetland basin (Fig. 1). They are

From Howell, J. A. & Aitken, J. F. (eds), *High Resolution Sequence Stratigraphy: Innovations and Applications,*
Geological Society Special Publication No. 104, pp. 87–113.

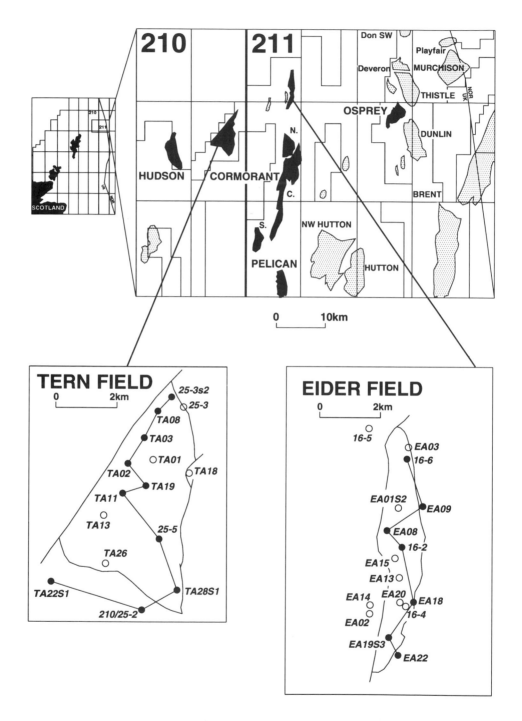

Fig. 1. Study area map highlighting the fields investigated. Insets show well locations and cross-section lines for Tern and Eider fields. Cormorant Block I cross-section line is shown in Fig. 14.

EIDER 211/16 - EA 18

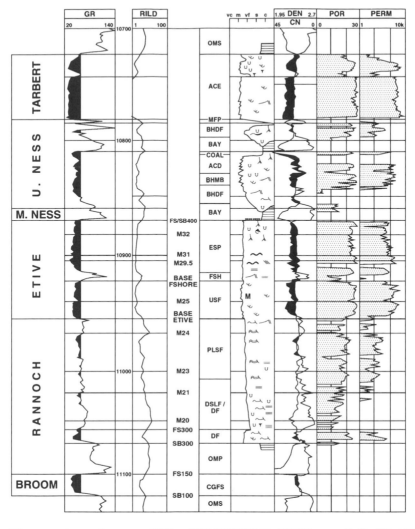

OMS	Offshore marine shelf	BAY	Mud-dominated bay / lagoon
OMP	Offshore marine prodelta	BHDF	Bay head delta front
CGSF	Coarse-grained shoreface	BHMB	Bay head mouth bar
DLSF	Distal lower shoreface	ACD	Active channel distributary
DF	Shallow marine delta front	ACE	Active channel estuarine
PLSF	Proximal lower shoreface		Trough cross bedding
USF	Upper shoreface		Foreset lamination
FSH	Foreshore		Planar lamination
ESP	Emergent strandplain		Hummocky cross bedding
MFP	Mud-dominated floodplain		Wave ripple
Coal	Coal swamp		Current ripple

Right column legend:

- Wavy lamination
- Pedogenic alteration
- U Burrows
- Churned
- M Massive
- Shale drape
- Roots
- oooo Pebble / coarse-grained lag

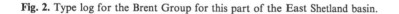

Fig. 2. Type log for the Brent Group for this part of the East Shetland basin.

Hudson, Osprey, Pelican, Cormorant, Eider and Tern fields and have combined recoverable reserves of 637 Mbbls. This study compiles data from over 140 wells and 3500 m (12 000 ft) of conventional core and summarizes over 64 km (40 miles) of stratigraphy in the basinward direction. Given the high density of subsurface data in the area, this study provides an excellent opportunity to refine the previous versions of the regional depositional facies and palaeogeographic framework for the lower Brent Group in this part of the basin (i.e. Budding & Inglin 1981; Livera & Caline 1990).

Lithofacies and depositional environments

The Brent Group is formally divided into five lithostratigraphic formations: the Broom, Rannoch, Etive, Ness, and Tarbert formations (Bowen 1975; Deegan & Scull 1977). Most workers divide the succession into four main genetic elements: (1) the basal coarse-grained and syntectonic Broom Formation, (2) sandy shoreface-coastal barrier complex of the Rannoch and Etive formations, (3) the heterolithic and cyclically arranged strata of the Ness backbarrier and coastal-plain system, and (4) the variably thick and erosively based Tarbert Formation (Johnson & Stewart 1985) (Fig. 2). Broom Formation deposition was dominated by coalesced fan deltas or coarse-grained proximal shoreface systems (Richards & Brown 1987; Graue et al. 1987; Livera & Caline 1990; Cannon et al. 1992; Mitchener et al. 1992). The basal surface of the Broom is interpreted as a low-order sequence boundary (SB100 on subsequent well-log cross sections) which locally erodes into the underlying Dunlin Group shale. A thin transgressive systems tract, consisting of one or two parasequences stacked in a retrogradational pattern, separates the main body of the coarse-grained Broom from the overlying shaley sandstone beds of the Rannoch. A low-order maximum flooding surface (FS150) separates the genetically distinct Broom from the overlying Rannoch–Etive complex (Graue et al. 1987; Helland-Hansen et al. 1989).

The following section describes the major lithofacies and stratigraphic relationships of the sandy nearshore-back barrier and interprets their depositional origin in the context of sequence stratigraphy. The discussion begins with the shallow marine strata associated with highstand and transgressive systems tracts. This is followed by a discussion of lowstand systems tracts, primarily focusing on 'Etive' facies associated with lowstand incised-valley fills.

Rannoch Formation

The Rannoch Formation is the distal, storm-dominated portion of the shallow-marine system (Budding & Inglin 1981; Brown et al. 1987; Graue et al. 1987; Livera & Caline 1990). Three different vertical facies successions of Rannoch occur in the study area, one of which has significantly better reservoir properties and fluid-flow characteristics than the others.

Normal lower shoreface (Pelican, Cormorant fields). In the southern part of the study area, the Broom Formation is overlain by a characteristic coarsening- and thickening-upward succession of wave-rippled and burrowed shaley sandstone. The trace-fossil suite is a high diversity and high abundance assemblage of open-marine, infaunal-feeding forms and includes *Teichichnus, Planolites, Palaeophycus, Rosselia, Thalassinoides, Asterosoma, Terebellina, Rhizocorallium* and *Skolithos.* A decrease in burrow intensity and shale content accompanies an upward increase in physical stratification. The most proximal facies in the succession is an unburrowed, amalgamated hummocky-bedded lithofacies which forms the upper part of the Rannoch Formation in most fields. Proximal lower shoreface facies range from 10–15 m thick (30–50 ft) and averages 18 m (60 ft) in thickness in Eider and Osprey fields.

The consistent vertical succession of bioturbated and wave-rippled strata into amalgamated hummocky strata is interpreted as the shoaling from distal to proximal facies of a storm-dominated lower shoreface (Mitchener et al. 1992; Scott 1992). The heterolithic, variably burrowed lower portion is interpreted as the distal part of the lower shoreface facies and records infrequent and relatively gentle storm-generated currents which disrupted fair-weather suspension sedimentation and infaunal burrowing. The onset of amalgamated hummocky facies reflects the intersection of the shoreline profile with effective storm wave base. From this point the relative strength and frequency of storm events was sufficient to erase all record of fairweather sedimentation and leave behind a thick succession of amalgamated hummocky beds.

Ebb tidal-delta (Tern, northern Cormorant fields). In the central part of the study area, an interval of relatively clean sandstone ranging from upper fine-grained to lower medium-grained sandstone is present in the lower Rannoch. Proximal and distal subfacies are recognized: (1) a shale-poor, cross-bedded sand-

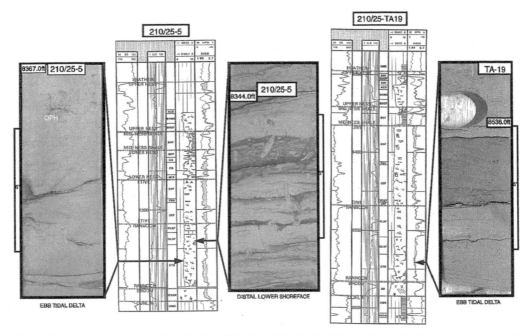

Fig. 3. Core photographs from Tern field, 210/25-5 and TA-19, highlighting lithofacies from the Rannoch ebb tidal-delta succession. Proximal facies are typified by trough cross-bedded sandstone with large *Ophiomorpha* (OPH) burrows. Note the response of the porosity logs for this interval in 25-5. Overlying the ebb-delta facies is typical distal lower-shoreface strata (DLSF, 8344 ft). Proximal facies interfinger with shalier distal ebb-delta facies in the more basinward TA-19, 8536 ft.

stone lithofacies containing a diagnostic low diversity but moderate to high abundance trace-fossil assemblage dominated by robust *Ophiomorpha*, and (2) a shalier and more micaceous sandstone lithofacies dominated by a mixed dwelling–feeding assemblage with *Thalassinoides*, *Arenicolites*, *Skolithos*, *Planolites* and rare *Cylindrichnus*. Although bioturbation has erased most of the physical sedimentary structures, planar laminations, low-angle trough cross-bedding and current ripples are found. The more proximal lithofacies has a distinct blocky gamma-ray log response and high reservoir quality is indicated by the considerable neutron and density log cross-over (Fig. 3). Core porosity and permeability values reach a maximum of 30% and 1400 md, respectively, which is an order of magnitude greater permeability observed in the lower part of a typical Rannoch succession.

In Tern field this sandstone body has been mapped as a northward-thinning wedge (Fig. 4). In the southern (landward) part of the field, the entire Rannoch interval is composed of this lithofacies; that is, no amalgamated hummocky strata are found beneath the upper shoreface deposits of the Etive (Fig. 4). North of the 25-2 well, a wedge of amalgamated hummocky strata

appears beneath the Etive and thickens as the underlying sandstone body thins. In the TA-19 core, the anomalous sandstone body occurs as a thin interval of high quality sandstone with faint cross-beds (Fig. 3). It is overlain by typical distal and proximal lower shoreface facies described earlier. The upper boundary of the sandstone body is difficult to correlate as a single chronostratigraphic surface.

A lithologically similar package of high-quality reservoir sandstone occurs in the lower Rannoch both in the landward direction (Hudson field) and along depositional strike (northern Cormorant field, southern block III and northern block IV) some 13 km (8 miles) to the east-southeast. The Cormorant example has similar reservoir quality and also displays pronounced basinward-thinning geometry. This lithofacies cannnot be related to the Broom because it onlaps the low-order transgressive flooding surface which caps the Broom (FS 150). Moreover, the sandstone is too fine-grained and well sorted to be considered a Broom Formation equivalent. In addition, the unit can be differentiated from the Etive (e.g. TA22) because of shale and mica content, stratification type and overall reservoir quality.

Based on geometry, trace fossil assemblage

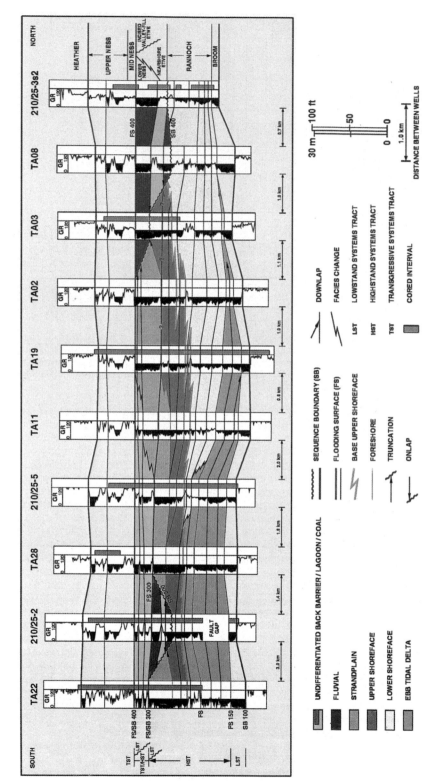

Fig. 4. Summary cross-section illustrating facies and sequence architecture of Tern field.

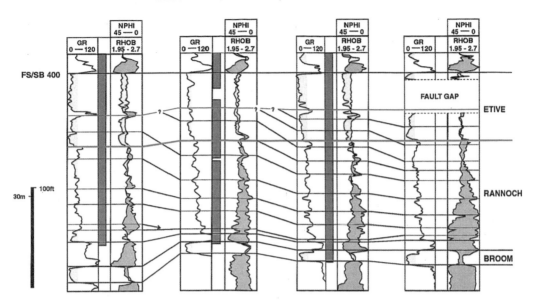

Fig. 5. Detail of correlation of bedsets boundaries which define seaward-dipping clinothems in the Rannoch Formation, Eider field.

and stratification types, this enigmatic unit is interpreted as the deposits of a large or possibly coalesced ebb-tidal delta complex. Because these strata occur transitionally beneath the Etive (TA22) and beneath the thickening wedge of proximal lower shoreface in Tern field (Fig. 4), the succession reflects deposition below fair-weather wave base in the lower shoreface realm. In contrast to the storm-wave-influenced strata found in typical successions (e.g. Pelican field), bedding in the ebb-delta sandstone body indicates sedimentation under sustained unidirectional or intermittent wave-modified currents (cf. Greer 1975; Hine 1975; Imperato et al. 1988). The absence of storm deposits suggests that these currents were active during fair-weather periods and were capable of erasing the higher energy but less frequent storm events. The low diversity but moderate to high abundance trace fossil assemblage suggests that the sand body was not fed by a fresh water deltaic system. Such a system would give rise to prodelta and delta-front facies both of which occur in Eider and Osprey fields and have markedly different trace fossil and bedding characteristics. The moderate to low levels of environmental stress suggest the sand body was more likely situated seaward of an estuary or expansive barrier inlet which debauched currents with fluctuating but near normal salinities. In addition turbidity was likely to have been low. The strongly seaward-prograding geometry and

the lack of bi-directional cross-bedding implies a system with strong ebb dominance (Imperato et al. 1988; Hayes 1989). The proximal subfacies may therefore represent deposition within the axis of the main ebb channel or by active swash bars. The distal facies may represent lower energy deposition in an abandoned ebb channel, on the sub-tidal platform or along the distal lobe-fringe (Greer 1975; Hine 1975; Imperato et al. 1988). The large areal dimension of the sandstone body (Tern, North Cormorant, Hudson fields) implies that the system was fairly long lived and was fed by active longshore currents (Greer 1975). Since the upper boundary of the ebb-tidal sandstone body is difficult to correlate as a single boundary or flooding surface, the boundary is considered to reflect the lateral migration or autocyclic abandonment of the feeder inlet channel and the return to normal wave-dominated sedimentation. The occurrence of this facies in the Rannoch Formation implies, that even during progradation, phases of tidally influenced deposition temporarily or at least locally dominated over the prevailing wave-influenced lower shoreface deposition.

Delta-front (Eider and Osprey fields). The lower-most Rannoch in Eider and Osprey fields is argillaceous and micaceous and contains thin siltstone layers with starved wave ripples and wave-modified current ripples. Infaunal reworking is largely subordinate to physical lamination

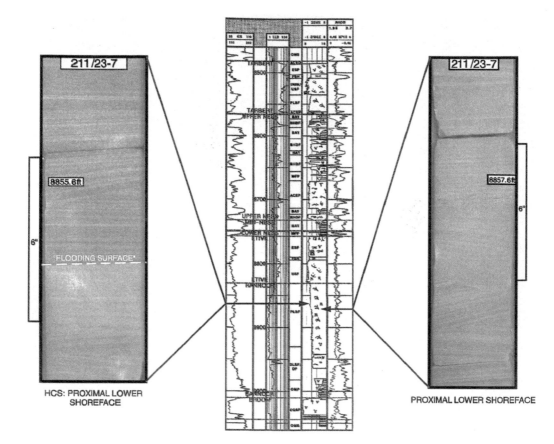

Fig. 6. Core photographs from Osprey 211/23-7, highlighting the proximal lower shoreface facies. Amalgamated hummocky lamination is evident at 8857.6 ft. Bedset boundaries are marked by the sharp colour change from lighter grey to darker grey in the sample from 8855.6 ft. This feature is attributed to an abrupt increase in mica content and accompanies a minor grain size reduction.

with ichnofauna restricted to delicate *Planolites*. The argillaceous and silty strata grade upward into micaceous, planar-laminated sandstone with current ripples, locally common detrital wood fragments and rare mudstone clasts. These graded beds occur at the transition between the argillaceous and micaceous strata and the overlying interval of amalgamated hummocky strata. Overall, the degree of burrowing in the sandier lithofacies is low and is limited to escape traces and simple vertical shafts at the top of fining-upward bedsets.

In the sandstone-rich upper portion of the Rannoch, small-scale asymmetric patterns are evident on the gamma-ray and density logs (Figs 2 and 5). The patterns reflect systematic decreases in the concentration of mica (Fig. 6). The units culminate with a sharp boundary and an abrupt increase in mica content with hum-

mocky stratification commonly occurring above and below the boundary. Carbonate cementation locally occurs along the boundaries. These bedding cycles are excellent field-wide markers for correlation (Fig. 5). In Eider and Osprey, the bounding surfaces dip gently to the north-northeast at 0.5°, a rate considerably higher than the 0.1° seen in Tern Field. Because the mica has a deleterious effect on permeability, these layers, coupled with the carbonate cement, are presumed to exert a strong influence on fluid flow within the Rannoch.

This Rannoch succession is interpreted to reflect storm dominated lower shoreface deposition, but under the influence of a fresh water deltaic system. The succession is composed of three genetically related depositional facies: prodelta, delta front and proximal lower shoreface. The association of abundant mica, un-

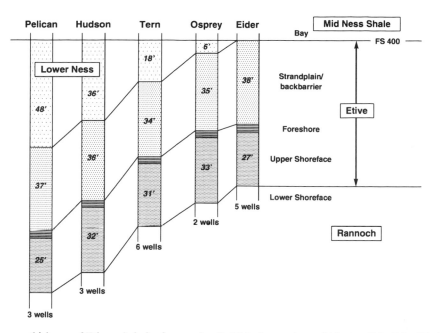

Fig. 7. Average thickness of Etive sub-facies for cored wells. Note the consistent thickness of the Etive lithofacies types across the area and the sequential loss of the Lower Ness in the seaward direction. These thickness relationships imply a depth-controlled facies boundary for the base of the nearshore Etive. This regional uniformity suggests that the depth of fairweather wave base was approximately 10 m (30 ft) (uncompacted) throughout deposition of the nearshore Etive in the study area.

burrowed mudstone and turbidites with the minor modification by wave-generated currents represents a prodelta (shale-prone) to delta-front (sand-prone) depositional facies (e.g. delta-slope facies of Cannon *et al.* 1992). The physical and biogenic structures found here imply deposition in relatively quiet but ecologically stressed nearshore conditions caused by frequent incursions of low salinity and turbid currents (cf. Pemberton & Wightman 1992). As with normal lower shoreface successions, the amalgamated hummocky beds represent the proximal lower shoreface and the onset of average storm wave base. The geographical occurrence of this Rannoch facies succession in Eider and Osprey implies that the northeastern part of the study area was closer to a fluvial–deltaic distributary system (cf. Cannon *et al.* 1992). The presence of the distributary system may be associated with the Northern Viking graben depositional axis which occurs along the UK–Norway border. This relationship may explain why the Rannoch has a more deltaic expression in Eider and Osprey fields, whereas away from the graben axis in Pelican, Tern and Cormorant fields a more typical lower shoreface succession prevails.

The clinothems shown in Fig. 5 clearly mark increments of shoreline progradation but the origin of these units is not clear. The patterns resemble parasequences on gamma ray and porosity logs but the bounding surfaces, particularly in the upper Rannoch, do not consistently mark obvious marine flooding events (cf. Van Wagoner *et al.* 1990). Although several widely correlative flooding events are recorded in the overlying beach–lagoonal system in the Lower Ness and locally in the Rannoch (e.g. Livera & Caline 1990), the closely juxtaposed clinothems are too numerous for each surface to mark significant sea-level rises which flooded the barrier complex. They resemble the sub-parasequence-scale compartments or shingles described by Valesek (1990) or the lower shoreface bedsets of O'Byrne & Flint (1993). O'Byrne & Flint (1993) cited no evidence for significant landward shift in facies tracts across bedset bounding surfaces and suggested that no significant time gap occurs between bedsets. Both studies postulate the role these surfaces may have in influencing fluid flow. The Rannoch clinothems corroborate this view as an order of magnitude permeability drop occurs across several of these boundaries. Although the three-dimensional geometry of the Rannoch clinothems is not known, they likely represent at least field wide, high frequency hiatal surfaces

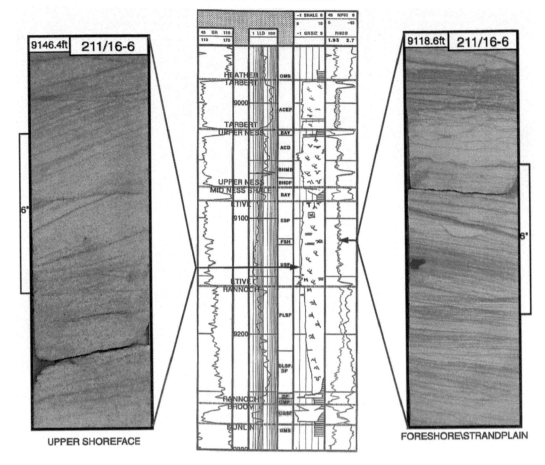

Fig. 8. Core photographs from the Eider 16-6, showing the unidirectional cross-bedding of the upper shoreface sub-facies and the low angle and wedging laminations of foreshore sub-facies.

which mark an abrupt decrease in sand-grade material and an increase in detrital mica and calcareous material. The surfaces may therefore be related to cyclic variations in sediment supply, possibly linked with short-term, climatically driven run-off cycles which influenced the relative influx of mica to the lower shoreface.

Nearshore-marine Etive and Lower Ness Formations

The onset of the nearshore-marine Etive Formation is marked by a subtle to abrupt increase in grainsize and a change in sedimentary structures. Above a variable basal interval, the Etive is found to have three distinct sub-facies which maintain relatively constant thickness across the area (Fig. 7). In ascending stratigraphic order the three sub-facies are upper

shoreface, foreshore and emergent strand plain. A discussion of the lower Ness Formations is included here because the Ness is the landward, time-equivalent facies of the Etive Formation. The discussion provides the comparison between the normal nearshore Etive facies succession and the much different 'Etive' lithofacies which occurs in lowstand incised valley fills.

Upper Shoreface. In the southern part of the study area in Pelican and Tern fields, the upper shoreface lithofacies is expressed as a medium-grained, moderately to well-sorted sandstone with trough cross-beds and plane-parallel lamination. Burrows are absent. Core porosity and permeability values commonly exceed 25% and 3000 md, respectively. The basal contact with the underlying Rannoch Formation varies from sharp and erosional to transitional and inter-bedded. Transitional contacts are expressed in

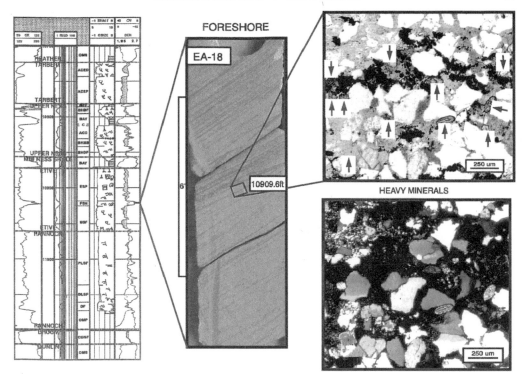

Fig. 9. Core photograph and thin-section micrographs from Eider EA-18 (deviated). The darker laminations are enriched with heavy minerals. This enrichment is easily detected by increased gamma ray values and higher bulk density and serves as a key recognition criterion for the foreshore in logs from uncored wells.

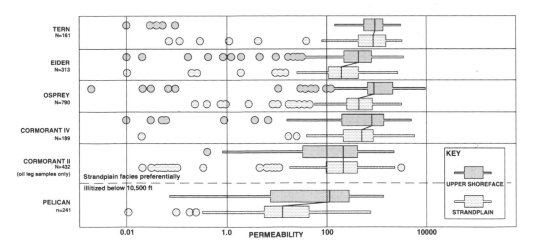

Fig. 10. Summary of Etive permeability values keyed to subfacies. Median values are indicated by the bold black lines and thick bars indicate data falling within the 25 and 75 percentile. The thin bars represent an envelope used to distinguish outliers (circles). The fields are arranged with increasing depth toward the bottom. On average, the strand plain shows similar to slightly reduced permeability values. This is also expressed in the permeability values for the Etive in Fig. 2. Preferential illitization of the strand plain significantly reduces permeability in Pelican Field.

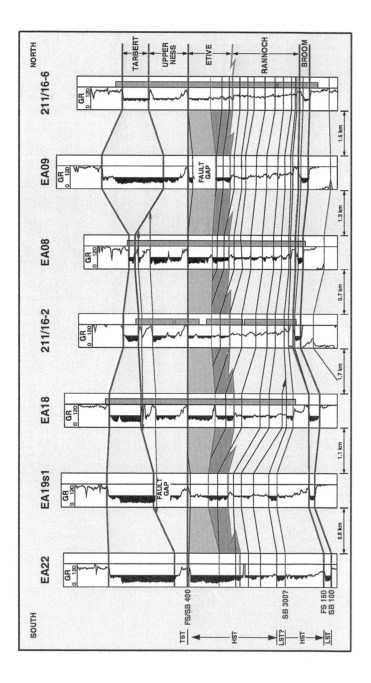

Fig. 11. Summary diagram of the facies and sequence architecture of the Eider field.

cores as intervals of micaceous, wave-rippled and hummocky laminated sandstone inter-bedded within the lower portion of the Etive (cored Pelican wells 211/26-13, 26-12; Tern wells 210/25-5, TA-19). Toward the northern part of the study area, in Eider and Osprey field, the lowermost Etive is very coarse-grained with thin layers of pure detrital mica. Above the coarser-grained beds, the Etive becomes better sorted with stratification dominated by small-scale, unidirectional trough cross-beds similar to examples from Tern and Pelican fields (Fig. 8). Thickness relationships are relatively constant across the study area for this as well as the other Etive subfacies.

Following previous authors, this subfacies is viewed as upper shoreface deposits (e.g. Budding & Inglin 1981; Johnson & Stewart 1985; Brown et al. 1987; Graue et al. 1987). A genetic relationship with the underlying Rannoch is supported by the local interbedding of Etive and Rannoch lithofacies. The lower contact repre-sents the water depth controlled transition from the lower-shoreface environment to the current dominated, fairweather deposits of the upper shoreface (cf. McCubbin 1982; Livera & Caline 1990). Much of the evidence gathered in this study supports the model of Scott (1992) which proposes that the Etive represents the prograda-tion of a storm influenced, barred nearshore system analogous to highly dissipative modern shoreline systems (Hunter et al. 1979; Wright et al. 1979). The locally sharp basal contact is interpreted to result from the temporal and spatial movement of longshore troughs and bars driven by rip currents and coastal jets. Studies of modern beaches show that such movements are driven by variations in wave climate which result from major storms as well as seasonally driven changes (Wright et al. 1979; Shipp 1984; Short 1984). In addition to storm activity, barred dissipative systems are commonly characterized by high sand supply (Wright et al. 1979). Evidence for a high sediment supply for the Rannoch–Etive system is given by Milton (1993), who showed the Brent ranks as one of the most significant Jurassic progradational successions in the Northern Viking Graben.

Foreshore. A discrete change in grain size and bedding style accompanies the transition from the upper shoreface to the foreshore. Cross-bedded sandstones are abruptly replaced by gently wedging to sub-parallel laminated sand-stone with shallow scour-and-fill and low amplitude current ripples (Fig. 8). Sandstones are slightly finer-grained (chiefly upper fine-grained) and are better sorted. Sub-facies thick-ness ranges from 1–2 m (3–6 ft). Core porosity and permeability in the foreshore are slightly lower than the upper shoreface. The unit is commonly associated with a slight to pro-nounced change in gamma ray and bulk density log character. Both gamma ray and bulk density values show an increase but neutron-porosity values remain unchanged. This characterisitic is a response to increased concentration of heavy minerals (Fig. 9).

These low angle planar to gently wedging laminations are interpreted as the product of variations in wave swash and the erosion and recovery episodes which follow energetic storm events. The low angle scour-and-fill and ripple lamination are attributed to deposition by bar and runnel systems and ephemeral creeks. The relatively restricted occurrence of zones enriched with heavy mineral lamination is consistent with a foreshore and berm interpretation (cf. Clifton 1969; McCubbin 1982).

Strand Plain. The upper part of the nearshore Etive Formation is typified by an extremely well sorted, upper fine-grained sandstone interval which has a characteristic fabric which alter-nates randomly between vaguely stratified to mottled to structureless. Stratification types include small-scale trough cross-bedding, gently wedging cross-lamination and planar-parallel lamination. Minor soft-sediment deformation and dewatering structures are locally observed. *Palaeophycus* burrows are common along dis-crete layers and become particularly abundant toward the top of the formation near the overlying Ness Formation. Locally unlined vertical shafts and U-shaped tubes are also observed toward the top of the interval. Vague root traces are evident with intense disruption occurring beneath coals. This lithofacies is capped by a coal or mudstone layer of the Lower Ness or the Mid-Ness Shale.

This lithofacies is interpreted as the subaerial portion of the shoreline complex. In contrast to other well-studied barrier-beach successions, the emergent part of the Etive strand plain or barrier-island system is well preserved (Budding & Inglin 1981; Livera & Caline 1990). The abundance of water-laid cross-lamination sug-gests that the dominant depositional process was storm-related washovers (Schwartz 1982). The indistinctly laminated to disrupted fabric is attributed to: (1) high degree of grain sorting, (2) post-depositional groundwater movement by a seasonally fluctuating water table, and (3) infaunal burrowing and rooting. The extreme uniformity in grain size implies that the source of the sand was probably the well sorted upper beach, foreshore and backshore. Wind rework-ing likely enhanced the sorting of the sediment but the absence of unequivocal wind-laid strata suggests that such deposits were reworked by later wash-over events or modified by post-depositional processes. The unlined burrows may record insect, amphipod or crustacean traces which are common in the subaerial portions of the strand plain (Budding & Inglin 1981; Frey & Pemberton 1987).

Etive subfacies influence on reservoir quality. Across the area, the Etive subfacies exhibit subtle but distinct variations in grainsize and clay content. Shoreface facies are slightly coar-ser-grained, have less silica cement and have lower depositional clay content. At shallow depths, above 3200 m (10 500 ft), these differ-ences translate into slightly reduced permeabil-ities for the beach-plain facies (Fig. 10). This variation is also illustrated in Fig. 2. Below 3200 m (10 500 ft), the reservoir quality contrast between the shoreface and strand plain is amplified because of preferential illitization of the more clay-rich strand-plain interval (Fig. 10). This relationship is a significant issue in the successful depletion of Pelican field which has its oil leg extending below the threshold depth of strong illitization (Kantorowicz *et al.* 1992).

Back-barrier/lagoon (Lower Ness). In the south-ern part of the study area, the Etive is overlain by the Ness Formation, a strongly layered and heterolithic interval composed of strata depos-ited in lagoonal, bay and coastal-plain environ-ments. The Ness Formation has been informally divided into lower, middle and upper members (Budding & Inglin 1981). The Lower Ness is commonly viewed as the back-barrier and lagoonal facies which are time-equivalent to the Etive Formation (Fig. 4) and includes an array of sub-environments: washover fans, tidal channels, lagoonal mudstones, tidal flats, coal swamps, flood-tidal deltas, crevasse splays and small bay-head deltas (Livera 1989). A thorough discussion of the Lower Ness in the nearby Cormorant field is found in Budding & Inglin (1981) and Livera & Caline (1990).

A diagnostic attribute of the Ness is highly continuous sandstone layers which are arranged as coarsening-upward successions capped by coals. These units are overlain by blanketing lagoonal mudstones (flooding surfaces) which are readily correlated between fields and accord-ingly enable a more accurate time-stratigraphic framework for the lower Brent system to be constructed.

The Lower Ness facies pass basinwards into emergent strand-plain deposits of the upper Etive. This is illustrated in Tern field (Fig. 4). A similar transition occurs along depositional strike in Cormorant field (Livera & Caline 1990). These Ness lagoonal flooding surfaces are correlated into the massive Etive strand plain lithofacies as zones of intense bioturbation, increased rooting and increased carbonate ce-ment. In several instances, coal beds correlate from the Ness into the strand plain deposits of the Etive Formation.

Nearshore facies stacking patterns. Stratigraphic and facies data support a genetically related shoreface/back-barrier interpretation for the Rannoch, nearshore Etive and Lower Ness Formation in this part of the East Shetland basin. Using examples from Tern and Eider field, two end-member shoreface geometries are evident: strongly progradational and aggrada-tional. An example of a progradational geo-metry occurs in Eider field where the Etive maintains a relatively uniform thickness. The foreshore log marker remains roughly 9 m (30 ft) above the base of the shoreface and 10.5 m (35 ft) below the top of the Etive (Figs 7 and 11). In addition, the interval of amalgamated hum-mocky cross-bedded sandstone remains a con-stant thickness of 18 m (60 ft) with the base of the proximal lower shoreface tracking parallel to the base of the Etive. This uniformity in thickness is interpreted as progradation during a time when little or no accommodation space is being created.

An aggradational stacking geometry occurs in the central part of Tern field. Across the southern half of Tern Field, the Etive and the Lower Ness formations maintain a uniform thickness toward the 25-5 well (Fig. 4). In the vicinity of TA11, several changes take place.

(1) Upper shoreface, foreshore, and strand plain subfacies and their associated bounding surfaces rise stratigraphically with respect to the Mid-Ness Shale datum.

(2) The Etive is interbedded with Rannoch facies in the cored TA19 and probably in the uncored TA02. The lower Etive has an irregular expression on the gamma ray log (TA11, TA 19, TA02) and contrasts with the blocky Etive pattern seen in TA28 (Fig. 4). In the vicinity of the TA02, well-log patterns in the Etive are easily correlated into the Rannoch in the seaward direction (cf. Livera & Caline 1990, fig. 6).

(3) An increase in stratiform carbonate cement is observed in the coeval Rannoch. The occurrence of cement has been recognized by previous workers and related to increased

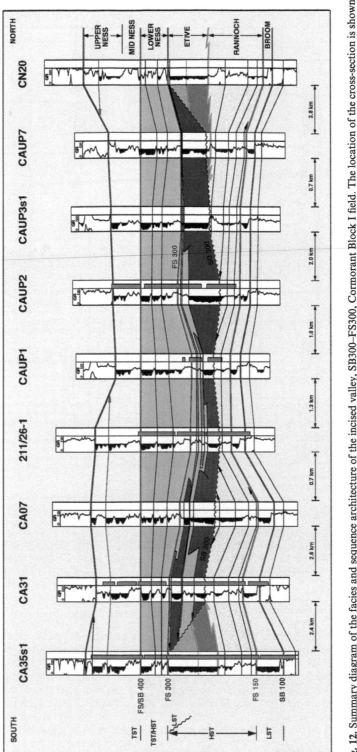
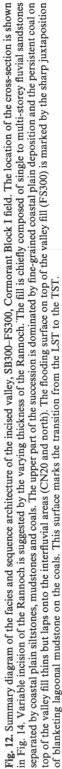

Fig. 12. Summary diagram of the facies and sequence architecture of the incised valley, SB300–FS300, Cormorant Block I field. The location of the cross-section is shown in Fig. 14. Variable incision of the Rannoch is suggested by the varying thickness of the Rannoch. The fill is chiefly composed of single to multi-storey fluvial sandstones separated by coastal plain siltstones, mudstones and coals. The upper part of the succession is dominated by fine-grained coastal plain deposition and the persistent coal on top of the valley fill thins but laps onto the interfluvial areas (CN20 and north). The flooding surface on top of the valley fill (FS300) is marked by the sharp juxtaposition of blanketing lagoonal mudstone on the coals. This surface marks the transition from the LST to the TST.

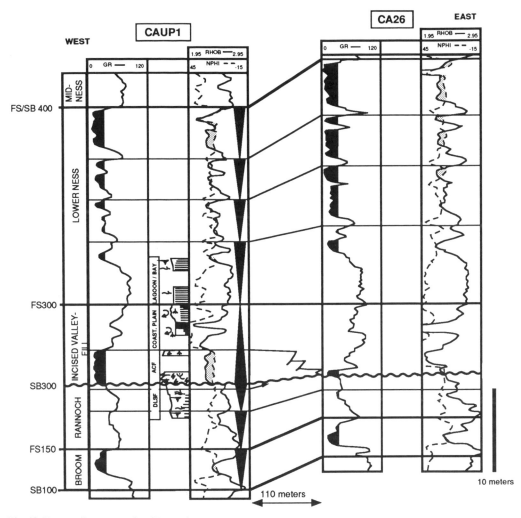

Fig. 13. Two-well cross-section illustrating the heterogeneity of the Etive incised valley fill deposits. The first well, CAUP1, encountered a 10 m thick channel sandstone overlain by a shale-prone and carbonaceous coastal plain interval. The channel sandstone was absent 110 m away in the CA26 well and the valley-fill succession is only composed of a thin sandstone bed and carbonaceous shale and coal. The Rannoch in the CA26 is interpreted to be thinned by truncation by SB300. Note the change from coastal plain to lagoonal deposition at FS300 and the minor flooding surfaces which define the parasequence boundaries in the highly layered Lower Ness.

detrital carbonate accumulation along hiatal surfaces (cf. Budding & Inglin 1981; Livera & Caline 1990; Cannon *et al.* 1992). The Rannoch is also very micaceous in the northern wells in the field and is characterized by lower than normal permeabilities and lower oil saturations.

(4) The Lower Ness back-barrier facies grade abruptly into the upper Etive strand-plain facies. This is expressed between wells TA28 to TA19. Flooding surfaces in the Lower Ness correlate with thin cemented zones in the upper Etive strand-plain lithofacies.

Incised valley-fill Etive

A considerably different Etive lithofacies assemblage occurs in discrete portions of Cormorant and Tern fields. These rocks contrast the uniform nearshore Etive lithofacies and are made up of a complex mix of sandstone and shale lithologies. The succession is commonly marked in core by an erosional surface consistently overlain by a layer of locally derived shale and sandstone clasts. The reservoir properties of the overlying strata vary laterally and vertically

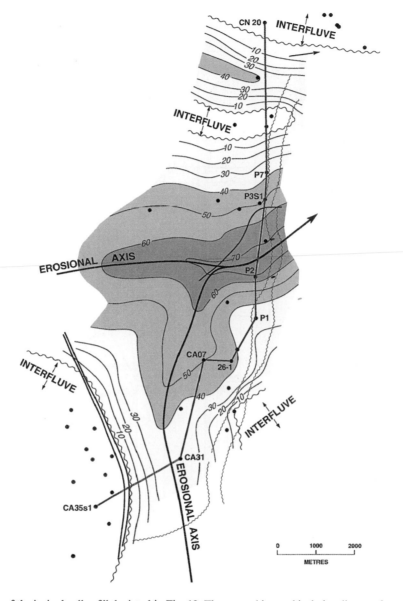

Fig. 14. Isochore of the incised valley fill depicted in Fig. 12. The mapped interval includes all strata from SB300 to the FS300. The cross-section from Fig. 14 is indicated. The interfluves indicate areas where the two surfaces, SB300 and FS300, have merged to form a single hiatal surface. The wells in these areas have a complete section of nearshore Etive subfacies.

from well to well both in terms of porosity, permeability and sandstone thickness (Fig. 12). Sandstones bodies are variably arranged as single, fining-upward stories separated by shale layers (CA31) or as thicker, amalgamated successions with abrupt changes in grain size and localized concentrations of coarse-grained carbonaceous detritus (CA20, CAUP2). The sandstone-poor intervals separating the single storey channels consist of coastal-plain/channel overbank facies with current-rippled sandstones, pedogenized mudstones and rooted coals. Lateral heterogeneity of this 'Etive' lithofacies is illustrated by the CA26 well which twinned the

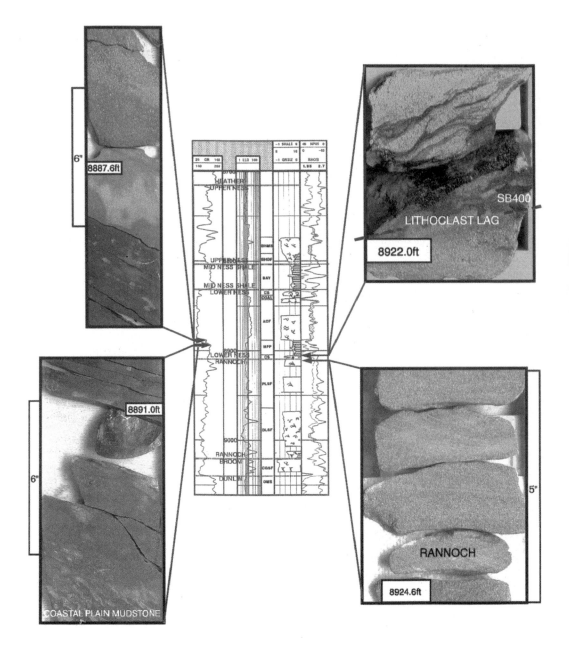

Fig. 15. Core photographs from the Tern 25-3s2, showing components of the incised valley (SB400–FS400). The sequence boundary is marked by a layer of poorly sorted sandstone with shale lithoclasts which sit on micaceous hummocky strata of the Rannoch Formation. The nearshore Etive formation has been completely truncated. A highly carbonaceous mudstone interval overlies the lag. The mudstone is devoid of burrows and is interpreted as a stressed lagoonal or lacustrine mudstone. Stacked channel deposits comprise the remaining valley-fill.

CA20 well. The well encountered a 1 m thick interval of sandstone which is abruptly overlain by coal-rich lower Ness strata (Fig. 13). Over a distance of 110 m, the Etive lithostratigraphic unit is virtually absent and renders a succession where the Lower Ness is effectively sitting on a thinned lower shoreface Rannoch. The variability of the fill also coincides with both a variable Rannoch formation. Rannoch well-log markers are truncated beneath the channelized succession with the maximum erosion occurring in the 211/26-1 where the Rannoch consists only of distal lower shoreface facies. Whereas the nearshore Etive consists of nearly all sandstone in a consistent vertical and lateral arrangement of subfacies, this type of Etive is characterized by a much more complex sandstone body stratigraphy.

In Cormorant field a fairly continuous interval of coal and carbonaceous mudstone occurs in the upper part of the channel complex. The coal bed can be traced northward to rest on the nearshore Etive deposits in the CN 20 (Fig. 12). A fissile, black lagoonal mudstone layer abruptly overlies the coastal-plain deposits as wells as on nearshore Etive deposits which occur to the south and north of the channel complex (Figs 12 and 13). This unit represents a flooding event and the onset of the readily correlatable mudstone–sandstone layers of the Lower Ness Formation on the fault block. The stratigraphic interval between the erosive base of the channelized complex to the lagoonal flooding surface defines a broad erosional feature which reaches 23 m (75 ft) thick and is 2.4 km (1.5 miles) in width (Fig. 14).

The broad channel complex appears to have regional extent. A similar succession of stacked channels and coastal plain facies occurs in the Etive Formation in NW Hutton Field, 11 km (7 miles) east of Cormorant Block I. Similar 'Etive' stratigraphy is reported in the 22/27-4a well where a few 'feet of Etive' sandstone is overlain by a heterolithic unit of coastal-plain siltstones and mudstones (Brown et al. 1987; Richards 1992). Since the Lower Ness stratigraphy can be correlated in both fields, the two features may be components of the same regional erosional and filling event.

Another erosional channel complex with similar lithologic and stratigraphic characteristics occurs in the Etive formation but occurs at a younger stratigraphic level. It is found directly beneath the Mid-Ness shale in the northern part of Tern field. In the vicinity of the 210/25-3st2 the shoreface and strand-plain facies of the nearshore Etive (cored in the TA19) are replaced in the basinward direction by stacked channel fill

and coastal-plain deposits (Fig. 4). In the 25-3st2 core a thin layer of lithoclasts separates crevasse-splay sandstones and very carbonaceous coastal-plain mudstones from proximal lower-shoreface strata of the Rannoch (Fig. 15). This succession is analogous to the example described from the CA26 well.

Sequence analysis

Lowstand systems tract: Cormorant Field (SB300–FS300), Tern Field (SB400–FS400). Facies and stratigraphic observations support the broadly channelized units in Cormorant and Tern fields to be parts of incised valley fill complexes which were cut into the Etive nearshore complex and backfilled with fluvial and coastal-plain strata. The two basal erosional surfaces which are mantled with shale lithoclasts (SB300 and SB400) are interpreted as sequence boundaries which separate nearshore Etive and lower shoreface Rannoch facies from the more heterogeneous channel deposits. The variable incision at the sequence boundaries contrasts with the planar and diachronous surface which marks the basal nearshore Etive seen in Eider and southern Tern field (Figs 11 and 4, respectively). Several lines of evidence suggest that these features are not distributary or barrier-inlet channels (e.g. Livera & Caline 1990; Scott 1992). The vertical and lateral suite of facies bears no resemblance to well documented barrier inlet channels (Kumar & Sanders 1974; Hubbard & Barwis 1976; Barwis & Makurath 1978; McCubbin 1982). The subaerial coastal-plain facies, such as pedogenized mudstones and coals, resting on the lower shoreface Rannoch implies that the cut and fill were not contemporaneous. This juxtaposition requires a relative fall in sea-level which is inconsistent with normal distributary migration. Individual channel bodies are much thinner than the 23 m (75 ft) thick erosional axis. For example, up to three fining-upward channels separated by overbank deposits with roots and coals occur within the valley fill in southern Block I. At least two discrete channel sandstones comprise the fill succession in Tern Field. This variable fluvial stratigraphy is expressed as the thin sandstone at the base of the valley fill in the 25-3st2 thickens to form a fining-upward sandstone body in the TA08.

Outside the limits of the incised valley on the interfluves, both sequence boundary and associated flooding surface are interpreted to merge and form a hiatal surface. In Cormorant Field a complete nearshore Etive succession (shoreface–foreshore–strandplain) occurs underneath the

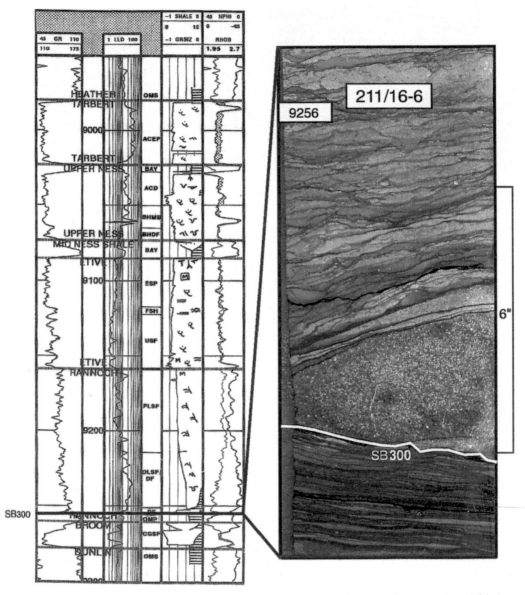

Fig. 16. Core photograph from the Eider 16-6, showing the coarse-grained wave and wavy laminated lithology resting on a surface inferred to be the equivalent of SB300. In the landward direction these overlying sandstone beds thicken to 8 m (25 ft) and contain trough cross-beds, shale lithoclasts and abundant carbonaceous material. This unit is provisionally correlated as the lowstand shoreline associated with the incised valley system from Cormorant field.

inferred hiatal surface on both margins of the erosional valley fill complex (Fig. 12). This relationship argues that the valley fill succession cannot be part of a genetically related inter-fingering between the Lower Ness and the Etive such as seen in the southern and central part of Tern field. The abrupt change from a coal-forming environment to sub-aqueous lagoonal

deposition in the upper valley fill is interpreted as the change from the lowstand to the transgressive systems tract (described below). This surface is regionally correlated as FS300. In Tern Field, interfluvial areas are mantled by a thin 1–5 cm thick coarse-grained lag which is significantly coarser than sandstone grain sizes from underlying Lower Ness or strand-plain

Etive. This thin layer is interpreted as a veneer of lowstand strata laid down on interfluvial areas. This boundary is also coincident with the base of the Mid-Ness Shale (FS400) which marks a regional flooding event for this part of the East Shetland Basin. All of the other fields have this thin, coarse-grained lag developed at the near-shore Etive/Mid-Ness shale boundary. This provides additional support for a regional hiatus at the base of the Mid-Ness shale.

Regional correlation argues for the two incised valley fill systems to be parts of different aged systems. The Cormorant Field example is overlain by several parasequences of Lower Ness-aged strata which can be correlated regionally (Figs 12 and 13). In Tern, the younger Mid-Ness shale serves as the transgressive surface which terminated valley-fill deposition. This age difference is illustrated within Tern field as the 210/25-2 well may have penetrated the margin of the older incised valley. An anomalous, coarser-grained sandstone and shale interval of fluvial and coastal-plain origin sits erosionally on finer-grained, trough cross-bedded sandstone of nearshore-Etive origin. This facies succession, together with the correlative Lower Ness stratigraphy, is interpreted as a penetration along the margin of the SB300 incised valley-fill complex mapped in Cormorant field. Away from the 25-2 well, the FS300 is correlated along the base of the Lower Ness and then into the nearshore Etive where it is eventually truncated by the younger SB400 in the northern part of the field. Since no other wells in Tern Field penetrate lowstand strata of this age, the valley-fill complex must have a west–east trend and probably flowed toward Cormorant field. This palaeocurrent pattern is consistent with the isochore of the incised-valley fill in Cormorant block I.

Unequivocal lowstand shoreline deposits linked to the incised valleys were not found within the limits of the highly drilled fields. The incised valley-fill isochore map predicts that associated lowstand shoreline deposits associated with SB300 should occur between Cormorant and Eider field to the north. A candidate for the distal expression of a lowstand shoreline deposits may exist in the lower Rannoch in Eider field. An enigmatic, basinward-thinning layer, 0.15–6 m (0.5–20 ft) thick is cored and consists of cross-bedded to flaser-bedded shaley sandstone with a locally developed coarse-grained lag at its base (Fig. 16). This stratal assemblage is very different from the distal, mud-prone Rannoch facies they rest on. This sandstone is unrelated to the Broom Formation because it rises systematically away from the top of the Broom (FS150)

in the landward (southwesterly) direction. The basal surface is too sharp and the basal lag too coarse-grained to have resulted from simple progradation. The upper surface is also abrupt and is overlain by micaceous siltstone of the lower Rannoch. In addition, the overlying Rannoch appears to prograde across the top of the unit in a downlapping relationship. The combined evidence suggests this unit was the high energy toe of a shoreline system which was rapidly emplaced during a relative fall in sea-level. The lack of intervening well data precludes a direct tie of these deposits to the sub-aerial portion of the shoreline system. However, if the candidate sequence boundary is projected from southern Eider (EA22) to northern Cormorant it has a dip of 0.4°. This value is consistent with the range of dips on Rannoch clinoforms. Therefore, this sandstone layer is provisionally interpreted to rest on the seaward expression of SB300 from Cormorant field.

Transgressive and highstand systems tract: (FS300–SB400). Regional variations in shoreface stacking patterns are evident away from the incised valleys. Widespread lagoonal flooding surfaces, FS300 and FS400, provide the stratal boundaries which enable the characterization of the nearshore environment's response to relative changes in sea-level. The aggradational shoreface stacking pattern described in Tern Field closely follows the transgressive flooding of the lowstand shoreline and incised valley fill complexes or FS300. This pattern correlates along depositional strike (southeast) with an identical geometry discussed by Livera & Caline (1990) from northern Cormorant field. The two examples reflect a synchronous response to a relative rise in sea-level. The aggradational barrier complex is therefore interpreted as the transgressive to early highstand systems tract. Several increments of relative rise in sea-level were recorded in the Lower Ness in the form of lagoonal and bay-fill mudstones which abruptly overlie coastal plain deposits and coals.

During the latter part of the HST the shoreface system rapidly evolved into one which stacked with strongly progradational geometries as the sediment supply far outpaced the decreasing rates of relative sea-level rise. Progradational deposition was terminated with the next major lowstand event, SB400, when the next phase of incised valley formation initiated the next sequence.

The volume of sediment supplied to the barrier was clearly sufficient to maintain at least a slight vector of progradation during the transgressive to early highstand systems tract.

This caused the barrier to rise stratigraphically without any apparent retrogradation or landward translation of shoreline position. In fact, no significant translation of the shoreline can be documented in any of the fields. This observation may help explain the preservation of the strand-plain facies. In a system which alternates between progradation and aggradation without retrogradation, transgressive ravinement of the proximal nearshore and strand plain would not occur and the subaerial part of the strand plain would be preserved.

Discussion

The integration of the field-scale sequence stratigraphy depicts a more complex lower Brent system than predicted by a simple model of a prograding shoreface system (e.g. Helland-Hansen et al. 1992; Eschard et al. 1993). The two incised valleys indicate that the system responded to at least two higher order falls in sea-level (Fig. 17). The sequence boundaries and the associated lowstand strata are readily identified in the vicinity of the incised valleys. On interfluves the expression of the sequence boundary is subtle and manifested by discrete changes in stacking patterns coupled with persistent coals or thin, coarse-grained lags. The area-wide flooding surfaces (FS300 and FS400) mark changes in Rannoch–Etive–Lower Ness stacking patterns from a progradational (late HST) below to aggradational (TST to early HST) above. Depending on the location, sequences therefore have a range of systems tract components: LST, TST and HST in the vicinity of the valley fills and lowstand shorelines, but are limited to TST and HST deposits away from the valley fills on interfluves.

The geometry and the organization of the systems tracts presented in this study may offer a model for sequence based interpretation in strongly progradational, sediment-laden nearshore systems. The high volume of sediment supplied to the shoreface was more than sufficient to keep pace with rising sea-level and increasing accommodation space. Accordingly, the barrier aggraded in place without any significant landward translation in shoreline position. Ravinement processes in this scenario would be minimized. The transgressive systems tract is a thin aggrading wedge of back-barrier and shoreface strata which presumably has a seaward thinning geometry because most of the longshore-delivered sediments within this systems tract would be trapped within the proximal nearshore environment. Only fine-grained sedi-

ment and hydrodynamically light mica were transported onto the distal lower shoreface during this time. As the rate of rising sea-level decreased, an equilibrium between sediment supply and accommodation was achieved and rapid progradation followed.

The base Mid-Ness shale (FS400) has been widely recognized as a major lithostratigraphic boundary and has at least regional time significance (Mitchener et al. 1992). This view is corroborated by this study. The Mid-Ness Shale is part of the transgressive systems tract which blanketed interfluves and the regional sequence boundary 400. This model would predict that the Mid-Ness shale is coeval with a phase of pronounced barrier aggradation north of the study area. This has been shown to occur in UK Quads 21/17 and 21/18 where several authors document an aggradational stack of proximal nearshore Etive Formation deposits with little or no Ness Formation (Brown & Richards 1989; Cannon et al. 1992; Mitchener et al. 1992; Helland-Hansen et al. 1992). Thus the Mid-Ness shale serves as a major, low-order change in the depositional stacking patterns of the lower Brent from strongly progradational below to aggradational above. Mitchener et al. (1992) and Partington et al. (1993) place a low order sequence boundary at the base of the Mid-Ness shale, citing both incision (SB400?) and major flooding along this surface (FS400).

Several recent studies have documented the higher resolution character of the Brent Group and emphasize the role which relative changes in sea-level had on facies and geometry. Many of these studies focus on the highly cyclic Ness Formation or its time-equivalent interval and show that several high frequency sequences make up the depositional record (e.g. Livera 1989; Van Wagoner et al. 1993; Crane et al. 1994; Kaas et al. 1994). For the highly layered Ness, most studies cite field-wide flooding surfaces as the principal fluid-flow barriers for the reservoir sandstones.

Less consensus has surrounded the depositional origin and sequence stratigraphy of the Etive. Although several studies have interpreted the Etive to reflect coalesced channel deposits (Johnson & Stewart 1985; Brown et al. 1987; Graue et al. 1987; Livera 1989; Brown & Richards 1989; Livera & Caline 1990; Helland-Hansen et al. 1992), only a few have described parts of the Etive as a lowstand systems tract. Van Wagoner et al. (1993) interpreted the Etive Formation in the Statfjord Field area to sit erosionally on the Rannoch. A sequence boundary, possibly synchronous with one of the surfaces described in this study, may have

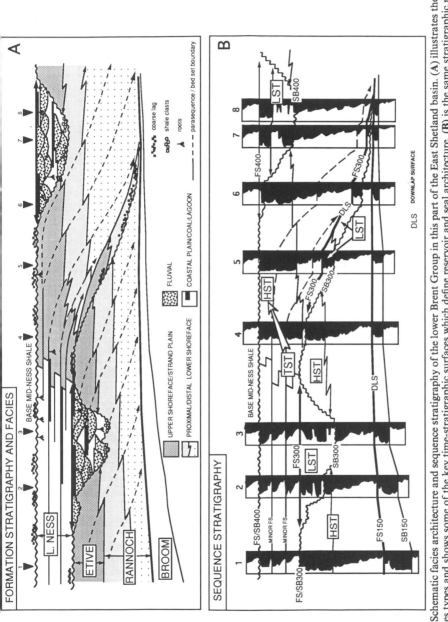

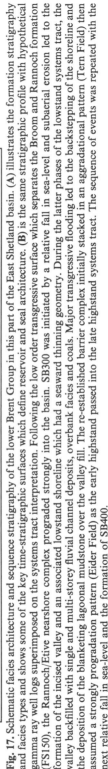

Fig. 17. Schematic facies architecture and sequence stratigraphy of the lower Brent Group in this part of the East Shetland basin. **(A)** illustrates the formation stratigraphy and facies types and shows some of the key time-stratigraphic surfaces which define reservoir and seal architecture. **(B)** is the same stratigraphic profile with hypothetical gamma ray well logs superimposed on the systems tract interpretation. Following the low order transgressive surface which separates the Broom and Rannoch formation (FS150), the Rannoch/Etive nearshore complex prograded strongly into the basin. SB300 was initiated by a relative fall in sea-level and subaerial erosion led to the formation of the incised valley and an associated lowstand shoreline which had a seaward thinning geometry. During the latter phases of the lowstand systems tract, the valley backfilled with single and multi-storey fluvial channel deposits, overbank facies and coals. Major transgressive flooding led to the backstepping of the shoreline and the deposition of the blanketing lagoonal mudstone over the valley fill. The re-established barrier complex initially stacked in an aggradational pattern (Tern Field) then assumed a strongly progradation pattern (Eider Field) as the early highstand passed into the late highstand systems tract. The sequence of events was repeated with the next relative fall in sea-level and the formation of SB400.

incised deeply toward the Tampen Spur and removed the underlying nearshore Etive across this relatively broad area. In addition, Milton (1993) deduced an intra-Etive sea-level fall on the basis of stratigraphic relationships from a regional Northern Viking graben study although the location and evidence for this feature was not stated. It is likely that as more field-based sequence-stratigraphic studies are conducted the complex nature of the Etive Formation will be unravelled.

Implications for reservoir management

The application of sequence stratigraphy in maturing fields has been particularly successful at providing a higher resolution characterization of reservoir and seal facies and has led to tangible rewards such as improved field management strategies, identification of stratigraphically controlled infill and work-over opportunities and more accurate geological input into reservoir engineering models (e.g. Kaas et al. 1994). A more accurate delineation of the stratigraphic elements presented in this study has led to an improved characterization of lower Brent reservoirs. Although the sequence-stratigraphic analysis of the more channelized Ness Formation provides a greater opportunity to resolve lithologic discontinuities and stratigraphically trapped hydrocarbons (e.g. Livera 1989; Kaas et al. 1994), several relationships are presented in this study which advance the understanding of the controls on fluid movement in the Rannoch and the Etive.

The ebb-tidal delta sandstone body and the gently clinoformal geometries have recognizable implications on reservoir performance. The ebb-tidal delta is characterized by very high sweep efficiencies and high injectivity compared with typical Rannoch strata. This injected volume provides considerable pressure support to updip production wells in both the Rannoch and Etive. The clinoforming geometries, largely defined by mica content, impart a strong permeability anisotropy to the upper Rannoch. Reservoir modelling studies by Wehr & Brasher (1996) show that these surfaces influence fluid movement and, under certain circumstances, may induce water to ride over and bypass oil-bearing Rannoch.

The nearshore Etive lithofacies is characterized by relatively high sweep efficiencies in all of the fields. The perpetual problem of water overrunning the Rannoch are discussed at length by others (e.g. Thomas & Bibby 1993). Owing to slight variations in permeability, a well recently drilled in Tern field showed water movement

controlled by the Etive subfacies; water is preferentially sweeping through the upper shoreface part of the Etive and both overrunning the Rannoch and underrunning oil in the foreshore and emergent strand plain.

In areas where stacking patterns are aggradational, a potential for bypassed oil exists where interbeds of much less permeable Rannoch vertically separate intervals of Etive from the main reservoir body (e.g. TA19 and TA02, Fig. 4). The extent of these tongues of Etive are unknown but the thickest in the study area is 3.3 m (10 ft) in Pelican field. The down-dip enrichment in mica of the Rannoch transgressive systems tract also impacts on producibility and recovery efficiency. In northern Tern field, the highly micaceous Rannoch has poorer than average flow rates. The mica may have the combined role of decreasing permeability, but may also adversely influence the composition of fault gouge and reduce cross-fault communication.

In each of the fields the complex channel sandstone geometries and variable porosity and permeability relationships contribute to well performance anomalies. A spectrum of reservoir management situations characterizes the incised valley-fill system in Cormorant. This spectrum includes very early water breakthrough and lower-than-predicted oil production volumes associated with thick, amalgamated channel sandstones to low injection volumes and poor pressure support where non-channelized sandstones such as crevasse splay sandstones comprise the entire 'Etive' interval. In Tern, individual channel complexes within the valley fill likely have good lateral continuity but the intervening coastal-plain mudstone is a barrier to vertical fluid movement between the channel sandbodies and the Rannoch Formation. This increased level of stratigraphic complexity is superimposed on an already complexly faulted part of the field and reservoir performance suffers accordingly. The wells on the northern flank of the field have poor pressure support and need additional water injectors to maintain economic production volumes.

Conclusions

In this part of the East Shetland Basin, the Rannoch, Etive and Lower Ness formations constitute a low order progradational succession of storm-dominated shoreface strata and coeval back-barrier facies. High resolution stratigraphic analysis indicates that the system also responded to higher frequency variations in sea-level. By recognizing the following relationships, a better insight into anomalous reservoir and seal units is gained.

Well-log markers in the Rannoch Formation have clinoformal geometries which dip gently to the north-east at a rate of 0.1–0.7°. Correlations in the unit are facilitated by cyclic variations in mica content and grain size. This characteristic of the Rannoch influences vertical and lateral permeability and impacts on fluid movement.

A large seaward-thinning, ebb-tidal delta sandstone body is mapped within the Rannoch in Tern and north Cormorant fields. It has much better reservoir quality than typical lower Rannoch and its occurrence helps explain anomalously high sweep efficiency and high water injection rates.

Three subfacies of the nearshore Etive-upper shoreface, foreshore and strand plain are associated with the prograding barrier complex. Two distinct stacking geometries are evident: a tabular geometry which represents a strongly progradational phase of deposition and an aggradational stacking pattern with strongly diachronous facies boundaries. With aggradational patterns the Etive interfingers with seaward Rannoch and landward lower Ness rocks. The geometry is attributed to the progradational barrier system stalling and taking on a vertical vector during a period of increased accommodation. Accordingly, the aggradational geometries are placed within the transgressive/early highstand systems tract and the progradational style is placed within the middle to late highstand systems tract.

The highstand and transgressive systems tracts are separated by a lowstand systems tract which records relative falls in sea-level. Two such events are seen in the study area and are expressed as incised valleys filled with channel and coastal-plain deposits which erode into and replace Etive shoreface and strand-plain facies. Incised valley lithologies add considerable vertical and lateral heterogeneity to the Etive lithostratigraphic interval and their recognition has been important in explaining and predicting well performance anomalies.

The authors express their thanks to Esso EXPRO UK Ltd and Shell UK EXPRO for the permission to publish this paper. Several colleagues are thanked for their contribution during the course of this study: R. Farmer, F. Wehr, C. Tenney, G. Pemberton and J. Van Wagoner. Geological Society reviewers, T. McKie and R. Steel, provided much appreciated technical commentary. J. Brine, F. Williams, A. Marsh, G. Muckle are thanked for exceptional drafting support.

References

BARWIS, J. H. & MAKURATH, J. H. 1978. Recognition of ancient tidal sequences: an example from the Upper Silurian Keyser Limestone in Virginia. *Sedimentology*, **25**, 61–82.

BOWEN, J. M. 1975. The Brent Oilfield. *In:* WOODLAND, A. W. (ed.) *Petroleum and the Continental Shelf of North-west Europe. I. Geology*. Applied Science Publishers, Barking, 353–362.

BROWN, S. & RICHARDS, P. C. 1989. Facies and development of the mid Jurassic delta near the northern limit of its progradation. *In:* WHATELEY, M. K. G. & PICKERING, K. T. (eds) *Deltas: Sites and Traps for Fossil Fuels*. Geological Society, London, Special Publication, **41**, 253–267.

——, —— & THOMSON, A. R. 1987. Patterns in the deposition of the Brent Group (Middle Jurassic), UK, North Sea. *In:* BROOKS, J. & GLENNIE, K. (eds) *Petroleum Geology of NW Europe*. Graham & Trotman, London, 899–913.

BUDDING, M. C. & INGLIN, H. F. 1981. A reservoir geological model of the Brent Sands in southern Cormorant. *In:* HOBSON, G. D. & ILLING, V. (eds) *Petroleum Geology of the Continental Shelf of NW Europe*. Institute of Petroleum, London, 326–334.

CANNON, S. J. C., GILES, M. R., WHITAKER, M. F., PLEASE, P. M. & MARTIN, S. V. 1992. A regional reassessment of the Brent Group, UK sector, North Sea. *In:* MORTON, A. C., HASZELDINE, R. S., GILES, M. R. & BROWN, S. (eds) *Geology of the Brent Group*. Geological Society, London, Special Publication, **61**, 81–108.

CLIFTON, H. E. 1969. Beach lamination: nature and origin. *Marine Geology*, **7**, 553–559.

CRANE, J., HARTLEY, A. & MAXWELL, G. 1994. Understanding reservoir architecture and production characteristics of the Middle Jurassic Beryl Formation using high resolution sequence stratigraphy (abstract). *In:* JOHNSON, S. D. (ed.) *High Resolution Sequence Stratigraphy: Innovations and Applications*. University of Liverpool, 59–62.

DEEGAN, C. E. & SCULL, B. J. 1977. *A Standard Lithostratigraphic Nomenclature for the Central and Northern North Sea*. Institute of Geological Sciences, Report, 77/36.

ESCHARD, R., TVEITEN, B., DESAUBLIAUX, G., LECOMTE, J. C. & VAN BUCHEM, F. S. P. 1993. High resolution sequence stratigraphy and reservoir prediction of the Brent Group (Tampen Spur Area) using an outcrop analogue (Mesa Verde Group. Colorado). *In:* ESCHARD, R. & DOLIGEZ, B. (eds) *Subsurface Reservoir Characterization from Outcrop Observations*. Editions Technip, Paris, 35–52.

FREY, R. W. & PEMBERTON, S. G. 1987. The *Psilonichnus* ichnocoenose along the Georgia coast. *Bulletin of Canadian Petroleum Geology*, **35**, 333–357.

GRAUE, E., HELLAND-HANSEN, S. G., JOHNSEN, J., LOMO, L., NOTTVEDT, A. *ET AL.* 1987. Advance and retreat of the Brent delta system, northern North Sea. *In:* BROOKS, J. & GLENNIE, K. (eds) *Petroleum Geology of NW Europe*. Graham & Trotman,

London, 915–937.

GREER, S. A. 1975. Estuaries of the Georgia Coast. USA: Sedimentology and Biology. III. Sandbody geometry and sedimentary facies at the estuary–marine transition zone, Ossabow Sound, Georgia, a stratigraphic model. *Senckenbergiana Maritima*, **7**, 105–135.

HAYES, M. O. 1989. *Modern Clastic Depositional Environments.*, 28th International Geological Conference, Field Trip, **T371**, 85.

HELLAND-HANSEN, W., ASHTON, M., LOMO, L. & STEEL, R. 1992. Advance and retreat of the Brent delta: recent contributions to the depositional model. *In:* MORTON, A. C., HASZELDINE, R. S., GILES, M. R. & BROWN, S. (eds) *Geology of the Brent Group*. Geological Society, London, Special Publication, **61**, 109–128.

——, STEEL, R., NAKAYAMA, K. & KENDALL, C. G. St. C. 1989. Review and computer modelling of the Brent Group stratigraphy. *In:* WHATELEY, M. K. G. & PICKERING, K. T. (eds) *Deltas: Sites and Traps for Fossil Fuels*. Geological Society, London, Special Publication, **41**, 237–252.

HINE, A. C. 1975. Bedform distribution and migration patterns on tidal deltas in the Chatham Harbor estuary, Cape Cod Massachusetts. *In:* CRONIN, L. E. (ed.) *Estuarine Research I. Geology & Engineering*. Academic Press, London, 235–252.

HUBBARD, D. K. & BARWIS, J. H. 1976. Discussion of tidal inlet sand deposits: examples from the South Carolina Coast *In:* HAYES, M. O. & KANA, T. W. (eds) *Terrigenous clastic depositional environments: some modern examples*. American Association of Petroleum Geologists, Field Course, University of South Carolina, Technical Report, **11-CRD**, II 128–II 142.

HUNTER, R. E, CLIFTON, H. E. & PHILLIPS, R. L. 1979. Depositional process, sedimentary structures and predicted vertical sequences in barred nearshore systems, southern Oregon coast. *Journal of Sedimentary Petrology*, **49**, 711–726.

IMPERATO, D. P., SEXTON, W. J. & HAYES, M. O. 1988. Stratigraphy and sediment characteristics of a mesotidal ebb-tidal delta. North Edisto inlet, South Carolina. *Journal of Sedimentary Petrology*, **58**, 950–958.

JOHNSON, H. D. & STEWART, D. J. 1985. Role of clastic sedimentology in the exploration and production of oil and gas in the North Sea. *In:* BRENCHLEY, P. J. & WILLIAMS, B. P. J. (eds) *Sedimentology: Recent Developments and Applied Aspects*. Geological Society, London, Special Publication, **18**, 249–310.

KAAS, I., SVANES, T., VAN WAGONER, J. C., HAMAR, G., JORGENVAG, S., SKARNES, S. & SUNDT, O. 1994. The use of high resolution sequence stratigraphy and stochastic modelling to reservoir management of the Ness Formation in the Statfjord Field, Offshore Norway (abstract). *In:* JOHNSON, S. D. (ed.) *High Resolution Sequence Stratigraphy: Innovations and Applications*. University of Liverpool, 57–58.

KANTOROWICZ, J. D., EIGNER, M. R. P., LIVERA, S. E., VAN SCHIJNDEL-GOESTNER, F. S. & HAMILTON,

P. J. 1992. Integration of petroleum engineering studies of producing Brent Group fields to predict reservoir properties in the Pelican Field, UK North Sea. *In:* MORTON, A. C., HASZELDINE, R. S., GILES, M. R. & BROWN, S. (eds) *Geology of the Brent Group*. Geological Society, London, Special Publication, **61**, 453–469.

KUMAR, N. & SANDERS, J. E. 1974. Inlet sequence: a vertical succession of sedimentary structures and textures created by lateral migration of tidal inlets. *Sedimentology*, **21**, 491–532.

LIVERA, S. E. 1989. Facies associations and sand-body geometries in the Ness Formation of the Brent Group, Brent Field. *In:* WHATELEY, M. K. G. & PICKERING, K. T. (eds) *Deltas: Sites and Traps for Fossil Fuels*. Geological Society, London, Special Publication, **41**, 269–286.

—— & CALINE, B. 1990. The sedimentology of the Brent Group in the Cormorant Block IV oilfield. *Journal of Petroleum Geology*, **13**, 367–396.

McCUBBIN, D. G. 1982. Barrier-island and strand plain facies. *In:* SCHOLLE, P. A. & SPEARING, D. (eds) *Sandstone Depositional Environments*. American Association of Petroleum Geologists, Memoir, **31**, 247–280.

MILTON, N. J. 1993. Evolving depositional geometries. *In:* PARKER, J. R. (ed.) *Petroleum Geology of Northwest Europe. Proceedings of the 4th Conference*. Geological Society, London, 425–442.

MITCHENER, B. C., LAWRENCE, D. A., PARTINGTON, M. A., BOWEN, M. B. J. & GLUYAS, J. 1992. Brent Group: sequence stratigraphy and regional implications. *In:* MORTON, A. C., HASZELDINE, R. S., GILES, M. R. & BROWN, S. (eds) *Geology of the Brent Group*. Geological Society, London, Special Publication, **61**, 45–80.

O'BYRNE, C. J. & FLINT, S. 1993. High-resolution sequence stratigraphy of Cretaceous shallow marine sandstones, Book Cliffs outcrops, Utah. USA – application to reservoir modelling. *First Break*, **11**, 45–459

PARTINGTON, M. A., MITCHENER, B. C., MILTON, N. J. & FRASER, A. J. 1993. Genetic sequence stratigraphy for the North Sea Late Jurassic and Early Cretaceous: distribution and prediction of Kimmeridgian–Late Ryazanian reservoirs in the North Sea and adjacent areas. *In:* PARKER, J. R. (ed.) *Petroleum Geology of Northwest Europe. Proceedings of the 4th Conference*. Geological Society, London, 347–370.

PEMBERTON, S. G. & WIGHTMAN, D. M. 1992. Ichnological characteristics of brackish water deposits. *In:* PEMBERTON, S. G. (ed.) *Applications of Ichnology to Petroleum Exploration*. SEPM, Core Workshop, **17**, 141–168.

RICHARDS, P. C. 1992. An introduction to the Brent Group: a literature review. *In:* MORTON, A. C., HASZELDINE, R. S., GILES, M. R. & BROWN, S. (eds) *Geology of the Brent Group*. Geological Society, London, Special Publication, **61**, 15–26.

—— & BROWN, S. 1987. *The nature of the Brent Delta, North Sea: a core workshop*. British Geological Survey, Open File Report, **87/17**.

SCHWARTZ, R. K. 1982. Bedform and stratification

characteristics of some modern small-scale wash-over sand bodies. *Sedimentology*, **29**, 835–849.

SCOTT, E. S. 1992. The palaeoenvironments and dynamics of the Rannoch–Etive nearshore and coastal successions, Brent Group, northern North Sea. *In:* MORTON, A. C., HASZELDINE, R. S., GILES, M. R. & BROWN, S. (eds) *Geology of the Brent Group.* Geological Society, London, Special Publication, **61**, 129–148.

SHIPP, R. C. 1984. Bedforms and depositional sedimentary structures of a barred nearshore system, eastern Long Island, New York. *Marine Geology*, **60**, 235–259.

SHORT, A. D. 1984. Beach and nearshore facies: southeast Australia. *Marine Geology*, **60**, 261–282.

THOMAS, J. M. D. & BIBBY, R. 1993. The depletion of the Rannoch–Etive Sand unit in Brent Sands reservoirs in the North Sea. *In:* LINFIELD, B. *Reservoir Characterisation III.* Pennwell, Tulsa, Oklahoma, 675–713.

VALESEK, D. W. 1990. Compartmentalisation of shoreface sequences in the Cretaceous Gallup sandstones west of Shiprock, New Mexico: Implications for reservoir quality and continuity (abstract). *American Association of Petroleum Geologists Bulletin*, **74**, 784.

VAN WAGONER, J. C., MITCHUM, R. M., CAMPION, K. M. & RAHMANIAN, V. D. 1990. *Siliciclastic Sequence Stratigraphy In Well Logs, Cores and Outcrops: Concepts for High-Resolution Correlation of Time and Facies.* American Association of Petroleum Geologists, Methods In Exploration Series, **7**.

——, JENNETTE, D. C., TSANG, P., HAMAR, G. P. & KAAS, I. 1993. Applications of high-resolution sequence stratigraphy and facies architecture in mapping potential additional hydrocarbon reserves in the Brent Group, Statfjord field (abstract). *Sequence Stratigraphy: Advances and Applications for Exploration and Production in Northwest Europe.* Norwegian Petroleum Society, Meeting, 2.

WEHR, F. L. & BRASHER, L. D. 1996. Impact of sequence-based correlation style on reservoir model behaviour, lower Brent Group, North Cormorant Field, UK North Sea. *This volume.*

WRIGHT, L. D., CHAPPELL, J., THOM, B. G., BRADSHAW, M. P. & COWELL, P. 1979. Morphodynamics of reflective and dissipative beach and inshore systems: southeastern Australia. *Marine Geology*, **32**, 105–140.

Impact of sequence-based correlation style on reservoir model behaviour, lower Brent Group, North Cormorant Field, UK North Sea

F. L. WEHR & L. D. BRASHER

Exxon Production Research Company, PO Box 2189, Houston TX 77001, USA

Abstract: An increasing percentage of remaining reserves in mature Brent Group producing fields is found in the Rannoch Formation, a lower shoreface succession characterized by low permeabilities, abundant mica and stratabound carbonate cements. The high contrast in permeability between the Rannoch and overlying Etive formations has resulted in severe water overrunning of oil-bearing Rannoch sandstones in a number of fields. Remaining Rannoch reserves will be difficult to recover and will require sophisticated reservoir management techniques. Differences in log correlation style which in the past may have been dismissed as inconsequential may now be important to consider in field depletion planning.

The objective of this study was to assess whether, in a high net-to-gross shoreface section such as the lower Brent Group at Cormorant Field, a sequence-based correlation would result in a significant difference in reservoir model behaviour, particularly in oil displacement patterns in lower shoreface sandstones of the Rannoch Formation. The modelled interval includes the prograding shoreline succession of the Rannoch (lower shoreface), Etive (upper shoreface–strandplain) and Ness (coastal plain) formations. There is a significant amount of facies interfingering between the Etive and Rannoch within the modelled area. Permeability models were constructed from both sequence-based and lithostratigraphic well-log correlations and used as input for 2-D reservoir simulations.

The results show a divergence in oil displacement behaviour between the two models, depending upon the degree of permeability anisotropy assumed. The principal reason for the difference lies in how the Rannoch layers are correlated. In the lithostratigraphic model, the Rannoch is treated as a tabular unit with layer boundaries parallel to the formation boundaries. Water injection into the upper Rannoch sweeps the upper Rannoch across the length of the model. In contrast, the sequence-stratigraphic approach treats the Rannoch–Etive as a series of clinoforms with layers offlapping toward the north in the direction of progradation. In this scenario, water injected into the upper Rannoch in the downdip area migrates parallel to layering into the lower Etive in the updip direction. The oil-bearing upper Rannoch in the updip areas correlates downdip to low-permeability facies which do not take water on injection; thus, the upper Rannoch remains unswept.

Predicted recovery efficiencies from the Rannoch lower shoreface sandstones are consistently lower in the sequence stratigraphic model, and more closely match observed reservoir behaviour at Cormorant Field than predictions based on lithostratigraphic correlation. This study suggests that, even in sand-rich systems such as the Rannoch–Etive, a sequence-based approach to reservoir zonation can improve predictions of reservoir performance.

The Brent Group is the most productive oil-bearing succession in western Europe, with total recoverable reserves estimated at over 30 billion oil-equivalent barrels (Morton *et al.* 1992). In most of the larger fields, production rates are currently near peak or in early decline. As these fields mature, remaining reserves are increasingly concentrated in reservoirs of poorer quality, characterized by relatively low permeabilities and/or internal barriers to flow. Economic recovery of these reserves is difficult and requires both a detailed understanding of reservoir architecture as well as sophisticated reservoir management. Geological subtleties that in the past may have been dismissed as inconsequential may now be significant in field depletion planning.

This study was designed to evaluate the effects of sequence stratigraphy-based well-log correlation style on reservoir model behaviour of the lower Brent Group, northern Cormorant Field. Specifically, the models assess the effect of low angle, clinoformal permeability barriers on predicted displacement efficiency of oil by injected water in a sandy shoreface depositional system. The modelled interval includes the prograding shoreface succession of the Rannoch (lower shoreface), Etive (upper shoreface–strandplain) and Ness (coastal plain) formations. Permeability models were constructed from both sequence-based and lithostratigraphic well-log correlations and used as input for two-dimensional fluid-flow simulations. As the study was designed as an analysis of model sensitivity to changes in stratigraphic parameters, production history matching was not a goal.

From Howell, J. A. & Aitken, J. F. (eds), *High Resolution Sequence Stratigraphy: Innovations and Applications*, Geological Society Special Publication No. 104, pp. 115–128.

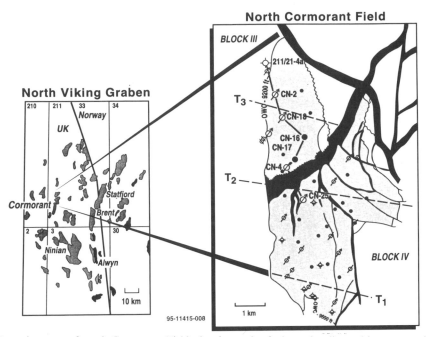

Fig. 1. Location map of north Cormorant Field, showing major faults and wells used in cross-section. Dashed lines T1–T3 show successive positions of the seaward limit of the upper shoreface within geological layers, illustrating the northward progradation of the Rannoch–Etive shoreline through time.

The significance of a sequence-based correlation to reservoir modelling has been demonstrated in the Brent Group elsewhere (e.g. Montmayeur *et al.* 1992; Crane *et al.* 1994; Kaas *et al.* 1994), mainly where the focus was on the distribution of discrete sand bodies such as incised-valley fill deposits. The objective in this study was to assess whether, in a high net-to-gross shoreface section like the Rannoch–Etive at Cormorant Field, a sequence-based correlation could result in a significant difference in reservoir model behaviour, particularly in oil displacement patterns in lower shoreface sandstones of the Rannoch Formation.

Geological setting. Cormorant Field is located 150 km northeast of the Shetland Islands in the western part of the Shetland Basin, UK North Sea (Fig. 1). The field is operated by Shell UK on behalf of the Shell/Esso joint venture. Total recoverable oil and gas reserves of approximately 623 million oil-equivalent barrels are trapped in Brent Group reservoirs (Taylor & Dietvorst 1991). Brent lithostratigraphy at Cormorant is typical of the Brent Group regionally, made up of (from base to top) the Broom, Rannoch, Etive, Ness and Tarbert formations (Deegan & Scull 1977; Budding & Inglin 1981; Livera & Caline 1990). The interval

of interest for this study includes the Rannoch, Etive and lower Ness formations, comprising a generally conformable, progradational succession of wave-dominated shoreface, strandplain and coastal-plain deposits. To the south in Cormorant and locally in surrounding fields, this succession includes fluvial-estuarine incised valleys that partly truncate the prograding shoreface (Jennette & Riley, 1996).

Oil production in most Brent Group fields, including Cormorant, relies on injection of water downdip to provide pressure support and to displace oil updip toward producing wells. The percentage of in-place oil that is ultimately produced (recovery factor) depends largely upon the efficiency of oil displacement by injected water (sweep efficiency), which itself is dependent upon an accurate understanding of the permeability structure of the reservoir.

Geological models

The basis for the study is a seven-well cross-section extending from south to north across the northern fault block (Block III) of Cormorant Field (Fig. 1). The cross-section is approximately 5 km in length and uses the base of the mid-Ness Shale, a prominent intra-Brent flooding surface, as a datum. Five of the seven wells

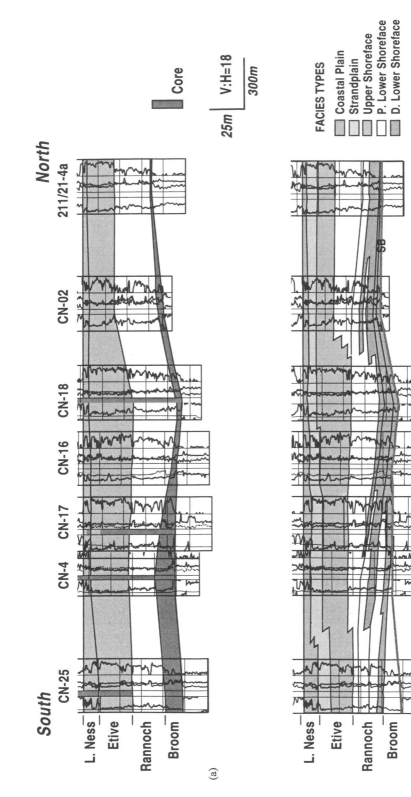
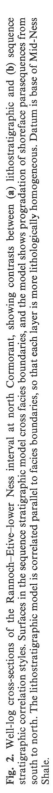

Fig. 2. Well-log cross-sections of the Rannoch–Etive–lower Ness interval at north Cormorant, showing contrasts between (**a**) lithostratigraphic and (**b**) sequence stratigraphic correlation styles. Surfaces in the sequence stratigraphic model cross facies boundaries, and the model shows progradation of shoreface parasequences from south to north. The lithostratigraphic model is correlated parallel to facies boundaries, so that each layer is more lithologically homogeneous. Datum is base of Mid-Ness Shale.

contain core in the modelled interval. Two contrasting correlation styles were used as input for the reservoir models: a lithostratigraphic zonation with model layers parallel to formation boundaries (Fig. 2a), and a sequence-stratigraphic interpretation (Fig. 2b) based on unpublished work by Wehr & Taylor (written comm. 1991) and similar in style to the interpretation of surrounding fields described in the previous paper by Jennette & Riley (1996). The main difference between the two models lies in the treatment of vertical facies changes such as the Rannoch–Etive contact: in the lithostratigraphic model facies changes are treated as model layer boundaries, whereas in the sequence model facies changes occur laterally within model layers. For example, the contact between the Rannoch and Etive formations represents a facies change between proximal lower shoreface and upper shoreface sandstones. This contact is treated as a single layer boundary in the lithostratigraphic model. In the sequence model the change from Etive to Rannoch occurs in successively younger layers from south to north, reflecting the northward progradation of the Rannoch–Etive shoreline system (Fig. 2b).

Permeability model

We have adopted a relatively unsophisticated, mainly deterministic approach to constructing a permeability model. For a study of this type, the well organized and gradational facies changes characteristic of the Rannoch–Etive in this area do not demand a geostatistical approach to permeability modelling. Furthermore, the strongly linear, shoreline-parallel facies trends characteristic of wave-dominated shorefaces such as the Rannoch–Etive lend themselves to a two-dimensional reservoir model.

Horizontal permeability. The reservoir description at the wells was obtained by arithmetically averaging porosity and permeability data from individual core analyses into the appropriate reservoir model layers. Porosity values were obtained from bulk-density curves calibrated to core analysis data. Permeability curves for individual wells were calculated from a linear bulk-density transform ($r^2 = 0.61$) calibrated to core-plug analyses. Permeabilities were then interpolated logarithmically between wells to build a two-dimensional matrix permeability model.

Vertical permeability. Reservoir model behaviour in a sand-rich system is highly sensitive to the treatment of vertical permeability. Un-

fortunately, vertical permeability (k_v) at the reservoir scale is difficult to constrain from subsurface data: in fact, it is sometimes used as a history-match parameter in reservoir simulation, adjusted during simulation runs to optimize the match to observed field behaviour.

Strata-bound, calcite-cemented horizons are abundant in the lower Brent Group. Such horizons in shallow-marine systems have been studied extensively due to their potential as vertical fluid-flow barriers (e.g. Walderhaug *et al.* 1989; Omre *et al.* 1990; Warrender & Spears 1991; Bjorkum & Walderhaug 1990*a, b*). Cemented horizons in the lower Brent most likely represent diagenetically-altered concentrations of shelly debris formed by winnowing of the shoreface during periods of nondeposition (Livera & Caline, 1990; Cannon *et al.* 1992). The lateral extent of calcite cements at Cormorant is not certain but is critical to fluid-flow behaviour. For the purposes of this study, cemented zones at the well bore were correlated between wells where possible, otherwise their lengths were modelled stochastically, using mean lengths of 350 or 700 m, depending on the thickness of the horizon at the wellbore. These lengths are considered a conservative estimate, based on published analogue studies (Bjorkum & Walderhaug, 1990*a*).

Other sources of permeability anisotropy in the Rannoch–Etive interval include internal lamination within beds, and highly micaceous or mud-rich layers at bed boundaries. Micaceous layers and muddy laminae tend to be concentrated at bedset boundaries such as flooding surfaces, and are typically more abundant in the lower shoreface (Rannoch) than in the upper shoreface (Etive).

In the absence of any hard data on the cumulative effect of these features on vertical permeability, a heirarchical model was assumed, based on qualitative core observations. Vertical permeabilities (k_v) at model layer boundaries were assumed to be 0.1 times the horizontal permeability (k_h) within geologic intervals, 0.01 md at the boundaries of geological intervals within the Etive reservoir, and 0.001 md at the boundaries of geologic intervals within the Rannoch. Calcite-cements are treated as vertical permeability barriers with zero transmissibility.

Reservoir models

Reservoir models and flow simulations were constructed using Exxon's proprietary reservoir simulation software (MARS). Lithostratigraphic and sequence stratigraphic models were constructed using the five central wells from the

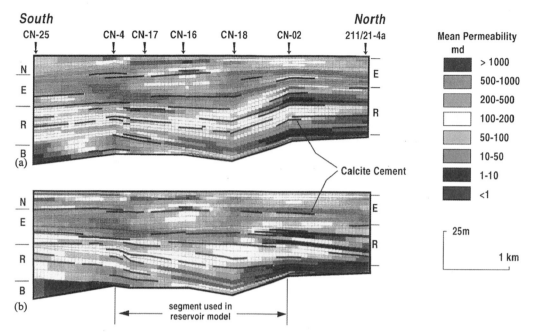

Fig. 3. Horizontal permeability models based on (a) lithostratigraphic and (b) sequence stratigraphic correlations. Note that the south–north progradation apparent on Fig. 2 can be seen in the permeability model. Only the central five wells were used to build the reservoir simulation.

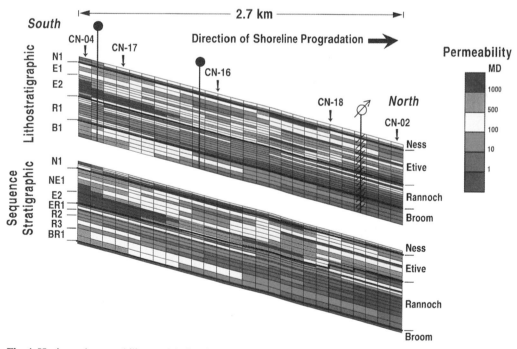

Fig. 4. Horizontal permeability models for the two reservoir simulations showing simulated well locations, north-dipping model. Note that due to graphical limitations of plotting software, layer thicknesses in the reservoir model plots appear to be fixed. Geological layer boundaries labelled on left (see Appendix), formations are indicated on right.

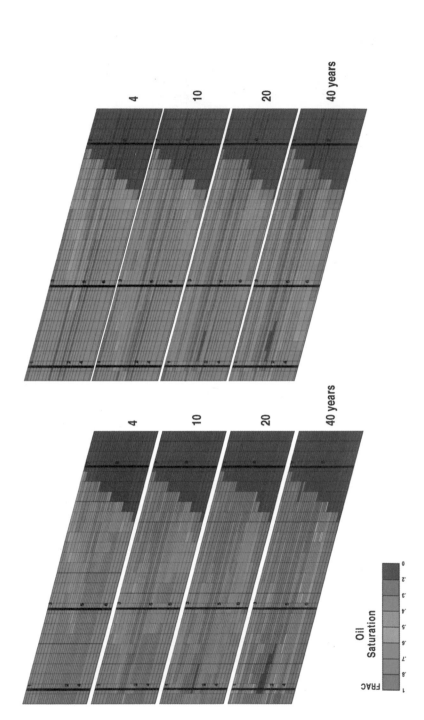

Fig. 5. Oil saturation plots through time at 4, 10, 20 and 40 years from start of simulated production for lithostratigraphic (left) and sequence stratigraphic (right) north-dipping models. Initial displacement patterns are similar, but the upper Rannoch Formation in the sequence model remains relatively unswept. See text for detailed discussion.

cross-section, with a total model length of 2.7 km. Correlations from the lithostratigraphic and sequence-stratigraphic cross-sections were subdivided into MARS cross-sectional model layers. Each MARS model was designed with 30 vertically stacked layers and 27 blocks in the horizontal direction, for a total of 810 grid blocks. The total length of each model was 2.7 km, with each grid block 100 m in the horizontal direction. The average layer thickness was about 2 m. Layers were grouped into geological intervals for the purpose of crossflow analysis (see Appendix).

Horizontal permeabilities for the two reservoir simulation runs are shown on the colour-coded plots (Fig. 4). Note that due to graphical limitations of the plotting software, layer thicknesses in the reservoir model plots are fixed: the thinning and downlap shown on the sequence stratigraphic well-log section and permeability plots (Figs 2 and 3) are not apparent. This does not affect the results but can be misleading when visually comparing these models.

To simulate the dip of a typical Brent producing field, the first set of models were tilted to a constant 8° toward the north from the oil column to the aquifer, in the direction of shoreline progradation. The oil–water contact was placed at a depth of 2720 m (9200 ft) subsea, between the CN2 and CN18 wells. Water was injected into the aquifer to displace oil toward two rows of producing wells. The displaceable oil column was approximately 2.0 km in length, and the two rows of producing wells are roughly equivalent to 160 acre well spacing.

Well management strategy

The well management program was designed to simulate a reasonable operating strategy, and to ensure consistency between cases. Specified well potentials were used for both water injection and producing wells. Water-injection wells were fixed to inject the amount of water necessary to maintain a constant potential of 5100 lb/in^2 (PSI). Producing wells for each model were specified to produce a total liquid rate of 8000 barrels/day (equivalent to a 10% per year depletion rate) during the first two years, and thereafter set to produce the volumes necessary (with limits on maximum liquid rate) to maintain a specified potential in the producing wells. In effect, all 30 layers in the reservoir models are perforated at each well location; however, for well management purposes, each location is treated as three separate wells (one in the Etive and two in the Rannoch), so that each model is built containing six producing wells. When the

water cut in a well reached 90%, that well was automatically shut in all layers producing above a 90% water cut. Forty years of production history were simulated.

Reservoir simulation results: north-dipping models

Plots of oil saturation as a function of time (Fig. 5) are used to demonstrate the oil displacement process in the key reservoir simulation cases. Invariably, these show rapid movement of the water front through the high permeability intervals of the Etive reservoir and much slower displacement of oil from the Rannoch. There appears to be good displacement of oil from all intervals that are effectively swept. This would be consistent with the favourable end-point mobility ratio of approximately 1.0, and the generally favourable permeability distribution with higher permeability intervals overlying intervals of lower permeability.

Lithostratigraphic model. The lithostratigraphic model is a relatively simple correlation with high continuity through the model layers. This simple description and high continuity result in efficient sweep and a high predicted recovery factor, with little variation in predicted recovery factors among the different geological layers. The simulator-predicted total recovery factor for the lithostratigraphic model was 55.7% of

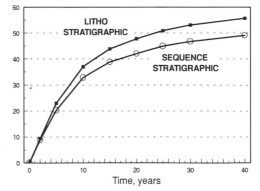

Recovery, % OOIP

Fig. 6. Comparison of recovery performance, lithostratigraphic and sequence stratigraphic reservoir simulations, north-dipping models. The sequence model predicts a recovery factor of 49.1% after 40 years, 6% lower than the lithostratigraphic model. Most of the difference between the two models can be accounted for by the amount of unswept oil in the upper Rannoch Formation (Fig. 7).

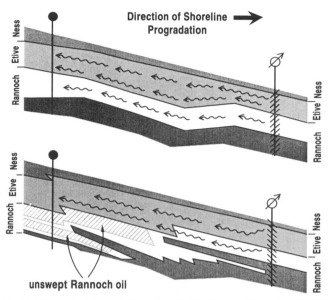

Fig. 7. Schematic interpretation of north-dipping model results. The lithostratigraphic model (top) predicts more efficient sweep of the Rannoch in part due to the lack of interfingering between the Rannoch and Etive. In the sequence model (bottom), injected water migrates parallel to layering from the upper Rannoch into the Etive, leaving unswept oil in the Rannoch updip.

original oil in place (OOIP). Recovery factors for the five geological intervals in the lithostratigraphic correlation varied over a narrow range from 50.4–60.3% of OOIP (see Appendix for summary of crossflow parameters). Distribution of water injection shows more than 61% of the total water injected into the wellbore was taken by the lower Etive (interval E2).

Sequence stratigraphic model. The sequence stratigraphic model differs from the lithostratigraphic model in the south-to-north clinoforming and interfingering of the lower Etive and Rannoch layers. Combined with the high degree of permeability anisotropy built into both models, the net effect is to decrease the recovery efficiencies of certain layers in the sequence stratigraphic model, leaving a pocket of poorly-swept oil in the updip Rannoch (Fig. 5).

The overall recovery factor predicted for the sequence stratigraphic model was 49.1% of OOIP. A more realistic range of recovery factors (19–59% of OOIP) was predicted for the seven major geological units in the sequence stratigraphic correlation. This result is more consistent with actual field performance, where displacement of oil from the Rannoch is slow and difficult. Over 50% of total water injected was taken at the wellbore by the Ness–Etive geological interval (NE1) with significant flow from these layers into adjacent geological units

within the model. The bottom three geological intervals in the Rannoch and Broom (R2, R3, RB1) took very little injected water at the wellbore, but water was able to make its way into the upper Rannoch interval R2 from the overlying Etive. Crossflows into the two bottom geological intervals were negligible, indicating that injection water was moving sequentially down through the lower geological layers. Both of the lower geological intervals R3 and RB1 had very low water–oil production ratios, confirming that very little water was injected.

Comparison of simulations. Figure 6 compares recovery performance of the two models over the 40-year simulation, measured as a percentage of OOIP. As outlined above, the predicted recovery factor for the sequence model is 6% lower than the lithostratigraphic model, largely because of a greater amount of unswept oil in the upper Rannoch. The degree to which the simulations differ is dependent upon the amount of permeability anisotropy built into the geological models: as it becomes more difficult for fluids to migrate across layer boundaries, the two models diverge. This is not surprising, as the only difference between the models is the geometry of the layers: if they are modelled as transparent to fluid flow, the models are indistinguishable.

A main reason for the difference in model

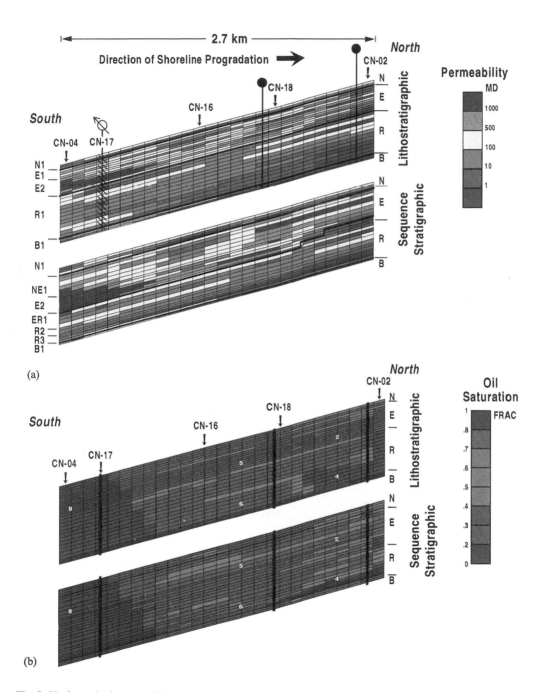

Fig. 8. Horizontal (**a**) permeability models and (**b**) oil saturations after 40 years (bottom) for the south-dipping reservoir simulations. In these models, water injection is in the more proximal shoreface facies (left), sweep is down depositional dip. Final displacement patterns in this set of models show less difference between the two models than in the cases with progradation in the direction of structural dip (Fig. 5). Geological layer boundaries labelled on left side (see Appendix), formations are indicated on right.

Recovery, % OOIP

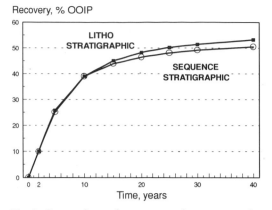

Fig. 9. Comparison of recovery performance, south-dipping lithostratigraphic and sequence stratigraphic reservoir simulations. Overall recovery factors are lower than in the previous models, and there is less difference between the two correlation styles.

results lies in how the Rannoch layers are correlated (Fig. 7). In both models, the high-permeability lower-Etive layers are swept early in the flow simulation, resulting in water over-running the oil-bearing layers in the Rannoch. However, in the lithostratigraphic model, the Rannoch is treated as a tabular unit with layer boundaries parallel to the formation boundaries. Injection into the upper Rannoch sweeps the upper Rannoch across the length of the model. In contrast, the sequence model treats the

Rannoch–Etive as a series of clinoforms that offlap toward the north in the direction of progradation (right). In this scenario, water injected into the upper Rannoch to the north migrates within the same layer into the lower Etive to the south, effectively decreasing the sweep efficiency in the Rannoch. The time-equivalent layers to the oil-bearing upper Rannoch updip (left) are low-permeability, lower-Rannoch facies at the injection well in the aquifer. These layers take on very little water during injection and are ineffective as conduits for sweep.

The process of gravity segregation can have a mitigating effect on the unswept oil in the upper Rannoch. Through time, water percolating downward from the Etive will displace oil upward from the unswept Rannoch, driven by the density contrast between oil and water. However, as the rate of gravity segregation is also dependent on the vertical permeability in the model, it is insignificant in these simulations.

Reservoir simulation results: south-dipping models

In the initial simulations, the direction of shore-line progradation was to the north and down depositional dip: that is, oil was produced from the higher permeability, more proximal facies to the south, and water was injected into the lower permeability, more distal facies to the north. As

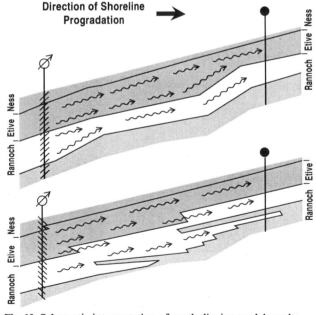

Direction of Shoreline Progradation

Lithostratigraphic

- 53% recovery efficiency
- Decrease due mainly to lower permeability in Rannoch at production well

Sequence Stratigraphic

- 50% recovery efficiency
- Increase due to more efficient sweep of Rannoch

Fig. 10. Schematic interpretation of south-dipping model results.

Fig. 11. Photomontage of well-exposed shoreface deposits from the lower Straight Cliffs Formation, southwestern Utah. (a) Original photo showing left to right progradation of shoreface, marked by resistant strata-bound carbonate cemented horizons. (b) Vertically exaggerated photo with prominent cement horizons traced and showing typical spacing of Brent injector-producer pair at Cormorant Field. Photos courtesy G. J. Moir.

a sensitivity test, the model dip was reversed so that the higher permeabilities were down-structure, rather than up-structure (Fig. 8). The objective of this sensitivity was to estimate the differences in predicted performance between the two models in a field where structural dip was the opposite of north Cormorant (for example, Tern Field).

The effect of dip reversal was to decrease the difference between the two correlation styles (Figs 9–10). In the south-dipping lithostratigraphic model, the overall oil recovery factor was reduced slightly from the original model, from 55.7–53.2% of OOIP. The predicted range of recovery factors for the five major geological layers was still relatively narrow, from 44.0–60.9% of OOIP. There was also a significant reduction in the share of total oil production from the up-structure row of wells, from 70.3–59.5% of total oil production. In the reversed sequence stratigraphic model, overall oil recovery factor increased slightly, from 49.1–50.5% of OOIP. The predicted range of recovery factors for the seven major geological intervals was from 27.6–55.7% of OOIP. There was a small decrease in the share of production from the up-structure wells, from 65.5–61.8% of total oil production. The share of total production from the Etive wells decreased from 74.3–56.2% of total oil production.

Relative to the original models, the structu-

rally reversed lithostratigraphic and sequence models behave more alike, borne out by comparisons of recovery performances (Figs 9–10). Overall recovery efficiencies are slightly lower, probably because the total permeability-thickness at the producer locations is lower, particularly in the Rannoch. Rannoch recovery efficiencies from the sequence model are higher in both models, due mainly to the increased ability to inject water downdip. Note that the area of high oil saturation at the updip end of both models (Fig. 8b) after 40 years corresponds to unrecoverable oil in low-permeability, lower Rannoch facies at CN-02 and does not represent a significant unswept reserve.

Discussion

The single most important cause of uncertainty in this study is the degree of permeability anisotropy in the system: initial runs with uniform $k_v : k_h$ values of $1:10$ showed no difference between the two models. Given the abundance of micaceous laminae and strata-bound calcite layers, and given the more realistic recovery factors resulting from the sequence model, we believe that the k_v values used in these models are geologically reasonable. However, it should be understood that these values are not based upon direct measurements from core analysis or engineering data at Cormorant:

reliable data on the effective permeability of intra-reservoir layering are extremely difficult to obtain from either subsurface or outcrop analogue studies.

Figure 11 is a photomosaic of an outcrop of Upper Cretaceous shallow-marine sandstones from southwestern Utah (Lower Straight Cliffs Formation: Moir 1974), showing a possible outcrop analogue for the Rannoch–Etive sequence model and illustrating the potential for highly layered reservoirs in a shoreface setting. The direction of shoreline progradation is from left to right. With no vertical exaggeration (Fig. 11a) a series of more resistant beds downlapping from left to right are clearly visible. These beds are carbonate cemented horizons, similar in morphology to those correlated in the Rannoch–Etive at Cormorant Field and elsewhere (Livera & Caline 1990; Jennette & Riley 1996). Vertical exaggeration and tracing of the cement horizons (Fig. 11b) highlights the downlapping geometry and the continuity of the beds. An injector–producer pair at typical Cormorant well spacing is also shown. When combined with other bed-parallel permeability baffles such as flooding surfaces and micaceous layers, it seems intuitive that such layering will play a role in controlling fluid flow patterns.

In summary, this study has shown that significant differences in model behaviour result from varying the style of well-log correlation, even within the sand-rich section of the Rannoch–Etive. Predicted recovery efficiencies from the Rannoch lower shoreface sandstones are lower in the sequence stratigraphic model, and more closely match observed reservoir behaviour at Cormorant Field. Reversing the dip of the model narrowed the difference in model behaviour but did not eliminate it. The results suggest that, even in sand-rich systems such as the Rannoch–Etive, a sequence-based approach to reservoir zonation could provide a significant improvement in reservoir management over the producing life of a field.

This study was funded as part of a collaborative research effort on North Sea reservoirs (NORTEC) by Esso Exploration and Production UK Ltd, Esso Norway as, and Exxon Production Research Company. Permission to publish from Esso and Shell UK is gratefully acknowledged. Earlier versions of this report were reviewed by Dave Jennette, Tom Jones and Dave Larue of Exxon Production Research Company. Careful reviews by Ian Bryant and Steve Flint substantially improved the final manuscript.

Appendix

Crossflow analysis

Simulation case results were analysed from the performance of the major geological intervals in the respective models. Crossflow is a measure of the net oil and water movement across the boundaries of the selected geological units. A positive crossflow indicates net flow of oil or water into a geological unit, and a negative crossflow indicates net flow out of the unit. Crossflow was not permitted within the wellbore. This appendix (Tables 1–4) contains a summary of the results of the reservoir simulation cases, both for the major geological units and for the models as a whole. The following conventions are used in the crossflow analysis tables:

oil production reflects the total oil actually produced from model layers and includes the effects of oil crossflow;

the water–oil ratio reflects cumulative water and oil production during the 40 year simulation period;

percentage of cumulative water injection column is the percentage of total water injected into a specific geological unit, and highlights the geological units taking proportionally large amounts of injection water at the wellbore.

The oil recovery column is based upon a comparison of the OOIP and OIP (oil in place) after 40 years of simulation. For these tables, oil recovery as a percent of OOIP is equal to (Oil Produced − Oil Crossflow) × 100/OOIP.

Table 1. *Crossflow summary by geological interval, lithostratigraphic model*

Fm.	Geol. int.	Layer	OOIP, MSTB	Oil Prod., MSTB	W–O Ratio	%Cumul. W. I.	Crossflow, Oil, MSTB	Crossflow, Water, MSTB	Oil Rec., % OOIP
Ness	N1	1–2	4.1	1.8	1.8	4.8	−0.4	2.3	53.8
U. Etive	E1	3–5	7.5	5.1	0.8	6.8	0.7	4.3	59.5
L. Etive	E2	6–14	22.8	12.7	1.2	61.1	−1.0	−13.4	60.3
Rannoch	R1	15–26	19.6	8.8	1.2	26.3	−1.0	3.0	50.4
Broom	B1	27–30	5.1	4.4	0.4	1.0	1.8	4.0	51.5
Total			**59.0**	**32.9**	**1.1**	**100.0**	**0.0**	**0.0**	**55.7**

Table 2. *Crossflow summary by geological interval, sequence stratigraphic model*

Fm.	Geol. int.	Layer	OOIP, MSTB	Oil Prod., MSTB	W–O Ratio	% Cumul. W. I.	Crossflow, Oil, MSTB	Crossflow, Water, MSTB	Oil Rec., %OOIP
Ness	N1	1–4	4.8	2.0	1.3	4.5	−0.4	2.6	50.1
NessEtiv	NE1	5–10	18.7	11.3	1.1	50.0	0.7	−6.6	57.0
EtivUs	E2	11–17	11.0	5.9	1.3	21.0	−0.6	2.0	59.1
Etiv/Rann1	ER1	18–20	6.1	2.4	1.4	17.4	−0.7	−4.1	51.2
Rann2	R2	21–24	8.8	4.8	0.6	3.6	0.9	5.1	43.7
Rann3	R3	25–27	4.9	0.9	0.0	0.2	0.0	0.9	18.8
RannBR	RB1	28–30	5.1	1.9	0.1	3.3	0.2	0.0	33.1
Total			**59.4**	**29.1**	**1.0**	**100.0**	**0.0**	**0.0**	**49.1**

Table 3. *Crossflow summary by geological interval, lithostratigraphic model (reversed)*

Fm.	Geol. int.	Layer	OOIP, MSTB	Oil Prod., MSTB	W–O Ratio	% Cumul. W. I.	Crossflow, Oil, MSTB	Crossflow, Water, MSTB	Oil Rec., %OOIP
Ness	N1	1–2	3.8	3.6	1.7	3.6	1.6	6.0	52.8
U. Etive	E1	3–5	6.5	2.7	0.5	12.1	−1.0	−2.9	57.0
L. Etive	E2	6–14	21.4	11.9	1.5	55.9	−1.1	−7.16	0.9
Rannoch	R1	15–26	20.1	10.0	1.2	23.2	1.2	6.2	44.0
Broom	B1	27–30	2.5	0.6	0.1	5.4	−0.6	−2.3	52.4
Total			**54.3**	**28.9**	**1.3**	**100.0**	**0.0**	**0.0**	**53.2**

Table 4. *Crossflow summary by geological interval, sequence stratigraphic model (reversed)*

Fm.	Geol. int.	Layer	OOIP, MSTB	Oil Prod., MSTB	W–O Ratio	% Cumul. W. I.	Crossflow, Oil, MSTB	Crossflow, Water, MSTB	Oil Rec., %OOIP
Ness	N1	1–4	4.7	2.2	1.8	5.5	−0.1	3.5	49.8
NessEtiv	NE1	5–10	17.3	11.2	0.9	33.8	1.6	1.3	55.3
EtivUs	E2	11–17	10.5	3.0	1.2	37.7	−2.8	−11.8	55.7
Etiv/Rann1	ER1	18–20	6.7	5.4	1.4	9.7	1.9	5.7	51.1
Rann2	R2	21–24	8.1	2.8	0.5	6.9	−0.7	1.5	43.1
Rann3	R3	25–27	3.1	1.0	0.2	1.1	0.1	0.6	27.6
RannBR	RB1	28–30	2.4	1.1	0.9	5.3	0.0	−0.8	47.1
Total			**52.9**	**26.7**	**1.0**	**100.0**	**0.0**	**0.0**	**50.5**

References

BJORKUM, P. A. & WALDERHAUG, O. 1990a. Lateral extent of calcite-cemented zones in shallow marine sandstones. *In:* BULLER, A. T., BERG, E., HJELMELAND, O., KLEPPE, J., TORSAETER, O. & AASEN, J. O. (eds) *North Sea Oil and Gas Reservoirs – II, Proceedings of the North Sea oil and gas reservoirs conference.* Graham and Trotman, London, 331–336.

—— 1990b. Geometrical arrangement of calcite cementation within shallow marine sandstones. *Earth Science Reviews*, **29**, 145–161.

BUDDING, M. C. & INGLIN, H. F. 1981. A reservoir

geological model of the Brent sands in Southern Cormorant: *In:* ILLING, L. V. & HOBSON, G. D. (eds) *Petroleum Geology of the Continental Shelf of North-West Europe.* Heyden, London, 326–334.

CANNON, S. J. C., GILES, M. R., WHITAKER, M. F., PLEASE, P. M. & MARTIN, S. V. 1992. A regional reassessment of the Brent Group, UK sector, North Sea: *In:* MORTON, A. C., HASZELDINE, R. S., GILES, M. R. & BROWN, S. (eds) *Geology of the Brent Group.* Geological Society, London, Special Publication, **61**, 81–107.

CRANE, J., HARTLEY, A. & MAXWELL, G. 1994. Understanding reservoir architecture and production characteristics of the Middle Jurassic Beryl Formation using high-resolution sequence stratigraphy. *In:* JOHNSON, S. D. (ed.) *High Resolution Sequence Stratigraphy: Innovations and Applications.* University of Liverpool, 59–62.

DEEGAN, C. E. & SCULL, B. J. 1977 *A standard lithostratigraphic nomenclature for the central and northern North Sea.* Report for the Institute of Geological Sciences, **77**/25.

JENNETTE, D. C. & RILEY, C. O. 1996 Influence of relative sea-level on facies and reservoir geometry of the Middle Jurassic lower Brent Group, UK North Viking Graben. *This volume.*

KAAS, I., SVANES, T., VAN WAGONER, J. C., HAMAR, G., JORGENVEG, S., SKARNES, P. I. & SUNDT, O. 1994. The use of high-resolution sequence stratigraphy and stochastic modelling to reservoir management of the Ness Formation in the Statfjord Field, offshore Norway. *In:* JOHNSON, S. D. (ed.) *High Resolution Sequence Stratigraphy: Innovations and Applications.* University of Liverpool, 57–58.

LIVERA, S. E. & CALINE, B. 1990 The sedimentology of the Brent Group in the Cormorant Block IV oilfield. *Journal of Petroleum Geology,* **13**, 367–396.

MOIR, G. J. 1974. *Depositional environments and stratigraphy of the Cretaceous Rocks, southwestern Utah.* PhD Thesis, University of California, Los Angeles.

MONTMAYEUR, H., RAE, S. F. & COOMBES, T. P. 1992. *Alwyn North Brent East panel: detailed reservoir analysis to improve reservoir management and recovery.* SPE paper 25009, 477–491.

MORTON, A. C., HASZELDINE, R. S., GILES, M. R. & BROWN, S. 1992. Geology of the Brent Group: Introduction. *In:* MORTON, A. C., HAZELDINE, R. S., GILES, M. R. & BROWN, S. (eds) *Geology of the Brent Group.* Geological Society, London, Special Publication, **61**, 1–2.

OMRE, H., SOLNA, K., TJELMELAND, H., CLAESSON, L. & HOLTER, C. 1990. *Calcite cementation: description and production consequences.* SPE paper 20607, 811–823.

TAYLOR, D. J. & DIETVORST, J. P. A. 1991. The Cormorant Field, Blocks 211/21a, 211/26a, UK North Sea. *In:* ABBOTTS, I. L. (ed.) *United Kingdom Oil and Gas Fields, 25 Year Commemorative Volume.* Geological Society, London, Memoir, **14**, 73–81.

WALDERHAUG, O., BJORKUM, P. A. & BOLAS, N. M. N. 1989. Correlation of calcite-cemented layers in shallow-marine sandstones of the Fensfjord Formation in the Brage Field: *In:* *Correlation in Hydrocarbon Exploration.* Graham and Trotman, London, 367–375.

WARRENDER, J. M. & SPEARS, A. 1991 Calcite cemented layers, their characterisation and use in improving reservoir recovery from Murchison Field, northern North Sea. *American Association of Petroleum Geologists Bulletin,* **75**, 1424.

A model for high resolution sequence stratigraphy within extensional basins

JOHN A. HOWELL & STEPHEN S. FLINT

STRAT Group, Department of Earth Sciences, University of Liverpool, PO Box 147, Liverpool L69 3BX, UK

Abstract: Existing sequence stratigraphic models imply that for any given time period in a sedimentary basin, the interplay between rates of eustatic sea-level change and subsidence will dictate the type of sequence boundary formed (type-1 boundary if rate of sea-level fall exceeds subsidence and type-2 if it does not). Basin margin profile will dictate whether lowstand fluvial systems feed submarine fans (shelf-slope break profile) or detached shorelines (forced regressions on ramp margin). A conceptual model presented here, proposes that in structurally active extensional domains the expressions of eustatic sea-level change are still present in the stratigraphic record. These expressions are modified by complex basin margin topography and by significant, along strike variations in subsidence and accommodation creation. This model also illustrates that tectonically complex basins can host a spectrum of time equivalent lowstand deposits previously considered to be spatially and mutually exclusive. This model is illustrated by an example from the Oxfordian of the UK North Sea Central Graben.

The concepts of sequence stratigraphy provide a framework within which clastic depositional responses to abrupt and gradual changes in accommodation space across a smooth shelf profile can be predicted (Posamentier & Vail 1988). Within this general model, basin margins exhibit either a shelf-slope break or a ramp profile. Sea-level fall across a shelf-slope break results in deep water turbidite deposition whilst ramp margins host lowstand shorefaces produced during forced regression (Posamentier *et al.* 1992). In both cases, incised valleys and interfluves form on the sub-aerially exposed shelf although they may not always be preserved (Posamentier & Allen 1993*a*). Within either basin margin type, if the rate of sea-level fall is slower than the rate of subsidence the shelf is not exposed and type-2 sequence boundaries are formed (Posamentier *et al.* 1988).

Within this paper the application of sequence stratigraphic concepts to extensional rift basins is considered. A theoretical model, combining the extensional fault geometry models of Walsh & Watterson (1988) with the accommodation models of Jervey (1988) is presented. Within this model a variety of accommodation regimes coexists across a complex, structurally derived, topographic template. These accommodation regimes reflect the interplay between changes in eustatic sea-level and subsidence patterns across the basin and produce varied but predictable stratigraphic expression.

Posamentier & Allen (1993*a, b*) discuss both basin physiography and subsidence profiles as local, controlling factors upon sequence architecture. When studying rift basins, it is necessary to develop further these concepts and consider subsidence variations in a depositional strike as well as dip direction and to consider more complex basin physiography than ramp v. shelf slope-break profiles. Prosser (1993) discussed rift basin fills within the context of four tectonic systems tracts equating to early-, syn- and two stages of post-rift deposition. Rift basin evolution was cited as the principal control upon sequence architecture. It is proposed herein that within these stages of basin filling, the depositional products of high frequency eustatic fluctuations in sea-level (third order and above, Van Wagoner *et al.* 1990) may still be recognized in the rock record. These signatures will vary from those originally described from passive margins (Posamentier *et al.* 1988). Recent work within the Gulf of Corinth (Gawthorpe *et al.* 1994), a recent extensional basin, has illustrated along strike changes in subsidence rates similar to those proposed here. These authors were able to quantify rates of fault movement and compare them to known Quaternary sea-level changes. Their findings support the model below, although the physiography of the Corinth faults are somewhat larger than those considered here and consequently the shallow marine shoreface systems included within the model are virtually absent in the Greek examples. Our model also considers the presence of a ductile sub-crop (such as the Zechstein salt in the North Sea) above which the creation of accom-

From Howell, J. A. & Aitken, J. F. (eds), *High Resolution Sequence Stratigraphy: Innovations and Applications,*
Geological Society Special Publication No. 104, pp. 129–137.

129

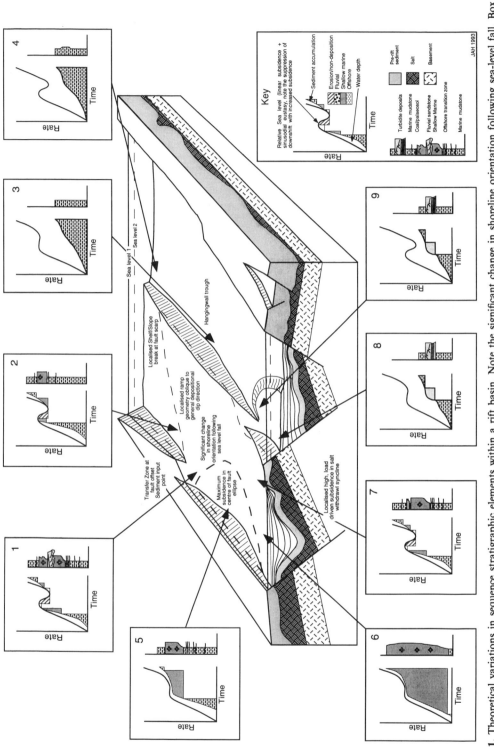

Fig. 1. Theoretical variations in sequence stratigraphic elements within a rift basin. Note the significant change in shoreline orientation following sea-level fall. Box numbers referred to in text, relative sea-level curves and idealized vertical sections adapted from the concepts of Jervey (1988), fault geometries from Walsh & Watterson (1988).

modation space may be load driven.

An application of the model is illustrated by an example from the Upper Jurassic of the UK North Sea Central Graben basin. The example illustrates the varied expression of a series of lowstand deposits, interpreted to lie above a single sequence boundary. A full discussion of these successions is detailed in Howell *et al.* (1995).

Sequence boundaries and lowstand sediments in rift basins

When considering the application of sequence stratigraphic concepts to extensional rift basins, two factors must be considered. Firstly, the topography of the basin will be extremely complex, unlike the smooth shelf profile of the theoretical model (Posamentier *et al.* 1988) and secondly, differential subsidence along depositional strike will modulate variations in accommodation which have previously been treated as constants.

Variations in the topography of the basin will dictate whether, following a base level fall, incised valley systems feed either lowstand shorelines or submarine fan systems (Posamentier *et al.* 1988; 1992). Within a rift setting (Fig. 1), if relative sea-level fall exposes the footwall block, lowstand fan deposits may be deposited within the deeper water, hangingwall trough segment. If the rate of fluvial incision matches the rate of sea-level fall, maximum incision would be anticipated on the crest of uplifted footwall blocks. However if the rate of sea-level fall were greater than the rate of incision, fluvial systems would be channelized through structural lows (transfer zones) and would exhibit comparable distribution to those observed in recent and modern rift basins (e.g. Leeder & Gawthorpe 1987). Elsewhere in the basin, deposition on gently sloping hangingwall blocks (simulating a ramp setting), would be characterized by time-equivalent lowstand shoreline deposits (forced regressions). Structurally generated complex topography may provide a tortuous route for fluvial systems supplying sediment to these lowstand shorelines and may also result in oblique shoreline orientations with respect to earlier depositional dip directions (Fig. 1). This will contrast with incised valley profiles on smooth shelf profiles which are typically straight (Van Wagoner *et al.* 1990).

Significant along strike variations in subsidence rates (and consequently accommodation creation) within different parts of the rift basin will also greatly affect the expression of a single

base level fall (Fig. 1). These variations are comparable to those modelled in a depositional dip direction away from a tectonic hinge point by Jervey (1988). Strike-variable accommodation will be caused by differing rates of fault propagation and between individual offset faults which bound the basin. An additional complication may be the presence of underlying, ductile salt. Localized salt withdrawal provides areas of very high subsidence as a direct response to sediment loading (within rim synclines; Johnson *et al.* 1986).

Jervey (1988) considered the effects of varying rates of subsidence upon a sinusoidal fluctuating eustatic sea-level. An increased subsidence rate results in progressively less relative sea-level fall associated with the falling limb of the eustatic curve, as more of the previously deposited sediment subsides below the level of potential erosion (Jervey 1988; Fig. 1). These variations in the model were implied to occur within different basins but these observations are equally valid when considering along-strike variations within a single extensional basin (Fig. 1).

In areas of slow to moderate subsidence (comparable to passive margins), the rate of eustatic sea-level fall may exceed the rate of subsidence and sub-aerial exposure occurs on the shelf (Fig. 1, cases 1, 2 and 7). In case 1, fluvial input through the adjacent transfer zone results in the cutting of incised valleys which are filled during subsequent transgression by fluvial and tidal deposits (Van Wagoner *et al.* 1990). Adjacent to the transfer zone (case 2), a ramp geometry exists towards the centre of the fault ellipse. Lowstand deposits are represented by sharp based shoreface sandstones (Posamentier *et al.* 1992) overlying offshore pelagic deposits of the previous highstand. Towards the centre of the basin, uplifted footwall blocks may be exposed (case 7); in such cases the lack of established fluvial drainage systems will preclude valley incision. Wave ravinement may remove some deposits during both sea-level fall and rise; however, the main lowstand deposits, if preserved, will be well drained soil horizons, similar to the interfluve sequence boundaries described by O'Byrne & Flint (1996).

Within the deeper parts of the basin where the major control upon the expression of sequences is largely sediment supply (controlled by relative sea-level at the basin margin and shelf topography), eustatic falls within deep marine grabens adjacent to moderately subsiding shelfal areas are expressed as submarine fans (Fig. 1, case 8). These are comparable to the lowstand fans of Posamentier *et al.* (1988), where sea-level falls beyond a shelf-slope break. In deeper water areas away from exposed shelves the stratigraphic expression of a drop in sea-level may be either absent (due to paucity of sand supply) or observed as a transition to less organic rich claystones (Fig. 1, case 4; cf. Price *et al.* 1993).

In areas where rapid subsidence exceeds the rate of sea-level fall (i.e. no sub-aerial exposure of the shelf occurs) a type-2 sequence boundary will be formed (Fig. 1, cases 5 and 6; Posamentier *et al.* 1988). Although type-2 sequences are rare on slowly subsiding passive margins, within a rift setting, ample opportunity exists for the rate of subsidence locally to exceed the rate of sea-level fall. In such cases the sequence boundary is typically picked at the base of an aggradational unit, the shelf margin systems tract, between the highstand and transgressive systems tracts (Posamentier & Vail 1988). However, the tectonic suppression of a relative sea-level fall (Fig. 1, case 5; Jervey 1988; Mitchum & Van Wagoner 1991) results in an increased sequence asymmetry with increasing subsidence rate. The extreme case occurs where the progradational component of the highstand systems tract is capped by an aggradational shelf margin system tract and subsequently the combination of rapid subsidence and rising base level produces a completely sediment starved, transgressive systems tract. The resultant vertical profile is a shallowing upward succession which, in a single vertical profile may be more akin to a thick parasequence than a sequence (Fig. 1, case 5).

Where the high subsidence is driven by sediment flux (e.g. sediment loading of salt within rim synclines), accommodation space creation will keep pace with sediment supply and aggradation will occur during the period of falling sea-level. The subsequent transgressive system tract will be retrogradational (Fig. 1, case 6) and the geometry of the sequence will resemble the predicted model of Posamentier *et al.* (1988).

Within deeper marine settings the lowstand fan deposits, if topographically permitted, will not vary significantly, irrespective of the relative rate of subsidence (Fig. 1, cases 8 and 9). The nature of these lowstand deposits is controlled by topography and the effects of eustatic fall across the shelf, not by subsidence within the basin. In clastic starved settings these downshifts may not be identifiable (Fig. 1, case 3).

Example from the North Sea Central Graben

The Upper Jurassic Fulmar Formation of the North Sea Central Graben (Fig. 2) is a succession of highly bioturbated, predominantly shallow marine sandstones overlain successively by offshore mudstone of the Heather Formation and the deep marine Kimmeridge Clay Formation (Johnson *et al.* 1986; Price *et al.* 1993). The Fulmar Formation comprises a succession of lithologically identical but geographically isolated sand bodies of different ages (cf. Price *et al.* 1993) which onlap the margins of the basin (Price *et al.* 1993; Underhill & Partington 1993; Howell *et al.* 1996). The relative ages and distributions of these sandbodies have been constrained within a stratigraphic framework based on correlation of flooding surfaces tied to ammonite biostratigraphy (Howell *et al.* 1996; cf. Partington *et al.* 1993).

The study from which these data are taken (Howell *et al.* 1996) was based upon the relogging of 1.7 km of core, supported by a new palynological study, from 20 wells within the area (Fig. 2). Wireline log suites (gamma ray, density/neutron and resistivity) were also studied from a further 66 released wells (Howell *et al.* 1996). Published seismic lines (Roberts *et al.* 1990; Price *et al.* 1993) illustrate that the Upper Jurassic sandstones were deposited within a synrift setting (cf. Ratty & Hayward 1993). The significant depths of burial within the centre of the Graben (around 6000 m in block 29/5) and the effects of salt pull-up mean that more detailed seismic is of limited value for understanding the internal geometries of individual reservoir sandbodies. Consequently workers are restricted to well-log data.

The work undertaken here (see also Howell *et al.* 1996) identified a sequence boundary and overlying lowstand deposits which illustrate most of the features described within the model proposed above. This sequence boundary lies between the *Glosense* and *Serratum* (J54a and J54b) maximum flooding surfaces (Partington *et al.* 1993) and is marked by a significant basinward shift of facies tracts in eight of the wells in which sandstones of this age occur. The sequence boundary has been documented from some of the study wells by previous workers (Donovan *et al.* 1993; Price *et al.* 1993). A sequence boundary of comparable age has also been described from the Moray Firth basin 200 km north (Partington *et al.* 1993; K. Stephens & R. Davies 1993 pers. comm.).

The basin physiography includes a major tilted fault block (the Puffin Terrace) which lies within the hangingwall of the basin margin faults to the south and west (Figs 2 and 3). The Puffin Terrace is bounded to the northeast by the Puffin Fault which down throws into the main

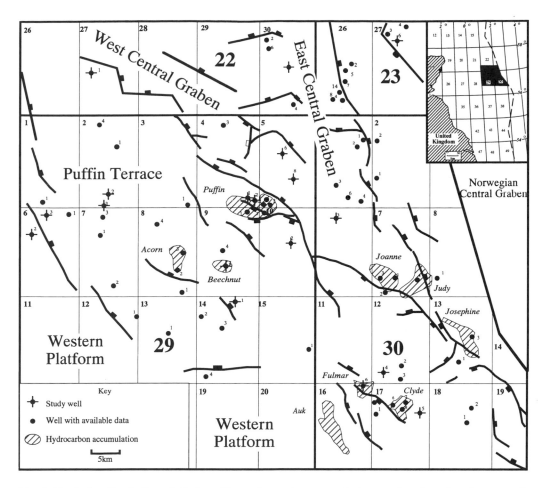

Fig. 2. North Sea South Central Graben, illustrating major structural elements, location of study area and released wells. Structure modified after Roberts *et al.* (1990) and Price *et al.* (1993).

Central Graben (Fig. 3; cf. Roberts *et al.* 1993). The hangingwall block of the Puffin Fault (the Central Graben) also dips, towards the southeast, parallel to the fault, consequently simulating a ramp setting in a NW–SE direction. Where the throw of the fault was greatest (SE), the structure acted as local shelf-slope break. Basin topography was also influenced by the presence of Zechstein salt which acted as a shallow decollement reducing the wavelength of individual structures and produced a greater number of small-scale features (Hodgson *et al.* 1992). The overall effect of this was to create an extremely complex but shallower dipping basin physiography (Figs 2 and 3) not directly comparable to that described from modern extensional basins (Leeder & Gawthorpe 1987;

Gawthorpe *et al.* 1994).

Footwall section

On the margins of the basin (Well A) the sequence boundary occurs as a sharp contact between lower shoreface argillaceous sandstones and mudstones and an overlying unit of clean, laminated sandstones with abundant terriginous, woody debris, 50 m thick (Fig. 4). The unit above the sequence boundary also contains lagoonal muds and is capped by two coal horizons. In an adjacent well, Well B (2.5 km south), at the same stratigraphic interval, the thick sandstone is absent and three coal horizons are interbedded with mudstones interpreted as coastal plain in origin. The thick sandstone

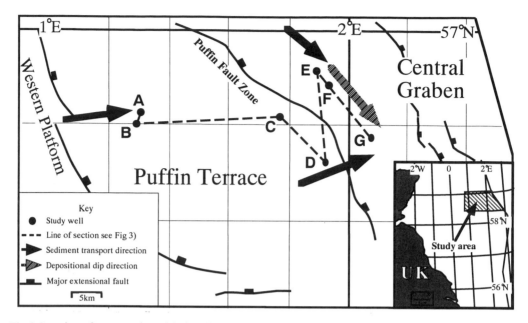

Fig. 3. Location of cross-section with simplified structure, including the relative dip of fault blocks. Map adapted from published sources (Roberts *et al.* 1990; Price *et al.* 1993; and others) and study wells. Note wells A and B on basin margin and wells C and D at crest of structure. Wells E, F and G lie in the hangingwall of the fault terrace.

interval in Well A is interpreted to represent the fill of an incised valley complex fed through a structural transfer zone (Figs 3 and 4) and the coals in the adjacent well represent the interfluvial expression of the same sea-level fall.

Towards the centre of the basin, on the uplifted footwall crest of the Puffin Fault block, significant incision is associated with the sequence boundary in Well C (Fig. 4). Clean, cross-stratified sandstones sharply overlie bioturbated argillaceous sandstones with a significant increase in grain size. Biostratigraphic data suggest the absence of an entire third order sequence below the surface (the entire *Glosense* ammonite zone is absent (Howell *et al.* 1996)) attributed to erosion caused by a drop in base level.

South of the fault block, a well (D) still within the footwall of the main Puffin structure, contains a thick succession of uniformly bioturbated sandstones, punctuated by 15 m of tidal sandstone and lagoonal mudstone, capped by two coal horizons (Fig. 4). Although the transition from bioturbated sandstone to sandstones with tractional structures may be progradational, the sharpness of the basal contact and parasequence stacking patterns above and below the surface support the interpretation of this unit as a lowstand to early transgressive,

estuarine/barrier island complex (cf. Howell *et al.* 1996).

Hangingwall section

Thickness trends within the Upper Jurassic interval, observed within the study wells, indicate that the hangingwall of the fault zone dips parallel to the fault, at *c.* 1°, towards the southeast. Wells E and F within the northern updip section of this ramp contain sharp based, massive, foreshore deposits overlying lower shoreface to offshore transition zone heteroliths (Fig. 4). A *Glossifungites* ichnofacies is interpreted beneath the surface which includes *Thalassinoides* and *Skolithos* trace fossils. Such burrowing has been observed associated with sequence boundary formation (Taylor & Gawthorpe 1993). Parasequence stacking patterns above and below this surface (Howell *et al.* 1995) also support the presence of a sequence boundary at this contact. The local ramp geometry produced a sequence boundary which is overlain by forced regression deposits (Posamentier *et al.* 1992).

In Well G, situated in a deeper water part of the hangingwall than F, 50 m of blocky turbidite sandstone lies within a 500 m thick basinal mudstone succession (Fig. 4). Although poor,

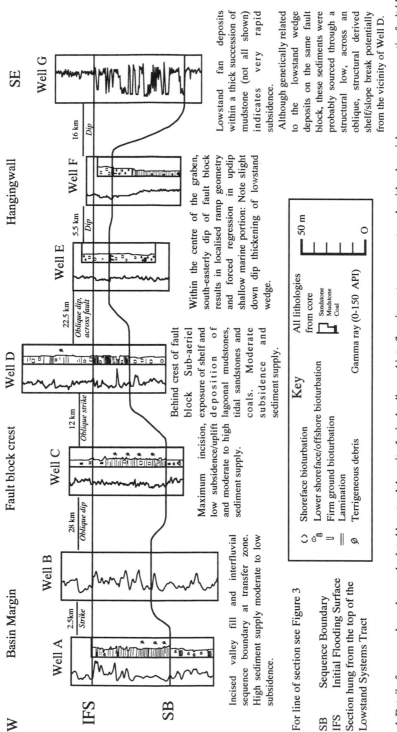

Fig. 4. Detail of sequence boundary and related lowstand deposits. Incised valley and interfluve in west, passes eastwards with sub-aerial exposure across entire fault block. Maximum incision occurred at the crest of the structure. Oblique dip in the hangingwall of the fault resulted in a shallow marine lowstand wedge (forced regression) passing downdip (SE) into deeper water turbidites. However, these hangingwall sediments are temporally but not spatially related (see text for full discussion). Section is datumed from overlying initial flooding surface (top lowstand systems tract).

micropalaeontological data indicate a comparable age to the lowstand complexes described above (cf. Price *et al.* 1993) and it is therefore proposed that these deposits represent coeval lowstand fan deposits associated with the same sequence boundary.

Discussion

Within the Upper Jurassic Central Graben rift basin, a drop in eustatic sea-level resulted in a variety of surfaces and related lowstand deposits. Adjacent to the basin margin sub-aerial exposure was associated with incised valley formation (Well A) at fluvial input points (structural transfer zones) and interfluvial sequence boundary development away from them (Well B). The thickest occurrence of estuarine deposits was topographically restricted, behind the uplifted footwall crest of the fault block (Well D) the most erosion occurred on the fault crest (Well C). During the early lowstand, sand bypassed the footwall shelf and was deposited as lowstand fan sediments within the deepest part of the hangingwall (Well G). In this case the fault zone acted as a local shelf slope break. Within the shallower water parts of the hangingwall a localized ramp geometry existed parallel to the fault zone. Following the drop in sealevel, forced regression deposits were laid down (Wells E and F). These were coeval but not genetically related to the lowstand fan deposits in well G.

Implications for stratigraphic modelling

A conceptual model has been presented which documents variations in accommodation creation within extensional basins. This model proposes that the stratal geometries observed within tectonically active basins will differ from those predicted by the basic sequence stratigraphic model for passive continental margins (cf. Posamentier & Allen 1993*a*, *b*). Within extensional rift basins, in addition to eustatic sea-level, it is also necessary to consider the effects of both strike variability in subsidence rates and the presence of tectonically generated basin physiography upon sequence architecture. Both of these parameters are predictable, given an adequate understanding of the structural configuration of the basin.

Elf Enterprise Caledonia are gratefully acknowledged for funding, data and permission to publish this work. The views expressed herein are those of the authors and do not necessarily reflect the opinions of the staff of E.E. Cal. John Keating of the University of Wales, Aberystwyth is acknowledged for biostratigraphical data and discussion. Henry Posamentier and Steve Cannon are thanked for constructive reviews on various stages of this manuscript.

References

DONOVAN, A. D., DJAKIC, A. W., IOANNIDES, N. S., GARFIELD, T. R. & JONES C. R. 1993. Sequence stratigraphic control on Middle and Upper Jurassic reservoir distribution within the UK Central Graben. *In:* PARKER, J. R. (ed.) *Petroleum Geology of Northwest Europe. Proceedings of the 4th Conference.* Geological Society, London, 251–270.

GAWTHORPE, R. L., FRASER, A. J. & COLLIER, R. E. Ll. 1994. Sequence stratigraphy in active extensional basins: implications for the interpretation of ancient basin-fills. *Marine and Petroleum Geology*, **11**, 642–658.

HODGSON, N. A., FARNSWORTH, J. & FRASER, A. J. 1992. Salt related tectonics, sedimentation and hydrocarbon plays in the Central Graben, North Sea, UKCS. *In:* HARDMAN, R. F. P. (ed.) *Exploration Britain: Geological insights for the next decade.* Geological Society, London, Special Publication, **67**, 31–63

HOWELL, J. A., FLINT, S. & KEATING, J. M. 1996. Sequence stratigraphic expressions of relative sea-level changes in extensional rift basins: an example from the Upper Jurassic of the UK Central Graben. *Journal of Sedimentary Research*, in press.

JERVEY, M. T. 1988. Quantitative modelling of siliciclastic rock sequences and their seismic expression. *In:* WILGUS, C. H., HASTINGS, B. S., KENDALL, C. G. St. C., POSAMENTIER, H. W., ROSS, C. A. & VAN WAGONER, J. C. (eds) *Sea Level Changes – an Integrated Approach.* SEPM Special Publication, **42**, 47–70.

JOHNSON, H. D., MACKAY, T. A. & STEWART, D. J. 1986. The Fulmar Oil-field (Central North Sea): geological aspects of its discovery, appraisal and development. *Marine and Petroleum Geology*, **3**, 99–125.

LEEDER, M. R. & GAWTHORPE, R. L. 1987. Sedimentary models for extensional tilt-block/half-graben basins. *In:* COWARD, M. P., DEWEY, J. F. & HANCOCK, P. L. (eds) *Continental Extensional Tectonics.* Geological Society, London, Special Publication, **28**, 139–152

MITCHUM, R. M. & VAN WAGONER, J. C. 1991. High-frequency sequences and their stacking patterns: sequence stratigraphic evidence of high-frequency eustatic cycles. *Sedimentary Geology*, **70**, 131–160.

O'BYRNE, C. J. & FLINT, S. 1996 Interfluve sequence boundaries from the Grassy Member, Book Cliffs, Utah: criteria for recognition and implications for subsurface correlation. *This volume.*

PARTINGTON, M. M., MITCHENER, B. C., MILTON, N. J. & FRASER, A. J. 1993. Genetic sequence stratigraphy for the North Sea Late Jurassic and Early Cretaceous: distribution and prediction of Kimmeridgian–Late Ryazanian reservoirs in the

North Sea and adjacent areas. *In:* PARKER, J. R. (ed.) *Petroleum Geology of Northwest Europe. Proceedings of the 4th Conference.* Geological Society, London, 347–370.

POSAMENTIER, H. W. & ALLEN, G. P. 1993*a*. Variability of the sequence stratigraphic model: effects of local basin factors. *Sedimentary Geology*, **86**, 91–109.

—— & —— 1993*b*. Siliciclastic sequence stratigraphic patterns in foreland ramp-type basins. *Geology*, **21**, 455–458.

—— & VAIL, P. R. 1988. Eustatic controls on clastic deposition II – sequence and systems tract models. *In:* WILGUS, C. H., HASTINGS, B. S., KENDALL, C. G. St. C., POSAMENTIER, H. W., ROSS, C. A. & VAN WAGONER, J. C. (eds) *Sea Level Changes – an Integrated Approach.* SEPM Special Publication, **42**, 125–154.

——, JERVEY, M. T. & VAIL, P. R. 1988. Eustatic controls on clastic deposition I – conceptual framework. *In:* WILGUS, C. H., HASTINGS, B. S., KENDALL, C. G. St. C., POSAMENTIER, H. W., ROSS, C. A. & VAN WAGONER, J. C. (eds) *Sea Level Changes – an Integrated Approach.* SEPM. Special Publication, **42**, 109–124.

——, ALLEN, G. P., JAMES, D. P. & TESSON, M. 1992. Forced regressions in a sequence stratigraphic framework: concepts, examples and exploration significance. *American Association of Petroleum Geologists Bulletin*, **76**, 1687–1709.

PRICE, J., DYER, R., GOODALL, I., McKIE, T., WATSON, P. & WILLIAMS, G. 1993. Effective stratigraphical subdivision of the Humber Group and the Late Jurassic evolution of the UK Central Graben. *In:* PARKER, J. R. (ed.) *Petroleum Geology of Northwest Europe. Proceedings of the 4th Conference.* Geological Society, London, 443–458.

PROSSER, S. 1993. Rift-related linked depositional systems and their seismic expression. *In:* WILLIAMS, G. D. & DOBB, A. (eds) *Tectonics and seismic sequence stratigraphy.* Geological Society, London, Special Publication, **71**, 35–66.

RATTEY, R. P. & HAYWARD, A. B. 1993. Sequence stratigraphy of a failed rift system: the Middle Jurassic to Late Cretaceous basin evolution of the Central and Northern North Sea. *In:* PARKER, J. R. (ed.) *Petroleum Geology of Northwest Europe. Proceedings of the 4th Conference.* Geological Society, London, Special Publication, 215–251.

ROBERTS, A. M., PRICE, J. D. & OLSEN, T. S. 1990. Late Jurassic half-graben control on the siting and structure of hydrocarbon accumulations: UK/Norwegian central Graben. *In:* HARDMAN, R. F. P. & BROOKS, J. (eds) *Tectonic events responsible for Britain's Oil and Gas reserves.* Geological Society, London, Special Publication, **55**, 229–257.

TAYLOR, A. M. & GAWTHORPE, R. L. 1993. Application of sequence stratigraphy and trace fossil analysis to reservoir description: examples from the Jurassic of the North Sea. *In:* PARKER, J. R. (ed.) *Petroleum Geology of Northwest Europe. Proceedings of the 4th Conference,* Geological Society, London, 317–336.

UNDERHILL, J. R. & PARTINGTON, M. A. 1993. Jurassic thermal doming and deflation in the North Sea: implications of the sequence stratigraphic evidence. *In:* PARKER, J. R. (ed.) *Petroleum Geology of Northwest Europe. Proceedings of the 4th Conference,* Geological Society, London, 337–345.

VAN WAGONER, J. C., MITCHUM, R. M., CAMPION, K. M. & RAHMANIAN, V. D. 1990. *Siliciclastic sequence stratigraphy in well logs, cores and outcrops.* American Association of Petroleum Geologists, Methods in Exploration, **7**.

WALSH, J. J. & WATTERSON, J. 1988. Analysis of the relationship between displacements and dimensions of faults. *Journal of Structural Geology*, **10**, 1039–1046.

Hierarchical stratigraphic cycles in the non-marine Clair Group (Devonian) UKCS

T. McKIE,[1] & I. R. GARDEN[1,2]

[1] *Badley, Ashton & Associates Ltd, Winceby House, Winceby, Horncastle LN9 6PB, UK*

[2] *Department of Petroleum Engineering, Heriot Watt University, Riccarton, Edinburgh EH14 4AS, UK*

Abstract: The Devonian lower Clair Group (*c.* 300–800 m thick) is dominantly composed of fluviatile sandstones and conglomerates, aeolian sandstones and minor floodplain and lacustrine shales. These facies are organized into stratigraphical cycles which have a three-fold hierarchy: an unconformity bounded cycle representing first order retreat and advance of the fluvial drainage system; three subsidiary second order cycles bounded by minor unconformities, and numerous third order cycles. These stratigraphical cycles exhibit common characteristics: they are bounded by sharp, commonly erosive surfaces; they are composed of a fining-upwards succession overlain by a variably developed coarsening-upward element; the fining-upward element records a progressive decrease in the fluvial component, and increasing proportion of aeolian, floodplain or lacustrine facies (this trend is reversed in the upper, coarsening-upward section). The grain size of the fluvial facies, and nature of the non-fluvial facies (aeolian vs floodplain/lacustrine) within any order of cycle is dependent on it's position within a lower order (larger scale) cycle.

The cycles are areally extensive, and are present in all wells within the Clair Basin, although spatial changes in internal facies make-up occur. The second order cycles show progressive onlap followed by offlap of the basin margin.

These hierarchical cycles are interpreted to be the product of changes in the accommodation:supply ratio of the depositional system. An increasing accommodation:supply ratio provided space to preserve the record of waning fluvial influence, enhanced aeolian reworking, preserved floodplain fines or lacustrine mudrocks. In contrast, a reducing accommodation:supply ratio resulted in filling of accommodation space and sediment bypassing. Accommodation space variations were dictated by tectonic variations in strain rate, which caused incremental uplift of the basin margin and subsidence of the basin. Sediment supply changes were influenced by source area uplift, which changed sediment yield, and climate change which affected runoff. Climatic changes are detectable within the non-fluvial elements of the Clair Group by stratigraphic variations between aeolian, floodplain and lacustrine facies. These vary systematically within and between second order unconformity bounded units, suggesting a tectonic influence on climate.

Within the second and third order cycles, those which were evolving towards, or were occurring within, semi-arid to arid climatic conditions are commonly (but not exclusively) biased towards fining-upward successions, whilst the more humid systems show a better developed coarsening-upward element. This may be the result of the changing impact of discharge versus sediment supply variations when one of these parameters became suppressed or enhanced under varying climatic conditions.

The advent of sequence stratigraphy has changed the way shallow marine and paralic successions are considered, from focusing on the autocyclic controls on depositional facies to the allocyclic controls on facies stacking patterns. This in turn offers the possibility of more deterministic reservoir descriptions and provides a more realistic framework within which 'random' autocyclic facies variability can be incorporated. Recently this philosophy has begun to be applied, albeit in a modified form, to the deposits of intracratonic basins isolated from a direct sea-level control, by the recognition of regional to basin-wide discontinuities in the stratigraphic record enclosing repetitive and predictable facies stacking patterns. Most notable have been the recognition of super surfaces in aeolian systems (Loope 1985; Kocurek & Havholm 1993) and to a lesser extent repetitive facies stacking patterns in fluvial depositional systems (e.g. Steele & Ryseth 1990; Legarreta *et al.* 1993; Aitken & Flint 1994, 1995).

The aim of this paper is to illustrate a hierarchy of stacking patterns in the Devonian Clair Group within a wide spectrum of fluvial, aeolian and lacustrine depositional systems. These cycles are areally widespread, and are interpreted to represent the depositional and preservational response to variations in climate and tectonism. Their recognition allows greater predictability in the subsurface and, because they also exhibit variations in detrital mineral-

From Howell, J. A. & Aitken, J. F. (eds), *High Resolution Sequence Stratigraphy: Innovations and Applications*, Geological Society Special Publication No. 104, pp. 139–157.

139

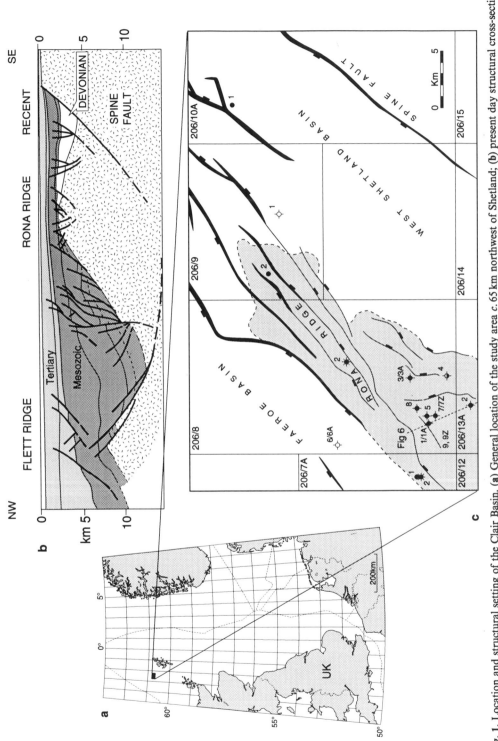

Fig. 1. Location and structural setting of the Clair Basin. (a) General location of the study area c. 65 km northwest of Shetland; (b) present day structural cross-section, showing the Devonian Clair Group forming an eastward thickening wedge from the Rona Ridge to the Spine Fault (modified after Duindam & van Hoorn 1987); (c) Clair Field area and well location map.

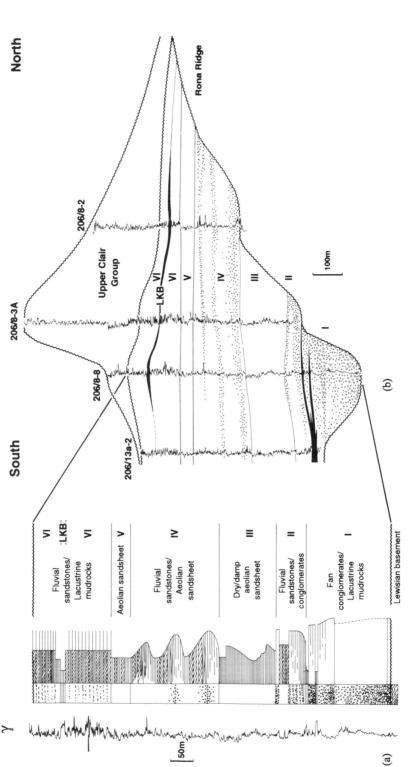

Fig. 2. Stratigraphic architecture of the Clair Group in the area of the Clair Field. (**a**) Core log and gamma profile from 206/8-8 illustrating the general vertical stratigraphy present in the lower Clair Group. (**b**) Well log correlation of the Clair Group and detailed lower Clair Group stratigraphy. Reservoir unit scheme after Allen & Mange-Rajetsky (1992).

ogy, grain sorting and clay infiltration, they offer the possibility of reservoir quality prediction.

Stratigraphic setting

The Clair Basin is situated *c.* 65 km northwest of Shetland, and forms aneastward thickening wedge (Duindam & van Hoorn 1987) bounded by the Spine Fault to the east and the Rona Ridge to the west (Fig. 1). The basin is believed to be extensional in origin (McClay *et al.* 1986), although the role of regional transtension is unresolved (Ziegler 1982). The Clair Field (Coney *et al.* 1993) occurs straddling the crest of the Rona Ridge, where the *c.* 1000 m thick Clair Group rests unconformably on Archaean Lewisian Gneiss (Fig. 2). In this area the Clair Group is informally subdivided into upper and lower groups by an angular-unconformity which was first recognized by Allen & Mange-Rajetsky (1992).

The lower Clair Group, which is the subject of this paper, has been tentatively dated as Givetian to Frasnian (potentially up to Dinantian) in age, whilst the upper part of the upper Clair Group has been dated as Dinantian to Visean (Ridd 1981; Blackbourn 1987; Allen & Mange-Rajetsky 1992). The entire section is dominated by sandstones, within which Blackbourn (1987) recognized a dominance of fluvial deposits, but with aeolian sandstones present within the lower parts of the section, and fluvio-deltaic facies in the upper parts. In a more detailed study, using a larger well database, Allen & Mange-Rajetsky (1992) used core, wireline logs and heavy mineral analysis to subdivide the Clair Group stratigraphy into ten sedimentologically distinct reservoir units. Units I–VI define the lower Clair Group and units VII–X comprise the upper Clair Group (Fig. 2).

Within the lower Clair Group these authors recognized two cycles of active fluvial deposition grading upwards into more aeolian dominated intervals characterized by intrabasinal reworking (Fig. 2, units I–III and IV–VI) each of which exhibited an increased rounding of detrital apatite grains and maturation of the detrital suite. The overall heavy mineral assemblage for the Lower Clair Group was interpreted to reflect a mainly Lewisian and metasedimentary (Moine and Dalradian) provenance, indicating a drainage network localized to within the Archaean and Neoproterozoic metamorphic Caledonides.

Lower Clair Group stratigraphy

For descriptive purposes the reservoir unit scheme of Allen & Mange-Rajetsky (1992) forms a robust, reproducible system which will be used in this section for describing the overall facies evolution. The depositional settings interpreted for these reservoir units largely concur with those of Allen & Mange-Rajetsky, to which the reader is referred for detailed facies descriptions. A summary of the facies association scheme used in this study is presented in Table 1. The emphasis in this paper will be placed on the stratigraphic architecture of the component facies, and concomitant textural and mineralogical variations.

Unit I

Unit I is the basal interval of the Clair Group, and is a 0–40 m thick, fining-upward succession of conglomerate, sandstone and mudrock which locally thickens to 120–200 m (Figs 2 and 3). The thickest intervals (e.g. 206/8-8, Fig. 3) comprise a clast-supported, boulder conglomerate dominated succession (Table 1). This contains subsidiary, sharply-based fining-upward cycles > 50 m thick which pass upwards into pebbly sandstones. The conglomerates are dominated by clasts of Lewisian gneiss, with subsidiary metasedimentary clasts (and rare sandstone clasts), and occur infilling a fault bounded topography on the Lewisian basement. The more areally widespread, upper parts of Unit I are composed of cross-stratified, fine- to very coarse-grained sandstone and pebble conglomerate (up to 20 m thick), intercalated with metre scale intervals of laminated mudrock (Fig. 3, Table 1). These mudrocks contain thin partings of fine-grained, current and wave ripple laminated sandstones with burrows locally present, and are widely correlatable across Unit I.

The boulder conglomerate sections are interpreted to be the product of local accumulation of fluvially transported basement material into topographic lows, with the large-scale fining-upward signatures representing waning-upward episodes of conglomerate supply, possibly in response to pulses in source uplift and denudation, or waning of fluvial discharge. The finer-grained, upper parts of this unit were interpreted as representing deposition within and along the margins of syn-rift lakes (Allen & Mange-Rajetsky 1992), and the section records the establishment of a lower gradient fluvial-lacustrine system. The wave rippled lacustrine mudrock packages are widely correlatable and separate equally widespread stratified, pebbly sandstone intervals, suggesting that the upper part of this interval represents fluctuations between widespread, subaqueous deposition

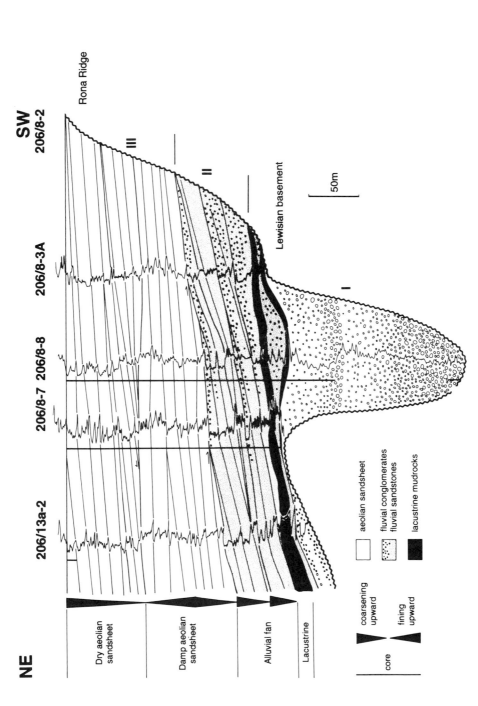

Fig. 3. Gamma log correlation (constrained by core observations) of Units I–III, illustrating the lateral persistence of lacustrine mudrocks in Unit I, offlap patterns in Unit II and on–offlap patterns in Unit III.

during lake flooding events, and deposition from bedload dominated, pebbly streams.

Unit II

This unit is 0–110 m thick and comprises interstratified pebble conglomerates (<8 m thick) and fine- to coarse-grained, cross-stratified and flat laminated sandstones (Table 1). The pebble grade material largely comprises basement Lewisian gneiss, quartz and rare reworked calcrete. These deposits are organised into <15 m thick coarsening-upward (cleaning upward gamma log motifs) cycles which, when correlated on well-logs, form an inclined, offlapping pattern dipping away from the Rona Ridge (Fig. 3). Wind ripple laminated sandstones form a very minor component (restricted to a single stratigraphic interval medial within Unit II), and mudrocks are largely absent.

These stratified sandstones and conglomerates represent the deposits of low sinuosity, bedload dominated streams with the record of interfluve deposition largely reworked, and represented by calcrete rip-up clasts. The offlapping patterns of the unit as a whole suggests alluvial fan progradation, with the component coarsening-upward cycles present within this unit recording pulses of progradation, possibly in response to discharge fluctuations or local channel avulsions up-dip. Local fan abandonment is suggested by the deposition of finer-grained sandstones with aeolian reworking into wind ripple laminated sands.

Unit III

This 0–160 m thick succession (typically >100 m) is largely composed of very fine- to fine-grained sandstone (Table 1) dominated by <10 m thick alternating packages of wind-ripple lamination and mud-prone, adhesion laminated sandstone (c.f. Kocurek & Dott 1981; Kocurek & Fielder 1982). Sandflow cross-stratification is rarely preserved as thin (<0.5 m) beds. Mudrocks are uncommon, but local concentrations of rip-up clasts occur in association with trough cross-stratified and structureless sandstones. Unit III forms an overall fining/coarsening-upward trend, with the central, finer-grained section containing a higher proportion of mud-prone, adhesion laminated sandstones with a correspondingly high gamma profile (Figs 2 and 3). The lower and upper parts are less mud-prone and, particularly in the upper part, are largely dominated by wind ripple lamination. In addition, the upper section is coarser-grained (fine to medium sand grade). Correlation of key marker horizons (the muddy adhesion laminated horizons) within this unit demonstrates that the fining-upward section shows an overall flat lying, onlapping pattern towards the Rona Ridge, whilst the coarsening-upward element displays an inclined offlapping stratal pattern dipping away from the Rona Ridge (Fig. 3).

Unit III is dominated by aeolian sandsheet deposits (Allen & Mange-Rajetsky 1992). The wind ripple laminated sandstones represent dry sandsheet conditions, while the adhesion laminated sandstones and rare, in situ mudrocks represent damp sandsheet environments with temporary ponding of surface water (c.f. Fryberger et al. 1979). Only the basal parts of dunes which may have been present on the sandsheet are preserved as thin, sandflow cross-stratified sandstones, possibly infilling dune scour pits. Preserved fluvial deposits are represented by the mudclast bearing, stratified and structureless sandstones, which may represent ephemeral flood events. These were commonly reworked by aeolian processes resulting in a low preservation potential.

The overall sandsheet architecture, comprising onlapping, 'wetting-upward' deposits succeeded by 'drying-upward', offlapping deposits suggests that there was an initially progressive diminution of sand supply and a water table close to the sediment surface matching the aggradation rate, followed by an enhanced sand supply of increasing calibre which caused sediment aggradation faster than the rate of water table rise. Fluvial deposits are very rare within the succession, indicating that even at the highest rate of sediment supply at the top of the unit, depositional events were sufficiently ephemeral to undergo almost total aeolian reworking.

Unit IV

Unit IV is a 50–200 m thick interval of pebbly, very coarse- to fine-grained sandstone (Table 1) which rests sharply (and locally unconformably; Allen & Mange-Rajetsky 1992) on the underlying fine-grained sandstones of Unit III. The unit onlaps the Rona Ridge beyond the northwest stratigraphic pinch out of the underlying Units I–III (Fig. 2). The top of this unit is locally lithologically gradational into the overlying Unit V, but is also marked by a sharp upward change into sandstones with markedly more common aeolian stratification.

Unit IV fines-upwards overall, and shows an upward reduction in the proportion of extrabasinal pebbly detritus (quartz and basement gneiss), gradual replacement of high angle cross-

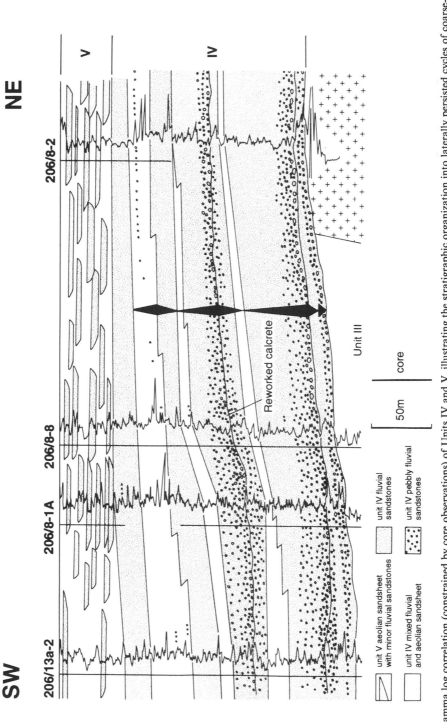

Fig. 4. Gamma log correlation (constrained by core observations) of Units IV and V, illustrating the stratigraphic organization into laterally persisted cycles of coarse-grained, pebbly fluvial sandstones and fine-grained aeolian sandsheet deposits within an overall fining-upward/drying-upward trend.

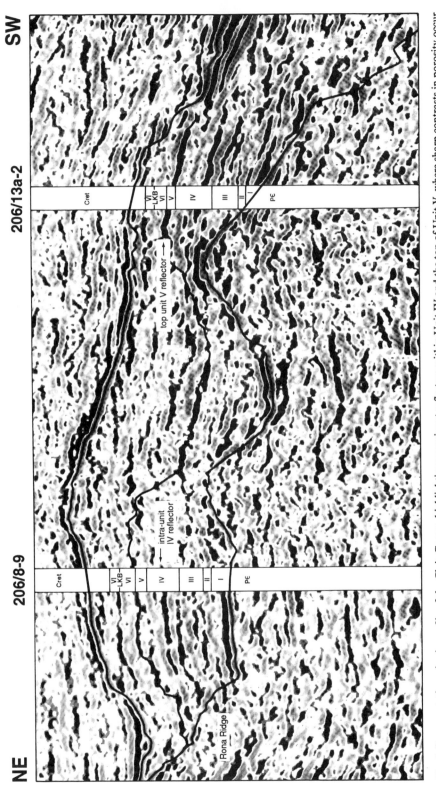

Fig. 5. Representative seismic profile of the Clair Group, highlighting prominent reflectors within Unit IV and the top of Unit V, where sharp contrasts in porosity occur between aeolian sandstones and overlying fluvial sandstones.

stratification with low angle and flat lying, mudclast bearing lamination, and an increase in the presence of adhesion and wind ripple laminae. This overall pattern is composed of at least three subsidiary cycles (Figs 2 and 4) characterized by fining/coarsening-upward trends. The lowermost cycle is poorly preserved, but forms a pebble-rich, < 15 m thick cycle with laminated and burrowed mudrocks locally preserved in the fine-grained medial part (Fig. 4). The upper part of this cycle is erosively truncated by the basal surface of the succeeding cycle, particularly towards the Rona Ridge. The topography along this surface is only of the order of a few metres, and no evidence of major incision has been detected in any of the Clair Basin wells. One notable feature of this surface (and the overlying one) is the presence of slightly higher gamma log values below the surface compared to the section above (Fig. 4). This is not due to clast type variation, but appears to be the response to infiltrated clay (from limited petrographic evidence). The overlying cycle is c. 50 m thick, and is dominated by a fining-upward succession, with a coarsening-upward component restricted to the upper 10 m. The coarse-grained, pebbly sections at the base and top are largely cross-stratified (with the local development of flat bedded conglomerates defining subtle, metre-scale fining-upward cycles), whilst the deposits at the top of the fining-upward section in the cycle show thinly bedded wind ripple and adhesion laminated sandstones interleaved with erosively based, structureless and trough cross-stratified sandstones. This fine-grained interval forms a prominent seismic reflector which is recognizable across the area (Fig. 5). The cycle thickens away from the Rona Ridge and is erosively truncated by the overlying cycle, which is comparable in overall character, but is markedly less pebbly in the coarsening-upward element. This cycle is succeeded by a third package of wind-ripple and adhesion laminated sandstones (up to 25 m thick), which is in turn sharply overlain by mudclast bearing, plane bedded sandstones, although in these uppermost sections there is only poorly developed coarsening/fining-upward trends within the fine-grained sandstones and granule grade detritus is uncommon.

Unit IV shows an overall waning-upward trend from cross-stratified, pebble bearing fluvial deposits of bedload dominated streams to largely plane bedded, mudclast bearing, unconfined fluvial sheet sandstones (Reid & Frostick 1989) and dry and damp aeolian sandsheet deposits (c.f. Fryberger et al. 1979). This trend is in turn composed of higher frequency cycles recording upward waning, succeeded by upward increasing fluvial discharge. The earliest cycle is poorly preserved due to erosional truncation, but shows the local preservation of burrowed lacustrine mudrocks in the low energy section. Succeeding cycles are increasingly dominated by aeolian sandsheet deposits in their low energy portions, and are finer-grained, less pebbly and less erosive during their peak fluvial discharge phases. Increased intrabasinal reworking is recorded by a progressive increase in sandstone textural and compositional maturity through the unit. The coarse-grained, pebbly sandstones are typically the most lithic (Fig. 6a) due to their coarser grain size. However, the fine-grained, mudclast-bearing fluvial sandstones (Fig. 6b), and the mixed fluvial–aeolian sandstones (Fig. 6c) are of comparable grain size, but the latter are generally more mature. This results from the protracted reworking of these sands by aeolian processes towards the top of the waning-upward trend and the destruction of less resistant lithic grains. The lack of evidence for significant feldspar destruction (c.f. Dutta et al. 1993) may indicate that episodes of aeolian reworking were relatively short lived.

Unit V

The base of Unit V is texturally gradational with Unit IV, but is marked by an abrupt increase in the proportion of wind ripple and adhesion laminated sandstones. The top is generally marked by a sharp, commonly erosional contact with Unit VI, but locally, e.g. 206/8-1A, the boundary is more transitional in character, with a progressive decrease in the proportion of aeolian stratification. On seismic, the top of Unit V is typically marked by a prominent reflector (Fig. 5).

The unit is typically 40–70 m thick and composed of fine-grained, well sorted sandstones depleted in lithic fragments (Fig. 6d), with well rounded apatite grains present (Allen & Mange-Rajetsky 1992). Wind ripple and adhesion laminated sandstones dominate, with subsidiary sandflow cross-stratification (Table 1). Up to 0.2 m thick, mudclast bearing, cross-bedded, plane bedded and current ripple laminated sandstones are sporadically distributed throughout, but are less common away from the Rona Ridge (Fig. 4). In general these sandstones typically have a comparable detrital mineralogy and texture to those of the wind ripple laminated sandstones, but are locally more lithic in composition (Fig. 6d). Granule lag deposits are also present as isolated horizons generally only a few grains thick.

Unit V is dominated by aeolian sandsheet deposits, with minor dune troughs preserved as sandflow cross-stratification. The mudclast bearing sandstones are interpreted as the deposits of ephemeral floods across the sandsheet and the overall setting appears to have been of a mixed fluvial–aeolian system analogous to examples described by Trewin (1993). The textural maturity of this unit indicates prolonged aeolian reworking, and the comparable maturity of both the fluvial and aeolian deposits is indicative of widespread fluvial reworking of aeolian sands. The granule lag deposits are interpreted as deflation lags and these, together with the proportion of sandsheet deposits is indicative of metasaturated sand supply conditions (Loope & Simpson 1992), with an insufficient sand budget to develop or preserve extensive climbing dune fields, resulting in thinly bedded, water table controlled aeolian deposits.

Unit VI

This unit is erosively truncated by the base of the upper Clair Group unconformity, and varies from 50–170 m thick. The basal contact is generally, but not exclusively sharp. Allen & Mange-Rajetsky (1992) recognized an areally widespread mudrock marker, the Lacustrine Key Bed (LKB) which forms a reliable correlation datum across the area (Figs 2 and 7). This datum is inclined relative to the underlying Units I–V (Fig. 2), indicating that tectonic tilting away from the Rona Ridge took place after deposition of Unit V. Using the top of this mudrock as a correlation datum shows that the base of Unit VI has a variable topography (Fig. 7). The presence of rounded apatite grains in the basal part of Unit VI is indicative of some reworking of Unit V, but the lack of significant thickness variations of Unit V within the study area indicates that major erosion of the top of this unit did not take place. The sandstones of Unit VI (Fig. 6e) are generally more feldspathic than the underlying units (apart from the burrowed sandstones), and above the basal section, detrital apatite is more angular than within Unit V (M. Mange-Rajetsky pers. comm. 1995). Instead, the regional tilting of Unit VI may have marked the onset of relatively gentle, local differential subsidence, with more rapidly subsiding areas e.g. 206/8-1A preserving a facies transition, whilst less rapidly subsiding areas were initially reworked, before being draped by later Unit VI deposits of markedly different character to Unit V, giving a more sharp facies change e.g. 206/8-2.

The dominant facies present within Unit VI comprise the following (Fig. 7, Table 1).

(i) Sharply based, fining-upward cycles up to c. 10 m thick with intraformational clast conglomerate lags and composed of fine- to medium-grained, cross-stratified sandstones with minor current ripple and planar laminated, fine-grained sandstones.

(ii) Very fine- to fine-grained, micaceous, current and plane bedded sandstones with local evidence of wave ripple laminae. These form decimetre scale beds with common mudclast lags. Calcrete and rooted palaeosol horizons are locally well developed within these sandstones at specific horizons (Fig. 7), but in general palaeosol development is limited, although more common above the LKB.

(iii) Bioturbated, clean, fine-grained sandstones forming metre scale intervals with indistinct bedding. The burrow assemblage comprises simple horizontal burrows with a structureless fill or with a backfilled, meniscate structure (possibly *Taenidium*).

(iv) Areally widespread mudrocks with wave ripple laminated sand streaks and root traces (e.g. the Lacustrine Key Bed).

The overall architecture of Unit VI is illustrated in Fig. 7. The LKB forms a field-wide marker, below which are several correlatable mud-prone, high gamma horizons which have associated wave ripples, bioturbation and organic remains (e.g. possible armoured fish remains in 206/8-8). These occur at the base of fining/cleaning-upward cycles up to 10 m thick. Between these areally widespread, mud-prone horizons there are more locally developed, sharply based fining-upward cycles composed of coarser-grained sandstone, with the correlative areas between these units marked by mature calcic palaeosols. The LKB itself appears to overlie an erosional surface which is locally floored by an erosively based, cross-stratified sandstone within the more deeply incised section. Away from the Rona Ridge the LKB passes into a burrowed sandstone (Fig. 7). The interval above the LKB is depleted in mudrock horizons, but has an overall cleaning-upward gamma profile (coarsening-upward in core), in turn composed of subsidiary cleaning/coarsening upward cycles. These are also locally interrupted by the sharply based, fining-upward sandstone units.

The presence of common wave formed structures within Unit VI led Allen & Mange-Rajetsky (1992) to interpret this interval within the context of a lacustrine depositional system. The results of this paper largely concur with

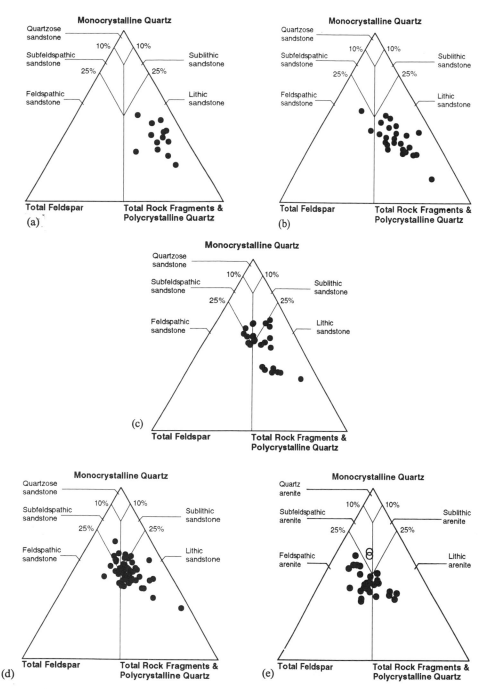

Fig. 6. Triangular plots of detrital mineralogy for Units IV, V and VI. (**a**) Pebbly, coarse-grained lithic deposits of Unit IV; (**b**) Fine-grained less lithic, mudclast bearing fluvial deposits of Unit IV. (**c**) Mixed fluvial–aeolian sandstones, Unit IV, with a bimodal detrital mineralogy. The more lithic sandstones are first cycle fluvial sandstones, whilst the less lithic examples are from both fluvial and aeolian facies in which extensive sediment reworking has occurred. (**d**) Unit V aeolian dominated sandstones principally comprising mature reworked sand from both fluvial and aeolian facies, and scarce, more lithic-rich first cycle fluvial sand. (**e**) Feldspathic Unit VI perilacustrine sandstones (open circles denote less feldspathic lake shoreline sandstones at the LKB stratigraphic level (206/13a-2)).

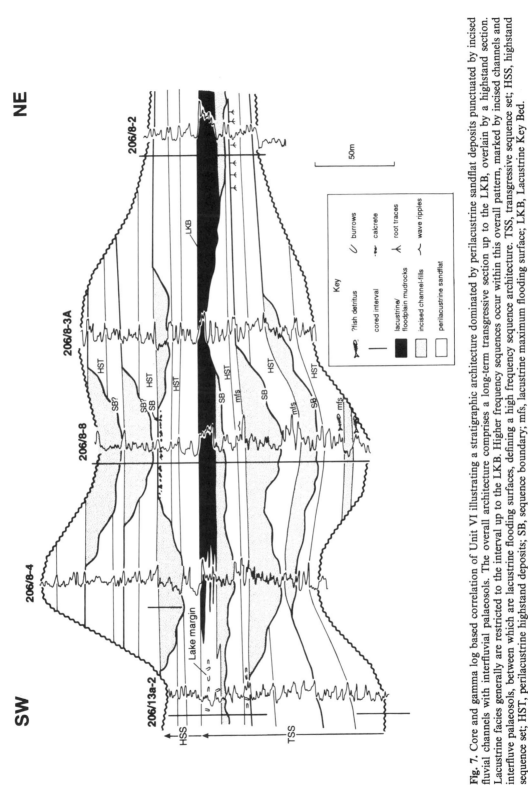

Fig. 7. Core and gamma log based correlation of Unit VI illustrating a stratigraphic architecture dominated by perilacustrine sandflat deposits punctuated by incised fluvial channels with interfluvial palaeosols. The overall architecture comprises a long-term transgressive section up to the LKB, overlain by a highstand section. Lacustrine facies generally are restricted to the interval up to the LKB. Higher frequency sequences occur within this overall pattern, marked by incised channels and interfluve palaeosols, between which are lacustrine flooding surfaces, defining a high frequency sequence architecture. TSS, transgressive sequence set; HSS, highstand sequence set; HST, perilacustrine highstand deposits; SB, sequence boundary; mfs, lacustrine maximum flooding surface; LKB, Lacustrine Key Bed.

Table 1. *Summary of the facies association scheme used in this study to subdivide the reservoir units of the lower Clair Group. Reservoir units follow those of Allen & Mange-Rajetsky (1992)*

Unit	Facies association	Lithology	Thickness	Principal characteristics	Depositional setting
VI	Incised channels	fine to medium grained, pebbly sandstones	intervals up to 10m thick	fining-upward cycles with intraformational conglomerate lags and dominant trough cross stratification	infilling of incised channels eroded during lake lowstands
	Sandflat	very fine to fine grained, micaceous sandstones	intervals up to 20m thick	dominantly planar and current ripple laminated (with rare calcic palaeosols and wave rippled horizons)	perilacustrine sandflat deposited during lake highstand
	Floodplain mudrocks	mudrocks with very fine grained sandstone partings	intervals up to 15m thick	dominantly root trace disrupted with locally common wave ripple laminae	floodplain and lacustrine mudrocks deposited during lake transgression
	Lake margin sandstones	fine grained sandstone	intervals up to 15m thick	clean and well sorted sandstone with primary structures largely destroyed by bioturbation	lake margin sandstones encountered at the mouth of an incised channel
V	Aeolian sandsheet	fine grained sandstone	intervals up to 10m thick	decimetre scale interbedding of wind ripple and adhesion laminated sandstone with locally developed sandflow cross bedding (occupying metre scale troughs in cored horizontal wells)	aeolian sandsheet with minor barchanoid dunes
	Fluvial sandstones	fine grained sandstone	0.2m thick intervals	mudclast bearing, cross stratified, current and planar laminated sandstone	ephemeral sheetflood deposits
IV	Mixed fluvial-aeolian sandstones	fine to medium grained sandstone	intervals up to 30m thick	up to 10m thick, mudclast bearing, plane bedded sandstone intervals interbedded with up to 4m thick packages of wind ripple and adhesion laminated sandstone	ephemeral fluvial system subject to aeolian reworking
	Fluvial sandstones	fine to medium grained sandstone	intervals up to 20m thick	trough cross stratification and common low angle to horizontal, planar lamination with common mudclasts	fleshy/ephemeral fluvial deposits
	Pebbly fluvial sandstones	medium to very coarse grained, pebbly sandstones	intervals up to 30m thick	dominated by trough cross stratification and minor horizontal lamination with common basement pebbly lags	bedload dominated stream deposits
III	Dry aeolian sandsheet	very fine to medium grained sandstone	intervals up to 10m thick	horizontal, bimodally sorted wind ripple lamination with rare sandflow cross bedding (and mudclast bearing, structureless sandstones)	dry sandsheet with minor dune bedform preservation (and local fluvial reworking)
	Damp aeolian sandsheet	very fine to fine grained, muddy sandstone	intervals up to 10m thick	horizontal, 'crinkly' lamination, isolated 'warts' and coarse grained lags (rare mud laminae)	damp sandsheet with minor ephemeral ponds and deflation lags
II	Fluvial conglomerates	granule to pebble grade, basement conglomerate	intervals <8m thick	flat laminated or structureless, rare trough cross stratification	coarse grained, bedload dominated stream deposits
	Fluvial sandstones	fine to coarse grained sandstones	<15m thick intervals	planar and trough cross stratified and laminated sandstones (rare wind ripple lamination)	bedload dominated stream deposits
I	Lacustrine mudrocks	mudrocks with fine grained, heterolithic partings	intervals up to 5m thick	well preserved primary lamination with wave and current ripple laminae within the heterolithic partings	lacustrine transgressive deposits
	Fluvial sandstones	fine to very coarse grained pebbly sandstones	intervals <20m thick	trough and planar cross stratified (and flat laminated), with common basement clasts and rare calcrete and mudclasts	bedload dominated stream deposits
	Fluvial conglomerates	pebble to boulder grade, basement conglomerate	intervals <50m thick	bedding locally defined by sharp clast size changes or thin sandstone partings	localised fan conglomerates

these authors, but greater emphasis is placed on the Unit VI stratigraphy being the product of lake level changes. The areally widespread LKB shows well developed features of lake level rise (burrowed sandstones, wave ripples, dominance of mudrock deposition) and can be regarded as a lacustrine maximum flooding surface (*sensu* Van Wagoner *et al.* 1990). The interval below, with a relatively high proportion of mudrock and lacustrine facies can be regarded as an overall transgressive section leading up to the LKB. The more sand-rich, upward-coarsening section above the LKB can in turn be regarded as a highstand deposit. However, the presence of up to 10 m thick, incised fluvial deposits, with mature palaeosols on their correlative interfluves suggests that higher frequency sequences are present, and the variable thickness of the base of LKB itself (Fig. 7) could be interpreted to reflect infilling of an incised topography. The sub-LKB section is therefore interpreted to be a transgressive sequence set composed of high frequency sequences, and the supra-LKB section as a highstand sequence set (*sensu* Mitchum & Van Wagoner 1991). The overall Unit VI stratigraphy is therefore interpreted to be controlled by lake level fluctuations, with lacustrine facies largely restricted to high and low frequency maximum lake flooding events, and fluvial conglomerates and sandstones representing the corresponding incised valley-fills cut during, and filled after lake lowstand conditions. Mature palaeosols occur almost exclusively on surfaces interpreted as interfluvial sequence boundaries representing soil formation during lowstand hiati. Comparable styles of stratigraphic architecture in fluvial successions, albeit controlled by marine base level fluctuations, have been described by Aitken & Flint (1994, 1995).

The reversion to fluvial and lacustrine deposition, together with the development of calcic palaeosols within this unit is consistent with more humid climatic conditions relative to the underlying Unit V. Under such conditions, feldspar and lithic grains are generally less resistant and more mature sandstone compositions would be favoured (Suttner *et al.* 1981; Suttner & Dutta 1984; Johnson 1993). However, the Unit VI sandstones are more feldspathic than those of Units IV and V, which suggests a rejuvenation of the source area and a change in provenance.

Stratigraphic architecture of the lower Clair Group

The lower Clair Group is an unconformity bounded interval of fluvial, aeolian and lacustrine deposits which shows a systematic hierarchy of facies stacking patterns. These can be rationalized into at least three orders of 'cyclicity' (Fig. 8), with the nature of the facies stacking within a cycle dependent upon its position within a larger-scale, lower frequency cycle.

The first order cycle is defined as the entire lower Clair Group (Fig. 8). This shows an overall waning and waxing of the discharge within the component fluvial systems marked by: an overall waning in Units I to III from fluvial conglomerates through fluvial sandstones (Units I–II) to a major aeolian sandsheet (Unit III); and an overall waxing through Units IV to VI marking a change from ephemeral fluvial/aeolian systems to perennial fluvio-lacustrine systems.

The development of a basal unconformity with significant relief, a local unconformity at the Unit III–IV boundary (Allen & Mange-Rajetsky 1992) and a more areally widespread angular unconformity at the V–VI boundary, prior to truncation by the lower/upper Clair Group unconformity indicate that the first order stratigraphic cycle is the product of tectonic activity. The retreat stage (Units I to III) represents a record of progressive reduction of fluvial gradient and discharge, infilling of the original accommodation space created by subsidence, and increased intrabasinal reworking by aeolian processes as sediment supply was reduced. The subsequent increase in fluvial activity (Units IV–VI) is punctuated by three progressively greater unconformities (the III–IV, V–VI and lower–upper Clair Group boundaries). This interval is characterized by increasingly perennial fluvial systems and a detrital mineralogical change between Units V and VI, suggesting that the increased fluvial runoff operated in conjunction with increasing tectonic instability. The facies make-up of the non-fluvial elements of this first order cycle indicate a corresponding climatic change; Units I–III record the progressive replacement of lacustrine environments with aeolian sandsheet environments, and this trend is systematically reversed in units IV–VI as aeolian sandsheet environments were replaced by lacustrine systems. There therefore appears to have been a climatic influence on fluvial runoff which operated in conjunction with the tectonic accommodation cycle and ultimately dictated fluvial character within the lower Clair Group. This suggests the possibility that climate within the area was influenced by the evolving hinterland topography, with times of tectonic uplift of the hinterland enhancing precipitation, thereby increasing fluvial discharge and sediment supply

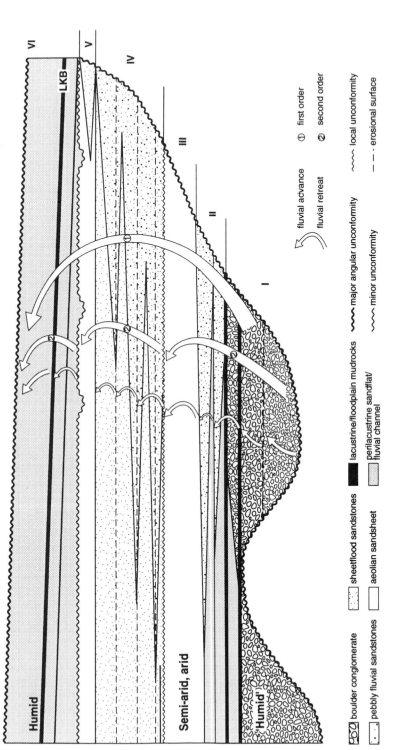

Fig. 8. A synthesis of the hierarchical stratigraphic cycles in the lower Clair Group. The high frequency (third order) cycles within Unit VI are illustrated schematically due to space limitation.

and elevating the palaeo-water table.

Three second order cycles are recognized (Fig. 8) and are interpreted to comprise: the 'drying/wetting-upward' facies cycles of Units I–III and Units IV/V, and the lacustrine transgressive–regressive cycle of Unit VI. Units I–III and IV/V record overall drying-upward from fluvial to aeolian deposits, with minor 'wetting-upward' trends preserved at the top of Unit III by offlap patterns of coarser-grained sandsheet facies (recording increased sediment flux of increasing calibre), and at the top of Unit V by the local preservation of an aeolian sandsheet to peri-lacustrine transitional facies below the V/VI unconformity. Units I–III occur on the first order waning-upward phase and show the most extreme spread of 'humid' to arid facies changes. Units IV/V occur in the early part of the first order increase in fluvial activity, and the low energy, maximum fluvial retreat phase of this second order cycle (Unit V) is less arid in this stratigraphic setting compared to Unit III. Unit VI comprises a fluvial retreat phase (up to the LKB) and advance of its fluvial systems, with the advancing phase well developed in this position within the larger-scale, first order advance. Unconformities bound these cycles and they are viewed as the product of higher frequency tectonic events. Concomitant climatic variation within these cycles are recorded by lacustrine to aeolian sandsheet facies trends within Units I–III and internally within Unit IV by locally developed lacustrine mudrocks at the base of the section which give way to aeolian deposits higher in the succession.

Third order cycles are recorded by: fining-upward cycles within the basal part of Unit I, fan progradational pulses interleaved with lacustrine and aeolian facies within Unit II, the dry/damp/dry sandsheet cycle in Unit III, 'drying/wetting-upward' fluvial/aeolian pulses in Unit IV and high frequency sequences within Unit VI. No correlatable cyclicity has been detected within Unit V.

Discussion

The above description of the lower Clair Group illustrates how this succession can be viewed as a series of 'nested' cycles which are the product of the genetically linked allocyclic forcing mechanisms of climate and tectonics. Previous studies on terrestrial systems emphasizing a sequence stratigraphic philosophy are few, with notable exceptions being Legarreta et al. (1993) and Steele & Ryseth (1990). Differentiation of the principal causes of such cycles remains enigmatic, but it is generally considered to be the depositional response to changes in the balance between the rate of accommodation creation and sediment supply through the influence of tectonics and climate (Blakey & Gubitosa 1984; Leeder & Gawthorpe 1987; Schlische 1991; López-Gómez & Arche 1993; Shanley & McCabe 1994). Forward modelling experiments by Paola et al. (1992) also suggest that the time span over which these changes occur, relative to the basin's response time (proportional to basin length and sediment transport rate) may also have an impact on stratigraphic architecture.

The Clair Group stratigraphy is interpreted to be the product of changes in the balance between tectonically induced accommodation and climatically modulated sediment supply, although the stratigraphic association of episodes of increased fluvial sediment flux and lacustrine facies with unconformities (of variable magnitude) suggests a tectonic control on basin climate. The initial phase of the first order cycle of the lower Clair Group was marked by the tectonic event which created the Clair Basin accommodation, followed by a progressive onlapping infill and source area denudation (Units I–III, second order cycle 1). Subsequent second order cycles 2 and 3 (Units IV/V and VI) reflect incrementally increasing tectonic uplift pulses which could reflect progressive uplift during fault block rotation. Hinterland uplift was associated with increased precipitation and enhanced runoff, but also resulted in less mature, initially lithic (second order cycle 2) and subsequently feldspathic sediment (second order cycle 3) being flushed into the basin. Maximum uplift generated the lower/upper Clair Group unconformity. Forward modelling experiments by Roberts et al. (1993) have demonstrated that such offlap patterns can be generated on a fault block by a progressively increasing strain rate. This provides a plausible explanation of the overall increase in activity of the fluvial systems of Units IV–VI as the product of successive pulses of increased strain rate causing uplift of the Rona Ridge hinterland and causing increased sediment to be flushed basinward, ultimately across proximal unconformities. Strain relaxation between these pulses allowed time for fluvial profiles gradually to equilibrate and generate a fining-upward signature as sediment supply decreased.

During the first order retreat (Units I–III) and early advance (Units IV/V) of the fluvial systems, the second order cycles show a general stratigraphic bias towards waning-upward deposits, with coarsening-upward elements recording increased sediment flux forming a volumetrically minor component. A similar trend is also reflected in the third order cycles within Unit IV. This stratigraphic bias indicates a general tendency for accommodation to exceed sediment supply as a result of reduced sediment flux as the system evolved towards increasingly arid climatic conditions. Similar stratigraphic bias within semi-arid to arid fluvial–aeolian systems has been reported by Chakraborty & Chaudhuri (1993), Clemmensen *et al.* (1989), Herries (1993) and Dreyer (1993), but also within a variety of climatic regimes by Legarreta *et al.* (1993). This does not appear to be the case within the wetting/drying upward cycle of Unit III, possibly because the top of this third order cycle also represents a second order sediment supply increase. In contrast, during the latter stages of the first order fluvial cycle (Unit VI) the stratigraphic bias is towards overall progradation, leading to a coarsening-upward bias. This may have been the result of enhanced runoff and sediment flux during climatically 'wet' phases together with more limited accommodation space. Again, the general tendency within such climatic conditions may be a greater stratigraphic bias towards coarsening upward cycles of fluvial advance, which may also be present in the general model proposed by Steele & Ryseth (1990) for fluvial megasequences in the North Sea Triassic.

Forward modelling experiments by Paola *et al.* (1992) have revealed the possibility for markedly different vertical grain size profiles in alluvial successions depending on variations in subsidence, water supply and sediment supply, and based on the time spans over which these parameters vary relative to the basin length and sediment transport rate. Whilst such experiments were simplified to isolate only one variable at a time, which is unlikely in real situations, they do give indications that the effects of each variable can generate very different stratigraphies. Their results from modelling rapid (*c.* 10^5 year) variations in sediment flux and water supply are of particular interest. Flux variations apparently generated a strong stratigraphic bias towards coarsening-upward cycles, whilst variations in water supply over comparable time spans produced a stratigraphic bias towards fining-upward successions. These results could be applied to the Clair succession to suggest that during semi-arid to arid conditions the fining-upward bias was largely influenced by changes in water supply, whilst in the more humid situation in Units II and VI, sediment supply variations were the dictating factor in sediment architecture. Clearly this is a simplification because climate and tectonics may have been related, and hence water supply,

sediment supply and accommodation were interdependent and varied simultaneously. However, depending on the overall balance of these variables at any one time, the effects of one may have been of greater impact on the stratigraphic architecture. Potentially while water supply was abundant, variations in sediment supply became the overriding variable in dictating stratigraphic architecture, whilst in semi-arid conditions with an adequate sand budget available, water supply variations became the critical factor in dictating sediment architecture.

Conclusions

Variations in the grain size, conglomerate fraction and mineralogy of fluvial, aeolian and lacustrine deposits within the lower Clair Group can be rationalized into three orders of stratigraphic cyclicity.

First order unconformity bounded cycle characterized by an overall decrease then increase in fluvial activity which shows evidence of a matching climatic change in non-fluvial facies from humid through semi-arid to arid conditions and back to humid conditions.

Second order cycles, bounded by minor unconformities, are characterized by a waning-upward facies in the semi-arid/arid parts of the first order cycle, and by a greater proportion of coarsening-upward deposits in the more humid end members of the first order cycle.

Third order cycles are commonly self similar in character to their enclosing second order cycles, but with the proportion and character of fluvial v. aeolian or lacustrine deposits dependent on the stratigraphic position within lower frequency cycles.

The controls on these stratigraphic cycles are multivariate and hence difficult to differentiate. Variations in tectonic strain rate (decreasing then increasing) probably created the first order cycle and had a sufficient impact on hinterland topography to influence sediment yield and precipitation. Pulses in tectonism may have resulted in the minor unconformities which bound the second order cycles. Variations in the style of stratigraphic architecture within the second and third order cycles are ascribed to changes in the accommodation : supply ratio of the alluvial system, in turn the product of variations in tectonic accommodation and genetically linked sediment and water supply variations. It may be speculated that during first order semi-arid/arid conditions, water supply variations were critical in generating a waning-upward stratigraphical bias, whilst in more humid conditions, sediment flux variations generated a greater stratigraphic bias towards waxing upward deposits.

In terms of subsurface reservoir characterization, the sequence stratigraphic approach described has provided a more detailed correlation scheme which is predictive of the distribution of porous, aeolian sandstones which form the main reservoir intervals. Within Unit IV these were previously regarded as random and discontinuous, but are now recognized to be the product of areally extensive aeolian reworking episodes, with their greater extent (and calibration with seismic) impacting on STOIIP calculations. In addition, the distribution, geometry and character of reservoir heterogeneities are more accurately defined for the reservoir. For example, the occurence of permeable, incised channels directly underlying the areally extensive LKB clearly presents the danger of very rapid fluid underrunning which will impact on well completions.

BP and the Clair Field partners are thanked for permission to publish this paper. Russell Smith in particular is thanked for his enthusiastic support. The manuscript has benefited from the reviews of J. Howell & S. Flint.

References

AITKEN, J. F. & FLINT, S. S. 1994. High frequency sequences and the nature of incised valley-fills in fluvial systems of the Breathitt Group (Pennsylvanian), Appalachian Foreland Basin, Eastern Kentucky. *In:* DALRYMPLE, R., BOYD, R. & ZAITLIN, B. (eds) *Incised Valley Systems: Origin and Sedimentary Sequences.* SEPM Special Publication, **51**, 353–368.

—— & —— 1995. The application of high resolution sequence stratigraphy to fluvial systems: a case study from the Upper Carboniferous Breathitt Group, eastern Kentucky. *Sedimentology*, **42**, 3–30.

ALLEN, P. A. & MANGE-RAJETSKY, M. 1992. Sedimentary evolution of the Devonian–Carboniferous Clair Field offshore northwestern UK: impact of changing provenance. *Marine and Petroleum Geology*, **9**, 29–52.

BLACKBOURN, G. A. 1987. Sedimentary environments and stratigraphy of the Late Devonian–Early Carboniferous Clair Basin, west of Shetland. *In:* MILLER, J., ADAMS, A. E. & WRIGHT, V. P. (eds) *European Dinantian Environments.* Wiley and Sons Ltd, 75–91.

BLAKEY, R. C. & GUBITOSA, R. 1984. Controls of sandstone body geometry and architecture in the Chinle Formation (Upper Triassic), Colorado Plateau. *Sedimentary Geology*, **38**, 51–86.

CHAKRABORTY, T. & CHAUDHURI, A. K. 1993. Fluvial–aeolian interactions in a Proterozoic alluvial plain: example from the Mancheral Quartzite, Sullavai Group, Pranhita-Godavari Valley, India. *In:* PYE, K. (ed.) *The Dynamics and Environmental Context of Aeolian Sedimentary Systems.* Geological Society, London, Special Publication, **72**, 127–141.

CLEMMENSON, L. B., OLSEN, H. & BLAKEY, R. C. 1989. Erg-margin deposits in the Lower Jurassic Moenave Formation and Wingate Sandstone, south Utah. *Geological Society of America Bulletin*, **101**, 759–773.

CONEY, D., FYFE, T. B., RETAIL, P. & SMITH, P. J. 1993. Clair Field appraisal: benefits of a co-operative approach. *In:* PARKER, J. R. (ed.) *Petroleum Geology of Northwest Europe. Proceedings of the 4th Conference.* Geological Society, London, 1409–1420.

DREYER, T. 1993. Quantified fluvial architecture in ephemeral stream deposits of the Esplugafreda Formation (Palaeocene), Tremp-Graus Basin, northern Spain. *In:* MARZO, M. & PUIGDEFABREGAS, C. (eds) *Alluvial Sedimentation.* IAS Special Publication, **17**, 337–362.

DUINDAM, P. & VAN HOORN, B. 1987. Structural evolution of the west of Shetland continental margin. *In:* BROOKS, J. & GLENNIE, K. (eds) *Petroleum Geology of Northwest Europe.* Graham and Trotman, 1131–1148.

DUTTA, P. K., ZHOU, Z. & DOS SANTOS, P. R. 1993. A theoretical study of mineralogical maturation of eolian sand. *Geological Society of America, Special Paper*, **284**, 203–209.

FRYBERGER, S. G., AHLBRANDT, T. S. & ANDREWS, S. 1979. Origin, sedimentary features and significance of low angle aeolian 'sand sheet' deposits, Great Sand Dunes National Monument and vicinity, Colorado. *Journal of Sedimentary Petrology*, **49**, 733–746.

HERRIES, R. D. 1993. Contrasting styles of fluvial–aeolian interaction at a downward erg margin: Jurassic Kayenta–Navajo transition, northeastern Arizona, USA. *In:* NORTH, C. P. & PROSSER, D. J. (eds) *Characterisation of Fluvial and Aeolian Reservoirs.* Geological Society, London, Special Publication, **73**, 199–218.

JOHNSON, M. J. 1993. The systems controlling the composition of clastic sediments. *Geological Society of America, Special Paper*, **284**, 1–19.

KOCUREK, G. & DOTT, R. H. 1981. Distinctions and uses of stratification types in the interpretation of aeolian sand. *Journal of Sedimentary Petrology*, **51**, 579–595.

—— & FIELDER, G. 1982. Adhesion structures. *Journal of Sedimentary Petrology*, **52**, 1229–1241.

—— & HAVHOLM. K. G. 1993. Eolian event stratigraphy – a conceptual framework. *In:* WEIMER, P. & POSAMENTIER, H. (eds) *Siliciclastic Sequence Stratigraphy: Recent Developments and Application.* American Association of Petroleum Geologists, Memoir, **58**, 393–410.

LEEDER, M. R. & GAWTHORPE, R. L. 1987. Sedimentary models for extensional tilt-block/half graben basins. *In:* COWARD, M. P., DEWEY, M. P. & HANCOCK, A. L. (eds) *Continental Extensional Tectonics*. Geological Society, London, Special Publication, **28**, 139–152.

LEGARRETA, L., ULIANA, M. A., LAROTONDA, C. A. & MECONI, G. R. 1993. Approaches to non-marine sequence stratigraphy – theoretical models and examples from Argentine Basins. *In:* ESCHARD, R. & DOLIGEZ, B. (eds) *Subsurface Reservoir Characterisation from Outcrop Observations*. Editions Technip, Paris, 125–143.

LOOPE, D. B. 1985. Episodic deposition and preservation of eolian sands: a late Palaeozoic example from south-eastern Utah. *Geology*, **13**, 73–76.

—— & SIMPSON, E. L. 1992. Significance of thin sets of eolian cross-strata. *Journal of Sedimentary Petrology*, **62**, 849–859.

LÓPEZ-GÓMEZ, J. & ARCHE, A. 1993. Architecture of the Cañuzar fluvial sheet sandstones, Early Triassic, Iberian Ranges, eastern Spain. *In:* MARZO, M. & PUIGDEFABREGAS, C. (eds) *Alluvial Sedimentation*. IAS Special Publication, **17**, 363–382.

McCLAY, K. R., NORTON, M. G., CONEY, P. & DAVIS, G. H. 1986. Collapse of the Caledonian orogen and the Old Red Sandstone. *Nature*, **323**, 147–149.

MITCHUM, R. M. & VAN WAGONER, J. C. 1991. High frequency sequences and their stacking patterns: sequence stratigraphic evidence of high frequency eustatic cycles. *Sedimentary Geology*, **70**, 131–160.

PAOLA, C., HELLER, P. L. & ANGEVINE, C. L. 1992. The large scale dynamics of grain size variation in alluvial basins, I: theory. *Basin Research*, **4**, 73–90.

REID, L. & FROSTICK, L. E. 1989. Channel form, flows and sediments in deserts. *In:* THOMAS, D. S. G. (ed.) *Arid Zone Geomorphology*. Belhaven Press, 117–135.

RIDD, M. F. 1981. Petroleum geology west of the Shetlands. *In:* ILLING, L. V. & HOBSON, C. D.

(eds) *Petroleum Geology of the Continental Shelf of North West Europe*. Heyden, London, 414–425.

ROBERTS, A. M., YIELDING, G. & BADLEY, M. E. 1993. Tectonic and bathymetric controls on stratigraphic sequences within evolving half-graben. *In:* WILLIAMS, G. & DOBB, A. (eds) *Tectonics and Seismic Sequence Stratigraphy*. Geological Society, London, Special Publication, **71**, 87–121.

SCHLISCHE, R. W. 1991. Half-graben basin filling models: new constraints on continental extensional basin development. *Basin Research*, **3**, 123–141.

SHANLEY, K. W. & McCABE, P. J. 1994. Perspectives on the sequence stratigraphy of continental strata. *American Association of Petroleum Geologists Bulletin*, **78**, 544–568.

STEELE, R. & RYSETH, A. 1990. The Triassic–early Jurassic succession in the northern North Sea: megasequence stratigraphy and intra-Triassic tectonics. *In:* HARDMAN, R. F. P. & BROOKS, J. (eds) *Tectonic Events Responsible for Britain's Oil and Gas Reserves*. Geological Society, London, Special Publication, **55**, 139–168.

SUTTNER, L. J. & DUTTA, P. K. 1986. Alluvial sandstone composition and palaeoclimate I. Framework mineralogy. *Journal of Sedimentary Petrology*, **56**, 329.

——, BASU, A. & MACK, G. H. 1981. Climate and the origin of quartz arenites. *Journal of Sedimentary Petrology*, **51**, 1235–1246.

TREWIN, N. H. 1993. Controls on fluvial deposition in mixed fluvial and aeolian facies within the Tumblagooda Sandstone (Late Silurian) of Western Australia. *Sedimentary Geology*, **85**, 387–400.

VAN WAGONER, J. C., MITCHUM, R. M., CAMPION, K. M. & RAHMANIAN, V. D. 1990. *Siliciclastic Sequence Stratigraphy in Well Logs, Cores and Outcrops*. American Association of Petroleum Geologists, Methods in Exploration Series, **7**.

ZIEGLER, P. A. 1982. Faulting and graben formation in Western and Central Europe. *Philosophical Transactions of the Royal Society of London*, **A305**, 113–143.

Marine and nonmarine systems tracts in fourth-order sequences in the Early–Middle Cenomanian, Dunvegan Alloformation, northeastern British Columbia, Canada

A. GUY PLINT

Department of Earth Sciences, University of Western Ontario, London, Ontario, N6A 5B7, Canada

Abstract: The Early to Middle Cenomanian Dunvegan Alloformation is a 3rd-order clastic sequence that was deposited on the western margin of the Alberta foreland basin. The alloformation, comprising ten allomembers (A–J) has been mapped over 33 000 km² using 1500 well logs and 40 outcrops. Regional flooding surfaces are used to define allomembers because they are more reliably identified in well logs than are Exxon-type unconformities (although in many instances unconformities and flooding surfaces are coplanar). The Dunvegan records episodic deltaic progradation from NW to SE along the axis of the foreland basin. This paper focuses on an 800 km² study area in NE British Columbia where both shallow marine and coastal plain strata are exposed in the valleys of the Peace and Beatton rivers. Marine strata comprise sandier-upward successions that grade from platy mudstone through cm-scale interbeds of wave rippled fine sandstone to dm-scale beds of HCS, commonly including large gutter casts. Successions may be capped by SCS and cross-bedded, fine-grained sandstone, sometimes with a rooted top. The succession of facies suggests a strongly wave-influenced deltaic shoreline. Gutter casts are consistently orientated shore-normal and appear to have been formed by oscillatory wave scour during storms. Sandstone distribution varies between successive progradational packages ('shingles') in allomember J, suggesting that delta switching was an important control on depositional cyclicity. Coastal plain deposits include small-scale (few metres) coarsening-upward successions of carbonaceous mudstone, siltstone and fine sandstone interpreted as lake and bay-fill deposits. Decimetre scale sharp-based sandstones may represent crevasse splays whereas lenticular, rippled and crossbedded sandstones were probably deposited in distributary channels. Rooted palaeosols are abundant. Both coastal plain and marine strata are locally replaced by erosive-based bodies of cross-bedded sandstone up to 25 m thick and 0.5–3 km wide, interpreted as incised valley fills underlain by type 1 sequence boundaries. Additional evidence of relative sea-level fall is provided by regressive surfaces of marine erosion beneath sharp-based shoreface sandstones. If attached to highstand deposits, these sandstones are assigned to the falling stage systems tract, whereas detached sandstones are assigned to the lowstand systems tract. During the early part of the transgressive systems tract, sedimentation was confined to incised valleys but later, interfluves were buried by sediments that locally contain dense *Thalassinoides* burrow systems, *Teredolites* and well-developed mud drapes, suggestive of brackish, tidally-influenced conditions. Overlying coastal plain strata lack evidence of marine influence and are attributed to the highstand systems tract.

Regional setting. The Dunvegan Formation, of Early to Middle Cenomanian age (Caldwell *et al.* 1978; Stott 1982; Singh 1983; Bhattacharya & Walker 1991*a*) represents an extensive deltaic complex that was deposited on the western side of the Alberta foreland basin. During the Late Cretaceous, this basin formed part of the Western Interior Seaway which extended from the Gulf of Mexico to the Arctic Ocean (Fig. 1). To the west, the seaway was bounded by the rising Cordillera, which yielded large volumes of clastic detritus, whereas to the east, the low-lying cratonic interior of North America yielded little sediment (Fig. 1). The bulk of the sediment forming the Dunvegan Formation was sourced from northeastern British Columbia and southern Yukon, and was transported to the southeast by rivers flowing parallel to the active margin of the basin (Stott 1982). Additional sediment was introduced by rivers flowing to the east and northeast, perpendicular to the adjacent Cordillera.

In northern British Columbia and southern Yukon, the Dunvegan consists dominantly of sandstones and conglomerates deposited in various alluvial environments (Stott 1982). Further south, the formation consists largely of fine- to medium-grained sandstone, siltstone and

From Howell, J. A. & Aitken, J. F. (eds), *High Resolution Sequence Stratigraphy: Innovations and Applications*, Geological Society Special Publication No. 104, pp. 159–191.

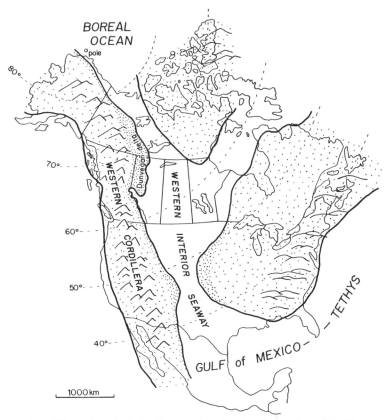

Fig. 1. Palaeogeography of North America in the Cenomanian, showing the location of the Dunvegan Formation delta complex in relation to the Western Interior Seaway and the Western Cordillera. After Bhattacharya (1993), modified from Williams & Stelck (1975). Palaeolatitudinal data from Irving *et al.* (1993).

mudstone, deposited in a range of deltaic and shallow marine environments.

Study area and database. The work reported here forms a small part of a regional study covering about 33 000 km², with stratigraphic control provided by 1500 well logs and about 40 major outcrop sections (Fig. 2). This database, in combination with the work of Bhattacharya & Walker (1991a), has been used to define and map ten allomembers. For practical purposes, regional flooding surfaces were chosen as the bounding surfaces of allomembers, rather than unconformities, as the former could be more reliably recognized in well logs. In reality, marine flooding surfaces and unconformities related to relative sea-level fall were coplanar in many wells. This paper focuses on a small area of northeastern British Columbia, north of the Peace River (Fig. 2). In this area, the Dunvegan Formation crops out extensively in the valleys of the Peace and Beatton rivers as a

result of the gentle southerly regional dip (Fig. 2). Although patchy exposure exists throughout these river valleys, thick, well exposed sections, suitable for regional stratigraphic correlation, are rare and widely spaced.

Purpose of paper. In addition to providing the first detailed stratigraphic analysis of the Dunvegan Formation in northeastern British Columbia, this paper illustrates the use of subsurface data to constrain correlations between outcrop sections. The allostratigraphic scheme employed herein allows the Dunvegan Formation to be divided into genetically related packages within which facies and palaeocurrent data can be interpreted in a temporal and palaeogeographic framework. The rocks have also been interpreted in sequence stratigraphic terms, although some difficulty was encountered in applying the existing terminology, particularly to rocks deposited during relative sea-level fall (cf. Ainsworth 1994; Ainsworth & Pattison

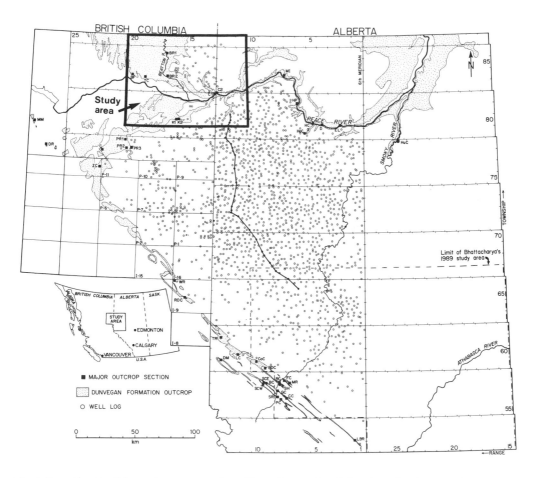

Fig. 2. Location of well log and outcrop stratigraphic sections used to define the regional allostratigraphy of the Dunvegan Formation on a regional basis. The study area partially overlaps study area of Bhattacharya (1989), shown by dashed line. The boundary of the detailed study area discussed in this paper is shown, as is the position of the cross-section shown in Fig. 4.

1994). The Dunvegan therefore provides an opportunity to discuss some problems with, and possible solutions to, the practical application of high resolution sequence stratigraphy.

Stratigraphy

Lithostratigraphy. The Dunvegan Formation has been mapped as a lithostratigraphic unit throughout the Foothills of Alberta and British Columbia, north of the Athabasca River (Fig. 2). A comprehensive account of the history of research on this unit is given by Stott (1982). The base of the formation is picked at the first thick sandstone overlying marine mudstone of the Shaftesbury Formation, and the top is marked

by an abrupt change from deltaic sandstone and siltstone to marine mudstone of the overlying Kaskapau Formation (Figs 3 and 4). South of the Athabasca River the formation grades laterally into mudstone and is mapped as a part of the Blackstone Formation (Fig. 3).

Allostratigraphy. Subsurface study of the Dunvegan in west-central Alberta (Fig. 2; Bhattacharya & Walker 1991a, b; Bhattacharya 1993; Plint 1994) showed that the formation comprises a series of aggradational and progradational deltaic packages that collectively form a third-order sequence. Bhattacharya & Walker (1991a) defined seven Dunvegan allomembers, designated A to G in descending order, each of which

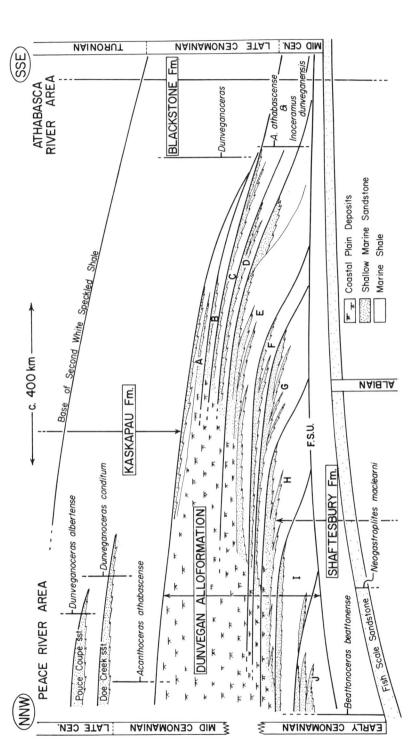

Fig. 3. Summary dip section across regional study area shown in Fig. 2, illustrating lithological and biostratigraphic relationships between the Shaftesbury, Dunvegan, Kaskapau and Blackstone formations. The upper and lower boundaries of the Dunvegan Alloformation, as used in this paper, are indicated. The base of the Dunvegan Formation, based on lithostratigraphic criteria, is defined at the first appearance of sandstone above the Shaftesbury shale. This is a highly diachronous contact and does not facilitate high resolution palaeogeographic analysis. Biostratigraphic data from Caldwell *et al.* (1978) and Stott (1963).

was bounded by a regional flooding surface, mappable in logs and core. Allomembers downlap from northwest to southeast onto a prominent condensed section, termed the 'Fish Scales Upper' or FSU marker, within the Shaftesbury Formation (Bhattacharya & Walker 1991*a*; Figs 3 and 4).

The new work reported here confirms the validity and mappability of Bhattacharya & Walker's (1991*a*) allomembers A–G, and reveals an additional three underlying allomembers, H, I and J that downlap and disappear to the north of Bhattacharya's (1989) study area (Fig. 2). Similarly, even older units must exist to the north of the present study area.

Bhattacharya & Walker (1991*a*) and Plint (1994) demonstrated that each Dunvegan allomember showed a lateral gradation from updip sandy ('Dunvegan') facies into downdip muddy ('Shaftesbury') facies, and that allomembers C–J downlapped onto the FSU marker. The top of the Dunvegan Formation (top of allomember A) is defined by a regionally-mappable flooding surface.

Although Bhattacharya & Walker (1991*a*) defined mappable Dunvegan *allomembers*, they did not extend their stratigraphic logic to define a Dunvegan *Alloformation*. Given the fact that the FSU marker is easily recognizable, both in every well log to depth, and in outcrop (Plint & Bhattacharya 1991; Plint unpublished data), and that the FSU marker forms the base of allomembers C–J, it is here proposed that the FSU marker be designated as the base of the *Dunvegan Alloformation*. The top of the Dunvegan Alloformation is placed at the flooding surface at the top of allomember A, a surface that can be traced throughout the subsurface study area (Figs 2, 3 and 4).

Bhattacharya & Walker (1991*a*) recognized that allomembers, bounded by flooding surfaces of regional extent and probable allogenic origin, commonly contained several, more localized shoaling-upward cycles, bounded by flooding surfaces. These they termed 'shingles' (broadly comparable to the parasequences of Van Wagoner *et al.* 1988). Shingles were interpreted to reflect autogenic processes of delta lobe switching. Sub-regional shoaling-upward successions have also been recognized in the new allomembers H, I and J, and to maintain consistency with the established terminology of Bhattacharya & Walker (1991*a*), the term 'shingle' is applied here too. Note, however, that in the present paper, it was thought more logical to number shingles in order of deposition, i.e. shingle J-1 underlies J-2 whereas Bhattacharya & Walker (1991*a*) numbered from youngest to oldest, i.e. shingle E-1

overlies E-2.

Bio- and chronostratigraphy. Because of its predominantly marginal- to nonmarine character, the Dunvegan Formation contains few ammonites and is consequently difficult to date. In the present study area, marine shales that would be ascribed to the Shaftesbury Formation on lithostratigraphic criteria (Figs 3 and 4) contain foraminifera of the *Textularia alcesensis* zone and the zonal ammonite *Beattonoceras beattonense*, indicative of a late Early Cenomanian age (Stelck *et al.* 1958; Caldwell *et al.* 1978). Figures 4 and 5 show that these 'Shaftesbury' shales can be traced updip into nearshore sandy facies, and belong to allomembers J and I of the Dunvegan Alloformation. In the same area, the basal 50 m of the marine Kaskapau Formation above the Dunvegan (Fig. 3) yields the ammonite *Acanthoceras athabascense* and foraminifera of the *Ammobaculites gravenori* subzone (Caldwell *et al.* 1978) indicative of a Middle to early Late Cenomanian age. Far to the south (approximately Township 52, Fig. 2) shales of the Blackstone Formation (Fig. 3) approximately equivalent to Dunvegan allomembers A–E have yielded the bivalve *Inoceramus dunveganensis* and ammonite *Acanthoceras athabascense* (Stott 1963) also indicative of a Middle to early Late Cenomanian age. Thus the Dunvegan Alloformation within the regional study area depicted in Fig. 2 appears to have been deposited during the *B. beattonense* and succeeding *A. athabascense* zones, representative of the late Early Cenomanian to perhaps the earliest Late Cenomanian. In terms of the most recent radiometric determinations (Obradovitch 1993) these zones probably represent an interval of about 1.5 ma. On this basis, each Dunvegan allomember represents about 150 000 years, making the simplistic assumption that each allomember represents an equal interval of time.

Outcrop to subsurface correlation. Although the measured Dunvegan outcrop sections are generally well-exposed, lateral thickness, facies changes and the lack of definitive stratigraphic markers made it impossible to correlate reliably between sections. It is, however, possible to correlate, with confidence, some outcrop sections with geophysical logs from wells a few kilometres distant. All available well logs (about 50) on the north side of the Peace River were correlated via a rigorously checked grid of cross sections. The subsurface correlation grid helped to establish regional thickness and facies trends within each allomember, made it possible to predict the stratigraphy likely to be encountered in outcrop, and provided a basis for projecting key surfaces from subsurface to outcrop.

The local stratigraphy, defined on the north side of the Peace River was correlated with the regional allostratigraphy, defined south of the river. Because regional allomember-bounding, and local shingle-bounding flooding surfaces are generally indistinguishable in individual logs or outcrops, it was essential to establish a correlation of local with regional stratigraphies. Only regional stratigraphic correlation permitted distinction of localized (few tens of kilometres), shingle-bounding surfaces from the more extensive (hundreds of kilometres) surfaces that bound allomembers. Figure 4 shows how the bounding surfaces of allomembers G, H, I and J can be correlated into the study area north of the Peace River. Figures 5 and 6 show detailed sections representative of the stratigraphy in the 'north Peace' study area.

Stratigraphy of allomembers G, H, I and J

Allomember J

Allomember J is divided into three, mappable, sandier-upward shingles, bounded by flooding surfaces (Figs 5 and 6). A more detailed correlation of facies successions in outcrop, constrained by the data in Figs 5 and 6, is presented in Fig. 7.

The J-1 shingle comprises several minor sandier-upward successions, typically 5–10 m thick, comprising cm- to dm-scale interbeds of very fine-grained wave-rippled sandstone and grey platy mudstone (Fig. 8). Thicker sandstone beds may show hummocky cross stratification (HCS), and gutter casts, either isolated in mudstone or extending from the base of sandstone beds (Fig. 9). The walls of gutters are commonly vertical, even undercut, and may be ornamented with small groove and flute marks which, however, rarely give a clear indication of flow polarity. Shingle J-1 is unusual in containing numerous sandstone beds that have undergone extensive loading and soft-sediment deformation. Shingle J-1 culminates in a fine-grained sandstone containing HCS, swaley cross-stratification (SCS), wave ripples and trough crossbedding. In places, roots are present at the top (Figs 5 and 7).

Between wells 6-1-85-16W6 and 14-1-84-15W6 (Fig. 5), shingle J-1 splits into several sandy tongues, below which are abrupt log deflections in the shaly part of the succession. The abrupt thickness and facies changes are interpreted to reflect facies changes over growth faults on the delta front, an interpretation supported by the observation of a growth fault in the same stratigraphic interval in outcrop at Tea Creek (Figs 7 and 10).

The distribution of shoreface/inner shelf sandstone in J-1, interpreted from well log response, forms an elongate lobe extending at least 60 km towards the NE (Fig. 11).

Shingle J-2 is best developed at Beatton River 1 (Figs 6 and 7) where it comprises an upward gradation from cm-scale interbeds of very fine-grained, wave-rippled sandstone and mudstone into SCS and trough cross-bedded fine-grained sandstone. The upper, sandy portion of J-2 is best developed in the north of the study area, and southward, the shingle is represented by thinly-interbedded mudstone and fine sandstone (e.g. Tea Creek, Bear Flat, Fig. 7). An elongate 'lobe' of sandstone apparently extends for about 20 km southeast of the main sandbody (Fig. 12).

Shingle J-3 at Beatton River 1 (Fig. 7) comprises three sandier-upward, wave-dominated successions characterized by wave ripples and HCS. The top of J-3 is marked by a 2 m thick unit of SCS sandstone with a scoured base marked by large gutter casts (Fig. 7). In outcrop at Bear Flat, Tea Creek and Beatton River 2 (Figs 5 and 7), shingle J-3 consists primarily of thinly-interbedded mudstone and wave-rippled sandstone with gutter casts. Most gutters are sandstone-filled although some are cut in sandstone and filled with laminated mudstone (Fig. 13; cf. Goldring & Aigner 1982). The distribution of sandstone in shingle J-3 is shown in Fig. 14.

Fig. 4. Representative regional well log dip section (located on Fig. 2) from south side of the Peace River to the northern part of Bhattacharya's (1989) study area. The section joins the south end of Fig. 5 to the north end of Bhattacharya & Walker's (1991a) fig. 4 (well 7-29-67-8W6). Note, however, that the top of allomember E shown here is at a slightly higher position than that picked by Bhattacharya & Walker. Regional correlation by the author supports the pick shown here. Stippled pattern denotes incised valley fill deposits. Logs are gamma ray (left) and resistivity (right).

Fig. 5. Detailed log cross-section A–A' linking outcrop sections on the Peace and Beatton Rivers. Location of cross-section shown in Fig. 11. Section joins the north end of Fig. 4. See text for detailed discussion.

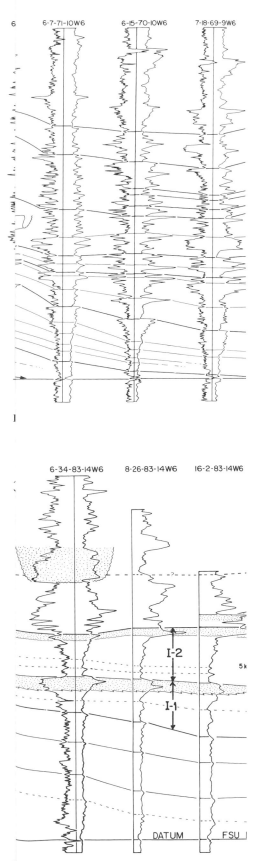

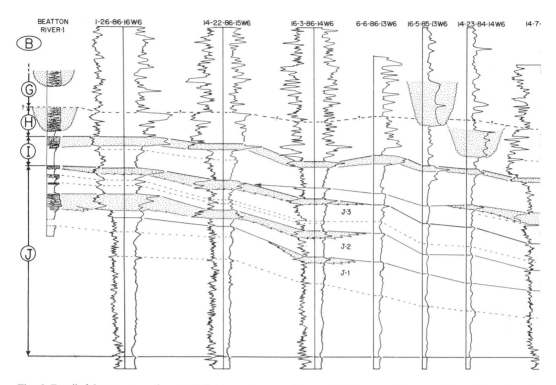

Fig. 6. Detailed log cross-section B–B′ linking outcrop sections on the Peace and Beatton Rivers. Location of cross-section shown in Fig. 11. Section joins the north end of Fig. 4. See text for detailed discussion.

interpreted onlapping relationship to J-3, shingle I-1 (Figs 5 and 8) is interpreted as a lowstand delta deposit that developed following a relative sea-level fall. A similar stratigraphic relationship was demonstrated between shingles 2 and 1 in allomember E, shingle E-1 being interpreted as a lowstand delta (Bhattacharya 1993). Relative sea-level rise drowned the I-1 lowstand delta and resulted in at least 70 km of shoreline backstep to a position west of Bear Flat (Figs 7 and 16). The fact that allomember I is dominated by sandstone at Bear Flat suggests that this locality lay close to the maximum limit of transgression.

Palaeogeography. The consistent orientation of wave ripples and orthogonal trend of gutter casts in shingle I-2 suggest a shoreline trending broadly SW–NE (Fig. 17). Southeasterly progradation of the delta shoreline is indicated by the southeasterly thickening of allomember I, preferential development of minor sandy successions to the northwest, and lateral pinchout of inferred shoreface sandstones towards the east and south (Fig. 17). Although sedimentary structures suggest a broadly SW–NE shoreline trend, the eastern margin of the I-2 sandstone,

inferred from well logs (Fig. 17), shows a lobate pattern suggesting local concentration of sand, perhaps around distributary mouths. This local observation is supported by regional mapping (beyond the 'north Peace' study area) of the I-2 sandstone which has a pronounced lobate margin suggesting coalescence of at least three major river deltas (Fig. 15B).

Allomember H

In the study area, allomember H is dominated by coastal plain deposits. Only in the south are marine shelf and shoreline deposits present, and are exposed at Clayhurst 1 and 2 (Figs 5, 6 and 7). Marine sandstone of allomember H can tentatively be traced northward from Clayhurst 1 for a few kilometres before the distinctive log response becomes unrecognizable (Figs 5 and 6). In exposures further to the north and west, shingle I-2 is directly overlain by fresh to brackish water deposits. At Bear Flat and Tea Creek, the basal 60 cm of allomember H consists of sandy siltstone, intensely burrowed by *Thalassinoides* (Fig. 20). Overlying beds at Bear Flat, Tea Creek and Beatton River 1 and 2 consist of

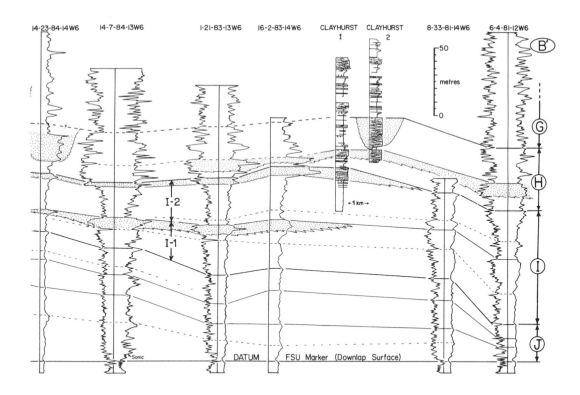

dark grey or brown, laminated mudstone rich in leaf debris and occasional *Teredo*-bored logs, interbedded with, or grading up into, very fine-grained sandstones with wave and current ripple cross-lamination. At Clayhurst 1, similar lithologies, including thin coals and rooted horizons, form the upper part of allomember H, above the marine shoreface sandstone (Fig. 7).

In most outcrop sections, the bulk of allomember H comprises a unit of fine- to medium-grained, trough cross-bedded sandstone up to 25 m thick (Figs 7 and 21). Palaeocurrents are directed broadly towards the southeast. The base of each sandbody is marked by an intraclast lag of mudstone and siderite pebbles. Near-continuous exposure along the walls of the Beatton River canyon (in the vicinity of locality Beatton River 1, Fig. 11) shows that the cross-bedded sandbodies are highly lenticular with steeply-incised margins (Fig. 22). In places, the sandstones are partitioned by large-scale, low angle accretion surfaces and may enclose mudstone lenses. Locally, the base of the sandstone cuts down into allomember I (Figs 21 and 22). Three-dimensional exposure afforded by the canyon shows that the sandbodies are several

hundred metres to perhaps 1–3 km wide in flow-transverse section, and trend NW–SE. The sandbodies can also be mapped in subsurface, based on their distinctive blocky log response (Fig. 15C). Preliminary work suggests that sandbodies have a highly sinuous planform.

Interpretation of allomember H

Facies successions. The restriction of marine facies to the southeast corner of the study area shows that marine transgression at the base of allomember H extended to some position between Clayhurst 1 and Beatton River 2 (Fig. 7). The presence of lagoonal/lacustrine deposits at the base of allomember H in sections north-west of Clayhurst 1 suggests that, behind the transgressing shoreline, accommodation space created on the coastal plain was filled by vertical aggradation in lakes and lagoons (Thorne & Swift 1991; Shanley & McCabe 1993). The presence of abundant *Thalassinoides* at the base of allomember H at Bear Flat, and *Teredolites* 4.5 m above the base, suggests that initial flooding of the coastal plain was by at least brackish water. The upward disappearance of marine

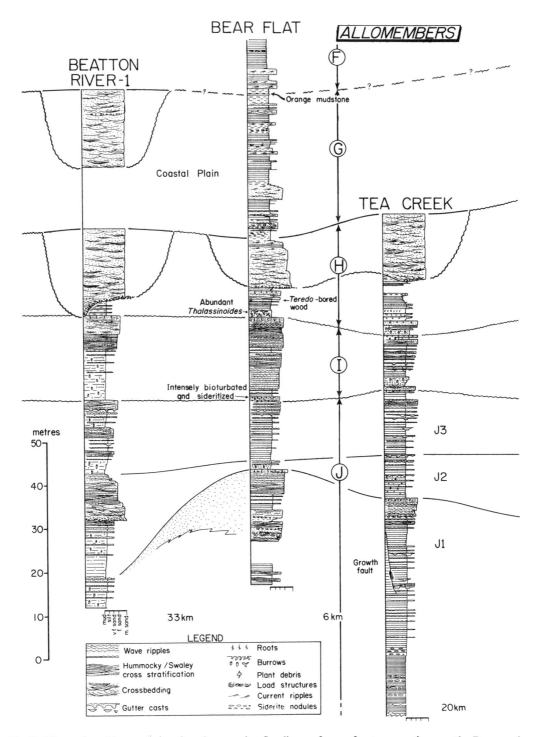

Fig. 7. Allostratigraphic correlation, based on marine flooding surfaces, of outcrop sections on the Peace and Beatton Rivers. Broad environmental interpretations are indicated. Inset map shows relative position of sections. A sequence stratigraphic interpretation of these sections is given in Fig. 23.

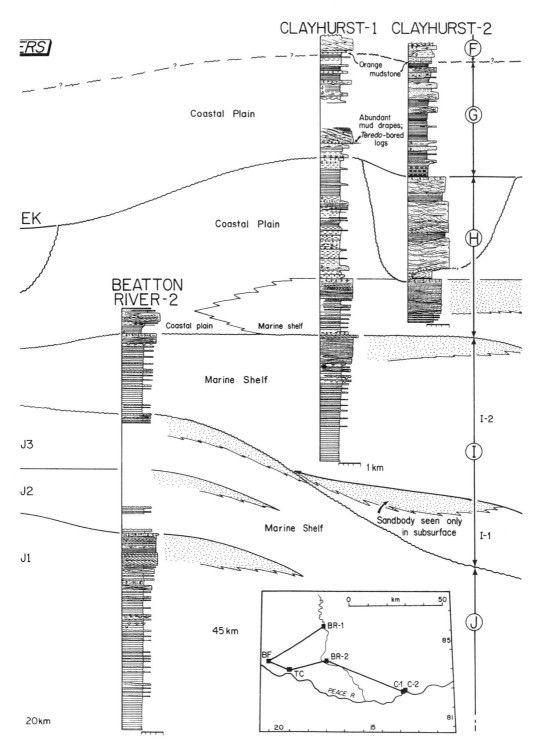

Fig. 8. Centimetre-scale interbeds of very fine-grained, plane-laminated and wave-rippled sandstone and platy grey mudstone, interpreted to represent storm deposition on a shelf several tens of kilometres from shore. Shingle J-1, Beatton River 1; scale bar 20 cm.

trace fossils suggests a reversion to freshwater lakes as transgression changed to regression. Subsurface data (Fig. 4) show that, to the southeast of the study area, allomember H contains several more shingles comprising transgressive marine mudstones and regressive marine shoreface sandstones.

Palaeogeography. The progradational limit of marine sandstone in allomember H, which lies 100 km SE of Clayhurst, shows an arcuate shoreline suggesting two coalesced, strongly wave-influenced delta lobes (Fig. 15C).

The upper 28 m of allomember H at Clayhurst 1 contain no field-observable evidence of marine conditions. It is therefore reasonable to infer that the shoreline never again transgressed this far north during deposition of the remainder of allomember H. Based on the lateral equivalence of marine and nonmarine strata in the lower part of allomember H, the upper 28 m of coastal plain deposits at Clayhurst 1 are interpreted to have aggraded behind an intermittently transgressive, but overall regressive shoreline that lay farther to the southeast.

The thick, lenticular, cross-bedded sandstone bodies that deeply incise coastal plain deposits of allomember H are interpreted, on the basis of their dimensions and complex multistorey fill, as incised valley fill deposits. Valley cutting is inferred to have resulted from relative sea-level fall near the end of allomember H time. Indeed, relative sea-level fall may have driven the later stages of coastal progradation as shoreface

Fig. 9. Nearly flow-parallel view of a gutter cast, indicated by arrows, showing undercut margin (u). Gutter fill shows well developed HCS. Shingle I-2, Tea Creek. Scale bar 20 cm.

Fig. 10. Growth fault exposed in lower part of shingle J-1 at Tea Creek. Surface labelled A is offset by 3.6 m whereas overlying sandstone (S) shows no offset.

sandstones near the progradational limit of allomember H (e.g. Fig. 4, well 10-12-72-12W6) are sharp based.

Incised valleys must have been separated by interfluves. A 1.6 m thick, pale grey, very blocky-textured and rooted siltstone containing compressed lumps of wood lies 70 m above the base of Clayhurst 1. This pale mudstone lies at almost the same stratigraphic level as the top of the incised valley fill sandstone in Clayhurst 2 (Fig. 23), and is tentatively interpreted as a palaeosol that developed on an interfluve, coeval with incision of valleys into allomember H.

Filling of the valleys probably took place in response to relative sea-level rise at the onset of transgression that initiated deposition of allomember G. Although these valley fills have not been examined in detail, preliminary inspection has not revealed obvious evidence of tidal processes.

Allomember G

Within the study area, the deposits of allomember G are entirely nonmarine and constitute a succession of very fine-grained sandstones and dark grey to brown, carbonaceous mudstones interbedded with occasional thin coals and rooted horizons. Mudstones may grade up, over 1–5 m, into wave and current rippled sandstones. Alternatively, decimetre-thick, current rippled sandstones may rest erosionally on mudstones. Rarely, erosionally-based sandstones are up to 2 m thick and are dominated by trough cross-bedding.

Well developed inclined heterolithic stratification (IHS; Thomas *et al.* 1987), 3.5 m thick, is present near the base of Clayhurst 1 (Figs 7 and 24A). Siderite pebbles, *Teredo*-bored logs of wood, very abundant fish bones and rare crocodile teeth are present at the base of the

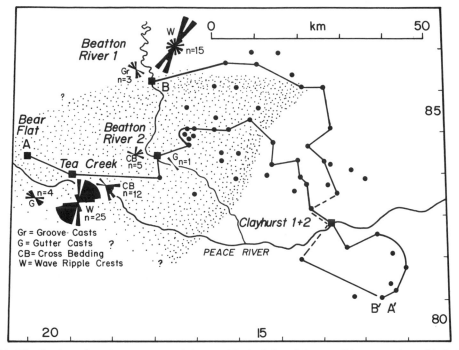

Fig. 11. Distribution of sandstone at top of the J-1 shingle, inferred from well-logs, together with facies and palaeocurrent data from outcrop. Sandstone distribution suggests a delta lobe fed from the west, and principal wave approach from the southeast.

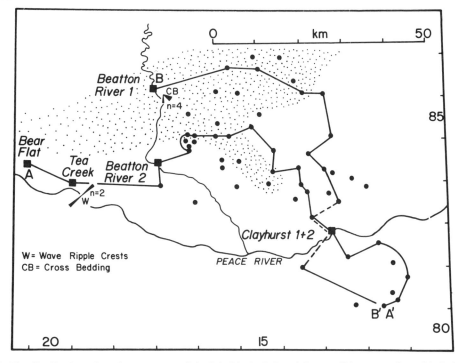

Fig. 12. The distribution of sandstone at top of the J-2 shingle, inferred from well-logs, together with facies and sparse palaeocurrent data, suggests a delta lobe prograding from the northwest.

Fig. 13. Mud-filled gutter cast, indicated by arrows, cut into an HCS sandstone bed, and erosively overlain by HCS sandstone. Shingle J-3 at Bear Flat. Scale bar 20 cm.

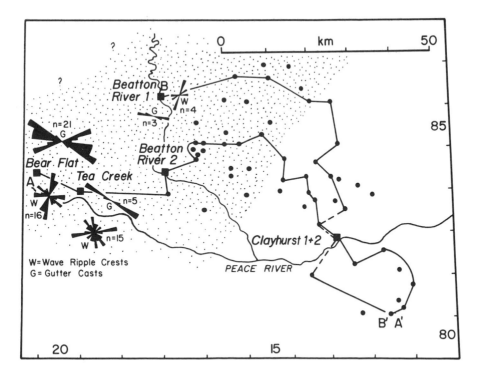

Fig. 14. Distribution of sandstone at top of J-3 shingle, inferred from well logs, plus facies and palaeocurrent data from outcrop. Data suggest southeastward progradation of a relatively straight, ?wave-dominated delta shoreline.

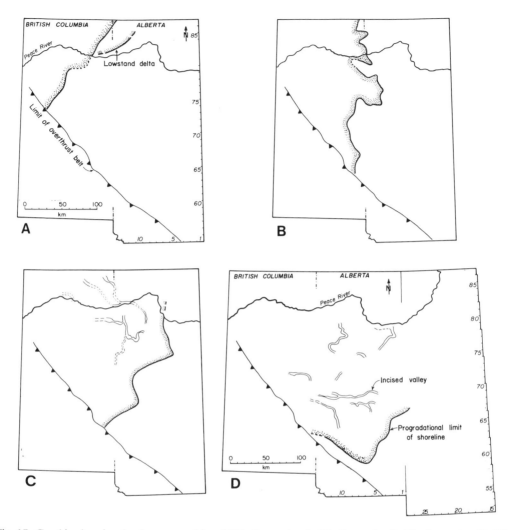

Fig. 15. Combined regional palaeogeographies of (**A**) allomember J; (**B**) allomember I; (**C**) allomember H; (**D**) allomember G. Each map shows the progradational margin of the shoreface sandstone of the youngest shingle of each allomember. Also shown are preliminary maps of incised valleys cut into the top of each allomember.

IHS unit (Fig. 24B) which consists mainly of centimetre- and decimetre-scale interbeds of fine- to medium-grained sandstone and mudstone. Bed contacts are commonly scoured, with dm-scale relief. Ripple cross-lamination and cross-bedding are the principal sedimentary structures. A distinctive, bright orange mudstone bed, about 20 cm thick is present at Clayhurst 1 and 2, and is tentatively equated with a similar unit at Bear Flat (Fig. 7). At Beatton River 1, cross-bedded sandstone bodies, hundreds of metres wide and up to 20 m thick are incised into nonmarine mudstones and sandstones (Fig. 21).

Interpretation of allomember G

Facies successions. Rocks of allomember G are interpreted to represent a low-lying coastal plain. Metre-scale sandier-upward successions are interpreted to represent prograding lacustrine delta mouth bars. Sharp based, dm-scale current rippled sandstone beds probably represent crevasse splays whereas thicker cross-bedded sandstones are interpreted as minor distributary channel fills. Evidence of desiccation is rare, and abundant preserved plant debris suggests that the sediments were waterlogged much of the time.

The IHS unit at Clayhurst 1 is interpreted as a point bar deposit. The occurrence of *Teredo*-bored logs in the channel-base lag is suggestive of a marine influence within the channel (Bromley *et al.* 1984). Abundant mud interbeds are also suggestive of a tidal influence (Bridges & Leeder 1976; Smith 1987, 1988; Shanley *et al.* 1992).

Thick cross-bedded sandstones at Beatton River 1 are interpreted as valley fill deposits and are similar to those at the top of allomember H. Several valley-fill sandstones in allomember G have also been mapped in subsurface (Fig. 15 D).

Sequence stratigraphy

Key surfaces

Allostratigraphic subdivision and correlation of the Dunvegan Alloformation has been based on recognition and mapping of key bounding surfaces. Dunvegan allomembers A–G, defined by Bhattacharya & Walker (1991*a*) are bounded by regionally-extensive marine flooding surfaces, recognizable in logs and core. Similar surfaces have been used to define allomembers H–J.

The use of flooding surfaces to define major genetic stratal packages differs from the sequence stratigraphic approach of the Exxon group (e.g. Posamentier *et al.* 1988) who define depositional sequences on the basis of unconformities and their correlative conformities. Although unconformities can be easily mapped beneath incised valley fills, they are less easily recognized in interfluve areas where they have been modified by transgressive marine erosion, resulting in a composite surface. Galloway (1989) recognized the utility of marine flooding surfaces for correlation, but emphasized the use of maximum flooding surfaces to define fundamental stratal packages that he termed genetic stratigraphic sequences. In the Dunvegan, maximum flooding surfaces do not have a sufficiently distinctive log response to make them useful stratigraphic markers, although in areas where core is available, Bhattacharya (1993) was able to reliably distinguish initial from maximum flooding surfaces.

Thus from a practical point of view, flooding surfaces have proved to be the most robust and useful markers in the Dunvegan, enabling allomember boundaries to be mapped from distal offshore areas to the lower reaches of the coastal plain (cf. Bhattacharya 1993). In dip sections, transgressive marine mudstones can be seen to thin and lose their identity at the landward limit of transgression (Fig. 4). How-ever, a recognizable high gamma ray and low resistivity log response can sometimes be traced for many tens of kilometres farther updip. This log response is interpreted to correspond to mudstones deposited in extensive coastal plain lakes and lagoons that developed behind the transgressive barrier shoreline (cf. Bhattacharya 1993). When exposed in outcrop, these deposits commonly comprise very dark grey and brown, platy and highly carbonaceous mudstones (cf. Johnson & Fitzsimmons 1994). The lagoonal/lacustrine mudstones and fine-grained sandstones at the base of allomember H at Bear Flat, Tea Creek and Beatton River 2 (Fig. 7) constitute exposed examples of these inferred 'transgressive' coastal plain deposits.

Mapping allomember boundaries in the Dunvegan Alloformation is very much more difficult than mapping allomembers in other units such as the Cardium (Plint *et al.* 1986), Marshybank (Plint 1990) and Viking (Boreen & Walker 1991) formations. In the latter formations, allomember boundaries are commonly marked by chert pebble beds that form transgressive lags derived from the reworking of shoreline conglomerates deposited at sea-level lowstand. In contrast, the Dunvegan is virtually devoid of chert pebbles in the regional study area (Fig. 2) and so distinctive lags are absent. As a result, Dunvegan allomember boundaries can confidently be recognized only when a particular surface can be shown to be of regional (hundreds of kilometres) extent. Such delineation is only possible through the use of the dense well-log control available in this basin.

Application of parasequence and systems tract terminology

Up to this point, depositional packages have been termed allomembers and have been discussed largely without the use of sequence stratigraphic terminology. However, stratigraphic relationships are sufficiently well established to permit a more interpretive terminology to be applied. In Fig. 7 the stratigraphic sections were correlated in terms of allomembers bounded by flooding surfaces. In Fig. 23 the same sections have been correlated and interpreted using sequence stratigraphic terminology (Posamentier *et al.* 1988; Posamentier & Vail 1988; Posamentier *et al.* 1992).

Parasequences. In this study, it was found that the application of parasequence terminology, as classically defined (Van Wagoner *et al.* 1988) was not particularly useful (cf. Martinsen 1993).

This was primarily because numerous upward shoaling facies successions (shingles) contain, at least locally, an internal erosion surface that separates offshore heterolithic strata from nearshore sandstones. Others, e.g. Swift *et al.* (1991) and Mitchum & Van Wagoner (1991), have also noted the lateral intergradation of classical parasequences into successions containing an internal erosion surface. In a pragmatic approach, Swift *et al.* (1991) grouped both types of succession together as 'parasequences' whereas Mitchum & Van Wagoner (1991) differentiated parasequences from high frequency sequences.

In the analysis of the Dunvegan, it was found more useful to differentiate shoaling-upward successions of strata (shingles) into component depositional systems tracts, separated by subaerial and marine erosion surfaces, initial flooding and maximum flooding surfaces. In this way, the terminological difficulties surrounding 'parasequences' and 'high frequency sequences' could be avoided.

Systems tracts. The term systems tract has been used in different ways by different workers. The original definition: 'a linkage of depositional systems' (Brown & Fisher 1977), was intended to provide a means of relating the deposits of numerous discrete depositional systems, each of which comprised 'an assemblage of process-related facies' (Fisher & McGowen 1967). Within a systems tract, it is possible to trace the deposits of one depositional system (e.g. a distributary channel) into the coeval deposits of other depositional systems (e.g. a sandy mouth bar or a muddy prodelta). A systems tract therefore encompasses the rocks deposited within a set of coeval depositional environments. Depositional systems tend to be progradational, and consequently deposit shoaling-upward successions of facies that, as Swift *et al.* (1991) point out, fit Van Wagoner *et al.*'s (1988) definition of a parasequence. Changes in factors such as accommodation space or sediment supply (the 'regime variables' of Swift & Thorne 1991) will 'reset' the depositional systems causing cessation of deposition and formation of a flooding surface upon which a new systems tract will build.

In their definition of sequence stratigraphic terms, Van Wagoner *et al.* (1988) described systems tracts in terms of relative sea-level (e.g. transgressive, highstand) and based their interpretation on the geometric stacking pattern of parasequence sets and their relationship to unconformities and maximum flooding surfaces.

In an attempt to differentiate facies-based and geometrically-based systems tracts, Swift *et al.*

(1991) suggested that the term 'geometric systems tract' be used when discussing the stacking pattern of parasequences (as seen in a lower-order sequence built of numerous parasequences), and that 'depositional systems tract' be used to describe a body of rock (such as a single parasequence) deposited under fixed regime variables.

The basis on which systems tracts are defined will depend both upon the scale of sequence, and on the quality of the stratigraphic data available. In general, as one moves to progressively higher levels of resolution, the basis for the interpretation of systems tracts shifts from geometric criteria (parasequence stacking pattern) to identification of key surfaces and facies criteria (depositional environments inferred from lithofacies).

Systems tracts (facies-based) and geometric systems tracts (geometrically-based) are differentiated in Fig. 23. In the following discussion, 'systems tract' refers to a facies-based, depositional systems tract.

Falling stage and lowstand systems tracts

Although they emphasize the importance of recognizing deposition during relative sea-level fall, Posamentier *et al.* (1992) do not follow their observations to a logical conclusion, namely the recognition of a discrete falling stage systems tract. Instead, 'late highstand' is followed by 'early lowstand' systems tracts. They interpret these systems tracts to be separated by a sequence boundary that forms at the onset of relative sea-level fall. In consequence, all strata deposited during relative sea-level fall, during lowstand and during relative rise prior to transgression are part of the lowstand systems tract. It seems illogical to recognize strata deposited during relative sea-level rise (the transgressive systems tract) but to ignore deposition during relative sea-level fall.

The Dunvegan provides good examples of sharp-based shoreface sandstones that are succeeded basinwards by a detached shoreface sandbody (e.g. J-3 and I-1 sandbodies, Figs 5 and 6). Also seen are sharp-based sandbodies that lack a detached terminal lowstand deposit (e.g. I-2 sandstone). Following the reasoning of Nummedal (1992), Hunt & Tucker (1992, 1995) and Ainsworth (1994), it appears more logical to assign strata deposited during relative sea-level fall to the falling stage systems tract whereas the lowstand systems tract should embrace those strata deposited after sea-level lowstand but prior to transgression.

At present, there is clearly a divergence of opinion regarding the position of 'the' sequence boundary in successions of rocks deposited during falling and low sea-level. Some workers place a sequence boundary at the subaerial erosion surface above a sharp based shoreface sandbody (e.g. Van Wagoner *et al.* 1990; Plint 1991, this paper; Ainsworth 1994; Hamberg *et al.* 1994). Others place the boundary at the marine erosion surface below the shoreface (e.g. Posamentier *et al.* 1992; Ainsworth & Pattison 1994; Johnson & Fitzsimmons 1994; Fitzsimmons 1994).

This question is addressed in Fig. 25 which is drawn with the assumption that the rates of subsidence and sediment supply remain invariant and that the shelf has a homoclinal ramp physiography.

In Figure 25A, there is no relative sea-level change and accommodation space and sediment supply are sufficient to allow development of a gradually upward-coarsening succession of shelf mudstone, through rippled and HCS sandstone of the lower shoreface into clean sandstones of the upper shoreface (cf. fig. 1b of Posamentier *et al.* 1992; fig 8A of Helland-Hansen & Gjelberg 1994). 'Fair weather wave base', although in reality a bathymetric zone rather than a sharply-defined depth, may be approximated, from a geological standpoint, as the depth above which mud is not preserved in the sedimentary succession. In Fig. 25B, relative sea-level begins to fall. Fair-weather wave base begins to impinge on muddy deposits of the lower shoreface and inner shelf, forming an erosion surface: a regressive surface of marine erosion (Plint 1988, 1991; Posamentier *et al.* 1992). In Fig. 25C, the rate of relative sea-level fall first accelerates, progressively eliminating accommodation space available for the prograding shoreface, then decelerates, allowing progressively more accommodation space to become available (due to bathymetry and subsidence). The resulting sharp-based shoreface sandbody becomes progressively thinner up to the point of fastest relative sea-level fall. Subsequently, as the rate of relative sea-level fall decelerates, the sandbody thickens and eventually accommodation space is sufficient to permit re-establishment of a gradationally-based shoreface (cf. fig. 9 of Helland-Hansen & Gjelberg 1994). Base level fall also results in fluvial incision. The fluvial erosion surface (i.e. the sequence boundary) is contiguous with the subaerial surface on the top of the prograding shoreface, which, as Van Wagoner *et al.* (1990, p. 36) point out, is the sequence boundary. If this reasoning is accepted, then the regressive surface of marine erosion below the

shoreface sandbody, although a manifestation of relative sea-level fall, cannot be 'the' sequence boundary in the accepted Exxon sense! The shoreface sandbody, which remains attached to the highstand shoreface throughout relative sea-level fall, would, in conventional sequence stratigraphic terminology, be assigned to the 'early lowstand systems tract' (Van Wagoner *et al.* 1990; Posamentier & Allen 1993; Kolla *et al.* 1995). However, rather than trying to split the lowstand systems tract, it seems more logical to assign the shoreface sandbody to the falling stage systems tract of Nummedal (1992) and Ainsworth (1994), or the forced regressive systems tract of Hunt & Tucker (1992) and Helland-Hansen & Gjelberg (1994), particularly if it is succeeded by a detached lowstand shoreface. The stratal geometry shown in Fig. 25C and D was termed an 'attached lowstand systems tract' by Ainsworth & Pattison (1994) who placed the sequence boundary at the regressive surface of marine erosion below the shoreface sandbody.

Figure 25D is a variant on Fig. 25C in which accelerating relative sea-level fall is maintained for longer, which causes the shoreface to be drawn out farther across the shelf, depositing a very thin, sharp-based sandbody. As in Fig. 25C, the shoreface remains attached, and is assigned to the falling stage systems tract. Subsequent marine transgression would result in a thin, erosive-based sheet of sandstone, typically lacking evidence of subaerial exposure as a result of transgressive erosion (e.g. units J and L in the Marshybank Formation; Plint & Norris 1991).

Figure 25E takes as its starting point Fig. 25B. However, in Fig. 25E, the rate of relative sea-level fall is so rapid (or the shelf gradient is so low) that, for a period, no accommodation space is left at the shoreline and a sediment bypass zone develops. As the rate of relative sea-level fall subsequently diminishes to zero (and/or shelf gradient steepens), new accommodation space becomes available basinward of the bypass zone and a detached shoreface sandbody develops (Plint 1988; Posamentier *et al.* 1992). The detached shoreface sandbody (or the most basinward if several detached shorefaces formed successively) is assigned to the lowstand systems tract because it formed at the lowest point of the relative sea-level curve (Hunt & Tucker 1992). In updip areas, the sequence boundary is formed by the subaerial erosion surface bounding incised valleys, and by the upper surface of the falling stage systems tract. The sequence boundary extends across the bypass zone and continues, as the regressive surface of marine erosion, below the detached lowstand systems tract.

In more basinward areas, beyond the seaward limit of regressive marine erosion, deposition of muddy sediments would be continuous throughout falling stage and lowstand systems tract time. It is possible that the onset of the falling stage systems tract may be marked by a rapidly-gradational boundary between mudstones of the highstand systems tract and more silt- or sand-rich sediments of the falling stage systems tract. This boundary would record the abrupt increase in the rate of progradation and more efficient seaward dispersal of sediment across the shelf. Hunt & Tucker (1995) suggest that it would be possible to distinguish falling stage from lowstand deposits in carbonate sediments in a basinal setting although such a distinction in a low gradient clastic ramp setting may not be possible (Helland-Hansen & Gjelberg 1994).

Example of falling stage and lowstand systems tracts. The sharp, scoured base to the SCS sandstone at the top of shingle J-3 at Beatton River 1 suggests that it is a shoreface sandbody deposited during a 'forced regression' resulting from relative sea-level fall (Figs 7, 23 and 25B, C, D). The erosion surface beneath the sandstone was cut into marine shelf muds as fair weather wave base impinged on the inner shelf surface cutting a regressive surface of marine erosion.

There is no preserved evidence that the J-3 shoreface extended much farther south than Beatton River 1, although thinly-bedded storm-deposited sandstones extend about 30 km farther southeast (Fig. 14). However, shingle I-1 contains a basinally-detached sandbody, interpreted as a lowstand delta, which, on geometric grounds, can be assigned to a lowstand systems tract (Fig. 25E). If this interpretation is correct, it suggests that the sharp-based shoreface in J-3 at Beatton River 1 records an initial response to (or at least the first lithological expression of), relative sea-level fall. It would therefore seem logical to assign this sandstone to the FSST because there is good evidence that continued relative sea-level fall exposed the delta top platform and moved the shoreline to the upper part of the J-3 delta front slope. The top surface of the J-3 sandstone is therefore assigned sequence boundary status; this surface extends as a correlative conformity beneath the LST formed by shingle I-1 (Figs 23 and 25E). Relative sea-level may have fallen by about 25 m from the level of the J-3 shoreface at Beatton River 1 to the top of the I-1 lowstand delta, assuming the top of the latter to approximate palaeo-horizontal, and making no correction for compaction.

The distribution of I-1 sandstone (Fig. 16) suggests that the lowstand delta shoreline prograded about 25 km to the southeast before being drowned by a relative sea-level rise that drove the shoreline beyond Bear Flat, at least 70 km to the WNW. Transgressive erosion (ravinement) modified the subaerial sequence boundary at the top of the J-3 sandstone, removing any physical evidence of emergence, resulting in a composite sequence boundary (i.e. a formerly subaerial surface subsequently reworked by marine processes; Posamentier & James 1993).

Unlike allomember E (extensively documented by Bhattacharya 1993) it is unfortunate that little log, and neither core nor outcrop data are available with which to determine better the geometry and lithology of the lowstand delta in shingle I-1. The lowstand delta of shingle E-1 was clearly related to, and fed by, an incised valley (the Simonette Channel; Bhattacharya 1993), and a similar conduit would be expected to have fed the I-1 lowstand delta. Unfortunately, the scant data in the study area provide no evidence for a feeder system.

It is worth noting that, despite the strong geometric similarities of the lowstand deltas in allomembers I and E, their stratigraphic terminology differs. Thus Bhattacharya & Walker (1991a) included the lowstand delta (the E1 shingle) within allomember E. In contrast, Plint (this paper) placed the allomember boundary at the top of the J-3 shingle, and assigned the lowstand delta to the basal part of allomember I. In part this reflects the influence of sequence stratigraphic philosophy, which dictates that a sequence boundary be placed beneath a lowstand systems tract, and is in part for practical reasons, because the sandy part of the lowstand delta is only well developed in the small 'north Peace' study area, whereas the surface marking the top of the J-3 shingle was mappable on a regional scale. Regardless of the differences in stratigraphic terminology between this paper and that of Bhattacharya & Walker (1991a), our interpretation of the rocks remains consistent.

Example of a falling stage systems tract. Within the regional study area shown in Fig. 2, shingle I-2 constitutes one principal shoaling-upward succession, generally capped by a sharp-based shoreface sandstone, the seaward limit of which is shown in Fig. 16B. Despite examination of hundreds of well logs, there is no evidence for the development of a discrete lowstand systems tract similar to that in shingle I-1. However, the widespread presence of a sharp-based shoreface suggests that progradation of much of the I-2

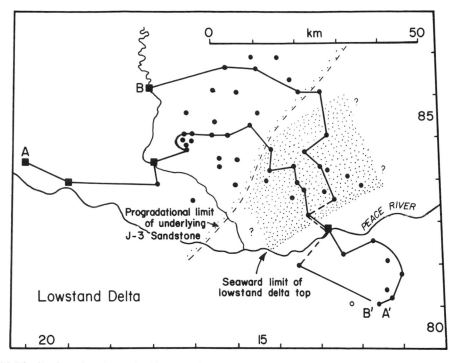

Fig. 16. Distribution of sandstone in shingle I-1, inferred to represent a lowstand delta. Onlap of shingle I-1 onto the front of shingle J-3 is illustrated in Figs 5 and 6.

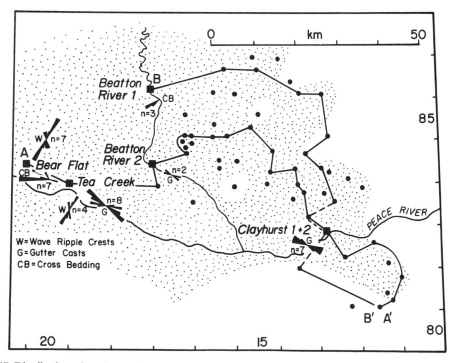

Fig. 17. Distribution of sandstone in shingle I-2, inferred from well logs, plus facies and palaeocurrent data from outcrop. Data suggest a broad southeastward prograding lobe which is the most northerly of three major lobes defined by regional facies mapping (see Fig. 15B).

Fig. 18. Sharp base (arrowed) of I-2 shoreface sandstone at Clayhurst 1. Marine mudstones below shoreface contain an example of a 'giant' gutter cast. These structures probably record intermittent storm-scouring of a muddy shelf during the early stages of relative sea-level fall (cf. Wright 1986; Plint & Norris 1991; Hadley & Elliott 1993). Figure is 1.7 m tall.

Fig. 19. Section at Bear Flat, showing upper part of shingle J-3 comprising thin-bedded, wave-rippled sandstone, sharply overlain at 'S' by a massive, burrowed sandstone with an intensely limonite cemented upper surface 'L'. Surface 'S' is inferred to represent the basal erosion surface of allomember I. The top of allomember I, 'I' can be seen in the distance, and the overlying sandstone 'V' is part of an incised valley fill in allomember H. Scale bar 1 m.

sandstone was driven by relative sea-level fall.

Following the reasoning explained in Fig. 25, the sandstone at the top of I-2 is assigned to a falling stage systems tract. The I-2 shoreface prograded about 100 km. If a progradation rate of about 2.5 m/year is estimated (see discussion of rates in Plint 1991), the I-2 sandstone may represent about 40 000 years of progradation. Development of a detached lowstand shoreface did not take place. This might reflect an invariant rate of relative sea-level fall, which might have favoured development of a continuous shoreface sandstone sheet, or possibly lack of a discrete shelf-slope topographic break that might have encouraged shoreface detachment. In contrast, initiation of the lowstand delta in shingle I-1, might have been triggered by an increase in the rate of relative sea-level fall, causing a 'quantum jump' of the shoreline onto the upper delta-front slope.

Development of incised valleys. Bhattacharya (1993) demonstrated a convincing relationship between valley cutting, to form the Simonette channel, and deposition of a lowstand delta, the E1 shingle, at the mouth of the valley. The onlap of the lowstand delta onto the front of the older E2 shingle, interpreted as a late highstand deposit, provides good evidence of relative sea-level fall to which can also be attributed fluvial incision and cutting of the Simonette Channel.

In a review of river response to base level change, Schumm (1993) and Leeder (1994) noted that eustatic changes of only a few tens of metres would be likely to produce valley

Fig. 20. (A) Plane-laminated beach sandstone with roots at the top of allomember I at Bear Flat. Beach sandstone is overlain by grey–green, silty sandstone, intensely burrowed by *Thalassinoides*. (B) Detail showing *Thalassinoides* burrow systems. Scale bar 20 cm.

incision for only a few tens of kilometres from the shoreline. Preliminary mapping of incised valleys in Dunvegan allomembers E, F, G and H (e.g. Fig. 15C, D; Plint 1994) suggests that some valleys extend over 200 km from the contemporaneous shoreline, but even at this distance from the coast, valleys have a depth of 20 m or more.

Geometric considerations (Plint 1994) suggest that the transgressions and regressions that generated Dunvegan allomembers were controlled primarily by eustasy. Although subsidence was an important factor, it appears to have controlled the geometry, but not the origin, of allomembers. Eustatic excursions are estimated to have been only a few tens of metres over periods of about 150 000 years. However, base level changes of this magnitude are, according to Schumm (1993), unlikely to cause valley incision on the scale observed, and it may therefore be necessary to appeal to additional mechanisms (?tectonic, climatic) to explain the observations. On the other hand, Shanley & McCabe (1993) suggest, on the basis of their observations on the Kaiparowits Plateau, that valley incision may extend 100–150 km inland, driven only by relative sea-level changes at the coast. Clearly, further investigation of this question is warranted.

Recognition of interfluves. At Clayhurst 1, a pale grey, apparently highly weathered palaeosol is interpreted to represent a subaerial surface that formed on an interfluve between valleys incised into Allomember H (Fig. 23). Similarly, an unusual but distinctive orange mudstone at Clayhust 1 and 2, and Bear Flat is tentatively interpreted as a weathered surface that formed when valleys were cut into Allomember G. The interpretations presented in Fig. 23 are tentative

at present, although regional thickness trends of allomembers predict that subaerial sequence boundaries, i.e. interfluve surfaces, should be present within a few metres of the horizons selected on the basis of field observations. Future work, using spectral gamma ray analysis of outcrop sections, is planned in an attempt to identify more accurately deeply-weathered interfluve surfaces.

Transgressive systems tract

Incised valley filling. Incised valley fills in the study area, although not the subject of exhaustive study, appear to be remarkably monotonous, consisting largely of decimetre-, and occasionally metre-scale trough cross-bedded upper fine- to medium-grained sandstone. Mudstone intraclasts are common, and decimetre- to metre-scale interbeds of carbonaceous mudstone are locally encountered. Valley filling was very probably related to relative sea-level rise and marine transgression from the southeast, and one would expect to find, especially in the upper part of the valley fill, features suggestive of tidal influence, (e.g. paired mud drapes, *Teredo*-bored wood, compound cross-bedding and reactivation surfaces; Shanley *et al.* 1992). Detailed exmination of these deposits is presently underway (J. A. Wadsworth, unpublished data).

Marine transgressive systems tract. Most marine mudstones in the Dunvegan lack benthic body fossils and are very weakly bioturbated to astrate. However, mudstones just above marine flooding surfaces are commonly more silt-rich and much more highly bioturbated, as, for example at the base of allomember I (Fig. 7). Bhattacharya (1993) documented a similar relationship in core and showed that the

---

Fig. 21. Panorama of west wall of Beatton River canyon (centred on NTS grid reference 437570 94/A7). At left can be seen a complete section through allomember J, with the tops of shingles J-1, J-2 and J-3 labelled. Note the relative insignificance of the J-3 sandstone in outcrop, despite the fact that subsurface correlation shows that it forms a major bounding surface (see Fig. 6). Shingle I-2 is capped by a prominent shoreface sandstone, the base of which is abruptly gradational. Shingle I-2 is overlain by coastal plain deposits of allomember H, 'H', which are deeply incised (in places cutting into the top of allomember I), by massive cross-bedded sandstone (base marked by arrows) forming a valley fill. The valley fill encloses a discrete mudstone lens 'M' and shows local lateral accretion surfaces 'LA'.

Fig. 22. Panorama of east wall of Beatton River canyon (centred on NTS grid reference 438574 94/A7, directly opposite cliff shown in Fig. 21). Prominent marine sandstones at top of shingles J-2 and I-2 are labelled. Coastal plain deposits of allomember H, 'H-CP' are incised by a valley, filled with cross-bedded sandstone of allomember H, 'H-VF'. The base of valley fill 'H-VF' locally cuts down into the top of Allomember I, and the valley fill consists of several mutually-erosive sandbodies. The base of allomember G is inferred to lie at the top of 'H-VF'. Allomember G comprises coastal plain deposits 'G-CP' locally incised by valley fill sandstones, 'G-VF', overlain by more coastal plain deposits inferred to belong to allomember F, 'F-CP'. Location of measured section Beatton River 1 shown by BR-1.

Fig.

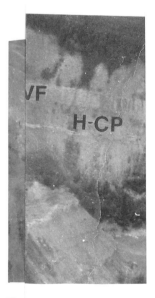

Fig.

bioturbated interval, commonly only 1–2 m thick, could be assigned to the transgressive systems tract whereas the overlying platy mudstones formed part of the highstand systems tract. The boundary between the two facies is the maximum flooding surface.

Marine benthic activity was probably controlled by a combination of rate of sediment flux, salinity and oxygenation. In the Dunvegan the first factor might have been the most important, with suspended sediment supply being relatively low during marine transgression, but increasing markedly in the regressive highstand systems tract. Fresh water discharge from rivers may also have caused significant reduction in salinity close to the shoreline, further inhibiting a normal marine fauna. Although the oxygen content of bottom water may have been reduced in the Dunvegan sea, the increase in bioturbation above marine flooding surfaces suggests that a pycnocline did not encroach into shallow shelf areas during transgression. This contrasts with the overlying Muskiki and Marshybank formations, in which mudstones just above flooding surfaces are highly pyritic and very weakly bioturbated, suggesting deposition from oxygen-deficient water below a pycnocline (Plint & Norris 1991; McKay et al. 1995).

Transgressive systems tract deposits have not been positively identified in shingles J-1, 2 and 3, although the lowest part of each shingle probably records a period of marine transgression.

Transgressive systems tract on the coastal plain. Shanley et al. (1992) and Shanley & McCabe (1993) documented a range of sedimentary and stratigraphic features that were suggestive of salt water and tidal influence on a coastal plain. They showed that these effects could be related to contemporaneous marine transgression. A similar relationship can be demonstrated in the Dunvegan Alloformation. Thus, at the base of sequence H at Bear Flat (Fig. 23), dense *Thalassinoides* bioturbation (Fig. 20) is interpreted to have formed in a brackish lagoon perhaps 20–30 km landward of the shoreline; *Teredolites* in the same unit provide additional evidence for saline conditions (Wright 1986). Similarly, the incursion of salt water onto the coastal plain in sequence G (Fig. 23) is inferred from *Teredo*-bored logs in tidally-influenced fluvial point bar deposits (Fig. 24). This indicates salt water penetration about 40 km behind the coeval shoreline which is inferred to have reached as far north as well 10-6-79-12W6 (Fig. 4; c.f. Shanley et al. 1992).

The apparent restriction of features indicative of marine influence to the basal part of

sequences H and G (Fig. 23) is analogous to that described by Shanley et al. (1993) and is similarly interpreted as evidence for coastal plain deposition during marine transgression. These deposits can therefore be assigned to a coastal plain transgressive systems tract (Fig. 23). Recognition of the maximum flooding surface (MFS) is equivocal. Shanley et al. (1993) suggested that the maximum landward extent of tidally influenced fluvial deposits was temporally equivalent to the MFS. Unfortunately, in the present study area, exposure is insufficient to determine accurately the maximum geographic extent of tidally influenced deposits, and hence pick the MFS. A similar difficulty was encountered by Aitken & Flint (1995). However, Shanley et al. (1993, fig. 4) indicate a zone of coals just above the coastal plain MFS and Aitken & Flint (1995) point out that well-developed coals occur up to, but not above the coastal plain MFS, reflecting changes in the rate at which accommodation space was generated. In light of these observations, a MFS is tentatively placed in sequence G (Clayhurst 2, Fig. 23) at a 10 cm coal (thick by Dunvegan standards!) 12 m above the top of the incised valley fill.

Highstand systems tract

Highstand systems tract deposits comprise facies successions that grade up from weakly bioturbated mudstone to cm-scale interbeds of very fine-grained, wave-rippled sandstone. Some successions culminate in dm-scale HCS sandstones or even SCS and cross-bedded sandstone of the upper shoreface. In sequence J (Fig. 23) each of the three shingles can be interpreted in terms of transgressive and highstand systems tracts, although it has not been possible to recognize a maximum flooding surface separating these systems tracts; the transgressive systems tract is probably very thin. The three major shingles in sequence J (plus several more minor sandier-upward successions, e.g. indicated by dashed lines in shingle J-3; Figs 5 and 6) when viewed together have an aggradational to progradational stacking pattern and can be classified as a highstand geometric systems tract (H-GST; Fig. 23). Sequence I-2 comprises a single highstand systems tract so, unlike sequence J, recognition of a highstand geometric systems tract on the basis of stacking pattern is not possible.

Conclusions

Ten allomembers, bounded by regional flooding surfaces have been recognized in the Early–Mid-

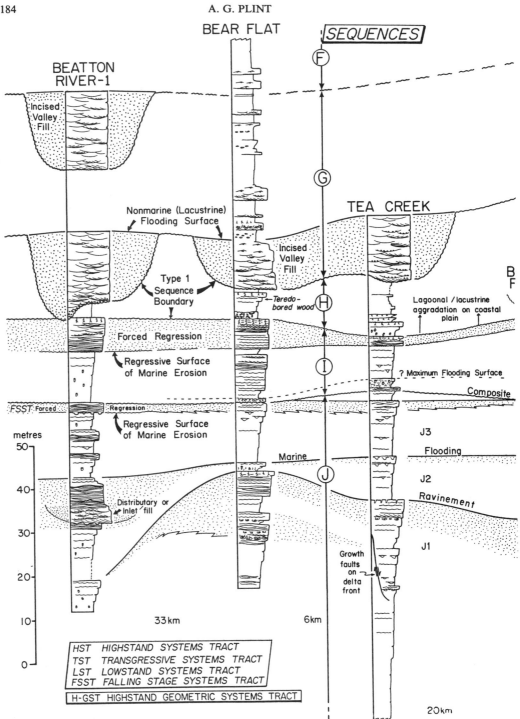

Fig. 23. Sequence stratigraphic interpretation of outcrop sections shown in Fig. 7. Systems tracts and sequence boundaries can be picked with confidence in marine rocks of allomembers J and I. Type 1 sequence boundaries can be picked at the base of incised valley fills in allomembers H and G. Maximum flooding surfaces separating coastal plain strata of the TST and HST in allomembers H and G are perhaps impossible to recognize, although a prominent coal near the middle of allomember G at Clayhurst 2 might be interpreted as a maximum flooding surface. Legend as in Fig. 7. See text for full discussion.

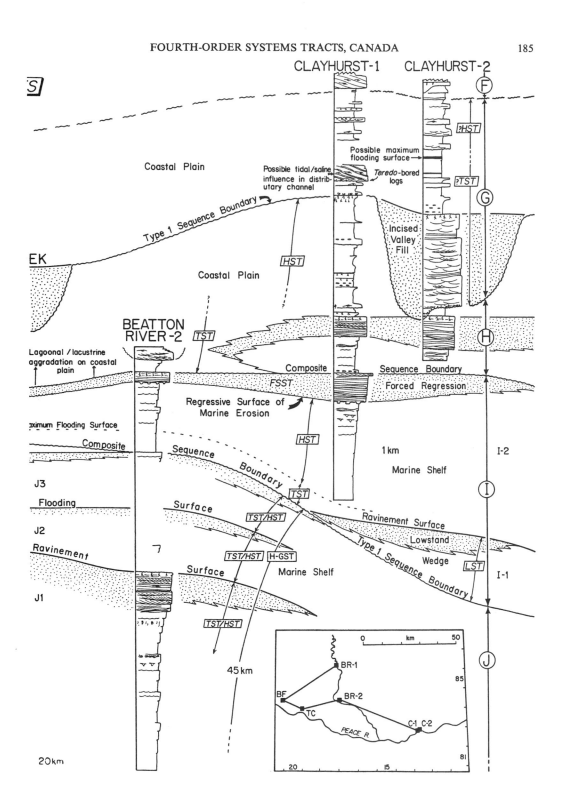

Fig. 24. (**A**) Overview of inclined heterolithic stratification, interpreted to represent a tidally-influenced point bar, near base of allomember G at Clayhurst 1. Base of IHS unit is arrowed. Grassy depression in distance (P) is underlain by black laminated mudstone rich in plant debris, inferred to represent an abandoned channel plug. (**B**) Detail of *Teredolites* (arrowed) in log from basal lag of IHS unit in (A).

Cenomanian Dunvegan Alloformation. Each allomember represents about 150 000 years and displays an overall shoaling-upward facies succession in which prodelta mudstones grade updip into shallow marine sandstones and coastal plain deposits. Allomembers downlap from NW to SE onto a prominent condensed section, which forms the base of the alloformation. The top of the Dunvegan Alloformation is defined by a regional flooding surface which marks a major marine transgression at the base of the Late Cenomanian–Turonian Kaskapau Formation. The Dunvegan crops out along the Peace and Beatton rivers in NE British Columbia. In the study area discussed here, attention is focused on the lowest four allomembers, G–J.

Fifty well-logs provide a reliable means of correlating six outcrop sections spaced tens of kilometres apart.

Marine strata comprise stacked, sandier-upward successions that grade from platy marine mudstone through cm-scale interbeds of fine-grained, wave-rippled sandstone; in places, HCS sandstone, commonly associated with gutter casts, and SCS and cross-bedded sandstones cap the succession. Integration of outcrop facies and palaeocurrent data with sandstone distribution mapped in subsurface show that gutter casts are consistently orientated shore-normal (NW–SE), perpendicular to wave ripple crests. Gutters were probably cut by oscillatory wave action during storms, and filled as the storm waned.

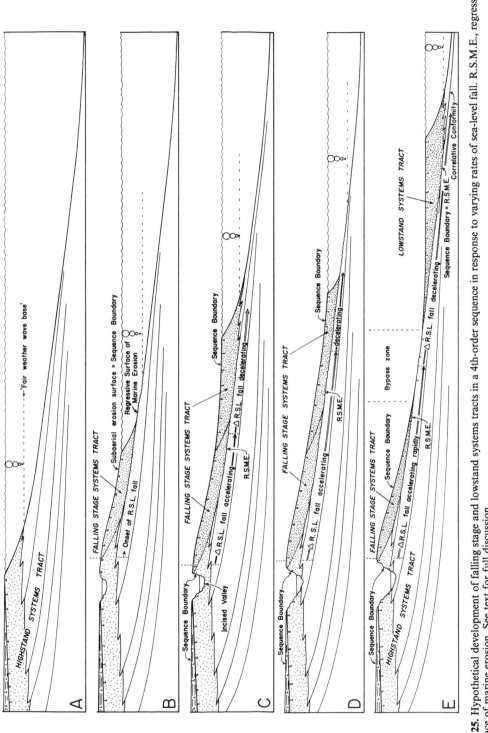

Fig. 25. Hypothetical development of falling stage and lowstand systems tracts in a 4th-order sequence in response to varying rates of sea-level fall. R.S.M.E., regressive surface of marine erosion. See text for full discussion.

Most allomembers comprise two or more sandier-upward successions of sub-regional extent, here termed shingles (broadly equivalent to parasequences). Allomember J comprises three shingles; each shingle has a different distribution of sandstone suggesting that depositional cyclicity was in part controlled by switching of delta distributaries. Regional facies mapping suggests that each shingle was formed by progradation of a wave-dominated linear delta shoreline.

Allomember I contains two shingles. The lower contains a basinally-isolated sandstone which onlaps the inclined front of shingle J-3, and is interpreted as a lowstand delta. The overlying shingle I-2 indicates over 70 km of shoreline backstep to the NW. Shingle I-2 is capped by a sheet-like sharp-based shoreface sandstone, the seaward margin of which comprises three distinct lobes of which the two more northerly suggest a strong river influence.

Allomember H contains marine shelf and shoreface deposits only in the SE. These grade laterally northwestward into carbonaceous mudstones, fine-grained sandstones, palaeosols, thin coals, etc., which represent a low-lying coastal plain. In the study area, allomember G consists entirely of coastal plain deposits but farther south these grade into marine strata. Coastal plain deposits of allomembers H and G are incised by cross-bedded sandstone bodies up to 25 m thick, 0.5–3 km wide and sometimes traceable for over 100 km. These units are interpreted as valley fill deposits.

By emphasizing unconformities, allomembers can be interpreted in terms of 4th-order sequences. Nevertheless, sequences are primarily defined on the basis of regionally extensive composite unconformity/flooding surfaces that record marine transgression over formerly subaerial interfluves. Sequences are built of shingles which, although similar to parasequences, differ in that they are sometimes (especially at the top of sequences) partitioned by regressive surfaces of marine erosion produced by relative sea-level fall. Shingles can be partitioned into systems tracts on the basis of facies successions and key surfaces. Extensive erosive-based shoreface sandstone sheets at the top of allomembers are most logically assigned to a falling stage systems tract and are overlain by the sequence-bounding subaerial unconformity.

A well developed delta top–delta front physiography in allomember J probably encouraged formation of a basinally-detached lowstand systems tract at the base of allomember I, although in the Dunvegan Alloformation as a whole, detached lowstand systems tracts are not common.

Valleys incised at sea-level lowstand were filled during the early transgressive systems tract by multistorey, sandstone-dominated channel deposits lacking clear tidal signatures. However, overlying transgressive systems tract deposits on the coastal plain contain evidence for both tidal action and brackish water. Identifying the boundary between transgressive and highstand systems tract deposits on the coastal plain is difficult, but might coincide with one or more poorly developed coal beds.

In general, the 4th-order sequences (allomembers) of the Dunvegan Alloformation contain too few component shingles to permit differentiation of systems tracts on the basis of geometric stacking pattern. At this relatively high level of stratigraphic resolution, partitioning of the strata into depositional systems tracts on the basis of initial and maximum flooding surfaces, and on regressive marine and subaerial erosion surfaces appears to offer the best means of reconstructing detailed chronostratigraphic and palaeogeographic relationships.

Work on the Dunvegan Formation began in 1984 as a post-doctoral project, supported by a Natural Sciences and Engineering Research Council (NSERC) strategic grant to Roger Walker. Subsequent research was supported by NSERC operating grants to AGP; Canadian Hunter Ltd, Home Oil Ltd and Husky Resources Ltd also provided generous financial and logistical support. I thank Patrick Elliott, Yuanxian Hu, Jennifer McKay, Darryl Reiter, Rick Vantfoort and, in particular, Annemarie Plint for tirelesss field assistance over nine seasons. I am grateful to Bruce Hart, Finn Surlyk and Fred Wehr whose incisive comments greatly improved the final version of the paper; responsibility for facts and interpretations nevertheless remains with the author. Photographs were prepared by Ian Craig and Alan Noon (UWO) and Jim Watkins (Reading). I am grateful to Dr A. Parker for provision of facilities at the Postgraduate Research Institute for Sedimentology, University of Reading, where this paper was prepared during sabbatical leave.

References

AIGNER, T. 1985. Storm depositional systems: dynamic stratigraphy in modern and ancient shallow marine sequences. *Lecture Notes in Earth Sciences*, 3, Springer-Verlag, Berlin.

AINSWORTH, R. B. 1994. Marginal marine sedimentology and high resolution sequence analysis; Bearpaw–Horseshoe Canyon transition, Drumheller, Alberta. *Bulletin of Canadian Petroleum Geology*, 42, 26–54.

—— & PATTISON, S. A. J. 1994. Where have all the lowstands gone? Evidence for attached lowstand systems tracts in the Western Interior of North America. *Geology*, 22, 415–418.

AITKEN, J. F. & FLINT, S. S. 1995. The application of high-resolution sequence stratigraphy to fluvial systems: a case study from the Upper Carboniferous Breathitt Group, eastern Kentucky. *Sedimentology*, **42**, 3–30.

BHATTACHARYA, J. P. 1989. *Allostratigraphy and river- and wave-dominated deltaic sediments of the Upper Cretaceous (Cenomanian) Dunvegan Formation, Alberta.* PhD Thesis, McMaster University, Ontario.

—— 1993. The expression and interpretation of marine flooding surfaces and erosional surfaces in core; examples from the Upper Cretaceous Dunvegan Formation, Alberta foreland basin, Canada. *In:* POSAMENTIER, H. W, SUMMERHAYES, C. P., HAQ, B. U. & ALLEN, G. P. (eds) *Sequence Stratigraphy and Facies Associations.* International Association of Sedimentologists, Special Publication, **18**, 125–160.

—— & WALKER, R. G. 1991*a*. Allostratigraphic subdivision of the Upper Cretaceous Dunvegan, Shaftesbury, and Kaskapau formations in the northwestern Alberta subsurface. *Bulletin of Canadian Petroleum Geology*, **39**, 145–164.

—— & —— 1991*b*. River- and wave-dominated depositional systems of the Upper Cretaceous Dunvegan Formation, northwest Alberta. *Bulletin of Canadian Petroleum Geology*, **39**, 165–191.

BOREEN, T. & WALKER, R. G. 1991. Definition of allomembers and their facies assemblages in the Viking Formation, Willesden Green area, Alberta. *Bulletin of Canadian Petroleum Geology*, **39**, 123–144.

BRIDGES, P. H. & LEEDER, M. R. 1976. Sedimentary model for intertidal mudflat channels with examples from the Solway Firth, Scotland. *Sedimentology*, **23**, 533–552.

BROMLEY, R. G., PEMBERTON, S. G. & RAHMANI, R. A. 1984. A Cretaceous woodground: the *Teredolites* ichnofacies. *Journal of Paleontology*, **58**, 488–498.

BROWN, L. F. & FISHER, W. L. 1977. Seismic-stratigraphic interpretation of depositional systems: examples from Brazil rift and pull-apart basins. *In:* PAYTON, C. E. (ed.) *Seismic Stratigraphy – Applications to Hydrocarbon Exploration.* American Association of Petroleum Geologists, Memoir, **26**, 213–248.

CALDWELL, W. G. E., NORTH, B. R., STELCK, C. R. & WALL, J. R. 1978. A foraminiferal zonal scheme for the Cretaceous System in the interior plains of Canada. *In:* STELCK, C. R. & CHATTERTON, B. D. E. (eds) *Western and Arctic Canadian Biostratigraphy.* Geological Association of Canada, Special Paper, **18**, 495–575.

DUKE, W. L. 1990. Geostrophic circulation or shallow marine turbidity currents? The dilemma of paleoflow patterns in storm-influenced prograding shoreline systems. *Journal of Sedimentary Petrology*, **60**, 870–883.

——, ARNOTT, R. W. C. & CHEEL, R. J. 1991. Shelf sandstones and hummocky cross stratification: new insights on a stormy debate. *Geology*, **19**, 625–628.

FISHER, W. L. & McGOWEN, J. H. 1967. Depositional systems in the Wilcox Group of Texas and their relationship to occurrence of oil and gas. *Transactions of the Gulf Coast Association of Geological Societies*, **17**, 105–125.

FITZSIMMONS, R. J. 1994. Identification of high order sequence boundaries and land attached forced regressions. *In:* JOHNSON, S. D. (ed.) *High resolution sequence stratigraphy: Innovations and applications* (Abstract volume). Department of Earth Sciences, University of Liverpool, 332–333.

GALLOWAY, W. E. 1989. Genetic stratigraphic sequences in basin analysis I: Architecture and genesis of flooding-surface bounded depositional units. *American Association of Petroleum Geologists Bulletin*, **73**, 125–142.

GOLDRING, R. 1971. Shallow water sedimentation as illustrated by the Upper Devonian Baggy Beds. *Geological Society, London, Memoir*, **5**.

—— & AIGNER, T. 1982. Scour and fill: The significance of event separation. *In:* EINSELE, G. & SEILACHER, A. (eds) *Cyclic and event stratification*, Springer-Verlag, Berlin, 354–362.

HADLEY, D. F. & ELLIOTT, T. 1993. The sequence-stratigraphic significance of erosive-based shoreface sequences in the Cretaceous Mesaverde Group of northwest Colorado. *In:* POSAMENTIER, H. W, SUMMERHAYES, C. P., HAQ, B. U. & ALLEN, G. P. (eds) *Sequence Stratigraphy and Facies Associations.* International Association of Sedimentologists, Special Publication, **18**, 521–535.

HAMBERG, L., NIELSEN, L. H. & KOPPELHUS, E. B. 1994. Dynamics and timing of shoreface deposition in the intracratonic Danish Basins: an example of Norian–Hettangian deposition controlled by high-frequency sea-level fluctuations. *In:* JOHNSON, S. D. (ed.) *High resolution sequence stratigraphy: Innovations and applications* (Abstract volume). Department of Earth Sciences, University of Liverpool, 335–339.

HELLAND-HANSEN, W. & GJELBERG, J. G. 1994. Conceptual basis and variability in sequence stratigraphy: a different perspective. *Sedimentary Geology*, **92**, 31–52.

HUNT, D. & TUCKER, M. E. 1992. Stranded parasequences and the forced regressive wedge systems tract: deposition during base-level fall. *Sedimentary Geology*, **81**, 1–9.

—— & —— 1995. Reply to Discussion on: Stranded parasequences and the forced regressive wedge systems tract: deposition during base-level fall. *Sedimentary Geology*, **95**, 147–160.

IRVING, E., WYNNE, P. J. & GLOBERMAN, B. R. 1993. Cretaceous palaeolatitudes and overprints of North American craton. *In:* CALDWELL, W. G. E. & KAUFFMAN, E. G. (eds) *Evolution of the Western Interior Basin.* Geological Association of Canada, Special Paper, **39**, 91–96.

JENETTE, D. C. & PRYOR, W. A. 1993. Cyclic alternation of proximal and distal storm facies: Kope and Fairview formations (Upper Ordovician), Ohio and Kentucky. *Journal of Sedimentary*

Petrology, **63**, 183–203.

JOHNSON, S. D. & FITZSIMMONS, R. J. 1994. High resolution sequence stratigraphy: The characterisation and genesis of key sequence stratigraphic surfaces in fluvial and shallow marine successions: Examples from the Mesaverde Group of the Big Horn Basin, Wyoming, U.S.A. *In:* JOHNSON, S. D. (ed.) *High resolution sequence stratigraphy: Innovations and applications* (Abstract volume). Department of Earth Sciences, University of Liverpool, 86–92.

KOLLA, V., POSAMENTIER, H. W. & EICHENSEER, H. 1995. Discussion on: Stranded parasequences and the forced regressive wedge systems tract: deposition during base-level fall. *Sedimentary Geology*, **95**, 139–145.

KOMAR, P. D. 1976. *Beach Processes and Sedimentation.* Prentice Hall, Englewood Cliffs, New Jersey.

LECKIE, D. A. & KRYSTINIK, L. F. 1989. Is there evidence for geostrophic currents preserved in the sedimentary record of inner to middle shelf deposits? *Journal of Sedimentary Petrology*, **59**, 862–870.

LEEDER, M. R. 1994. Fluvial systems and sequence stratigraphy. *In:* JOHNSON, S. D. (ed.) *High resolution sequence stratigraphy: Innovations and applications* (Abstract volume). Department of Earth Sciences, University of Liverpool, 19.

MCKAY, J. L., LONGSTAFFE, F. J. & PLINT, A. G. 1995. Early diagenesis and its relationship to depositional environment and relative sea-level changes (Upper Cretaceous Marshybank Formation, Alberta and British Columbia). *Sedimentology*, **42**, 161–190.

MARTINSEN, O. J. 1993. Namurian (late Carboniferous) depositional systems of the Craven–Askrigg area, northern England: implications for sequence stratigraphic models. *In:* POSAMENTIER, H. W., SUMMERHAYES, C. P., HAQ, B. U. & ALLEN, G. P. (eds) *Sequence Stratigraphy and Facies Associations.* International Association of Sedimentologists, Special Publication, **18**, 247–281.

MITCHUM, R. M. & VAN WAGONER, J. C. 1991. High-frequency sequences and their stacking patterns: sequence stratigraphic evidence of high-frequency eustatic cycles. *Sedimentary Geology*, **70**, 131–160.

MYROW, P. M. 1992. Pot and gutter casts from the Chapel Island Formation, southeast Newfoundland. *Journal of Sedimentary Petrology*, **62**, 992–1007.

NUMMEDAL, D. 1992. The falling sea-level systems tract in ramp settings. *Mesozoic of the Western Interior*. SEPM Theme meeting, Fort Collins, Colorado, Abstract volume, 50.

OBRADOVITCH, J. D. 1993. A Cretaceous time scale. *In:* CALDWELL, W. G. E. & KAUFFMAN, E. G. (eds) *Evolution of the Western Interior Basin.* Geological Association of Canada, Special Paper, **39**, 379–396.

PLINT, A. G. 1988. Sharp-based shoreface sequences and 'offshore bars' in the Cardium Formation: Their relationship to relative changes in sea-level. *In:* WILGUS, C. K., HASTINGS, B. S., KENDALL, C. G. St. C., POSAMENTIER, H. W., ROSS, C. A.

& VAN WAGONER, J. C. (eds) *Sea-Level Changes: An Integrated Approach.* Society of Economic Paleontologists and Mineralogists, Special Publication, **42**, 357–370.

—— 1990. An allostratigraphic correlation of the Muskiki and Marshybank formations (Coniacian–Santonian) in the Foothills and subsurface of the Alberta Basin. *Bulletin of Canadian Petroleum Geology*, **38**, 288–306.

—— 1991. High-frequency relative sea-level oscillations in Upper Cretaceous shelf clastics of the Alberta Foreland Basin: Evidence for a Milankovitch- scale glacio-eustatic control? *In:* MACDONALD, D. I. M. (ed.) *Sedimentation, Tectonics and Eustacy.* International Association of Sedimentologists, Special Publication, **12**, 409–428.

—— 1994. Sheep from Goats: Separating eustatic from tectonic controls on 4th order sequence development: Cenomanian Dunvegan Formation, Alberta foreland basin. *In:* 2nd Research Symposium on High-Resolution Sequence Stratigraphy. Tremp, Spanish Pyrenees, 21–26 June, 1994, Abstract volume, 153–158.

—— & BHATTACHARYA, J. 1991. Allostratigraphy in outcrop: initial results of subsurface to outcrop correlation in the Upper Cretaceous Dunvegan Formation, Alberta and British Columbia (Abstract). *Bulletin of Canadian Petroleum Geology*, **39**, 221.

—— & NORRIS, B. 1991. Anatomy of a ramp margin sequence: Facies successions, paleogeography and sediment dispersal patterns in the Muskiki and Marshybank formations, Alberta Foreland Basin. *Bulletin of Canadian Petroleum Geology*, **39**, 18–42.

—— & WALKER, R. G. 1987. Cardium Formation 8. Facies and environments of the Cardium shoreline and coastal plain in the Kakwa field and adjacent areas, northwestern Alberta. *Bulletin of Canadian Petroleum Geology*, **35**, 48–64.

——, —— & BERGMAN, K. M. 1986. Cardium Formation 6: Stratigraphic framework of the Cardium in subsurface. *Bulletin of Canadian Petroleum Geology*, **34**, 213–225.

——, ——, & DUKE, W. L. 1988. An outcrop to subsurface correlation of the Cardium Formation in Alberta. *In:* JAMES, D. P. & LECKIE, D. A. (eds) *Sequences, Stratigraphy, Sedimentology: Surface and Subsurface.* Canadian Society of Petroleum Geologists, Memoir, **15**, 167–184.

POSAMENTIER, H. W. & ALLEN, G. P. 1993. Variability of the sequence stratigraphic model: effects of local basin factors. *Sedimentary Geology*, **91**, 91–109.

—— & JAMES, D. P. 1993. An overview of sequence-stratigraphic concepts: uses and abuses. *In:* POSAMENTIER, H. W., SUMMERHAYES, C. P., HAQ, B. U. & ALLEN, G. P. (eds) *Sequence Stratigraphy and Facies Associations.* International Association of Sedimentologists, Special Publication, **18**, 3–18.

—— & VAIL, P. R. 1988. Eustatic controls on clastic deposition II – sequence and systems tract models. *In:* WILGUS, C. K., HASTINGS, B. S., KENDALL,

C. G. St. C., POSAMENTIER, H. W., ROSS, C. A. & VAN WAGONER, J. C. (eds) *Sea-Level Changes: An Integrated Approach*. Society of Economic Paleontologists and Mineralogists, Special Publication, **42**, 125–154.

——, JERVEY, M. T., & VAIL, P. R. 1988. Eustatic controls on clastic deposition I – conceptual framework. *In:* WILGUS, C. K., HASTINGS, B. S., KENDALL, C. G. St. C., POSAMENTIER, H. W., ROSS, C. A. & VAN WAGONER, J. C. (eds) *Sea-Level Changes: An Integrated Approach*. Society of Economic Paleontologists and Mineralogists, Special Publication, **42**, 109–124.

——, ALLEN, G. P., JAMES, D. P. & TESSON, M. 1992. Forced regressions in a sequence stratigraphic framework: concepts, examples and explorations significance. *American Association of Petroleum Geologists Bulletin*, **76**, 1687–1709.

SCHUMM, S. A. 1993. River response to baselevel change: implications for sequence stratigraphy. *Journal of Geology*, **101**, 279–294.

SHANLEY, K. W. & MCCABE, P. J. 1993. Alluvial architecture in a sequence stratigraphic framework: a case history from the Upper Cretaceous of southern Utah, U.S.A. *In:* FLINT, S. S. & BRYANT, I. D. (eds) *Quantitative description and modelling of clastic hydrocarbon reservoirs and outcrop analogues*. International Association of Sedimentologists, Special Publication, **15**, 21–56.

——, —— & HETTINGER, R. D. 1992. Tidal influence in Cretaceous fluvial strata from Utah, U.S.A.: a key to sequence stratigraphic interpretation. *Sedimentology*, **39**, 905–930.

SINGH, C. 1983. *Cenomanian microfloras of the Peace River area, northwestern Alberta*. Alberta Research Council, Bulletin, **44**, Edmonton.

SMITH, D. G. 1987. Meandering river point bar lithofacies models: modern and ancient examples compared. *In:* ETHRIDGE, F. G., FLORES, R. M. & HARVEY, M. D. (eds) *Recent developments in fluvial sedimentology*. Society of Economic Paleontologists and Mineralogists, Special Publication, **39**, 83–91.

—— 1988. Modern point bar deposits analagous to the Athabasca Oil Sands, Alberta, Canada. *In:* DE BOER, P. L., VAN GELDER, A. & NIO, S. D. (eds) *Tide-influenced sedimentary environments and facies*, D. Riedel, Dordrecht, 417–432.

STELCK, C. R., WALL, J. H. & WETTER, R. E. 1958. Lower Cenomanian foraminifera from the Peace River area, western Canada. *Research Council of Alberta, Geological Division, Bulletin*, **2**, 1–25.

STOTT, D. F. 1963. *The Cretaceous Alberta Group and equivalent rocks, Rocky Mountain Foothills, Alberta*. Geological Survey of Canada, Memoir, **317**.

—— 1982. *Lower Cretaceous Fort St. John Group and Upper Cretaceous Dunvegan Formation of the Foothills and Plains of Alberta, British Columbia, District of Mackenzie and Yukon Territory*. Geological Survey of Canada, Bulletin, **328**.

SWIFT, D. J. P. & THORNE, J. A. 1991. Sedimentation on continental margins, I: a general model for shelf sedimentation. *In:* SWIFT, D. J. P., OERTEL, G. F., TILLMAN, R. W. & THORNE, J. A. (eds) *Shelf sand and sandstone bodies*. International Association of Sedimentologists, Special Publication, **14**, 3–31.

——, PHILLIPS, S. & THORNE, J. A. 1991. Sedimentation on continental margins, V: parasequences. *In:* SWIFT, D. J. P., OERTEL, G. F., TILLMAN, R. W. & THORNE, J. A. (eds) *Shelf sand and sandstone bodies*. International Association of Sedimentologists, Special Publication, **14**, 153–187.

THOMAS, R. G., SMITH, D. G., WOOD, J. M., VISSER, J., CALVERLEY-RANGE, E. A. & KOSTER, E. H. 1987. Inclined heterolithic stratification – terminology, description, interpretation and significance. *Sedimentary Geology*, **53**, 123–179.

THORNE, J. A. & SWIFT, D. J. P. 1991. Sedimentation on continental margins, VI: a regime model for depositional sequences, their component systems tracts and bounding surfaces. *In:* SWIFT, D. J. P., OERTEL, G. F., TILLMAN, R. W. & THORNE, J. A. (eds) *Shelf sand and sandstone bodies*. International Association of Sedimentologists, Special Publication, **14**, 189–225.

VAN WAGONER, J. C., MITCHUM, R. M., CAMPION, K. M. & RAHMANIAN, V. D. 1990. *Siliciclastic sequence stratigraphy in well logs, cores, and outcrops*. American Association of Petroleum Geologists, Methods in Exploration Series, 7.

——, POSAMENTIER, H. W., MITCHUM, R. M., VAIL, P. R., SARG, J. F., LOUTIT, T. S. & HARDENBOL, J. 1988. An overview of sequence stratigraphy and key definitions. *In:* WILGUS, C. K., HASTINGS, B. S., KENDALL, C. G. St. C., POSAMENTIER, H. W., ROSS, C. A. & VAN WAGONER, J. C. (eds) *Sea-Level Changes: An Integrated Approach*. Society of Economic Paleontologists and Mineralogists, Special Publication, **42**, 39–45.

WILLIAMS, G. D. & STELCK, C. R. 1975. Speculations on the Cretaceous palaeogeography of North America. *In:* CALDWELL, W. G. E. (ed.) *The Cretaceous System in the Western Interior of North America*. Geological Association of Canada, Special Paper, **13**, 1–20.

WRIGHT, R. 1986. Cycle stratigraphy as a paleogeographic tool: Point Lookout sandstone, southeastern San Juan basin, New Mexico. *Geological Society of America, Bulletin*, **96**, 661–673.

Variable expressions of interfluvial sequence boundaries in the Breathitt Group (Pennsylvanian), eastern Kentucky, USA

JOHN F. AITKEN[1] & STEPHEN S. FLINT[2]

[1]*Geology and Cartography Division, Oxford Brookes University, Gipsy Lane Campus, Headington, Oxford OX3 0BP, UK*
[2]*STRAT Group, Department of Earth Sciences, University of Liverpool, Jane Herdman Laboratories, PO Box 147, Brownlow St, Liverpool L69 3BX, UK*

Abstract: Interfluvial sequence boundaries are a largely neglected aspect of sequence stratigraphy. They are subtle lowstand exposure surfaces which in marginal and nonmarine environments are often difficult to distinguish from exposure surfaces developed during the transgressive and highstand systems tracts. Three examples of interfluvial sequence boundaries, correlated with incised valley fills along strike, are described from Pennsylvanian delta plain deposits of the Breathitt Group of eastern Kentucky. The most common interfluvial palaeosols in the Breathitt Group are gleys developed under poorly drained conditions. This is in contrast to much of the published literature which suggests that soils on interfluves ought to be freely drained. Previously published geochemical analyses suggest that interfluvial palaeosols within the Breathitt Group are composite, forming initially under freely drained conditions and subsequently becoming gleyed due to a rise in the water table during the late lowstand and early transgressive systems tracts. This is a good criterion for differentiating lowstand palaeosols from those developed during the transgressive and highstand systems tracts.

Interfluvial sequence boundaries in the Breathitt Group have low preservation potential, being readily eroded by fluvial activity in the transgressive systems tract, which may totally remove all evidence of their former existence. In addition, if incised valleys overspill their margins during major flood events, interfluves may be represented by aggradational successions comprising thin coal seams, rootlet horizons and mature palaeoforests interbedded with crevasse splay sandstone. In such circumstances the presence of an interfluve can only be determined by the presence of correlative incised valley fills along strike.

Sequence stratigraphy has proven itself useful as a predictive tool to assess the occurrence and geometry of sedimentary strata and to reconstruct a relative chronological framework. In recent years many of the early models of high resolution sequence stratigraphy from well logs, cores and outcrops (Jervey 1988; Posamentier *et al.* 1988; Posamentier & Vail 1988; Van Wagoner *et al.* 1990) have been extensively discussed (e.g. Miall 1991, 1992; Hunt & Tucker 1992; Posamentier & Allen 1993*a, b*; Posamentier & James 1993; Schlager, 1993; Helland-Hansen & Gjellberg 1994; Shanley & McCabe 1994), but one aspect, the interfluvial sequence boundary, has received little attention. Interfluves are important because: (1) they indicate the presence of incised valleys, commonly thick sandbodies, along strike, which may be useful in subsurface hydrocarbon exploration; (2) they may be more areally extensive than their correlative incised valley fills; and (3) the recognition of interfluve sequence boundaries is one of the key criteria for distinguishing incised valley fills from major channel deposits.

This paper briefly reviews current descriptions of interfluvial sequence boundaries and subsequently describes three interfluves from Pennsylvanian delta plain deposits of the Breathitt Group, eastern Kentucky, USA, illustrating the variable nature of such surfaces. Each of these examples is correlatable with incised valley fills along strike. Although each example is considered as a single entity for this paper, they are representative of a number of similar examples occurring throughout the studied stratigraphic interval. The reader is referred to Aitken & Flint (1994, 1995) for detailed descriptions of the regional geological context and sequence stratigraphic interpretation of the Breathitt Group.

Interfluvial sequence boundaries

Van Wagoner *et al.* (1990, p. 31) state that 'adjacent to incised valleys the erosional surface passes into a correlative subaerial exposure surface marked by soils or rooted horizons'; these surfaces may be removed by erosion associated with the subsequent sea level rise.

From Howell, J. A. & Aitken, J. F. (eds), *High Resolution Sequence Stratigraphy: Innovations and Applications,* Geological Society Special Publication No. 104, pp. 193–206.

This theoretical statement has recently been reaffirmed by both Wright & Marriott (1993) and Shanley & McCabe (1994), but few well documented examples exist, with many authors simply stating that mature palaeosols, with tenuous if any direct correlation to incised valley fills, represent interfluves (e.g. Krystinik & Blakeney-DeJarnett 1991; Wood & Hopkins 1992; Gibling & Bird 1994; Tandon & Gibling 1994).

O'Byrne & Flint (1993, 1995, 1996) report that interfluvial sequence boundaries in the Blackhawk Formation of the Book Cliffs, Utah, USA, are characterized by thin coals and carbonaceous shales in up dip areas (cf. Aitken & Flint 1994) and down dip by thin oxidized, haematitic layers or sub-rounded haematized mud or siltstone chips and rare sideritized clasts representing transgressively reworked soil horizons. O'Byrne & Flint (1993, 1995, 1996) further note that interfluves are commonly cut by narrow, steep-sided tributary channels, with dendritic drainage patterns, which fed the incised valleys.

Leckie *et al.* (1989) provide a detailed documentation of lower Cretaceous discrete palaeosols and palaeosol complexes from British Columbia, Canada. These palaeosols developed as a consequence of base level fall, resulting in stream rejuvenation and incision and entrenchment into the floodplain. Consequently most sediment was flushed through the system and raised terraces and 'interfluves' were starved of sediment. Many of these palaeosols appear to be gleysols (i.e. formed in poorly drained conditions) because of their grey colour and the occurrence of slickensides and mottling. Thin sections indicate that the mottles are siderite spherules that are superimposed onto translocated clay. Leckie *et al.* (1989) interpret this to imply that gleization was a secondary process and that the palaeosols initially developed under well drained conditions with a lowered water table. Similarly, Shanley & McCabe (1994) suggest that previously well drained soils on valley margins may become poorly drained as transgression raises the water table.

Carboniferous interfluvial sequence boundaries have been described from County Clare, Western Ireland, by Davies & Elliott (1996). Here interfluvial palaeosols comprise light coloured sandstone beds with abundant rootlets, *in situ Stigmaria* and polygonal desiccation cracks which extend approximately half way through the profile. In contrast, Hampson *et al.* (1996) describe silica-enriched, variably bleached palaeosols displaying rootlets characterized by siliceous tubules with carbonaceous linings from the Namurian of northern England. These are interpreted as gannisters and are correlated with incised valley fills.

Regional setting of the Breathitt Group, eastern Kentucky

The Middle to Upper Pennsylvanian Breathitt Group, up to 950 m thick, comprises delta plain facies of siltstone, claystone, sandstone, bituminous coal and rare ironstone and limestone, deposited in a foreland basin setting. Traditional models of deposition for Carboniferous strata in eastern Kentucky equate all the units from early Mississippian to late Pennsylvanian age as broadly time equivalent. All adjacent lithologies within this stratigraphic interval were thought to represent separate facies in a single environmental continuum reflecting Walther's law. Within this framework, the Breathitt Group was interpreted as the deposits of a fluvially-dominated, shallow water deltaic environment associated with periodic, widespread marine incursions (Horne *et al.* 1978). Nonetheless, traditional lithostratigraphic analysis fails to recognize regionally significant sequence boundary unconformities similar to those described by Van Wagoner *et al.* (1990). Such surfaces were first described from the Breathitt Group by Aitken & Flint (1991), Davies & Burn (1991) and Davies *et al.* (1992). Aitken & Flint (1994, 1995) describe the sequence stratigraphy of the outcrops in detail and recognize 18 sequence boundaries in the constituent Pikeville, Hyden and Four Corners Formations (Fig. 1). The Breathitt Group is exposed in a series of roadcuts for *c.* 110 km along United States Highway 23 and for *c.* 100 km along Kentucky Highway 80 and the Daniel Boone Parkway (Fig. 1). Stratigraphic correlation between outcrops was based on the coal correlation framework of Horne (1978, 1979*a, b*) and Baganz & Horne (1979) and the lithostratigraphic correlation of Chesnut (1991).

Example 1: Magoffin Shale member interfluve, Four Corners outcrop, near Hazard

The Magoffin Shale is the most widespread and persistent marine unit in eastern Kentucky and is commonly incised by thick, multistorey, multilateral, low sinuosity to braided channel complexes interpreted as incised valley fills (Aitken & Flint 1994, 1995). At the Four Corners outcrop near Hazard (outcrop a, Fig. 1) the Magoffin Shale is some 9 m thick and comprises a medium grey siltstone with calcar-

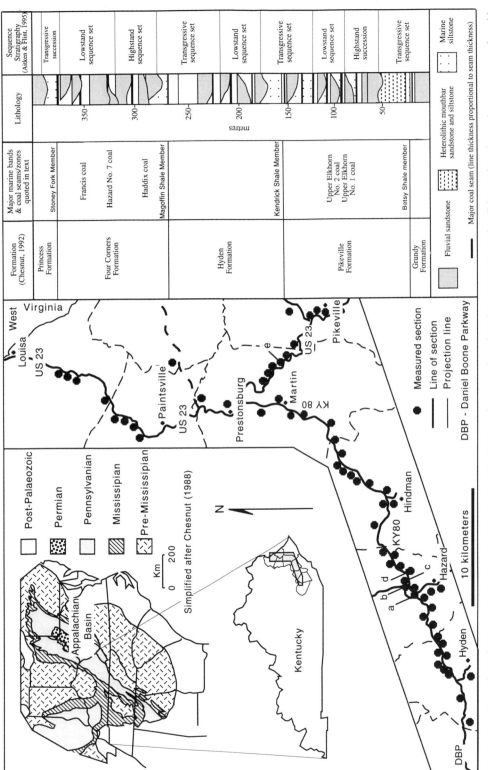

Fig. 1. Location of the field area in the Appalachian Basin, distribution of measured sections in eastern Kentucky, lithostratigraphy and sequence stratigraphic interpretation of the Pikeville, Hyden and Four Corners Formations. a, Four Corners; b, Perry County milepost 9; c, Meadow Branch; d, Darb Fork; e, HR9.

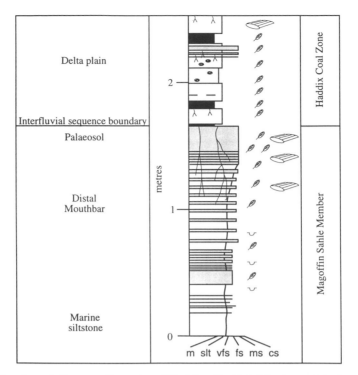

Fig. 2. Lithological log through the upper few metres of the Magoffin Shale Member and the base of the Haddix coal zone at Four Corners (outcrop a, Fig. 1).

Fig. 3. The interfluvial sequence boundary overlying the Magoffin Shale Member at Four Corners (outcrop a, Fig. 1).

eous concretions. In its upper 2 m the Magoffin Shale coarsens-upward through very thinly interbedded pale grey sandstone and siltstone beds, within which sandstone beds rarely exceed 1 cm in thickness and thin siderite beds are common (Figs 2 and 3). This is gradationally overlain by 50 cm of silty, fine-grained sandstone with small, woody fragments and carbonaceous streaks of plant debris along bedding planes and simple, unlined vertical burrows up to 5 cm deep. This is in turn gradationally overlain by pale grey, finely bedded, upper fine-grained sandstone with rare sandy siltstone partings. Sandstone beds, some of which are normally graded, thicken upward to a maximum of 2 cm (Figs 2 and 3). Roots, characterized by carbonaceous films, penetrate from the unit above. This upper unit comprises horizontally bedded fine- to medium-grained sandstone, with beds up to 4 cm thick, with thin siltstone partings towards the base. This unit is dominantly pale grey but is locally mottled reddish-brown and orange, contains slickensides and abundant plant debris and is heavily rooted and passes upward into very heavily rooted, bleached and mottled, very hard sandstone, containing a considerable amount of plant debris. This is overlain by the first thin

(4 cm thick) coal seam of the Haddix coal zone (Figs 1, 2 and 3).

The Magoffin Shale at the Four Corners is interpreted as a marine siltstone which coarsens-upward into a distal mouthbar deposit, reflecting maximum water depths of up to 10–15 m. The uppermost portion of this distal mouthbar has clearly undergone pedogenesis, recording a basinward shift in facies and fall in relative sea level. The mottled and grey colouration is generally indicative of originally poorly drained conditions where both oxidation and reduction processes were active (Retallack 1976), hence the palaeosol is interpreted as a gleysol. The occurrence of pedogenized strata at the top of a marine unit and correlative to incised valley fills indicates that this is an interfluvial sequence boundary. The occurrence of a gleyed soil on an interfluve is in contradiction to much of the published literature which suggests that interfluvial palaeosols should be well drained. Nonetheless, this is the most common type of interfluvial palaeosol in the Breathitt Group (Aitken & Flint 1994, 1995).

Example 2: Upper Elkhorn # 1 and 2 coal seam interval, HR9 outcrop, near Betsy Layne

Between the Upper Elkhorn # 1 and 2 coal seams (Fig. 1) at least three thick, multistorey, multilateral low sinuosity fluvial sandstone complexes, interpreted as incised valley fills, occur at the same stratigraphic level (Aitken & Flint 1994, 1995). On the United States Highway 23 near Betsy Layne in outcrop HR9 (outcrop e, Fig. 1) this stratigraphic interval is 6–10 m thick and comprises a heterolithic succession of grey, laminated siltstone interbedded with rippled and trough cross-bedded sandstone deposited in crevasse splay and sub-aerial levee environments (Figs 4 and 5b). A 60–70 cm-thick orange–yellow, heavily rooted, structureless silty claystone lies c. 2 m above the Upper Elkhorn # 1 seam (Figs 4 and 5a). This claystone is bounded at its base by a 5–9 cm-thick discontinuous siderite bed beneath which roots do not penetrate. The top of the clay is marked by *Lepidodendron* and *Calamites* stumps (up to 40 cm in diameter) in growth positions with abundant fossilized branches and a very thin (< 3 mm thick), discontinuous highly carbonaceous shale. Organic material within the claystone is preserved along clay-coated peds, whilst the over- and underlying laminated siltstones contain disseminated plant fragments along bedding planes. The contact between the orange–yellow claystone and overlying heterolithic strata is locally horizontal and appears to be non-erosional (Fig. 4). Elsewhere, the upper surface is erosional and in the middle of the outcrop the claystone has been entirely removed by a complex interbedded package of sandstone and siltstone (Figs 4 and 5c).

The orange–yellow claystone is interpreted as a well drained palaeosol on the basis of its uniform colour. The absence of well developed horizon zonation typical of modern well-developed soils is, however, unusual (see below). The occurrence of a well drained soil in a stratigraphic interval at which incised valley fills are known to occur, and where no other palaeosols are developed, strongly implies that this palaeosol marks an interfluvial sequence boundary.

Geochemical analysis of this palaeosol by Gardner et al. (1988) supports this interfluvial sequence boundary interpretation. Gardner et al. (1988) showed that the laminated siltstones above and below the palaeosol have a constant kaolinite/illite ratio and a constant chlorite component, whilst the palaeosol itself contains no detectable chlorite and an increasing kaolinite/illite ratio with depth. Gardner et al. (1988) noted similar patterns from age-equivalent palaeosols in western Pennsylvania, and developed a two-stage pedogenic model to explain the observed clay mineral profiles. In the first stage kaolinization occurred under conditions of good drainage with the downward movement of acidic, oxidizing meteoric water which promoted the downward movement of dissolved constituents, especially silica, alumina and base cations from the dissolution of feldspars, chlorite and detrital dolomite. The kaolinization resulted in the development of a podsol, probably with soil horizons. Subsequently the water table rose eventually to the land surface, such that pedogenesis by the downward movement of water ceased and gleying processes occurred. Since gleying mainly involves the removal of iron under reducing conditions, it cannot account for the observed vertical variation in clay minerals. The poorly drained gleying processes removed the iron oxide zonation produced during podsolization (Gardner et al. 1988).

This two-stage model of development of interfluvial soils is logical when base level dynamics are considered. During the lowstand, water tables are lowered as a consequence of base level fall and therefore water would flow freely downwards through soils developing on raised interfluves. During the subsequent transgression, water tables rise until both the incised valley fills and interfluves are flooded and

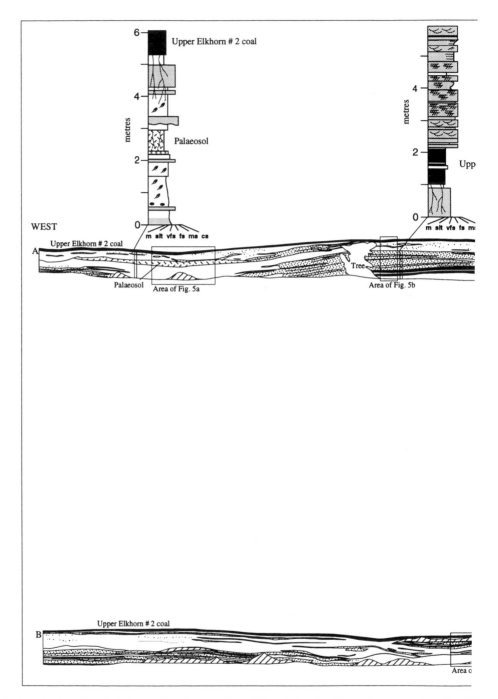

Fig. 4. Architectural variability between the Upper Elkhorn # 1 and 2 coals at HR9 (outcrop e, Fig. 1), with detailed lithological logs. The palaeosol midway between the two seams marks an interfluvial sequence boundary correlatable with incised valley fills along strike. This interfluvial palaeosol developed initially under freely drained conditions and subsequently became gleyed following a rise in the water table (see text for full discussion). The preservation potential of the interfluve is low as is illustrated by its erosion by overlying heterolithic fluvial deposits.

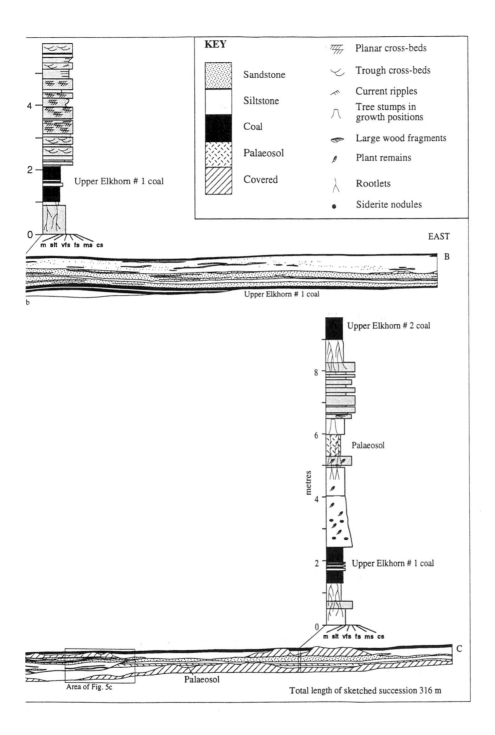

(a)

(b)

Fig. 5. The strata between the Upper Elkhorn # 1 and 2 coals at HR9 (outcrop e, Fig. 1). Location of the photographs illustrated in Fig. 4. (a) The interfluvial palaeosol and its surrounding strata; (b) details of the crevasse splay/levee facies underlying the interfluvial palaeosol; (c) erosion of the palaeosol by overlying heterolithic fluvial deposits of the transgressive systems tract, illustrating the low preservation potential of interfluves.

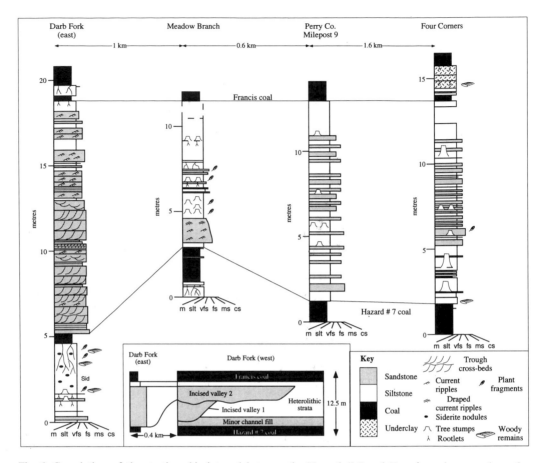

Fig. 6. Correlation of the stratigraphic interval between the Hazard # 7 and Francis coal seams along the Kentucky Highway 80 between Darb Fork (east) and Four Corners (outcrops d to a, Fig. 1). Two incised valley fills pinch out at Darb Fork (west), the upper of which has entirely cut out the lower, 0.4 km to the east (inset). Although two interfluvial sequence boundaries are expected only a thick crevasse splay succession is apparent, with numerous horizons of tree stumps in growth positions (see text for full discussion).

progressive water logging of previously well drained palaeosols will occur (cf. Leckie *et al.* 1989).

Example 3: The Hazard # 7 to Francis coal interval near Hazard

The exposed interval between the Hazard # 7 and Francis coal seams (Fig. 1) is most commonly characterizd by two vertically superimposed, laterally continuous, multilateral, multistorey, low sinuosity channel complexes (Salter 1992; Aitken & Flint 1994, 1995). These sandstone bodies are interpreted as incised valley fills, each marking a sequence boundary

(Aitken & Flint 1994, 1995). However, at the southwestern end of Kentucky Highway 80, near Hazard (Fig. 1), between the Darb Fork and Four Corners outcrops (Figs 1 and 6), large stacked sand bodies are absent and the margins of both sand bodies can be seen at Darb Fork west (Fig. 6 inset). Between Darb Fork and Four Corners the Hazard # 7 to Francis coal interval is characterized by complexly interbedded fine- to medium-grained sandstone and laminated siltstone (Fig. 6). Individual sandstone beds have sharp or erosive bases and sole marks are occasionally present. Internal sedimentary structures include abundant small-scale trough crossbeds, climbing ripple cross-lamination, current ripples, parallel lamination and rare dewatering

Fig. 7. The crevasse splay succession overlying the Hazard # 7 coal at Four Corners (outcrop a, Fig. 1). Particularly note the three horizons of large, mature tree stumps.

structures. Rootlet horizons, *in situ* tree stumps and well preserved plant fossils are common. This succession is interpreted as periodic sub-aerial overbank deposits dominated by crevasse splays.

The character of this overbank succession changes between the outcrops (Fig. 6). At Four Corners, crevasse splays are commonly amalgamated into single sandstone bodies up to 2.1 m thick interbedded with thinner crevasse splay sandstone and at least three thin (< 6 cm thick) coal seams (Fig. 6). Small tree stumps are scattered throughout the outcrop, but three horizons of very large tree stumps, up to 70 cm in diameter and 90 cm high, occur (Fig. 7), and represent mature tree growth (Jennings 1981). Up to 20 individual stumps occur at these horizons (cf. Cobb *et al.* 1981; Chesnut *et al.* 1986). Approximately 1.6 km to the north-east, in an outcrop by Perry County milepost 9, the succession is characterized by thinner, less amalgamated crevasse splay sandstone, and the

siltstone interbeds are thicker (Fig. 6). Abundant tree stumps, with diameters of 10–20 cm, in growth positions, occur scattered throughout the succession. 0.6 km farther to the north-east the Hazard # 7 seam is overlain by a 1.4 m thick, fining-upward channel sandstone overlain by a laminated silt-dominated succession with thin (< 20 cm) crevasse splay sandstone beds within which there are at least 5 horizons of small (< 15 cm in diameter) tree stumps (Fig. 6). This pattern of thinning of crevasse splay sandstones to the northeast, in conjunction with palaeoflow data (Aitken & Flint 1995), is taken to indicate that the channel responsible for these deposits lay to the south or southwest.

Since two superimposed incised valley fills are known to occur between the Hazard # 7 and Francis coals (Fig. 6 inset), interfluvial sequence boundaries should be expected where these incised valley fills are absent. The occurrence of tree stumps and rootlet horizons is indicative of exposure and pedogenesis (Retallack 1988). However, no well developed palaeosols, either of poorly drained or freely drained origin, are apparent between the Hazard # 7 and Francis seams, with primary sedimentary structures being preserved in both the siltstone and sandstone facies. Nor are there simply two horizons of mature palaeoforest which could mark the interfluves. Instead a number of palaeoforests appear to have grown throughout the deposition of the Hazard # 7 to Francis coals interval. Furthermore, Jennings (1981) suggests that close proximity to a source of terrigenous clastic input was necessary for plant growth in these deposits.

The absence of clear interfluves can be explained by high sediment supplies during periods of low accommodation space. In such circumstances incised valleys may overspill their margins during periodic major flood events (> 100 year flood event?) due to excess sediment in relation to available accommodation. This will result in the creation of widely spaced (in time) crevasse splay successions inundating interfluvial areas and the suppression of pedogenesis in interfluvial areas at the margins of the incised valleys. The relatively long time periods between crevassing events are indicated by the development of thin mires and/or mature palaeoforests. Hence interfluvial sequence boundaries need not be recognized as simple exposure surfaces but may be represented by lowstand sediments deposited on the margins of the incised valleys. One or more sequence boundaries may be present within these deposits but will be extremely difficult to locate precisely. It is probable that incised valleys will not overspill their margins until the later stages of filling such that interfluve palaeosols may develop during the early stages of filling. However, the high energy crevassing events described above have the potential to erode and totally remove any palaeosol deposits developed prior to overspilling.

Discussion

Gleysols, similar to that described for the Magoffin interfluve (example 1 above), are the most common type of interfluvial palaeosol in the Breathitt Group. The gleyed character of interfluvial palaeosols may indicate that the period of base level lowstand was relatively short in relation to the period required for mature palaeosol development. However, the two stage development of the interfluvial palaeosol in the HR9 outcrop (example 2 above) illustrates that originally well drained interfluvial soils may become gleyed during transgression. The occurrence of this pattern in both eastern Kentucky and western Pennsylvania shows that this is a widespread phenomenon in this part of the Appalachian Basin (Gardner et al. 1988), and supports the hypothesis that many of the gleyed interfluvial palaeosols may have originally developed under freely drained conditions. This hypothesis is strongly supported by the common occurrence of thick coal seams overlying the top of incised valley fills and interfluves in eastern Kentucky, which indicate poor drainage conditions, with a high and rising water table, immediately above the interfluves as base level rose at an increasing rate (Aitken 1994; Aitken & Flint 1994, 1995).

It is important to note that, taken in isolation, interfluvial palaeosols are difficult to distinguish from similar sediments deposited in the transgressive and highstand systems tracts (Aitken & Flint 1994, 1995). This difficulty is enhanced by the fact that many of the sequence boundaries in the delta plain deposits of eastern Kentucky are encased in non-marine strata (Aitken & Flint 1994, 1995), hence a basinward shift in facies may not necessarily be apparent at the base of interfluvial palaeosols (e.g. Fig. 3). It is only by mapping the relationship of palaeosols with incised valley fills along strike that their significance as interfluvial sequence boundaries can be clearly defined.

Palaeosols on interfluves have low preservation potential with reworking during transgression having commonly been reported for down dip areas (e.g. Van Wagoner et al. 1990; O'Byrne & Flint 1993, 1995, 1996). In alluvially dominated successions it is likely that interfluvial palaeosols may be eroded, with all evidence of their former existence removed, by fluvial activity in the transgressive systems tract. This case is illustrated by the incision of the interfluvial palaeosol between the Upper Elkhorn # 1 and 2 coals (e.g. Figs 4 and 5c), leaving a discontinuous erosional remnant. The interfluve would be very easy to miss in core. Clear interfluves will also not be apparent in situations where incised valleys overspill their margins resulting in the deposition of interfluvial lowstand deposits comprising aggradational crevasse splay packages, rootlet horizons and palaeoforests (Fig. 6).

Conclusions

Interfluvial sequence boundaries in the Pennsylvanian delta plain deposits of eastern Kentucky are subtle and difficult to identify other than by the correlation of palaeosols with incised valley fills along strike. This potential for correlating along strike from incised valley fills into areas of no incision creates an ideal opportunity to assess the variability of interfluve expressions and to discuss the controls on their development and preservation. The most common forms of interfluvial palaeosol in Kentucky are gleysols, which probably developed in two stages, initially in a freely drained environment during the lowstand. The palaeosols were subsequently modified under transgressive conditions of poor drainage. Gleying superimposed onto previously well drained palaeosols (e.g. Gardner et al. 1988; Leckie et al. 1989) is potentially a powerful tool to identify interfluvial palaeosols. Whilst it is clear that interfluvial palaeosols can be recognized, their preservation potential is, however, low and they may be entirely removed either by transgressive erosion or the activities of fluvial channels in the transgressive systems tract. As an additional complication clear surfaces may not develop if incised valley fills overspill their margins during filling such that an interfluvial zone comprising stacked crevasse splays interbedded with palaeoforests may accumulate.

The research of which this paper forms a part was funded by the Stratigraphic Prediction Unit and Reservoir Description Section of British Petroleum, London, while JFA was a Research Associate at the University of Liverpool. The work has profited from discussions with John Ferm (University of Kentucky) and Don Chesnut (Kentucky Geological Survey). Sarah Davies and Gary Hampson commented on a previous version of the manuscript, which was reviewed by David Oliver and Bill Read.

References

AITKEN, J. F. 1994. Coal in a sequence stratigraphic framework. *Geoscientist*, **4**, 9–12.

—— & FLINT, S. S. 1991. Reservoir characterisation and sequence stratigraphy of Upper Carboniferous (Pennsylvanian) deltaic sequences of eastern Kentucky, USA. *In: British Sedimentology Research Group 30th Annual Meeting Programme and Abstracts*. University of Edinburgh.

—— & —— 1994. High-frequency sequences and the nature of incised-valley fills in fluvial systems of the Breathitt Group (Pennsylvanian), Appalachian Foreland Basin, eastern Kentucky. *In:* DALRYMPLE, R., BOYD, R. & ZAITLIN, B. A. (eds), *Incised Valley Systems: Origin and Sedimentary Sequences*. Society of Economic Paleontologists and Mineralogists, Special Publication, **51**, 353–368.

—— & —— 1995. The application of high resolution sequence stratigraphy to fluvial systems: a case study from the Upper Carboniferous Breathitt Group, eastern Kentucky, U.S.A. *Sedimentology*, **42**, 3–30.

BAGANZ, B. B. & HORNE, J. C. 1979. *Geologic cross-section along US Highway 23 between Prestonsburg and Pikeville, Johnson and Floyd Counties, Kentucky*. Carolina Coal Group Geologic Cross-Section GCS-4, Carolina Coal Group, University of South Carolina, 2 sheets.

CHESNUT, D. R. 1988. *Stratigraphic analysis of the Carboniferous rocks of the central Appalachian Basin*. PhD Thesis, University of Kentucky.

—— 1991. *Geologic highway cross section: Kentucky Highway 80, Hazard to Prestonsburg*. Map and Chart Series 2 (Series XI), Kentucky Geological Survey, Lexington, 1 sheet.

—— 1992. *Stratigraphic and structural framework of the Carboniferous rocks of the Central Appalachian Basin in Kentucky*. Kentucky Geological Survey, Bulletin 3 (series XI), Lexington.

——, COBB, J. C. & KIEFER, J. D. 1986. 'Four Corners': A 3-dimensional panorama of stratigraphy and depositional structures in the Breathitt Formation (Pennsylvanian), Hazard, Kentucky. *In:* NEATHERY, T. L. (ed.) *Geological Society of America Southeastern Section*. Centennial Field Guide, **6**, 47–48.

COBB, J. C., CHESNUT, D. R., HESTER, N. C. & HOWER, J. C. 1981. *Coal and coal-bearing rocks of eastern Kentucky (Field trip road log)*. Kentucky Geological Survey, Lexington.

DAVIES, A. H. & BURN, M. J. 1991. High resolution sequence stratigraphy of fluvio-deltaic cyclothems: the Upper Carboniferous Breathitt Group, east Kentucky. *In: British Sedimentology Research Group 30th Annual Meeting Programme and Abstracts*. University of Edinburgh.

——, ——, BUDDING, M. C. & WILLIAMS, H. 1992. High resolution sequence stratigraphic analysis of fluvio-deltaic cyclothems: the Pennsylvanian Breathitt Group, east Kentucky. *In: American Association of Petroleum Geologists 1992 Annual Convention Programme*. 27.

DAVIES, S. J. & ELLIOTT, T. 1996. Spectral gamma ray characterization of high resolution sequence stratigraphy: examples from Upper Carboniferous fluvio-deltaic systems, County Clare, Ireland. *This volume*.

GARDNER, T. W., WILLIAMS, E. G. & HOLBROOK, P. W. 1988. Pedogenesis of some Pennsylvanian underclays: ground water, topographic, and tectonic controls. *In:* REINHARDT, J. & SIGLEO, W. R. (eds) *Paleosols and weathering through geologic time: Principles and applications*. Geological Society of America, Special Paper, **216**, 81–102.

GIBLING, M. R. & BIRD, D. J. 1994. Late Carboniferous cyclothems and alluvial palaeovalleys in the Sydney Basin, Nova Scotia. *Bulletin of the Geological Society of America*, **106**, 105–117.

HAMPSON, G., ELLIOTT, T. & FLINT, S. 1996. Critical application of high resolution sequence stratigraphic concepts to the Rough Rock Group (Upper Carboniferous) of northern England. *This volume*.

HELLAND-HANSEN, W. & GJELBERG, J. G. 1994. Conceptual basis and variability in sequence stratigraphy: a different perspective. *Sedimentary Geology*, **92**, 31–52.

HORNE, J. C. 1978. *Geologic cross-section along the Daniel Boone Parkway between Hazard and Buckhorn Reservoir, Perry and Leslie Counties, Kentucky*. Carolina Coal Group Geologic Cross Section GCS-1, Carolina Coal Group, University of South Carolina, Columbia, 2 sheets.

—— 1979a. *Geologic cross-section along US Highway 23 between Louisa and Paintsville, Lawrence and Johnson Counties, Kentucky*. Carolina Coal Group Geologic Cross-Section GCS-2, Carolina Coal Group, University of South Carolina, Columbia, 3 sheets.

—— 1979b. *Geologic cross-section along US Highway 23 between Paintsville and Prestonsburg, Floyd and Pike Counties, Kentucky*. Carolina Coal Group Geologic Cross Section GCS-3, Carolina Coal Group, University of South Carolina, Columbia, 2 sheets.

——, FERM, J. C., CARUCCIO, F. T. & BAGANZ, B. P. 1978. Depositional models in coal exploration and mine planning in Appalachian region. *American Association of Petroleum Geologists Bulletin*, **62**, 2379–2411.

HUNT, D. & TUCKER, M. E. 1992. Stranded parasequences and the forced regressive wedge systems tract: deposition during base level fall. *Sedimentary Geology*, **81**, 1–9.

JENNINGS, J. R. 1981. Pennsylvanian plants of eastern Kentucky: compression fossils from the Breathitt Formation near Hazard, Kentucky. *In:* COBB, J. C., CHESNUT, D. R., HESTER, N. C. & HOWER, J. C. (eds) *Coal and coal-bearing rocks of eastern Kentucky*. Annual Geological Society of America Coal Division Field Trip 1981, Kentucky Geological Survey, Lexington, 147–159.

JERVEY, M. T. 1988. Quantitiative modelling of siliciclastic rock sequences and their seismic expression. *In:* WILGUS, C. K., HASTINGS, B. S., KENDAL, C. G. St. C., POSAMENTIER, H. W., ROSS,

C. A. & VAN WAGONER, J. C. (eds) *Sea-level changes: an integrated approach*. Society of Economic Paleontologists and Mineralogists, Special Publication, **42**, 47–70.

KRYSTINIK, L. F. & BLAKENEY-DEJARNETT, A. 1991. Sequence stratigraphy and sedimentologic character of valley fills, Lower Pennsylvanian Morrow Formation, Eastern Colorado and western Kansas. *In: Programme, proceedings, guidebook 1991 NUNA conference on high resolution sequence stratigraphy*. Global Sedimentary Geology Programme, 24–26.

LECKIE, D., FOX, C. & TARNOCAI, C. 1989. Multiple paleosols of the late Albian Boulder Creek Formation, British Columbia, Canada. *Sedimentology*, **36**, 307–322.

MIALL, A. D. 1991. Stratigraphic sequences and their chronostratigraphic correlation. *Journal of Sedimentary Petrology*, **61**, 497–505.

—— 1992. Exxon global cycle chart: an event for every occassion? *Geology*, **20**, 787–790.

O'BYRNE, C. J. & FLINT, S. S. 1993. High-resolution sequence stratigraphy of Cretaceous shallow marine sandstones, Book Cliffs outcrops, Utah, USA – application to reservoir modelling. *First Break*, **11**, 445–459.

—— & —— 1995. Sequence, parasequence and intraparasequence architecture of the Grassy Member, Blackhawk Formation, Book Cliffs, Utah, USA. *In:* VAN WAGONER, J. C. & BERTRAM, G. (eds) *Sequence stratigraphy of foreland basin deposits; examples from the Cretaceous of N. America*. American Association of Petroleum Geologists, Memoir, **65**, in press.

—— & —— 1996. Interfluve sequence boundaries in the Grassy Member, Book Cliffs, Utah: criteria for recognition and implications for subsurface correlation. *This volume.*

POSAMENTIER, H. W. & ALLEN, G. P. 1993a. Siliciclastic sequence stratigraphic patterns in foreland ramp-type basins. *Geology*, **21**, 455–458.

—— & —— 1993b. Variability of the sequence stratigraphic model: effects of local basin factors. *Sedimentary Geology*, **86**, 91–109.

—— & JAMES, D. P. 1993. An overview of sequence stratigraphic concepts: uses and abuses. *In:* POSAMENTIER, H. W., SUMMERHAYES, C. P., HAQ, B. U. & ALLEN, G. P. (eds) *Sequence stratigraphy and facies associations*. International Association of Sedimentologists, Special Publication, **18**, 3–18.

—— & VAIL, P. R. 1988. Eustatic controls on clastic deposition II – sequence and systems tract models. *In:* WILGUS, C. K., HASTINGS, B. S., KENDAL, C.

G. St. C., POSAMENTIER, H. W., ROSS, C. A. & VAN WAGONER, J. C. (eds) *Sea-level changes: an integrated approach*. Society of Economic Paleontologists and Mineralogists, Special Publication, **42**, 125–154.

——, JERVEY, M. T. & VAIL, P. R. 1988. Eustatic controls on clastic deposition I – conceptual framework. *In:* WILGUS, C. K., HASTINGS, B. S., KENDAL, C. G. St. C., POSAMENTIER, H. W., ROSS, C. A. & VAN WAGONER, J. C. (eds) *Sea-level changes: an integrated approach*, Society of Economic Paleontologists and Mineralogists, Special Publication, **42**, 109–124.

RETALLACK, G. J. 1976. Triassic palaeosols in the upper Narrabeen Group of New South Wales. Part I. Features of the palaeosols. *Journal of the Geological Society of Australia*, **23**, 383–399.

—— 1988. Field recognition of paleosols. *In:* REINHARDT, J. & SIGLEO, W. R. (eds) *Paleosols and weathering through geologic time: Principles and applications*. Geological Society of America, Special Paper, **216**, 1–20.

SALTER, T. 1992. *Facies, geometrical and palaeocurrent analysis of well exposed alluvial reservoir analogues and applications to sub surface studies*. PhD Thesis, University of Leeds.

SCHLAGER, W. 1993. Accommodation and supply – a dual control on stratigraphic sequences. *Sedimentary Geology*, **86**, 111–136.

SHANLEY, K. W. & McCABE, P. J. 1994. Perspectives on the sequence stratigraphy of continental strata. *American Association of Petroleum Geologists Bulletin*, **78**, 544–568.

TANDON, S. K. & GIBLING, M. R. 1994. Calcrete and coal in Late Carboniferous cyclothems of Nova Scotia, Canada: climate and sea-level changes linked. *Geology*, **22**, 755–758.

VAN WAGONER, J. C., MITCHUM, R. M., CAMPION, K. M. & RAHMANIAN, V. D. 1990. *Siliciclastic sequence stratigraphy in well logs, cores and outcrops*. American Association of Petroleum Geologists, Methods in exploration 7.

WOOD, J. M. & HOPKINS, J. C. 1992. Traps associated with palaeovalleys and interfluves in an unconformity bounded sequence: Lower Cretaceous Glauconitic Member, southern Alberta, Canada. *American Association of Petroleum Geologists Bulletin*, **76**, 904–926.

WRIGHT, V. P. & MARRIOTT, S. B. 1993. The sequence stratigraphy of fluvial depositional systems: the role of floodplain sediment storage. *Sedimentary Geology*, **86**, 203–210.

Interfluve sequence boundaries in the Grassy Member, Book Cliffs, Utah: criteria for recognition and implications for subsurface correlation

CIARAN J. O'BYRNE[1] & STEPHEN FLINT[2]

[1]*Amoco Production Company, PO Box 3092, Houston TX 77253, USA*
[2]*STRAT Group, Department of Earth Sciences, University of Liverpool, Liverpool L69 3BX, UK*

Abstract: The Late Campanian Grassy Member of the Blackhawk Formation, Mesaverde Group, Utah consists of two unconformity-bounded fourth-order sequences within a third-order highstand sequence set. Several orders of bounding surface define the internal architecture of each sequence and component parasequences. Incised valley fill deposits, dominated by braided channel-fills updip and laterally accreted fluvial and estuarine channel-fills down dip, overlie the fourth-order sequence boundaries, truncating shoreface and shelf facies of underlying sequences. Incised valleys are separated by extensive interfluves with evidence of subaerial exposure and minor tributary channel systems. The interfluve gullies are interpreted to have transported very little bedload during lowstand and acted primarily as rainwater run-off or bypass channels lateral to the main incised valley system. They are filled with bioturbated fine-grained sandstone and siltstone and capped by marine shale. Interfluve areas provide subtle but crucial evidence for base level falls, including rare examples of roots penetrating formerly offshore marine shale, partially eroded hematitic hardpan sediments and sand or silt filled gullies containing coal and shell debris. Interfluve sequence boundaries have powerful predictive application in respect to recognition of along-strike incised valley fills. Recognition of these surfaces is commonly hampered by removal of subaerial indicators as a result of subsequent transgressive erosion. There is indirect stratigraphic evidence of an emerging salt or basement fault driven structural high, orthogonal to the Late Campanian palaeoshoreline, which controlled the locus of valley incision and associated interfluves in two successive high frequency sequences.

Application of high resolution sequence stratigraphy to sub-surface reservoir characterization over the past decade has highlighted the need for more detailed datasets describing the internal geometry of incised valley fill complexes and the regional expression of sequence boundaries. Identification of sequence boundaries and associated valley fill deposits has important implications for reservoir prediction and exploration in deeper basinal settings and in the recognition of potential high permeability zones susceptible to water breakthrough (e.g. Jennette & Riley 1996). However, sequence boundaries are commonly expressed as interfluve surfaces modified by subsequent transgressve erosion (Van Wagoner et al. 1990). The subtlety and thus potential for non-recognition of these surfaces may lead to the incorrect interpretation of unconformities as having limited extent (Cartwright et al. 1993). This paper focuses on the variable expression of sequence boundaries within the Grassy Member of the Book Cliffs, east-central Utah and documents in detail the recognition criteria for and predictive significance of stacked interfluve sequence boundaries. The mesa topography and dissection of the Book Cliffs by canyons allows for detailed 3-D mapping of stratal surfaces over tens to hundreds of square kilometres (Fig. 1) and tracing of interfluve surfaces into the time equivalent incised valleys along strike. Furthermore, while recent estuarine facies models provide criteria to assess the up-dip to down-dip variability in incised valley fills (e.g. Dalrymple et al. 1994), no such studies have been undertaken on the corresponding interfluve surfaces; in this paper we describe the proximal to distal expressions of interfluve sequence boundaries.

Regional geological setting and stratigraphy

The Late Campanian Blackhawk Formation (Mesaverde Group), exposed in the Book Cliffs, consists of six lithostratigraphic units: the Spring Canyon, Aberdeen, Kenilworth, Sunnyside, Grassy and Desert Members (Balsley 1980; Fig. 2), comprising a regressive set of clastic deposits up to 400 m thick, dominated by marine shoreface sand bodies. Recent studies have focused on building a high resolution sequence

From Howell, J. A. & Aitken, J. F. (eds), *High Resolution Sequence Stratigraphy: Innovations and Applications,*
Geological Society Special Publication No. 104, pp. 208–220.

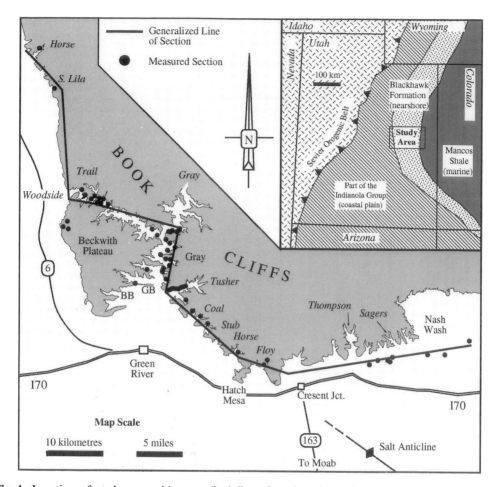

Fig. 1. Location of study area with generalized line of section shown in Fig. 3. The inset shows the palaeogeographic setting of the Blackhawk Formation during the Late Campanian.

stratigraphic framework for the formation, recognizing a series of unconformity bounded high frequency sequences within an overall highstand sequence set (Van Wagoner et al. 1990; Taylor & Lovell 1991; O'Byrne & Flint 1993, 1995; Kamola & Van Wagoner 1995).

Sequence stratigraphy of the Grassy Member

The age of the Grassy Member is c. 78–79 Ma (middle to late Campanian). Balsley (1980) defined the Grassy Member as a barrier island–lagoonal complex with associated tidal inlet and flood-tidal delta deposits. More recent high resolution sequence stratigraphic analysis (Fig. 2) has identified two genetically unrelated,

unconformity bounded sequences within a single lithostratigraphic unit, the sedimentology and sequence stratigraphy of which are described in detail elsewhere (O'Byrne & Flint 1993, 1995). Thus only the key data are summarized herein.

The down-dip extension of the Grassy Member allows further regional mapping of sequence boundaries and lowstand surfaces of subaerial shelf exposure (Figs 2 and 3). Grassy Sequence Boundary 1 (GSB1) is not discussed here. Grassy sequence boundary 2 (GSB2) truncates the GPS2 parasequence and defines the base of Grassy Sequence 2 (Fig. 2). Desert sequence boundary 1 (DSB1), as defined in this paper, truncates Grassy parasequence 4 (GPS4) and the up-dip extension of the GPS3 parasequence. This unconformity is difficult to trace in the up-dip section being obscured by scree slopes

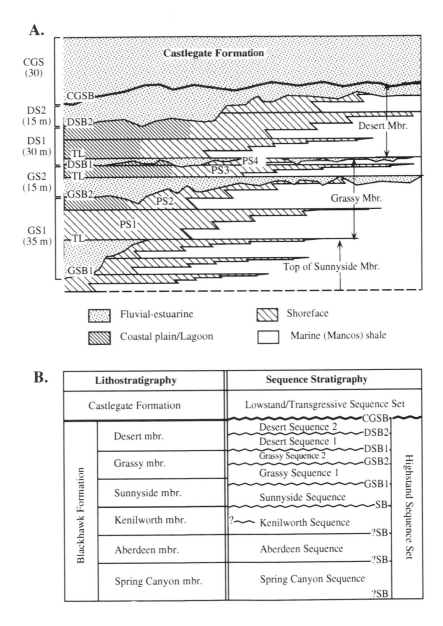

Fig. 2. (A) Schematic sequence stratigraphy for the Grassy Member. Scales are distorted. GPS1 and GPS2 represent the highstand systems tract of Grassy Sequence 1 (GS1), GPS3 and GPS4 define the highstand systems tract of Grassy Sequence 2 (GS2). Grassy Sequence boundaries (identified in this study) are labelled GSB1, GSB2 and DSB1. The last is the unconformity at the base of the newly defined Desert sequence 1. Desert sequence 2 is bounded below by the Desert sequence boundary described by Van Wagoner *et al.* (1990). DSB, desert sequence boundary; DS1: desert sequence 1; DS2, desert sequence 2; CGSB, castlegate sequence boundary; CGS, castlegate sequence set; TL, transgressive lag. (B) Provisional sequence stratigraphic interpretation of the Blackhawk Formation. At least eight high frequency sequences comprise a highstand sequence set truncated by the Castlegate sequence boundary. SB, sequence boundary.

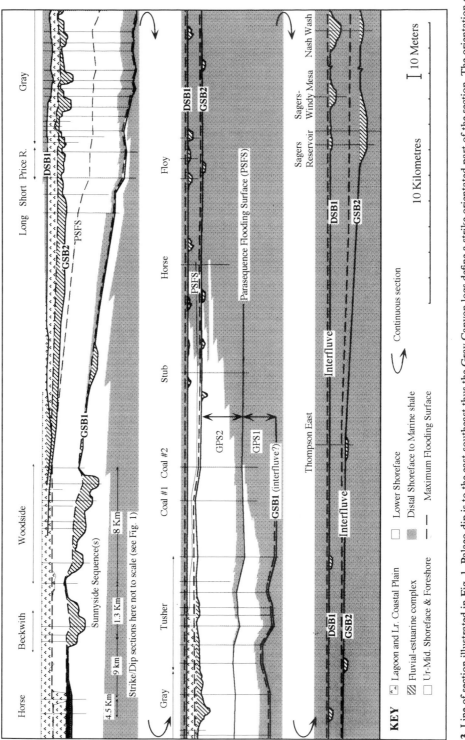

Fig. 3. Line of section illustrated in Fig. 1. Palaeo-dip is to the east-southeast thus the Gray Canyon logs define a strike orientated part of the section. The orientation of the section overall is such that the valley fill facies in the Sagers Reservoir area correlate directly with valley fill deposits in Tusher and Gray Canyons. GSB1, Grassy Sequence Boundary 1; GSB2, Grassy Sequence Boundary 2; DSB1, Desert Sequence Boundary 1.

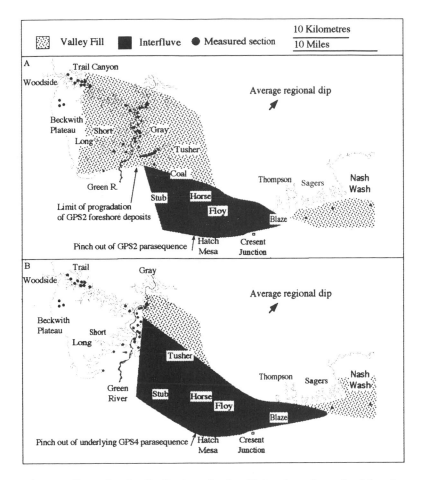

Fig. 4. (A) Facies map illustrating the distribution of valley fill deposits and associated interfluve area on the GSB2 sequence boundary. The distance from the GPS2 foreshore to the most basinward mapped outcrop of valley fill deposits is at least 37 km. Using an erosional depth of up to 20 m for the incised valleys gives a shelf slope of approximately 0.03°. The main area of valley fill consists of a complex of interconnected channel systems with localized interfluve areas. The area between Hatch Mesa and Stub Canyon was primarily a subaerially exposed surface with numerous small (< 10 m wide and < 4 m deep) tributary channels. (B) Outcrop facies map illustrating the distribution of valley fill deposits and associated interfluve area on the DSB1 sequence boundary. This valley system is believed to have migrated north relative to the incised valley associated with GSB2 due to an emerging structural high in the area of Hatch Mesa (see text and Fig. 12). The basinward shift in facies tracts associated with the DSB1 sequence boundary is a minimum of 35 km, based on the distance from the GPS4 pinch out to the basinward limit of mapped valley fill deposits.

extending from the base of the overlying, cliff-forming Desert Member.

The Grassy and Desert sequence boundary unconformities record a minimum (observable) basinward shift in facies tracts of 40 km (Fig. 4). No lowstand fan deposits were found in outcrop. However, thin bedded 'Mancos B' sandstones of in part Late Campanian age form hydrocarbon reservoirs further to the east in the subsurface (Cole & Young 1991) and may represent lowstand shoreline deposits time equivalent to more proximal facies in the study area

Incised valley fills: up-dip to down-dip facies development

Valley fill sands of both Grassy sequences have both lensoid and sheet geometries with no

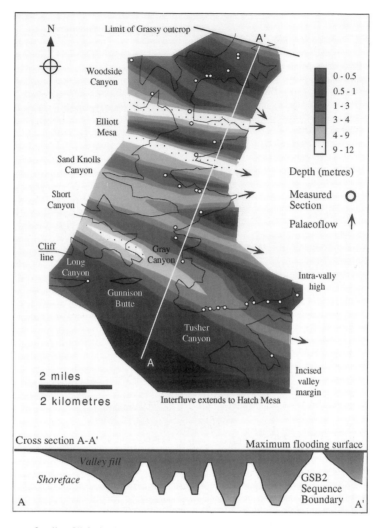

Fig. 5. Isopach map of valley fill facies in Gray (Green River) Canyon and adjacent areas showing the depth of incision on Grassy Sequence Boundary 2. Palaeocurrents indicate flow to the east and southeast. The incised valley fill consists of fluvial to marginal marine facies of the lowstand and transgressive systems tracts. Intravalley highs contain soil horizons overlain by peat swamp deposits laterally correlatable across the incised valley fill.

consistent vertical stacking pattern of internal facies. In general, however, multistorey sand bodies are characterized by a basal sand-rich (high net:gross) section overlain by an upper heterolithic (low net:gross) section. The basal section contains amalgamated, erosive based lenses and sheets of unidirectionally trough cross-bedded, non-fossiliferous sandstone with minor, discontinuous shales. The units are interpreted as low sinuosity, fluvial, braided stream deposits. The upper and more basinward parts of valley fills generally consist of mixed

sandstone and shale, commonly with well developed lateral accretion bedding and are interpreted as high sinuosity fluvial deposits. The upper parts of the high sinuosity facies have variably developed indicators of tidal influence such as inclined heterolithic facies, terridolites-bored logs, brackish water microfaunas, clay drapes on foresets and sigmoidal cross-bedding with doublets. These deposits are interpreted as estuarine shoals and channels (O'Byrne & Flint 1995).

In down-dip sections the fluvial association

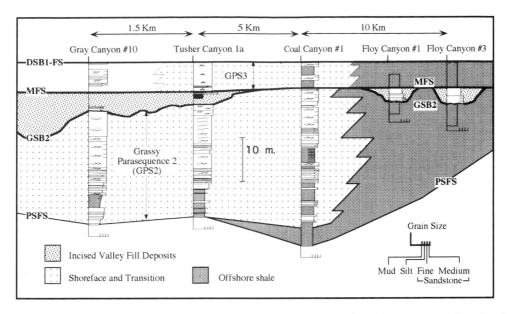

Fig. 6. Continuous oblique strike section showing the laterally variable expression of Grassy sequence boundary 2 (GSB2). A major incised valley fill in the Gray canyon area can be correlated laterally to an interfluve. For locations, see Fig. 1. For legend, see Fig. 3.

Fig. 7. Outcrop appearance of up-dip interfluve sequence boundary GSB2 in Gray canyon (Fig. 1). Note the slightly erosional relationship between the rooted zone and the top of the underlying HST shoreface; the surface dips off to the left of view (north). Vertical field of view is c. 1.5 m.

comprises current rippled and planar bedded heterolithic facies rich in terrigenous carbonaceous material. A more pervasive estuarine influence is indicated by bi-directional ripples, clay drapes and increased evidence of marine bioturbation. In some cases channels are abruptly overlain by marine shales or fine-grained sands and silts. These in turn are capped by flaser bedded muddy heterolithics with restricted trace fossil assemblages, interpreted as tidal flat deposits. The upper surface of the incised valley fill is usually heavily bioturbated, calcite cemented and contains whole shelled bivalves including *Inoceramus*.

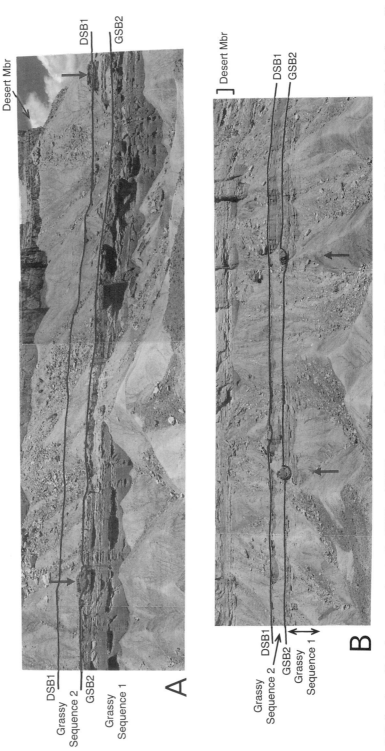

Fig. 8. Strike view of interfluve expressions to the GSB2 and DSB1 sequence boundaries (Fig. 2) between Horse Canyon and Hatch Mesa (Fig. 1). The interfluve gullies (arrows) indicate drainage to the E or NE. View B is some 500 m to the north of A.

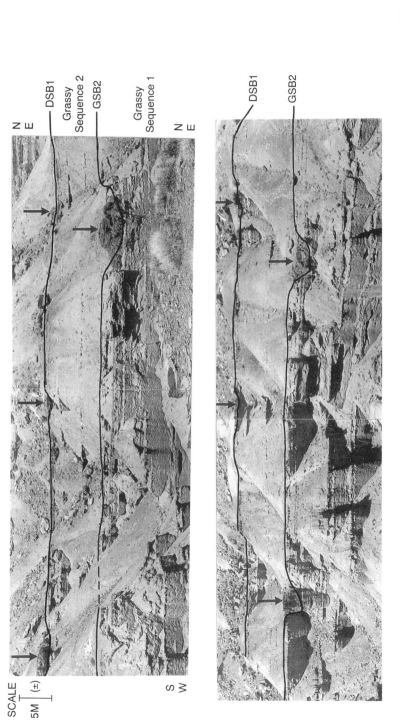

Fig. 9. Continuous oblique strike section showing numerous tributary drainage gullies (arrowed) on two interfluve sequence boundary surfaces (GSB2 and DSB1) within the offshore marine Mancos shale, Horse Canyon area (Fig. 1).

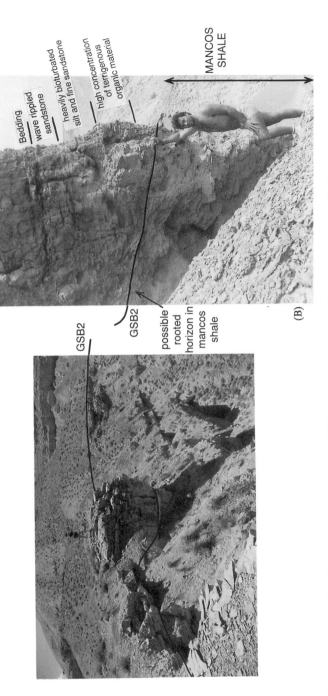

Fig. 10. Down-dip characteristics of interfluve sequence boundary to GSB2 in Floy Canyon (Fig. 1), some 25km down dip from the same surface as in Fig. 7. (A) Interfluvial drainage channel lying on the GSB2 surface, which truncates offshore shales, down dip of the seaward pinchout of GPS2. (B) Close-up view of the pod base showing slight angular relationship with underlying offshore marine Mancos shale and a basal concentration of terrigenous detrital carbonaceous material.

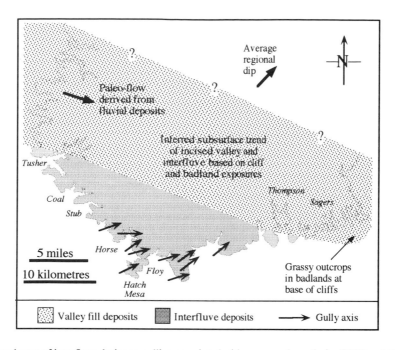

Fig. 11. Channel axes of interfluve drainage gullies associated with sequence boundaries GSB2 and DSB1. Of the ten localities shown there are *c.* 28 gullies. Orientation of channel axes, however, are often approximate due to the 2D nature of the exposures. The gullies appear to be concentrated around the area between Horse and Floy Canyons along the northerly trend of the presently exposed Salt Valley Anticline (Fig. 12).

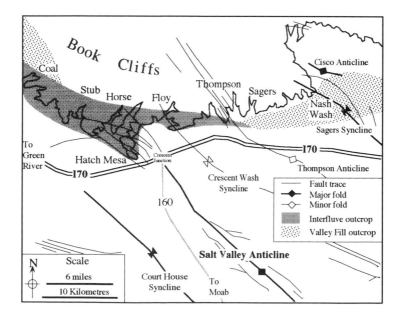

Fig. 12. Generalized structural map of the SE part of the study area (see Fig. 1 for location). The Salt Valley Anticline is believed to extend northward in the subsurface approximately parallel to the interfluvial area shown on the map. See text for details.

Palaeocurrent and palaeo-channel orientation data were recorded from 21 localities within the fluvial-estuarine deposits. Data taken from trough and planar cross-beds and current ripple lamination show a spread about a mean vector of 100°. This orientation parallels the east-southeast trend of the incised valleys (Fig. 5) and is orthogonal to the Grassy palaeo-shore-lines (O'Byrne & Flint 1995). Movement vectors from bank collapse structures indicate transport at a high angle to the predominant valley palaeocurrent direction. Localized erosional remnants or terraces are present within incised valley complexes. These intra-valley highs (Fig. 5) are similar to features described in flume experiments by Wood et al. (1994).

Interfluve sequence boundaries

Exposure surface expression

In up-dip areas, interfluves are represented by rooted soil horizons on top of underlying highstand carbonaceous shales or upper shore-face deposits (Figs 6 and 7), which themselves contain sphaero-siderite, indicative of fresh or brackish water interaction. Thin (10–200 cm) preserved coastal plain and/or lagoonal deposits overlying the sequence boundary aggraded during subsequent early base level rise. Farther down-dip, soil horizons which developed on interfluves have been partially or completely reworked by subsequent transgressive erosion, leaving only a remnant oxidized (hematite stained) surface overlain by a lag containing siderite nodules and hematite clasts (evidence of a reworked soil horizon) and rare dinosaur bone fragments. Clasts indicative of an open marine environment include phosphatic nodules, sharks' teeth and marine shell fragments.

Interfluvial drainage expression

Curvilinear narrow, steep-sided channels 2–10 m wide and 1–2.5 m deep are present on interfluves (Figs 6, 8, 9 and 10). When aligned from canyon to canyon the channels are seen to drain into adjacent deeper erosional lows cut in the shelf strata (Fig. 11). These minor channels appear to form a dendritic drainage network across an extensive interfluve area between Stub and Floy Canyons (Fig. 4). The channels are cut into distal shoreface sandstone and siltstone or off-shore shale (Figs 6 and 8) and are interpreted as headward eroding, water run-off gullies incised during lowstand shelf exposure. Interfluve chan-

nel-fills comprise fine-grained and heavily bio-turbated, (?)reworked estuarine facies and are interpreted as being coeval with deposition of estuarine facies within the laterally adjacent incised valleys during subsequent transgression (Fig. 10). A major flooding surface can be traced across both these tributary channels and the main incised valley system (Fig. 6).

Tectonic controls on the location of incised valleys and interfluves

The extensive area of interfluve associated with sequence boundaries GSB1 and DSB1 coincides with a northwesterly extension of a present day structural high, the Salt Valley Anticline (Fig. 12). There is evidence from a number of sources to suggest that this structure may have been active during the late Campanian. These include local tilting and formation of minor angular unconformities (related structures have been shown to affect the Castlegate sequence bound-ary; Van Wagoner et al. 1991), localization of lowstand deposits coincident with a 'rim-syn-cline' (Kenilworth Member; Balsley 1980; Tay-lor & Lovell 1991) and variation in the thickness of the Desert Member and Castlegate Forma-tion across these structures (J. K. Balsley pers. comm. 1992). The fold structures trend generally northwest, parallel to established basement lineaments in the area (Fig. 12). We have four lines of evidence for the influence of an emerging structural high at the time of deposition of the Grassy Member.

(1) The small scale tributary river systems adjacent to Grassy incised valleys are consistent with drainage off the flanks of a structure coincident with the Salt Valley Anticline (Figs 11 and 12).
(2) Grassy parasequences GPS3 and GPS4 (Fig. 2) thin across this area.
(3) The incised valleys are confined to lows or rim synclines parallel to this structural high.
(4) The principal axis of each successively younger valley system, still contained by the structural low, migrates to the northeast with growth of the anticlinal structure over time.

There is no evidence to suggest that this emerging high had significant relief; rather we envision a very gentle doming related to plastic cover deformation, without fault piercement, along the present day trend of the Salt Valley Anticline sufficient to influence the site of valley incision.

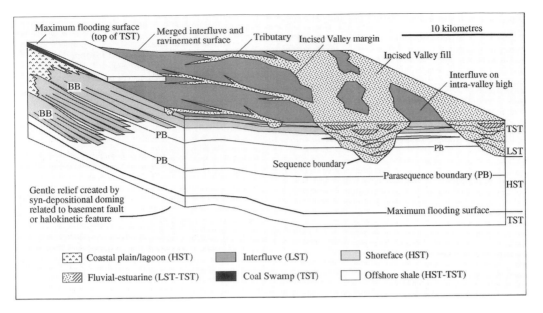

Fig. 13. Schematic isometric diagram of a Grassy highstand parasequence set cut by a sequence boundary. Cemented lags can occur on the interfluve areas and at the base of the incised valley fill. Identification of the maximum flooding surface or parasequence set boundary, which merges with interfluves on the sequence boundary, provides information on seal distribution only. Identification of the interfluve (see text) is critical in predicting out of section valley fill deposits, possible deeper water lowstand deposits and for determining the lateral (strike) connectivity of shoreface parasequences. PB, parasequence boundary; BB, bedset boundary.

Implications for subsurface correlation

Recognition of interfluves in the sub-surface is important in identifying 'out of section' incised valleys (Fig. 13). Identification of incised valley drainage systems is critical to hydrocarbon field development and water injection programmes allowing, for example, prediction of early water breakthrough in high permeability valley fill facies. Furthermore, lowstand valley incision could affect the lateral strike continuity of truncated shoreface/mouthbar facies with implications for sand body connectivity (Fig. 13). Formation of hematite and other cements on interfluves during lowstand subaerial exposure and later carbonate cements during subsequent transgression can create significant vertical permeability barriers within a field. Recognition of the interfluve surface thus becomes important away from the incised valley. Stacking patterns may be useful in this regard, with the possibility of an interfluve sequence boundary lying between forward-stepping and superjacent back-stepping parasequence sets. However, on a basin margin with strike-variable sediment supply and subsidence rates, stacking patterns alone may not be sufficient to ensure correct correlation of

the surface throughout the field (Howell & Flint 1996).

A number of criteria have now been established based on outcrop work by Van Wagoner *et al.* (1990, 1991); Walker & Eyles (1991); Jennette *et al.* (1991) and by the present authors, which may aid recognition of interfluves in core. These data are meant to be used where other, more conclusive evidence of a sequence boundary is known or suspected, such as a regional change in stacking patterns above a common horizon or where a field-wide significant or abrupt change in facies has occurred. Criteria include the following.

(1) The presence of siderite clasts indicative of a reworked soil horizon or a hard pan formed beneath a swamp deposit. Evidence of well developed hematite cements overlain by hard-pans or calcite cements associated with a transgressive lag are common indicators of interfluves in outcrop and could also be identified in core.

(2) A further possible indicator is the presence of sphaero-siderite nodules which, in the case of the Grassy sequences, always occur in proximal

to distal shoreface and shoreface transition sandstones immediately below sequence boundaries. Sphaero-siderite, as a product of fresh water interaction, is useful in these facies as an indicator of base level change and should be easily recognizable in core. Sphaero-siderite is, however, common in marginal marine environments such as lagoon edge deposits including washover fans and flood tidal deltas and thus is not of stand-alone use in determination of subaerial exposure of the shelf.

We acknowledge the financial support of Shell UK Ltd for this project while the senior author was in receipt of a Shell Postdoctoral Fellowship at Liverpool University (1990–92).

References

BALSLEY, J. K. 1980. *Cretaceous wave dominated delta systems, Book Cliffs, east-central Utah.* American Association of Petroleum Geologists Continuing Education Course, Field Guide.

CARTWRIGHT, J. A., HADDOCK, R. C. & PINHEIRO, L. M. 1993. The lateral extent of sequence boundaries. *In:* WILLIAMS, G. D. & DOBB, A. (eds) *Tectonics and seismic sequence stratigraphy.* Geological Society, London, Special Publication, **71**, 15–34.

COLE, R. D. & YOUNG, R. G. 1991. Facies characterization and architecture of a muddy shelf-sandstone complex: the Mancos B interval of Upper Cretaceous Mancos shale, Northwest Colorado–Northeast Utah. *In:* MIALL, A. D. & TYLER, N. (eds) *The 3-dimensional facies architecture of terrigenous clastic sediments and its implications for hydrocarbon discovery and recovery.* SEPM, Concepts in Sedimentology and Palaeontology, **3**, 277–287.

DALRYMPLE, R. W., BOYD, R. & ZAITLIN, B. A. 1994. History of research, types and internal organisation of incised valley systems: introduction to the volume. *In:* DALRYMPLE, R. W., BOYD, R. & ZAITLIN, B. A. (eds) *Incised Valley Systems: Origin and Sedimentary Sequences.* SEPM, Special Publication, **51**, 3–10.

HOWELL, J. & FLINT, S. 1996. Sequence stratigraphic expressions of relative sea level changes in tectonically active rift basins. *Journal of Sedimentary Research,* in press.

JENNETTE, D. C., JONES, C. R., VAN WAGONER, J. C. & LARSEN, J. E. 1991. High resolution sequence stratigraphy of the upper Cretaceous Tocito Sandstone: the relationship between incised valleys and hydrocarbon accumulation, San Juan Basin, New Mexico. *In:* VAN WAGONER, J. C., NUMMEDAL, D., JONES, C. R., TAYLOR, D. R., JENNETTE, D. C. & RILEY, G. W. (eds) *Sequence Stratigraphy Applications to Shelf Sandstone Reservoirs: Outcrop to Subsurface Examples.*

American Association of Petroleum Geologists, Field Conference Guide.

—— & RILEY, C. O. 1996. Influence of relative sea level on facies and reservoir geometry of the Middle Jurassic lower Brent Group, UK North Viking Graben. *This volume.*

KAMOLA, D. L. & VAN WAGONER, J. C. 1995. Stratigraphy and facies architecture of parasequences with examples from the Spring Canyon Member, Blackhawk Formation, Utah. *In:* VAN WAGONER, J. C. & BERTRAM, G. (eds) *Sequence stratigraphy of foreland basin deposits: examples from the Cretaceous of North America.* American Association of Petroleum Geologists, Memoir, **65**, in press.

O'BYRNE, C. J. & FLINT, S. 1993. Sequence stratigraphy of Cretaceous shallow marine sandstones, Book Cliffs, Utah: application to reservoir modelling. *First Break,* **11**, 445–459.

—— & —— 1995. Sequence, parasequence and intraparasequence architecture of the Grassy Member, Blackhawk Formation, Book Cliffs, Utah. *In:* VAN WAGONER, J. C. & BERTRAM, G. (eds) *Sequence stratigraphy of foreland basin deposits: examples from the Cretaceous of North America.* American Association of Petroleum Geologists, Memoir, **65**, in press.

TAYLOR, D. R. & LOVELL, R. W. 1991. Recognition of High-Frequency Sequences in the Kenilworth Member of the Blackhawk Formation, Book Cliffs, Utah. *In:* VAN WAGONER, J. C., NUMMEDAL, D., JONES, C. R., TAYLOR, D. R., JENNETTE, D. C. & RILEY, G. W. (eds) *Sequence Stratigraphy Applications to Shelf Sandstone Reservoirs: Outcrop to Subsurface Examples.* American Association of Petroleum Geologists, Field Conference Guide.

VAN WAGONER, J. C., MITCHUM, R. M., CAMPION, K. M. & RAHMANIAN, V. M. 1990. *Siliciclastic sequence stratigraphy in well logs, cores and outcrops: concepts for high resolution correlation of time and facies.* American Association of Petroleum Geologists, Methods in Exploration Series, 7.

——, NUMMEDAL, D., JONES, C. R., TAYLOR, D. R., JENNETTE, D. C. & RILEY, G. W. (eds) 1991. *Sequence Stratigraphy Applications to Shelf Sandstone Reservoirs: Outcrop to Subsurface Examples.* American Association of Petroleum Geologists, Field Conference Guide.

WALKER, R. G. & EYLES, C. H. 1991. Topography and significance of a sequence-bounding erosion surface in the Cretaceous Cardium Formation, Alberta, Canada. *Journal of Sedimentary Petrology,* **61**, 473–496.

WOOD, L. J., ETHRIDGE, F. G. & SCHUMM, S. A. 1994. An experimental study of the influence of subaqueous angles on coastal plain and shelf deposits. *In:* WEIMER, P. & POSAMENTIER, H. W. (eds) *Siliciclastic Sequence Stratigraphy: Recent Developments and Applications.* American Association of Petroleum Geologists, Memoir, **58**, 381–391.

Critical application of high resolution sequence stratigraphic concepts to the Rough Rock Group (Upper Carboniferous) of northern England

GARY J. HAMPSON, TREVOR ELLIOTT & STEPHEN S. FLINT

Department of Earth Sciences, University of Liverpool, PO Box 147, Liverpool L69 3BX, UK

Abstract: The Rough Rock Group (Upper Carboniferous, Namurian G_1) of northern England comprises fluvio-deltaic strata exhibiting high frequency cyclicity. Key elements of high-frequency sequences are recognized in these strata, including lowstand braided fluvial incised valley fills and their correlative interfluve palaeosols. The geometry of these incised valley fills was controlled by the local, synsedimentary tectonic setting. Regionally-extensive coals overlie these incised valley fills and their interfluve palaeosols and are interpreted as initial flooding surface correlatives. These coals are in turn overlain by shales containing faunal-concentrate condensed horizons (marine bands) which are interpreted as maximum flooding surfaces. A previously unrecognized marine band which overlies the Sand Rock Mine Coal and underlies the Rough Rock Flags is documented, implying a major revision of the published stratigraphy and sequence stratigraphy. These sequence stratigraphic elements define four high frequency siliciclastic sequences, which are integrated into a sequence stratigraphic framework for the Rough Rock Group. This framework allows sequence architecture, palaeogeographies and stacking patterns to be described in detail and also provides a basis for identifying potential controls on these variables.

The Rough Rock Group (Upper Carboniferous, Namurian G_1) comprises the uppermost strata of the Millstone Grit Series in the Pennine Basin of northern England. The Millstone Grit Series is characterized by siliciclastic fluvio-deltaic successions exhibiting high frequency cyclicity. The Rough Rock Group is exposed across northern England in natural streams, crags and quarries and has been cored in oil exploration boreholes from the East Midlands Oilfield (Fig. 1). The combination of widespread exposure, high frequency cyclicity and an integrated, high resolution biostratigraphic scheme makes this stratigraphic interval an ideal testing ground for the concepts of high resolution sequence stratigraphy, using detailed sedimentological logging. This paper aims (1) to document, describe and interpret key lithostratigraphic elements within the Rough Rock Group in a sequence stratigraphic context; and (2) to integrate these elements into a coherent sequence stratigraphic framework for the Rough Rock Group. This framework is then used to discuss sequence stacking patterns and palaeogeographies.

Geological setting

The Pennine Basin was one of a series of linked, intracratonic basins created by late Devonian and early Carboniferous rifting in Britain and northwestern Europe (Leeder 1988). This rifting also created sub-basins within the Pennine Basin (Collinson 1988; Lee 1988; Fig. 2). During the late Carboniferous the basin underwent passive, thermal subsidence with contemporaneous minor active extensional tectonism confined to the East Midlands Shelf (Edwards 1967; Smith *et al.* 1973; Fig. 2). The Pennine Basin and its constituent sub-basins were not infilled during syn-rift times. Instead, inherited rift bathymetry was largely infilled during deposition of the post-rift, Millstone Grit Series (Collinson 1988) by a series of southerly-advancing turbidite-fronted delta systems (*sensu* Collinson 1988) sourced from the northern Scottish–Scandinavian Caledonide metamorphic terrain (see reviews in Collinson 1988; Leeder 1988), prior to the deposition of the Rough Rock Group. Hence, during Rough Rock Group deposition the Pennine Basin had gently-dipping ramp margins, although the limited lateral distribution of some lithostratigraphic units suggests that some rifted sub-basins were still defined, probably by minor bathymetric features.

Millstone Grit cyclicity has classically been defined using cyclothems, each of which comprises a fossiliferous shale horizon, passing upward into progressively shallower, delta front facies and culminating in fluviatile and coal-bearing facies. The fossiliferous shale horizons which bound the cyclothems are faunal-concentrate condensed horizons, many of which bear

From Howell, J. A. & Aitken, J. F. (eds), *High Resolution Sequence Stratigraphy: Innovations and Applications,*
Geological Society Special Publication No. 104, pp. 221–246.

221

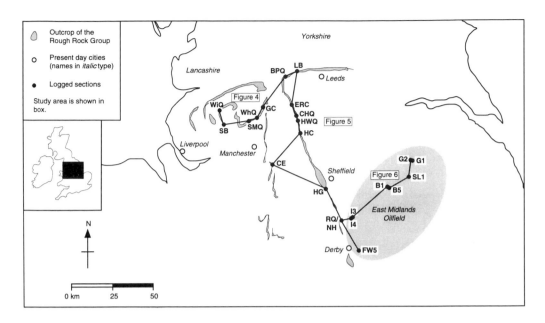

Fig. 1. Simplified outcrop map of the Rough Rock Group showing the location and data coverage of the study area. Logged section localities are abbreviated as follows: WiQ, Withnell Quarry [National Grid Reference SD 642 218]; SB, Stepback Brook [SD 670 214]; SMQ, Scout Moor Quarry [SD 815 187]; WhQ, Whitworth Quarry [SD 872 203]; GC, Greens Clough [SD 890 256]; BPQ, Buck Park Quarry [SE 070 352]; LB, Loadpit Beck [SE 128 390]; ERC, Elland Road Cutting [SE 103 214]; CHQ, Crossland Hill Quarry [SE 118 144]; HWQ, Honey Wood Quarry [SE 117 114]; HC, Harden Clough [SE 147 042]; CE, Cracken Edge [SK 036 838]; HG, Harewood Grange [SK 311 684]; RQ, Ridgeway Quarry [SK 359 515]; NH, Nether Heage borehole [SK 360 512]; FW5, Farleys Wood No. 5 [SK 471 372]; 14, Ironville No. 4 [SK 432 519]; 13, Ironville No. 3 [SK 433 523]; B1, Bothamsall No. 1 [SK 659 737]; B5, Bothamsall No. 5 [SK 666 734]; SL1, South Leverton No. 1 [SK 793 804]; G2, Gainsborough No. 2 [SK 817 908]; G1, Gainsborough No. 1 [SK 832 902]. The lines of cross-section illustrated in Figs 4 (WiQ–LB), 5 (LB–NH) and 6 (FW5–G1) are also shown.

distinctive goniatite faunas, and are termed 'marine bands'. Marine bands approximate to time lines and provide an important correlation tool within a goniatite biostratigraphic framework with an average time resolution of *c.* 180 ka (Holdsworth & Collinson 1988).

Calver (1968) and Wignall (1987) document a series of faunal phases which can be distinguished within marine bands, from fully marine, nektonic faunas (goniatites, pectinoid bivalves) in the basin centre to restricted marine, benthic faunas (brachiopods including *Lingula*) and non-marine, benthic faunas (non-marine bivalves, '*Estheria*') towards the basin margin. The lateral distribution of these faunal phases records palaeoecological changes within the marine bands, such as increasing distance from the shoreline and corresponding variations in salinity, water depth and oxygenation, substrate type and sedimentation (Calver 1968). The palaeoenvironments occupied by the different

faunal phases are visualized as a series of time-equivalent, palaeogeographic facies belts extending from non-marine swamps to the fully marine basin (Calver 1968, fig. 5).

An extensive polar ice cap was situated over southern Gondwanaland during the Carboniferous (see review in Leeder 1988). The ice cap is inferred to have behaved in an analogous way to its Quaternary counterparts: Milankovitch climatic cycles controlled the growth and decay of polar ice, producing high-frequency and high-magnitude glacio-eustatic sea-level fluctuations. Using conservative estimates of Gondwanan ice extent, Crowley & Baum (1991) calculated isostasy-corrected glacio-eustatic sea-level fluctuations during the late Carboniferous as 60 ±15 m. These glacio-eustatic changes are widely interpreted as a driving mechanism for Millstone Grit cyclicity (Ramsbottom 1977; Holdsworth & Collinson 1988; Leeder 1988; Maynard & Leeder 1992).

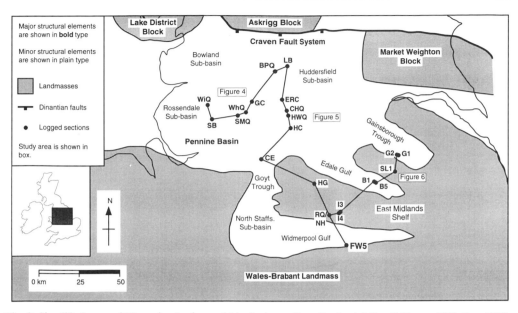

Fig. 2. Simplified map of Namurian basins and blocks in northern England (after Collinson 1988; Lee 1988), showing logged sections in bold type. Tilted rift blocks/half graben defining landmasses are particularly prominent along the southeastern basin margin (East Midlands Shelf), where the dominant fault trend is NW–SE. The lines of cross-section illustrated in Figs 4 (WiQ–LB), 5 (LB–NH) and 6 (FW5–G1) are also shown.

Rough Rock Group stratigraphy

The Rough Rock Group is defined at its base and top by the *Gastrioceras cancellatum* and *G. subcrenatum* marine bands, respectively (Fig. 3). In addition, the *G. cumbriense* marine band is documented within the Group (Fig. 3). At outcrop, in the northern sub-basins of the Pennine Basin (Fig. 2), there are also several widespread lithological units within the Rough Rock Group which have been used previously to define lithostratigraphy. One of these widespread lithological units is the Rough Rock, which is a regionally-extensive, coarse-grained sandstone sheet. This sandstone sheet is associated with two regionally-extensive coals: the Six Inch Mine or Pot Clay Coal, which caps the Rough Rock; and the Sand Rock Mine Coal, which lies within the Rough Rock and splits it into a lower leaf and an upper leaf in the Rossendale sub-basin (Fig. 3). Previous workers have used these marine bands and widespread lithological units to define a widely used stratigraphy for the Rough Rock Group (Fig. 3a). Here we present the evidence for a 'new' marine band which significantly revises Rough Rock Group stratigraphy (Fig. 3b), with reference to three correlation panels showing logged sections in the Rough Rock Group (Figs 4, 5 and 6).

A 'new' marine band

A fossiliferous, condensed shale horizon containing a distinctive, nonmarine bivalve fauna (including *Carbonicola, Anthraconaia*) is commonly recorded in shales directly above the Sand Rock Mine Coal in boreholes from west Lancashire (Eagar 1951, 1952; Table 1), in the Rossendale sub-basin (Fig. 2). This condensed shale horizon is inferred to comprise the most marginal faunal phase (*sensu* Calver 1968) of a marine band. These shales are absent in well exposed outcrop sections (e.g. Scout Moor Quarry and Whitworth Quarry; Figs 1, 2 and 4) lying farther east, and are presumed to have been cut out by the overlying, upper leaf of the Rough Rock.

A comparable nonmarine bivalve fauna is observed in a fossiliferous shale horizon in a key outcrop section at Harewood Grange (Calver *in* Smith 1967, p. 107; Figs 1, 2, 5 and 7; Table 1), which lies in the south of the Huddersfield sub-basin (Fig. 2), almost 100 km southeast of the boreholes in west Lancashire. This horizon at Harewood Grange directly underlies the Rough Rock Flags, but occurs (within the limits imposed by published marine band biostratigraphy; Fig. 3a) at the same stratigraphic horizon as those described above in west Lancashire. Further north, in the north of the Huddersfield

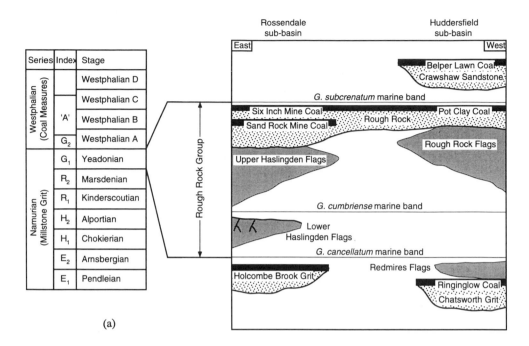

(a)

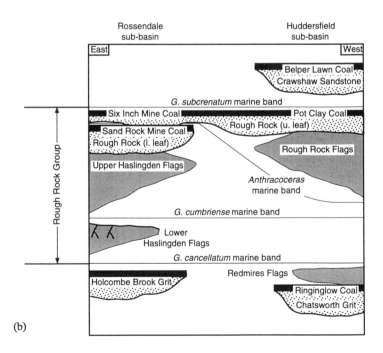

(b)

Fig. 3. Schematic stratigraphy of the Rough Rock Group in the northern Pennine Basin; (**a**) prior to recognition of a 'new' *Anthracoceras* marine band (after Bristow 1988) and (**b**) incorporating a 'new' *Anthracoceras* marine band (see text for details). The horizontal scale is *c.* 70 km and no vertical scale is implied.

Table 1. *Occurrence and faunal content of a 'new Anthracoceras marine band*

Section	Location	Fauna	Reference
Rossendale sub-basin			
Beechwood Colliery	SD 515 135	nonmarine bivalves (*Anthraconaia, Carbonicola*)	Eagar (1951, 1952)
Mountain Mine Colliery	SD 514 035		
Pimbo Lane Colliery	SD 512 040		
Upholland Railway Cutting	SD 509 037		
Big Wood boring	SD 538 167	nonmarine bivalves (*Carbonicola, Naiadites, Geisina*)	pp. 29, 105 in Price *et al.* (1963)
Huddersfield sub-basin			
Sandoz Chemical Co. boring, Bradford	SE 163 343	goniatites, fish, brachiopods (*Lingula*)	p. 134 in Hudson & Dunnington (1939)
		nonmarine bivalves (*Anthracomya, Carbonicola*)	p. 65 in Stephens *et al.* (1953)
Sydney Works boring, Bradford	SE 134 333	nonmarine bivalves (*Anthracomya, Carbonicola*)	p. 212 in Hudson & Dunnington (1940)
			p. 65 in Stephens *et al.* (1953)
Meanwood Tannery boring, Leeds	SE 290 375	brachiopods (*Lingula*)	Gilligan (1921)
			p. 12 in Edwards *et al.* (1950)
Roundhay Park, Leeds	SE 336 386	goniatites (*Anthracoceras*), fish brachiopods (*Lingula*)	Slinger (1936)
			p. 12 in Edwards *et al.* (1950)
Harewood Grange	SK 311 683	nonmarine bivalves (*Anthraconaia, Carbonicola, Naiadites, Geisina*)	p. 107 in Smith (1967)
			p. 76–77 in Smith *et al.* (1967)
Stonehay Farm	SK 332 684	nonmarine bivavles (*Carbonicola, Geisina*)	p. 77 in Smith *et al.* (1967)
Smeekley No. 3 borehole	SK 297 766	nonmarine bivalves (*Carbonicola, Naiadites, Geisina*)	p. 23, 218 Eden *et al.* (1957)
East Midlands Shelf			
Nether Heage borehole	SK 360 512	goniatites, fish, brachiopods (*Lingula*)	
Ironville No. 3	SK 433 523	gamma ray peak	
Ironville No. 4	SK 432 519		
Gainsborough No. 1	SK 832 902	brachiopods (*Lingula*)	

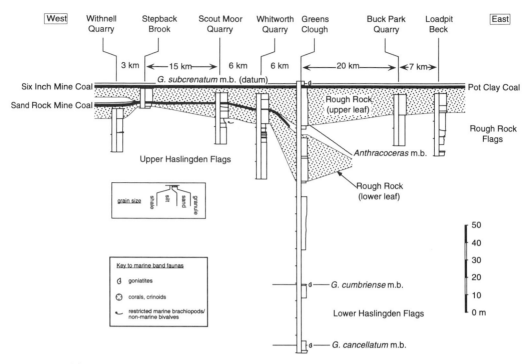

Fig. 4. Depositional strike cross-section through the Rough Rock Group in the Rossendale and Huddersfield subbasins. The locations of the cross-section and logged sections are shown in Figs 1 and 2.

sub-basin, several boreholes record the same fossiliferous shale horizon, directly underlying the Rough Rock Flags (Table 1). In different boreholes, the fauna is nonmarine, restricted marine or fully marine. These various faunas are interpreted as different faunal phases (*sensu* Calver 1968) within the same marine band. Where present, the fully marine fauna contains *Anthracoceras* cf. *arcuatilobum*, a non-diagnostic goniatite (Slinger 1936; Table 1) and accordingly this 'new' marine band is termed an *Anthracoceras* marine band (Figs 3b, 4, 5 and 6). In one section (Sandoz Chemical Co. boring; Table 1), this 'new' marine band comprises a fossiliferous horizon bearing nonmarine bivalves and '7 ft above this, a *Lingula* band with fish teeth and indeterminate goniatite impressions' (Stephens *et al.* 1953, p. 65). This succession records increasing marine influence as transgression progressed.

In a continuously-cored section from Nether Heage (Figs 1, 2, 5 and 6; Table 1), a condensed shale horizon bearing goniatites, fish scales and brachiopods (*Lingula*) is documented to occur approximately 3 m above the *G. cumbriense* marine band. This horizon is interpreted as a distinct marine band and is correlated with the

'new' marine band described above (Fig. 5). Further correlation of this marine band in the East Midlands Shelf and adjacent sub-basins (Fig. 2) is more subjective because individual marine bands are difficult to distinguish owing to the scarcity of recovered core intervals containing marine bands, and the non-diagnostic faunas within many of these marine bands.

Frost & Smart (1979, p. 169) document the *G. cumbriense* marine band in Ironville No. 4 oil bore and an overlying gamma ray peak in unrecovered shales is interpreted as the expression of this 'new' marine band in this section (Fig. 6, Table 1). In this interpretation, the *G. cumbriense* marine band and overlying 'new' marine band are separated by *c.* 3 m of pyrite-rich siltstones. Gamma ray peaks corresponding to both of these marine bands are identified in Ironville No. 3 oil bore (Fig. 6, Table 1). However, the initial identification of *G. cumbriense* within Ironville No. 4 oil bore is disputed by Church & Gawthorpe (1994), who identify goniatites from the same horizon as *G. cancellatum*.

Farther east on the East Midlands Shelf (Fig. 2), the *G. cumbriense* and 'new' marine bands are interpreted to merge (Fig. 6). This is consistent

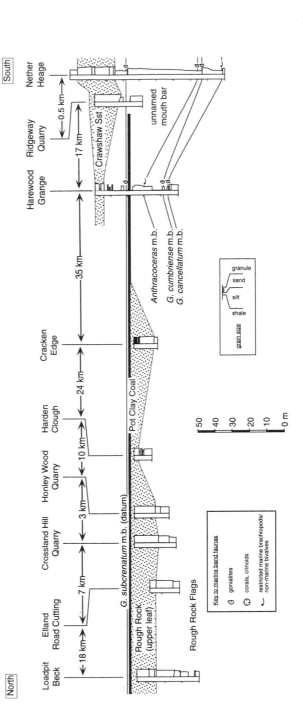

Fig. 5. Depositional dip cross-section through the Rough Rock Group in the Huddersfield sub-basin and Edale Gulf. The locations of the cross-section and logged sections are shown in Figs 1 and 2.

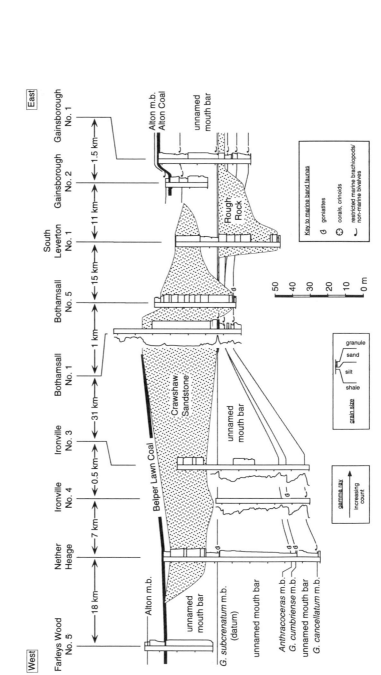

Fig. 6. Depositional strike cross-section through the Rough Rock Group on the East Midlands Shelf. The locations of the cross-section and logged sections are shown in Figs 1 and 2.

with a sub-regional trend recorded by the merging of the *G. cancellatum* and *G. cumbriense* marine bands in this direction (Church & Gawthorpe 1994, fig. 7). Also, the entire Namurian section thins to the east in this region onto a carbonate platform which defines the eastern margin of the Pennine Basin. In the Gainsborough Trough (Fig. 2), most Upper Carboniferous marine bands are represented by *Lingula*-rich shales. Based on Ramsbottom's interpretation (BGS-held notes) of the *G. cancellatum* and *G. subcrenatum* marine bands, the 'new' marine band is identified as one of these *Lingula*-rich shales (Fig. 6, Table 1).

Lateral variation

Net palaeocurrent orientations indicate that the northern margin of the Pennine Basin was orientated west–east in the Rossendale and Huddersfield sub-basins. Therefore, the lines of cross-section illustrated in Figs 4 and 5 constitute a depositional strike section and an oblique depositional dip section, respectively (Fig. 2). The cross-section line illustrated in Fig. 6 is orientated across the Widmerpool Gulf, East Midlands Shelf and Gainsborough Trough and is inferred to be an oblique depositional strike section (Fig. 2).

The Rough Rock Group is thickest in the Rossendale sub-basin and attains a maximum thickness of approximately 160 m in the depocentre of this sub-basin, near Greens Clough (Figs 2 and 4). The Group thins to the east and south in the Huddersfield sub-basin onto the East Midlands Shelf and London–Brabant Massif (Figs 2 and 5), respectively. The Rough Rock Group is particularly thin (< 20 m) in the Widmerpool Gulf, East Midlands Shelf and Gainsborough Trough (Figs 2 and 6). Also, several stratigraphic elements are confined to particular sub-basins, suggesting a bathymetric control on sedimentation. The Haslingden Flags and lower leaf of the Rough Rock are present only in the Rossendale sub-basin, while the Rough Rock Flags are confined to the Huddersfield sub-basin (Figs 2, 3 and 4).

Sedimentology

The sedimentology of the lithostratigraphic units within the Rough Rock Group has been described comprehensively by previous workers (see review in Hampson 1995) and is summarized in Table 2. Key aspects of the sedimentology used in the sequence stratigraphic reinterpretation of these units are described in detail below.

Key sequence stratigraphic elements

Incised valley fills (IVFs)

Major fluvial sandstone complexes in the Millstone Grit Series, including the Rough Rock, are candidates for incised valley fills (IVFs). The base of an incised valley defines a sequence boundary and is a key component of sequences in delta plain settings. Criteria for evaluating major fluvial sandstone complexes as candidate IVFs include (1) the context of the fluvial complex in relation to underlying and overlying systems tracts, (2) the relief on the basal erosional surface of the complex relative to the thickness of individual channel storeys within the complex, (3) the lateral, regional extent of the complex, and (4) the presence of contemporaneous interfluve surfaces at the margins of the complex.

In the Rossendale and Huddersfield sub-basins (Fig. 2), the Rough Rock Group contains two major, coarse-grained, braided fluvial sheet sandstone complexes, the lower and upper leaves of the Rough Rock. Both sandstone complexes are overlain by regionally-extensive coal seams and marine bands (Figs 3, 4 and 5), indicative of relative sea-level rise (see later discussion of regionally-extensive coals). It is probable, therefore, that both complexes and their basal surfaces represent sedimentation as relative sea-level fell, slowed to stillstand and started to rise, thus satisfying criterion (1). Both sandstone complexes are multistorey with negligible basal erosional relief on a local, outcrop scale, but exhibit up to 35 m of basal erosional relief on a regional scale (Figs 4 and 5; Table 3). Channel storeys within these complexes are strongly erosive (Bristow 1988, 1993; Bristow & Myers 1989), but fully preserved channel storeys including fine-grained channel plugs attain a maximum thickness of 15 m (e.g. upper leaf of the Rough Rock at Cracken Edge in Fig. 5; Hampson 1995), which is significantly smaller than the regional, basal relief of these sandstone complexes (up to 35 m), thus satisfying criterion (2). In addition, the upper leaf of the Rough Rock eroded through the newly-documented *Anthracoceras* marine band (Fig. 4), proving significant erosion. Both the lower and upper leaves of the Rough Rock are significantly more regionally-extensive and coarser grained than underlying delta systems (the Upper Haslingden Flags and Rough Rock Flags, respectively), thus satisfying criterion (3). Contemporaneous interfluves (criterion 4) are not exposed. Both the lower and upper leaves of the Rough Rock satisfy criteria (1), (2) and (3) and are interpreted

Table 2. *Summary sedimentology of lithostratigraphic units in the Rough Rock Group*

Lithostratigraphic units	Description	Interpretation	Previous descriptions
marine bands	fossiliferous, flat-laminated, very dark grey, fissile shale horizons	faunal-concentrate condensed horizons	Calver (1968) Spears & Sezgin (1985) Wignall (1987) Church & Gawthorpe (1994)
Lower Haslingden Flags Upper Haslingden Flags Rough Rock Flags	(a) coarsening-upwards interbedded siltstones and fine-grained sandstones with wave- and current-ripples and bioturbation, overlain by (b) low angle cross-bedded micaceous, medium-grained sandstones with current ripples and freshwater bioturbation (*Pelecepodychnus*), overlain by (c) channelized, cross-bedded, coarse-grained sandstones, interbedded siltstones and cross-bedded/current-rippled medium-grained sandstones with freshwater bioturbation, locally-restricted coals, and unfossiliferous silty shales	(a) distal distributary mouth bar, overlain by (b) proximal distributary mouth bar, overlain by (c) delta plain deposits: fluvial distributary channels, crevasse-splay deposits and ?interdistributary bay deposits	Collinson & Banks (1975) Miller (1986) Bristow & Myers (1989) Church & Gawthorpe (1994)
Rough Rock (lower leaf) Rough Rock (upper leaf) Crawshaw Sandstone	cross-bedded, coarse- to very coarse-grained sandstones in sets with intraformational conglomerate lags and current-rippled tops	major braided fluvial complexes	Bristow (1988) Bristow & Myers (1989) Bristow (1993) Church & Gawthorpe (1994)
Sand Rock Mine Coal Six Inch Mine/Pot Clay Coal	regionally extensive coal and carbonaceous shale horizons	low-lying peat mires	Spears & Sezgin (1985) Hampson (1995)

as IVFs. The base of the lower leaf of the Rough Rock also marks a change in sedimentary provenance, with the introduction of a much wider range of clast compositions relative to the underlying Upper Haslingden Flags. This provenance change is source-related, but is likely across a sequence boundary.

Bristow & Maynard (1994) attributed Rough Rock thickness variations in the Rossendale and Huddersfield sub-basins to differential subsidence, rather than fluvial incision. Hence they questioned the interpretation of the Rough Rock as a lowstand IVF(s) here. While the Rough Rock does not incise through either the Upper Haslingden Flags or Rough Rock Flags, both leaves of the Rough Rock satisfy criteria 2 and 3 above. In particular, both leaves of the Rough Rock are significantly coarser-grained and more regionally extensive than underlying mouth bar lobes in the Upper Haslingden Flags and Rough Rock Flags. Locally, Rough Rock sandstones overlie either distal delta front sediments (e.g. the Rough Rock Flags underlying the upper leaf of the Rough Rock at Greens Clough, Fig. 4; Hampson 1995) or delta abandonment facies (e.g. the Upper Haslingden Flags underlying the lower leaf of the Rough Rock at Scout Moor Quarry and Whitworth Quarry, Fig. 4; Hampson 1995). These facies juxtapositions strongly suggest that both leaves of the Rough Rock do not record continued progradation of underlying, genetically-linked mouth bar successions. The apparent absence of deep incision at the base of either leaf of the Rough Rock in the Rossendale and Huddersfield sub-basins may be attributed to high subsidence rates in these sub-basins, which enabled deposition of thick (up to 60 m) delta front successions underlying each leaf of the Rough Rock, and suppressed fluvial incision (cf. Jervey 1988; Van Wagoner et al. 1990).

Alternatively, Rough Rock deposition in the Rossendale and Huddersfield sub-basins may be interpreted to record a significant increase in sediment supply triggered by extrabasinal tectonics (Bristow & Maynard 1994, fig. 5). The interpretation of an extrabasinal tectonic control is partly supported by the provenance change at the base of the lower leaf of the Rough Rock (see above), but tectonically-driven changes in sediment supply cannot solely account for the coherent, cyclical and basinwide changes in sedimentation observed in the Rough Rock Group.

On the East Midlands Shelf (Fig. 2), one major, coarse-grained, braided fluvial sandstone complex is recognized in core within the Rough Rock Group, correlating to the upper leaf of the Rough Rock in the Rossendale and Huddersfield sub-basins (Figs 4, 5 and 6). This sandstone complex is overlain by an impersistent coal seam and the *Gastrioceras subcrenatum* marine band (Fig. 6), indicative of relative sea-level rise (see later discussion of regionally-extensive coals). Hence it is probable that this complex and its basal surface represent sedimentation as relative sea-level fell, slowed to stillstand and started to rise, thus satisfying criterion (1). This sandstone complex is interpreted as multistorey with up to 3 m of local basal erosional relief (Fig. 6), although channel storeys within it are poorly defined. Additionally, this sandstone body removes up to four underlying cyclothems (Church & Gawthorpe 1994; Fig. 6), proving significant incision and thus satisfying criterion (2). This Rough Rock sandstone complex is 5–15 km wide (Steele 1988; Church & Gawthorpe 1994), which is wider than its constituent channel storeys and therefore proves its multilateral character. This sandstone complex is comparable in width to the underlying delta front, but it is correlative to the upper leaf of the Rough Rock in the Rossendale and Huddersfield sub-basins, thereby proving regional incision at the same stratigraphic level and satisfying criterion (3). Also, this fluvial sandstone complex is laterally offset from the thickest section of the underlying delta front, which implies that it does not represent a simple distributary to this delta front. Contemporaneous interfluve palaeosols are recognized (see later discussion), thus satisfying criterion (4). This sandstone complex satisfies criteria (1) to (4) and is therefore interpreted as an IVF. In addition, the Crawshaw Sandstone, a regionally-extensive Westphalian A sandstone body which is lithologically and sedimentologically identical to the Rough Rock, is also identified as an IVF using criteria (1) to (4). The Crawshaw Sandstone cuts down to the *G. cancellatum* marine band in Bothamsall No. 5 core (Fig. 6), thereby demonstrating deep (up to 40 m) incision, and is overlain by the Belper Lawn Coal (Fig. 6).

In summary, three IVFs are recognized in the Rough Rock Group; the lower and upper leaves of the Rough Rock in the Rossendale and Huddersfield sub-basins, the latter correlating to the Rough Rock IVF on the East Midlands Shelf. Each of these IVFs comprises vertically-stacked channel storeys, which implies that these incised valleys were filled as accommodation space increased during rising base level. However, sediment flux was sufficiently high to keep pace with this base level rise, thereby precluding transgression, and hence these IVFs are interpreted as deposits in lowstand systems tracts

Table 3. *Incised valley fill (IVF) dimensions in the Rough Rock Group and overlying strata*

	Rossendale and Huddersfield sub-basins		East Midlands Shelf	
	Rough Rock (lower leaf)	Rough Rock (upper leaf)	Rough Rock (upper leaf)	Crawshaw Sandstone
Width	> 50 km	> 70 km	5–15 km	< 15 km
Thickness	< 25 m	< 35 m	< 35 m	< 40 m
Aspect ratio	1 : 2000	1 : 2000	1 : 150–400	1 : 400
Area	> 2700 km^2	> 6500 km^2	unconstrained	unconstrained

Dimensions for the upper Rough Rock IVF on the East Midlands Shelf taken from Steele (1988) and Church & Gawthorpe (1994)

(LSTs). Each of these IVFs is also capped by a palaeosol, implying that valleys were infilled to emergence prior to the transgression associated with the initial flooding surface. Evidence of emergence at the top of Rough Rock IVFs appears contrary to the interpretation that these units were deposited during rising base level, but is attributed to high sediment flux. This high flux may also have contributed to fluvial channel storeys remaining sand-rich and braided throughout valley infilling.

The poor distribution of exposed outcrops precludes recognition of down-dip lowstand depositional systems, including lowstand shorelines, which correspond to these lowstand IVFs. Consequently, regional stratal geometries cannot be defined and the assignment of Rough Rock IVFs to LSTs remains unproven. Also, it is not possible to define the extent of basinward facies shifts recorded by these IVFs.

IVF geometry

Evidence for active tectonism is absent within the Rossendale and Huddersfield sub-basins and, accordingly, the contemporaneous northern margin of the Pennine Basin is interpreted as tectonically-stable. In the Rossendale sub-basin, localized synsedimentary growth faults are recognized in the Upper Haslingden Flags (Bristow 1988, fig. 11.4). These synsedimentary growth faults sole out within the Flags (Bristow 1988) and are attributed to gravity-driven slides. The Upper Haslingden Flags have a flat top, as defined by overlying, local interdistributary bay/lake fill shales (e.g. Scout Moor Quarry in Fig. 4; Hampson 1995), which indicates that fault scarp relief had been completely infilled by the time of shale deposition and local abandonment of the Upper Haslingden Flags delta system. Subsequent Rough Rock deposition occurred on a broad, flat delta plain with very subdued topography. The sheet-like geometry of Rough

Rock IVFs in these sub-basins (Figs 4 and 5; Table 3) is inferred to have resulted from relatively uniform fluvial incision during falling relative sea-level, which is consistent with their formation on a broad, gently-dipping, tectonically-stable ramp with very subdued topography. In addition, the large widths of these IVFs (> 50 km and > 70 km for the lower and upper leaves of the Rough Rock, respectively; Table 3) is partly attributed to long-lived fluvial incision, which allowed extensive lateral migration of actively incising fluvial systems. This interpretation implies a prolonged period of low base level and is supported by the highly weathered character of a palaeosol which caps the upper leaf of the Rough Rock and formed during prolonged subaerial exposure (Spears & Sezgin 1985).

On the East Midlands Shelf, the Rough Rock IVF is up to 40 m thick (Church & Gawthorpe 1994; Table 3) and only 5–15 km wide (Steele 1988; Church & Gawthorpe 1994; Table 3), suggesting that fluvial incision was laterally confined. The Crawshaw Sandstone IVF exhibits abrupt lateral thickness changes (e.g. 30 m erosional relief over a lateral distance of 800 m is observed between Bothamsall Nos. 1 and 5 boreholes in Fig. 6), which also implies laterally confined fluvial incision. In this region, Namurian R_2 strata, which underlie the Rough Rock Group, contain significant thicknesses of volcanic tuffs (Edwards 1967; Smith *et al.* 1973), recording a history of late Namurian volcanism and implied, associated tectonism in this region. Workers in similar, tectonically-active, Upper Carboniferous strata in the Durham and South Wales Coalfields document small-scale, synsedimentary growth faults with fault scarp relief of never more than a few metres which focused adjacent fluvial systems (Elliott & Lapido 1981; Fielding & Johnson 1987; Hartley 1993). We invoke similar processes to explain the confined geometry of the Rough Rock IVF on the East Midlands Shelf: the palaeotopography gener-

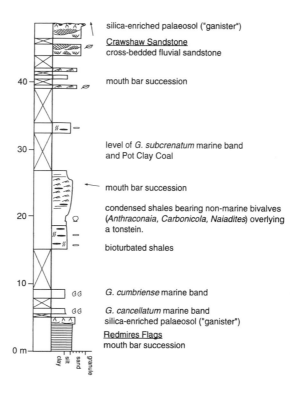

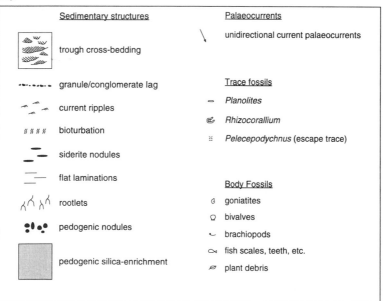

Fig. 7. Sedimentary log through the Rough Rock Group and surrounding strata at Harewood Grange (Figs 1, 2 and 5). Marine band palaeontology is taken from Smith (1967).

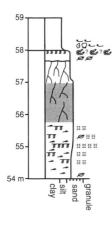

(a)

(b)

Fig. 8. (**a**) Sedimentary log of an interfluve palaeosol underlying the *G. subcrenatum* marine band, and correlative to the upper Rough Rock IVF, from the Nether Heage borehole (Fig. 6). Key to logging symbols is the same as that for Fig. 7. (**b**) Quartz-enriched horizon (ganister) from this interfluve palaeosol, which is indicative of leaching and free drainage conditions during pedogenesis. Preservation of carbonaceous rootlet traces is attributed to overprinting by later, 'wetter' (transgressive) palaeosols (see text for details).

ated by small-scale synsedimentary growth faults was sufficient to focus the Rough Rock fluvial system, which then incised into underlying strata during relative sea-level falls, producing a laterally confined incised valley. Abrupt increases in the thickness of the Crawshaw Sandstone IVF are associated with basal lags of anomalously angular intraclasts, which are interpreted as reworked breccias. In addition, documented late Namurian IVFs, including the Ashover Grit, the Chatsworth Grit and the Rough Rock (Church & Gawthorpe 1994;

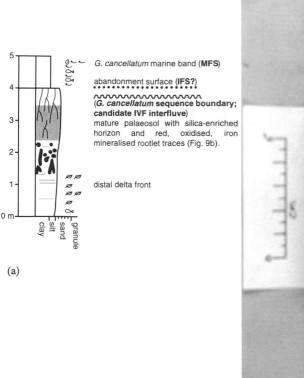

(a)

(b)

Fig. 9. (a) Sedimentary log of an interfluve palaeosol underlying the *G. cancellatum* marine band, and correlative to an undetermined incised valley, from the Nether Heage borehole (Fig. 6). Note the basinward facies shift at the base of this palaeosol. Key to logging symbols is the same as that for Fig. 7. **(b)** Red, iron-stained rootlet traces from this interfluve palaeosol. These structures are indicative of oxidation and free drainage conditions during pedogenesis.

O'Beirne 1994), exhibit a persistent NE–SW directional trend (Steele 1988, fig. 10.7) which is also consistent with a minor structural control on IVF geometry and orientation. This IVF directional trend cross-cuts basement structures defining sub-basins on the East Midlands Shelf (Fig. 2).

Interfluve palaeosols

Palaeosols which correlate as interfluves to the Rough Rock IVF on the East Midlands Shelf (Fig. 2) are recognized in Nether Heage and Bothamsall No. 1 cores (Figs 6 and 8). These palaeosols are not thick (< 1 m) or composite, indicating relatively short-lived pedogenesis, but do contain a variably bleached and silica-enriched upper horizon. These horizons contain downward-branching vertical and sub-vertical siliceous tubules, some of which have carbonaceous linings and attached carbonaceous streaks (e.g. Fig. 8), and are comparable to 'ganisters' (*sensu* Percival 1983). Silica-enrichment in these horizons indicates that clays and other fine particles were partially removed (leached) from the upper part of the soil profile and translocated to a lower position in the profile by gravity-driven percolation under freely-drained conditions, leading to characteristic silica enrichment of the surface horizon (Percival 1983; Wright 1989). This partial leaching indicates that the contemporaneous water table was relatively low and a degree of free drainage occurred. Such an interpretation precludes the accumulation of thick peat deposits (Wright 1989).

In addition, correlative interfluves to the Crawshaw Sandstone IVF on the East Midlands Shelf (Fig. 2) are recognized in Farleys Wood No. 5, Gainsborough No. 1 and Gainsborough No. 2 cores (Fig. 6). In Farleys Wood No. 5 and Gainsborough No. 2 cores, these interfluves comprise dark grey, carbonaceous palaeosols which form seatearths to the Belper Lawn Coal. These palaeosols are interpreted as gleys or gleysols which formed under permanent or semi-permanent waterlogged drainage conditions. Their grey colour reflects the dominantly reduced, ferrous state of iron within the palaeosol profile, which in turn reflects waterlogged conditions and the presence of abundant plant material (Duchaufour 1982; Besly & Fielding 1989). Reduced conditions also preserved carbonaceous material within these palaeosols. Waterlogged conditions were probably enhanced by the fine-grained, relatively impermeable character of the substrate.

Using spectral gamma ray analysis, Davies & Elliott (1996) characterize interfluve palaeosols in Upper Carboniferous strata in western Ireland as possessing anomalously high Th/K ratios (> 18), due to leaching of potassium-bearing, clay minerals from the upper horizons of these palaeosols. Despite the presence of leached upper horizons in some interfluve palaeosols recognized within the Rough Rock Group, there is wide variation in interfluve palaeosol character, such that extreme Th/K enrichment may not be observed. This variation is interpreted to reflect variability in substrate grain-size and sorting, and associated depositional permeability; variable palaeosol maturity; and/or overprinting and reworking by later, 'wetter' (transgressive) palaeosols. The last is well documented in Upper Carboniferous strata by Gardner *et al.* (1988) and may be particularly important in obscuring the freely-drained character of some interfluve palaeosols.

Other palaeosols within the Rough Rock Group contain bleached, upper horizons, some of which display red staining. In particular, on the East Midlands Shelf, the *G. cumbriense* and *G. cancellatum* marine bands overlie such palaeosols in the Nether Heage, Bothamsall No. 1 and Bothamsall No. 5 cores (Figs 6 and 9). These bleached and red-stained palaeosols are interpreted to record pedogenesis under conditions of free drainage. Due to their freely-drained character and stratigraphic occurrence underlying transgressive marine bands (see later discussion), these palaeosols are interpreted as interfluve correlatives to, as yet unrecognized, IVFs. Similarly, Maynard *et al.* (1991, 1992) document a palaeosol capping the Lower

Haslingden Flags and underlying the *G. cumbriense* marine band at Withnell Quarry (Figs 1, 2 and 4), which they interpret as a subaerial exposure surface marking a sequence boundary.

Regionally-extensive coals

In the northern Pennine Basin (Rossendale and Huddersfield sub-basins, Fig. 2), the Sand Rock Mine Coal and Six Inch Mine/Pot Clay Coal overlie the lower and upper leaves of the Rough Rock, respectively (Figs 3 and 4). Thus both Rough Rock IVFs are directly overlain by regionally-extensive coals. The Sand Rock Mine Coal and the Six Inch Mine/Pot Clay Coal generally occur as thin (< 50 cm), uniform single seams and cover exceptionally large areas (> 2600 km² and > 8400 km², respectively; Fig. 10). Both coals also thin to the south and east (Fig. 10). In particular, the Pot Clay Coal thins to a single seam of poor quality, clastic-rich coal ('cannel') up to 20 cm thick in the southern Huddersfield sub-basin and is absent on the East Midlands Shelf (Wray *et al.* 1930; Edwards *et al.* 1950; Stephens *et al.* 1953; Eden *et al.* 1957; Frost & Smart 1979; Aitkenhead *et al.* 1985; Hampson 1995). Significant local thickening of any of these coals is recorded only in the northern Rossendale sub-basin (Fig. 10), where the Sand Rock Mine Coal reaches a maximum thickness of 1.5 m and splits (Price *et al.* 1963, p. 23). Here the two leaves of the seam are separated by 'argillaceous measures up to 21 feet thick' over a lateral extent of *c.* 7 km (Price *et al.* 1963, p.23, fig. 6).

The thin, uniform, regionally-extensive character of these coal seams suggests that peat swamps were established in relatively uniform, regionally-extensive environments (Hampson 1995) such as blanket bogs (*sensu* Haszeldine 1989) or low-lying mires (*sensu* McCabe 1991). This is consistent with the tectonically-stable ramp setting of the northern Pennine Basin with its inferred subdued topography. The local thickening and splitting of the Sand Rock Mine Coal represents deposition in a locally-restricted environment (Hampson 1995), which is apparently confined to the upper delta plain. The absence of the Pot Clay Coal from the East Midlands Shelf is consistent with evidence for syn-sedimentary tectonism, producing an unstable substrate for peat swamp formation, in this area. Alternatively, drainage conditions may have been better on the East Midlands Shelf, but this alternative interpretation is not favoured because it contradicts the preservation of carbonaceous material in interfluve palaeosols which

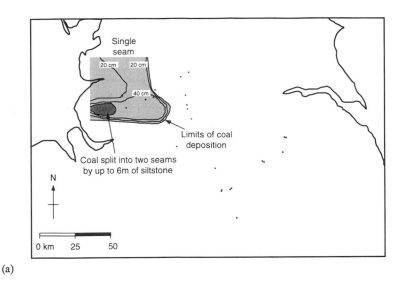

(a)

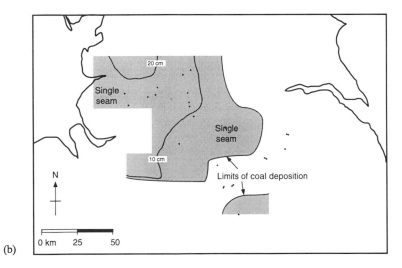

(b)

Fig. 10. Isopach maps of regionally-extensive coals in the Rough Rock Group; (**a**) the Sand Rock Mine Coal and (**b**) the Six Inch Mine/Pot Clay Coal. These maps are constructed using logged sections (shown as small filled circles) supplemented by descriptions in the relevant British Geological Survey Memoirs (Wright *et al.* 1927; Wray *et al.* 1930; Bromehead *et al.* 1933; Mitchell *et al.* 1947; Edwards *et al.* 1950; Stephens *et al.* 1953; Eden *et al.* 1957; Earp *et al.* 1961; Price *et al.* 1963; Edwards 1967; Smith *et al.* 1967, 1973; Evans *et al.* 1968; Stevenson & Gaunt 1971; Frost & Smart 1979; Aitkenhead *et al.* 1985; Chisholm *et al.* 1988).

correlate to the upper Rough Rock IVF here (e.g. Fig. 8).

The Sand Rock Mine Coal and Six Inch Mine/Pot Clay Coal both record the establishment of a regionally-extensive mire after infilling of an underlying fluvial IVF and represent, on a regional scale, clastic sediment starvation, and a synchronous rise in the regional water table (taken to approximate to regional base level),

creating accommodation for mires to develop and peat to accumulate and be preserved within these mires (cf. Aitken 1994). Rising base level during IVF deposition is documented by the vertical stacking of channel storeys within these IVFs and peat deposition in overlying mires may have occurred under the same rate of base level rise. Clastic sediment starvation during peat deposition may have been achieved as a result of sediment dispersal over a much wider area after incised valleys were completely filled and sediment flux overspilled onto their adjacent interfluves (Shanley & McCabe 1994). However, the great width of Rough Rock IVFs (> 50 km and > 70 km for the lower and upper leaves of the Rough Rock, respectively; Table 3) implies that their interfluves occupied a correspondingly small area. Accordingly, sediment starvation resulting from sediment overspill onto the interfluves of these Rough Rock IVFs is regarded as insufficient to account for the establishment of regionally-extensive mires. Instead, sediment supply is inferred to have been trapped up-dip as a result of rising base level, which reduced fluvial gradients and sediment transport efficiencies. Given these base level and sediment supply constraints, combined with the stratigraphic occurrence of these coals overlying lowstand IVFs and underlying condensed marine shales (marine bands), both the Sand Rock Mine Coal and the Six Inch Mine/Pot Clay Coal are interpreted as initial flooding surfaces (IFS) correlatives (Hampson 1995). In the absence of exposed down-dip strata, which precludes the recognition of initial flooding at the lowstand shoreline, and regional stratal geometries, this interpretation is somewhat speculative. However, similar arguments have been used by Arditto (1987), Hartley (1993), Aitken & Flint (1994, 1995) and Flint *et al.* (1995) in other examples of coal-bearing strata.

Marine bands

The *G. cancellatum, G. cumbriense,* newly documented *Anthracoceras* and *G. subcrenatum* marine bands are regionally-extensive horizons recording condensed sedimentation, and the most marine influence within their respective cyclothems. Accordingly, these horizons are intepreted as marine flooding surfaces (FSs). The interpretation of different faunal phases, which record palaeoecological changes, within marine bands (Calver 1968, Wignall 1987) is important in the correlation of these horizons. Furthermore, the interpretation of marginal marine and nonmarine faunal phases within marine bands as FS components has significant implications for the interpretation of similar restricted marine and nonmarine, fossiliferous, condensed shale horizons in other coastal plain successions (e.g. Flint *et al.* 1995).

Maynard *et al.* (1991) document a radioactive condensed shale horizon with an impoverished marine, nektonic fauna (Owd Bett's Horizon) which underlies the *G. cumbriense* marine band. These two condensed shale horizons define

onlap onto an underlying palaeosol which caps the Lower Haslingden Flags and is interpreted to define a sequence boundary (Maynard *et al.* 1991). Consequently, Maynard *et al.* (1991) interpret the Owd Bett's Horizon as a FS within a transgressive systems tract (TST) and the *G. cumbriense* marine band, which records maximum marine influence, as a maximum flooding surface (MFS). Work of similar detail has not been undertaken on other marine bands and associated shales in the Rough Rock Group. However, the newly documented *Anthracoceras* and *G. subcrenatum* marine bands both overlie regionally-extensive coals (the Sand Rock Mine Coal and Six Inch Mine/Pot Clay Coal, respectively) which are interpreted as IFS correlatives. Consequently, these marine bands are considered to be MFSs. In addition, O'Beirne (1994) interprets the *G. cancellatum* marine band as a MFS which caps a TST comprising the regionally-extensive Ringinglow Coal overlain by the Redmires Flags delta front succession (Fig. 3).

Sequence stratigraphic framework

The recognition of key sequence stratigraphic elements, particularly sequence boundaries defined by IVFs and their interfluve palaeosols, permits the construction of a sequence stratigraphic framework for the Rough Rock Group (Fig. 11). Four high-frequency sequences are documented in the Rough Rock Group and surrounding strata, each of which is named according to the marine band which defines the MFS within the sequence (according to the convention of Church & Gawthorpe 1994). Sequence boundaries are named according to the overlying marine band. Maps of key stratigraphic units within these sequences (Fig. 12) are constructed using logged sections supplemented by the maps of Calver (1968; *G subcrenatum* marine band), Collinson & Banks (1975; Upper Haslingden Flags), Bristow (1988, 1993; Rough Rock), Steele (1988; Rough Rock) and Church & Gawthorpe (1994; Rough Rock), and descriptions in the relevant British Geological Survey Memoirs (Wright *et al.* 1927; Wray *et al.* 1930; Bromehead *et al.* 1933; Mitchell *et al.* 1947; Edwards *et al.* 1950; Stephens *et al.* 1953; Eden *et al.* 1957; Earp *et al.* 1961; Price *et al.* 1963; Edwards 1967; Smith *et al.* 1967, 1973; Evans *et al.* 1968; Stevenson & Gaunt 1971; Frost & Smart 1979; Aitkenhead *et al.* 1985; Chisholm *et al.* 1988).

G. cancellatum *sequence*

The base of this sequence lies below the Rough

High frequency sequence stacking patterns are most clearly developed along the northern basin margin (Figs 4 and 11). Here the *G. cancellatum* and *G. cumbriense* sequences are sand-poor, progradationally-stacked and exhibit suppressed lowstand incised valley development. These sequences compare in internal architecture and stacking pattern to a highstand sequence set (*sensu* Mitchum & Van Wagoner 1991). In contrast, the *Anthracoceras* and *G. subcrenatum* sequences are sand-rich, progradationally-stacked and contain thick (25–35 m), well developed IVFs. These sequences compare in internal architecture and stacking pattern to a lowstand sequence set (*sensu* Mitchum & Van Wagoner 1991). This interpretation of high frequency sequence stacking patterns implies a late Namurian relative sea-level curve similar to that of Church & Gawthorpe (1994, fig. 11). However, these stacking patterns are not observed on the East Midlands Shelf, implying that local sediment supply is the predominant control on the distribution of the *Anthracoceras* sequence boundary, which defines the base of the lower Rough Rock IVF.

Discussion

While it is important to consider alternative scenarios of tectonics, climate, sediment supply and relative sea-level to explain Rough Rock Group stratigraphy (Bristow & Maynard 1994), these variables cannot be treated independently. The cyclicity and basin-wide extent of high frequency sequences in the Rough Rock Group discount a purely tectonic origin for relative sea-level fluctuations which controlled the formation of these sequences. There is abundant evidence of high magnitude (60 ±15 m; Crowley & Baum 1991), glacio-eustatic sea-level fluctuations throughout the Upper Carboniferous (see reviews in Ramsbottom 1977; Holdsworth & Collinson 1988; Leeder 1988; Maynard & Leeder 1992). Such glacio-eustatic sea-level fluctuations are climatically-forced and partly control the surface area available for weathering and, consequently, sediment flux derived from fluvial run-off. These fluctuations alone are considered to have been sufficient to account for observed changes in facies and sediment supply within the Rough Rock Group stratigraphy, with the exception of the change in provenance at the base of the lower Rough Rock IVF in the Rossendale sub-basin. This provenance change is inferred to be source-determined and may be driven by tectonic changes in the drainage basin of the Rough Rock river system.

Conclusions

The concepts of high resolution sequence stratigraphy can be successfully applied to the Rough Rock Group, which comprises four high frequency sequences. Biostratigraphy is a key tool to aid recognition and correlation of these sequences and the documentation of a 'new' *Anthracoceras* marine band within the Rough Rock Group significantly revises the sequence stratigraphic interpretation of these strata.

LSTs comprise major braided fluvial sandstone complexes, interpreted as IVFs, and correlative interfluve palaeosols. The character of these palaeosols is variable, reflecting variations in substrate, maturity and transgressive overprinting. Incised valleys are sheet-like in the tectonically-stable Rossendale and Huddersfield sub-basins and laterally confined on the East Midlands Shelf, which underwent contemporaneous minor active extensional tectonism. Regionally-extensive coals cap IVFs and represent the suppression of clastic sediment supply and synchronous rising base level. These coals are interpreted as IFS correlatives and define the bases of TSTs, each of which typically comprises several metres of marine/marginal marine shale capped by a faunal-concentrate condensed horizon (marine band) interpreted as a MFS. Point-sourced lobate deltas characterize HSTs.

A change in palaeogeography within the Rough Rock Group occurs across the newly documented *Anthracoceras* marine band and is controlled by lateral accommodation space distribution. Also, in the Rossendale sub-basin a change in sandstone provenance observed at the base of the lower Rough Rock IVF is source-determined and may relate to drainage basin tectonics.

The authors thank J. F. Aitken, K. Church, S. J. Davies and J. R. Maynard for fruitful discussions regarding the work presented here. We thank Evered Quarries Ltd., Johnsons Wellfield Ltd., Marshalls Mono and Ridgeway Quarry for access to quarry sites and the British Geological Survey for access to core material. A. J. Hartley and P. J. McCabe are thanked for their constructive reviews. This work was carried out while GJH was in receipt of a NERC studentship. Additional funding for fieldwork was supplied by Amoco, B.P. and Mobil.

References

AITKEN, J. F. 1994. Coal in a sequence stratigraphic framework. *Geoscientist*, **4**, 9–12.
—— & FLINT, S. S. 1994. High frequency sequences and the nature of incised valley fills in fluvial systems of the Breathitt Group (Pennsylvanian),

Appalachian foreland basin, eastern Kentucky. *In:* DALRYMPLE, R. W., BOYD, R. & ZAITLIN, B. A. (eds) *Incised-valley systems: origins and sedimentary sequences.* Society of Economic Paleontologists and Mineralogists, Special Publication, **51**, 353–368.

—— & —— 1995. The application of high resolution sequence stratigraphy to fluvial systems: a case study from the Upper Carboniferous Breathitt Group, eastern Kentucky, USA. *Sedimentology*, **42**, 3–30.

AITKENHEAD, N., CHISHOLM, J. I. & STEVENSON, I. P. 1985. *Geology of the country around Buxton, Leek and Bakewell.* British Geological Survey, Sheet Memoir, **111**.

ARDITTO, P. A. 1987. Eustasy, sequence stratigraphic analysis and peat formation: a model for widespread late Permian coal deposition in the Sydney Basin, N.S.W. *Advanced studies of the Sydney Basin: 21st Newcastle Symposium proceedings*, 11–17.

BESLY, B. M. & FIELDING, C. R. 1989. Palaeosols in Westphalian coal-bearing and red bed sequences, central and northern England. *Palaeogeography, Palaeoclimatology, Palaeoecology*, **70**, 303–330.

BRISTOW, C. S. 1988. Controls on the sedimentation of the Rough Rock Group (Namurian) from the Pennine Basin of northern England. *In:* BESLY, B. M. & KELLING, G. (eds) *Sedimentation in a synorogenic basin complex: the upper Carboniferous of northwest Europe*, Blackie, Glasgow, 114–131.

—— 1993. Sedimentology of the Rough Rock: a Carboniferous braided river sheet sandstone in northern England. *In:* BEST, J. L. & BRISTOW, C. S. (eds) *Braided Rivers.* Geological Society, London, Special Publication, **75**, 291–304.

—— & MAYNARD, J. R. 1994. Alternative sequence stratigraphic models for the Rough Rock Group: a Carboniferous delta in the Pennine Basin, England. *In:* JOHNSON, S. D. (ed.) *High resolution sequence stratigraphy: innovations and applications.* Abstract volume, University of Liverpool, 353–357.

—— & MYERS, K. J. 1989. Detailed sedimentology and gamma-ray log characteristics of a Namurian deltaic succession I: sedimentology and facies analysis. *In:* WHATELEY, M. K. G. & PICKERING, K. T. (eds) *Deltas: sites and traps for fossil fuels.* Geological Society, London, Special Publication, **41**, 75–80.

BROMEHEAD, C. E. N., EDWARDS, W. N., WRAY, D. A. & STEPHENS, J. D. 1933. *Geology of the country around Holmfirth and Glossop.* British Geological Survey, Sheet Memoir, **86**.

CALVER, M. A. 1968. Distribution of Westphalian marine faunas in northern England and adjoining areas. *Proceedings of the Yorkshire Geological Society*, **37**, 1–72.

CHISHOLM, J. I., CHARSLEY, T. J. & AITKENHEAD, N. 1988. *Geology of the country around Ashbourne and Cheadle.* British Geological Survey, Sheet Memoir, **124**.

CHURCH, K. D. & GAWTHORPE, R. L. 1994. High resolution sequence stratigraphy of the late Namurian in the Widmerpool Gulf (East Midlands, UK). *Marine and Petroleum Geology*, **11**, 528–544.

COLLINSON, J. D. 1988. Controls on Namurian sedimentation in the Central Province Basins of northern England. *In:* BESLY, B. M. & KELLING, G. (eds) *Sedimentation in a synorogenic basin complex: the Upper Carboniferous of northwest Europe*, 85–101.

—— & BANKS, N. L. 1975. The Haslingden Flags (Namurian G_1) of south-east Lancashire: bar finger sands in the Pennine Basin. *Proceedings of the Yorkshire Geological Society*, **40**, 431–458.

CROWLEY, T. J. & BAUM, S. K. 1991. Estimating Carboniferous sea-level fluctuations from Gondwanan ice extent. *Geology*, **19**, 975–977.

DAVIES, S. J. & ELLIOTT, T. 1996. Spectral gamma characterization of high resolution sequence stratigraphy: examples from Upper Carboniferous fluvio-deltaic systems, County Clare, Ireland. *This volume.*

DUCHAUFOUR, P. 1982. *Pedology.* Allen & Unwin, London.

EAGAR, R. M. C. 1951. A revision of the sequence and correlation of the Lower Coal Measures west of Wigan. *Journal of the Geological Society of London*, **107**, 23–48.

—— 1952. Growth and variation in the non-marine lamellibranch fauna above the Sand Rock Mine Coal of the Lancashire Millstone Grit. *Journal of the Geological Society of London*, **107**, 339–373.

EARP, J. R., MAGRAW, D., POOLE, E. G., LAND, D. H. & WHITEMAN, A. J. 1961. *Geology of the country around Clitheroe and Nelson.* British Geological Survey, Sheet Memoir, **68**.

EDEN, R. A., STEPHENSON, I. P. & EDWARDS, W. N. 1957. *Geology of the country around Sheffield.* British Geological Survey, Sheet Memoir, **100**.

EDWARDS, W. 1967. *The geology of the country around Ollerton.* British Geological Survey, Sheet Memoir, **113**.

——, MITCHELL, G. H. & WHITEHEAD, T. H. 1950. *Geology of the country around Leeds.* British Geological Survey, Sheet Memoir, **70**.

ELLIOTT, T. & LAPIDO, K. O. 1981. Syn-sedimentary gravity slides (growth faults) in the Coal Measures of South Wales. *Nature*, **291**, 220–222.

EVANS, W. B., WILSON, A. A., TAYLOR, B. J. & PRICE, D. 1968. *Geology of the country around Macclesfield, Congleton, Crewe and Middlewich.* British Geological Survey, Sheet Memoir, **110**.

FIELDING, C. R. & JOHNSON, G. A. L. 1987. Sedimentary structures associated with extensional fault movement from the Westphalian of NE England. *In:* COWARD, M. P., DEWEY, J. F. & HANCOCK, P. L. (eds) *Continental extensional tectonics.* Geological Society, London, Special Publication, **28**, 511–516.

FLINT, S. S., AITKEN, J. F. & HAMPSON, G. J. 1995. The application of sequence stratigraphy to coal-bearing, fluvial successions: implications for the U.K. Coal Measures. *In:* WHATELEY, M. K. G. & SPEARS, D. A (eds) *European Coal Geology.*

Geological Society, London, Special Publication, **82**, 1–16.

FROST, D. V. & SMART, J. G. O. 1979. *Geology of the country around Derby*. British Geological Survey, Sheet Memoir, **124**.

GARDNER, T. W., WILLIAMS, E. G. & HOLBROOK, P. W. 1988. Pedogenesis of some Pennsylvanian underclays; ground-water, topographic, and tectonic controls. *In:* REINHARDT, J. & SIGLEO, W. R. (eds) *Paleosols and weathering through geologic time: principles and applications*. Geological Society of America, Special Paper, **216**, 81–101.

GILLIGAN, A. 1921. A borehole for water at Meanwood, Leeds. *Transactions of the Leeds Geological Association*, **4**, 15–16.

HAMPSON, G. J. 1995. Discrimination of regionally-extensive coals in the Upper Carboniferous of the Pennine Basin, U.K. using high-resolution sequence stratigraphic concepts. *In:* WHATELEY, M. K. G. & SPEARS, D. A (eds) *European Coal Geology*. Geological Society, London, Special Publication, **82**, 79–97.

HARTLEY, A. J. 1993. A depositional model for the Mid-Westphalian A to late Westphalian B Coal Measures of South Wales. *Journal of the Geological Society, London*, **150**, 1121–1136.

HASZELDINE, R. S. 1989. Coal reviewed: depositional controls, modern analogues and ancient climates. *In:* WHATELEY, M. K. G. & PICKERING, K. T. (eds) *Deltas: sites and traps for fossil fuels*, Geological Society, London, Special Publication, **41**, 289–308.

HOLDSWORTH, B. K. & COLLINSON, J. D. 1988. Millstone Grit cyclicity revisited. *In:* BESLY, B. M. & KELLING, G. (eds) *Sedimentation in a synorogenic basin complex: the Upper Carboniferous of northwest Europe*, 132–152.

HUDSON, R. G. S. & DUNNINGTON, H. V. 1939. A boring in the Lower Coal Measures and Millstone Grit at Bradford. *Proceedings of the Yorkshire Geological Society*, **24**, 129–136.

—— & —— 1940. A borehole section in the Carboniferous at Fairweather Green, Bradford. *Proceedings of the Yorkshire Geological Society*, **24**, 206–218.

JERVEY, M. T. 1988. Quantitative geological modelling of siliciclastic rock sequences and their seismic expression. *In:* WILGUS, C. K., HASTINGS, B. S., KENDALL, C. G. St. C., POSAMENTIER, H. W., ROSS, C. A. & VAN WAGONER, J. C. (eds) *Sea-level changes – an integrated approach*. Society of Economic Paleontologists and Mineralogists, Special Publication, **42**, 47–69.

LEE, A. G. 1988. Carboniferous basin configuration of central and northern England modelled using gravity data. *In:* BESLY, B. M. & KELLING, G. (eds) *Sedimentation in a synorogenic basin complex: the Upper Carboniferous of northwest Europe*. Blackie, Glasgow, 69–84.

LEEDER, M. R. 1988. Recent developments in Carboniferous geology: a critical review with implications for the British Isles and northwest Europe. *Proceedings of the Geologists' Association*, **99**, 73–100.

McCABE, P. J. 1991. Geology of coal; environments of deposition. *In:* GLUSKOTER, H. J., RICE, D. D. & TAYLOR, R. B. (eds) *The geology of North America, Vol. P-2, Economic geology*, Geological Society of America, 469–482.

MAYNARD, J. R. 1992. Sequence stratigraphy of the Upper Yeadonian of northern England. *Marine and Petroleum Geology*, **9**, 197–207.

—— & LEEDER, M. R. 1992. On the periodicity and magnitude of Late Carboniferous glacio-eustatic sea level changes. *Journal of the Geological Society, London*, **149**, 303–311.

——, WIGNALL, P. B. & VARKER, W. J. 1991. A 'hot' new shale facies from the Upper Carboniferous of northern England. *Journal of the Geological Society, London*, **148**, 805–808.

MILLER, G. D. 1986. The sediments and trace fossils of the Rough Rock Group on Cracken Edge, Derbyshire. *Mercian Geologist*, **10**, 189–202.

MITCHELL, G. H., STEPHENS, J. V., BROMEHEAD, C. E. N. & WRAY, D. A. 1947. *Geology of the country around Barnsley*. British Geological Survey, Sheet Memoir, **87**.

MITCHUM, R. M. & VAN WAGONER, J. C. 1991. High-frequency sequences and their stacking patterns: sequence-stratigraphic evidence of high-frequency eustatic cycles. *Sedimentary Geology*, **70**, 131–160.

O'BEIRNE, A. 1994. The progradation of a delta during transgression: from the Namurian of the Central Province, England. *In:* JOHNSON, S. D. (ed.) *High resolution sequence stratigraphy: innovations and applications*. Abstract Volume. University of Liverpool, 389–391.

PERCIVAL, C. J. 1983. A definition of the term ganister. *Geological Magazine*, **120**, 187–190.

PRICE, D., WRIGHT, W. B., JONES, R. C. B., TONKS, L. H. & WHITEHEAD, T. H. 1963. *Geology of the country around Preston*. British Geological Survey, Sheet Memoir, **75**.

RAMSBOTTOM, W. H. C. 1977. Major cycles of transgression and regression (mesothems) in the Namurian. *Proceedings of the Yorkshire Geological Society*, **41**, 261–291.

SHANLEY, K. W. & McCABE, P. J. 1994. Perspectives on the sequence stratigraphy of continental strata. *American Association of Petroleum Geologists Bulletin*, **78**, 544–568.

SLINGER, F. C. 1936. The succession in the Rough Rock Series north of Leeds. *Transactions of the Leeds Geological Association*, **5**, 188–196.

SMITH, E. G. 1967. The Namurian and basal Westphalian rocks of the Beeley–Holymoorside area. *In:* NEVES, R. & DOWNIE, C. (eds) *Geological excursions in the Sheffield area*, University of Sheffield Press. 103–108.

——, RHYS, G. H. & EDEN, R. A. 1967. *Geology of the country around Chesterfield, Matlock and Mansfield*. British Geological Survey, Sheet Memoir, **112**.

——, —— & GOOSENS, R.F. 1973. *Geology of the country around East Retford*. British Geological Survey, Sheet Memoir, **101**.

SPEARS, D. A. & SEZGIN, H. I. 1985. Mineralogy and geochemistry of the subcrenatum marine band

and associated coal-bearing sediments, Langsett, south Yorkshire. *Journal of Sedimentary Petrology*, **55**, 570–578.

STEELE, R. P. 1988. The Namurian sedimentary history of the Gainsborough Trough. *In:* BESLY, B. M. & KELLING, G. (eds) *Sedimentation in a synorogenic basin complex: the Upper Carboniferous of northwest Europe*, Blackie, Glasgow, 102–113.

STEPHENS, J. V., MITCHELL, G. H. & EDWARDS, W. 1953. *Geology of the country between Bradford and Skipton*. British Geological Survey, Sheet Memoir, **69**.

STEVENSON, I. P. & GAUNT, G. D. 1971. *Geology of the country around Chapel-en-le-Frith*. British Geological Survey, Sheet Memoir, **99**.

VAN WAGONER, J. C., MITCHUM, R. M., CAMPION, K. M. & RAHMANIAN, V. D. 1990. *Siliciclastic sequence stratigraphy in well logs, cores and outcrops*. American Association of Petroleum Geologists, Methods in Exploration Series, **7**.

WIGNALL, P. B. 1987. A biofacies analysis of the *Gastrioceras cumbriense* marine band (Namurian) of the central Pennines. *Proceedings of the Yorkshire Geological Society*, **46**, 111–128.

WRAY, D. A., STEPHENS, J. V., EDWARDS, W. N. & BROMEHEAD, C. E. N. 1930. *Geology of the country round Huddersfield and Halifax*. British Geological Survey, Sheet Memoir, **77**.

WRIGHT, W. B., SHERLOCK, R. L., WRAY, D. A., LLOYD, W. & TONKS, L. H. 1927. *Geology of the Rossendale anticline*. British Geological Survey, Sheet Memoir, **76**.

WRIGHT, V. P. 1989. Palaeosols in deltaic settings. *In:* *Palaeosols in siliciclastic sequences*, Postgraduate Research Institute for Sedimentology, University of Reading, 70–79.

Response to high frequency sea-level change in a fluvial to estuarine succession: Cenomanian palaeovalley fill, Bohemian Cretaceous Basin

DAVID ULIČNÝ & LENKA ŠPIČÁKOVÁ

Department of Geology, Charles University, Albertov 6, 128 43 Praha 2, Czech Republic

Abstract: A high resolution sequence stratigraphic analysis of large-scale exposures in the Pecínov quarry (west–central Bohemia, Czech Republic) revealed a complex record of high frequency sea-level change in fluvial to estuarine deposits of late middle to late Cenomanian age. Within the investigated third-order sequence, four parasequences make up a transgressive systems tract and the highstand systems tract is represented by a single parasequence. The parasequences formed on a slowly subsiding basin margin in response to incremental eustatic sea-level rise. The parasequences, composed successively of fluvial, tide-dominated fluvial, supratidal marsh, tidal flat, and estuarine ebb tidal delta deposits, show extensive channelling and erosive features both within the parasequences and at the parasequence boundaries. Many of the channels incise into the underlying parasequences and superficially resemble sequence boundaries. However, sedimentological evidence indicates that the erosion took place on marine-flooding surfaces and their updip equivalents, due to landward translation of highly erosive environments. The most intense erosion was caused by subtidal currents.

Because of the low subsidence rate, the preservation of parasequences was greatly affected by the channelling at most parasequence boundaries. Lateral and vertical extent of parasequences vary over distances of tens of metres. Architectural analysis of sedimentary bodies proved essential in sequence stratigraphic interpretation of the complicated and highly erosive fluvial to tide-dominated estuarine facies succession.

In recent sequence stratigraphic studies, increasing attention is paid to the record of relative sea-level change in fluvial and estuarine depositional systems and to specific features of sequences and their building blocks in such settings (Shanley *et al.* 1992; Dalrymple & Zaitlin 1994). Most descriptions of parasequences and parasequence boundaries in recent literature deal with examples from coastal to shelf depositional settings. Probably the most common are examples of shallowing-upward successions from wave-dominated shelf–shoreface systems and from deltaic settings, whereas estuarine examples are less common (cf. Van Wagoner *et al.* 1990; Swift *et al.* 1991; Hadley & Elliott 1993; McCabe 1994).

On most basin margins (except for foreland basins, cf. Posamentier & Allen 1993), the potential for preservation of the sedimentary record of relative sea-level change decreases in the landward direction, and the signal of sea-level change gets increasingly modified updip by erosion and by autocyclic phenomena, particularly in fluvial and estuarine environments. In such environments, the character and preservation of parasequences may differ significantly from the established parasequence models.

Our study focuses on the important characteristics of sedimentation and preservation of parasequences formed in a fluvial to estuarine setting on a slowly subsiding, ramp-type basin margin. Laterally extensive exposures of Cenomanian strata in the Pecínov quarry (west-central Bohemia, Czech Republic) permitted detailed investigation of the sedimentary body geometries within the parasequences, as well as of the variability of parasequence boundaries. Our results may provide important information for the prediction of reservoir heterogeneity in fluvial to estuarine sedimentary successions.

Regional framework

The Bohemian Cretaceous Basin

The Bohemian Cretaceous Basin began to form in the Bohemian Massif (Central Europe) during the mid-Cretaceous in response to compressional phases in the Alpine–Carpathian collision zone (Fig. 1). Intra-plate stresses were transmitted throughout the Alpine–Carpathian foreland and led to reactivation of pre-existing faults in the Variscan basement of the Bohemian Massif (cf. Ziegler 1987). Unpublished preliminary studies indicate that the Bohemian Cretaceous Basin was a strike-slip basin with greatest subsidence rates close to its northeastern margin, where the Lusatian Fault Zone and adjacent fault systems acted as divergent wrench faults with a dextral sense of motion (Fig. 1).

From Howell, J. A. & Aitken, J. F. (eds), *High Resolution Sequence Stratigraphy: Innovations and Applications*, Geological Society Special Publication No. 104, pp. 247–268.

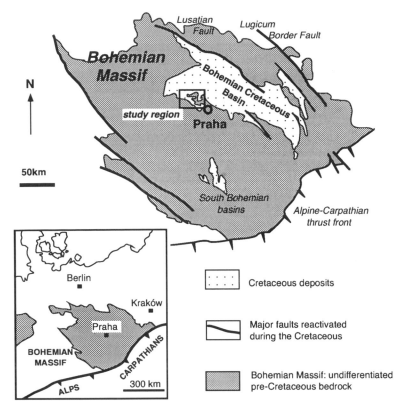

Fig. 1. The Bohemian Cretaceous Basin in the tectonic framework of the Bohemian Massif. Cenozoic sedimentary cover omitted. Inset: geographical setting of the Bohemian Massif.

The study area is situated in west-central Bohemia, on the soutwestern flank of the basin. During the mid-Cretaceous this region was characterized by a low subsidence rate of not more than 30 m/Ma, as inferred from quantitative subsidence analysis (Uličný 1992). Deposition took place on a generally flat, ramp-type basin margin with low sediment input from a low-relief drainage area of the southwestern part of the Bohemian Massif. Due to low subsidence rates and the ramp margin physiography, the nature and preservation of the Cenomanian stratigraphic record in this part of the basin was principally affected by eustasy, local and regional topography, and by autocyclic sedimentary processes.

Cenomanian palaeogeography and depositional environments

In west–central Bohemia, the initial Late Cretaceous transgression during the late Middle Cenomanian created an estuarine depositional setting by flooding a system of broad, northeast-directed palaeovalleys (Fig. 2). The palaeovalleys formed during a long period of pre-Cenomanian weathering and erosion (Triassic–Early Jurassic) and were cut into Late Palaeozoic and Late Precambrian rocks. The valleys were separated from each other by low-relief palaeohighs: the 'Unhošt'–'Tursko High' and the 'Jedomělice High' (Matějka 1936; Jelen & Malecha 1988). The strike of the valleys and highs was inherited from structural elements of the Variscan basement, such as the general strike of fold axes in the Proterozoic metamorphic rocks, and the strike of major outcrop belts of Late Palaeozoic clastics. During the Cenomanian sedimentation, the palaeohighs were gradually lowered by denudation before they were finally buried by late Cenomanian deposits. Thus, the estimates of palaeorelief, obtained from thicknesses of strata covering the buried topography, give only minimum values (Fig. 2a).

The sedimentary environments formed in response to the stepwise sea-level rise of the late Middle to Late Cenomanian, gave rise to a range

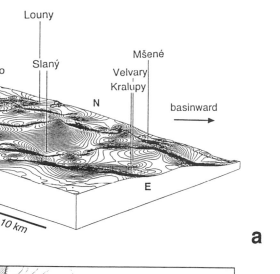

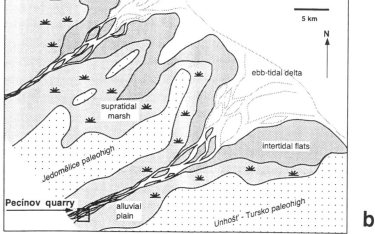

Fig. 2. Palaeogeographic framework of the Cenomanian basin margin in the study region of west-central Bohemia. (**a**) Palaeorelief image, based on inverted values of preserved thicknesses of the Peruc and Korycany members in 229 boreholes. Geographical locations of the Pecínov quarry and other localities used in constructing the sequence stratigraphic scheme (Fig. 4) are indicated. Contour interval is 2 m, vertical exaggeration 50×, maximum relief 70 m. (**b**) Palaeogeographic reconstruction of the region approximately during the deposition of parasequence 2 of the ZC 2.4 sequence (for details see text).

of fluvial, estuarine and coastal deposits (Fig. 2b). The Cenomanian strata, exposed in the Pecínov quarry, approximately 40 km west of Prague, were deposited near the axis of the broad, shallow palaeovalley bounded by the Jedomělice and Unhošť'–Tursko palaeohighs, and represent the most landward part of the preserved basin fill. The cross-section of the palaeovalley fill reveals a typical pattern of fluvial deposits in its axial part which are overlain successively by estuarine marsh, intertidal and subtidal deposits (Fig. 3).

Stratigraphy of the Cenomanian deposits

The Cenomanian Peruc–Korycany Formation (Čech *et al.* 1980) is made up of fluvial, marginal marine (Peruc Member), shallow marine (Korycany Member) siliciclastic deposits, and offshore shales to marls (the Pecínov Member, proposed by Uličný 1992; Figs 3 and 4). The Peruc–Korycany Formation is the oldest part of the Bohemian Cretaceous basin fill and reflects the initial Cenomanian transgression. In the study region, the oldest Cretaceous deposits are of late

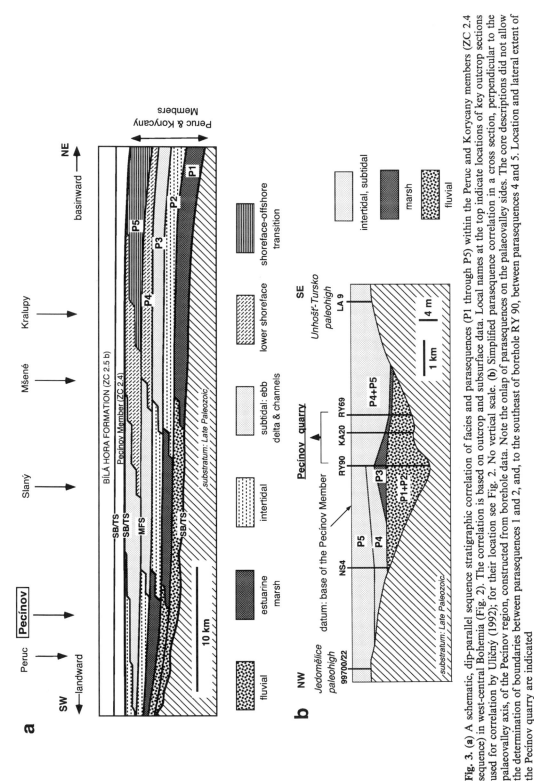

Fig. 3. (a) A schematic, dip-parallel sequence stratigraphic correlation of facies and parasequences (P1 through P5) within the Peruc and Korycany members (ZC 2.4 sequence) in west-central Bohemia (Fig. 2). The correlation is based on outcrop and subsurface data. Local names at the top indicate locations of key outcrop sections used for correlation by Uličný (1992); for their location see Fig. 2. No vertical scale. **(b)** Simplified parasequence correlation in a cross section, perpendicular to the palaeovalley axis, of the Pecínov region, constructed from borehole data. Note the onlap of parasequences on the palaeovalley sides. The core descriptions did not allow the determination of boundaries between parasequences 1 and 2, and, to the southeast of borehole RY 90, between parasequences 4 and 5. Location and lateral extent of the Pecínov quarry are indicated

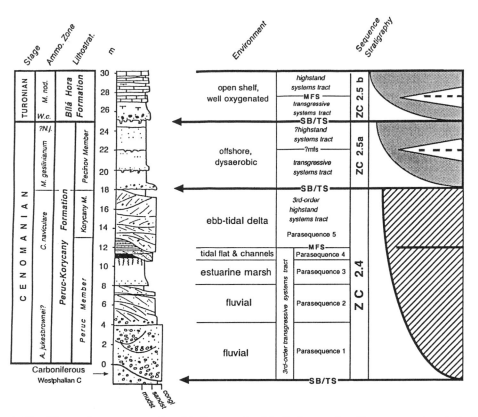

Fig. 4. Bio-, litho- and sequence- stratigraphic framework of the Pecínov section. 'Sequence stratigraphy' column shows the 3rd-order sequences, correlated to the scheme of Haq *et al.* (1988) and Juignet & Breton (1992), and the parasequences within the ZC 2.4 sequence. Further explanation in text.

Middle Cenomanian age, based on palynologic correlation (Svobodová 1995). During the late middle Cenomanian (*Acanthoceras jukesbrownei* Zone), a rapid sea-level rise, coeval with the ZC 2.4 transgression of Haq *et al.* (1988), caused the flooding of most of the basin. In west-central Bohemia the deposits of the Peruc and Korycany Members correspond to the ZC 2.4 sequence of Haq *et al.* (1988) (cross-hatched in Fig. 4). After another major sea-level rise (*Metoicoceras geslinianum* Zone; beginning of the ZC 2.5 cycle), dark offshore shales of the Pecínov Member were deposited over most of the basin, and the facies of the Peruc and Korycany Members were confined to the basin margin. The Peruc–Korycany Formation is overlain by hemipelagic marls and micritic limestones of the Bílá Hora Formation of early Turonian age.

Sequence stratigraphic and biostratigraphic correlations of the upper Cenomanian and Turonian strata in Bohemia to the record of German Mesozoic basins (Hilbrecht 1986), the Anglo-Paris Basin (Juignet & Breton 1992), northern Africa (Hardenbol *et al.* 1991) and the Western Interior of North America (e.g. Mellere 1994; Elder *et al.* 1994) show remarkable coincidences in the record of 3rd-order sea-level change. Apart from these data, the comparison of the Bohemian record with the 3rd-order sea-level cycles of the Haq *et al.* (1988) chart, as updated by Juignet & Breton (1992), also indicates that eustasy was the most important control on the relative sea-level change in this part of the Bohemian Cretaceous basin.

This paper focuses on the high resolution sequence stratigraphy of the deposits of the Peruc and Korycany Members, which correspond to the ZC 2.4 3rd-order sequence of Haq *et al.* (1988).

Methods and terminology

This paper illustrates the important characteristics of parasequences within a 3rd-order sequence, as exposed in a single quarry. How-

Table 1. *List of abbreviations used for lithologies and sedimentary geometries*

Parasequence	Sedimentary geometry	Lithofacies
1	CH: channels of different size, arranged in multi-storey pattern and/or laterally shifting	Gp planar cross-bedded conglomerate Gm massively bedded conglomerate Gt trough cross-bedded conglomerate Gh horizontally bedded conglomerate Sp planar cross-bedded sandstone
	channel infill style: massive or lateral GB, SB: gravel or sandy bedforms	Sm massively bedded sandstone St trough cross-bedded sandstone Ml laminated mudstone Mm massive mudstone
2	LA–DA: laterally and downstream accreted macrofoms	Gp planar cross-bedded conglomerate Sp planar cross-bedded sandstone St trough cross-bedded sandstone
	G–SB: gravel to sandy bedforms	Sr sandstone with ripples Ml laminated mudstone
	CH: channels	Mm massive mudstone Mr mudstone with rootlets H heterolithic, interlaminated sandstone and mudstone
3	CH: channels	Mc coaly mudstone Mr mudstone with rootles Mm massive mudstone
	channel infill style: LA: lateral accretion	H heterolithic lithologies; interlaminated sandstone and mudstone
4	CH: channels	Stb cross-bedded sandstone with tidal bundles Msl mudstone with sandstone laminae Sp planar cross-bedded sandstone
	channel infill style: LA: lateral accretion SB: sandy bedforms	Gp planar cross-bedded conglomerate
5	CH: channels	Sp planar cross-bedded sandstone St trough cross-bedded sandstone Stb cross-bedded sandstone with tidal bundles
	channel infill style: LA: lateral accretion DA: down stream accretion	H heterolithic, interlaminated sandstone and mudstone Ml laminated mudstone

ever, some of the sequence stratigraphic conclusions presented would be impossible to prove based only on evidence from one locality. The observations from the Pecínov quarry were compared with results of field studies in west-central Bohemia (Uličný 1992), supplemented with the re-evaluation of abundant borehole data, used for constructing the pseudo-3D images of the pre-Cenomanian topography (Fig. 2a) of west-central Bohemia.

A standard lithofacies study, based on vertical measured sections, was insufficient because the deposits show a high degree of vertical and lateral variability. Therefore, much of the high resolution sequence stratigraphic analysis is based on evaluation of photomosaics of laterally extensive exposures, which were supplemented by local measured sections.

Descriptions of the sedimentary units in this paper have two main aspects. (1) Lithofacies: a standard description of clastic rocks, using the classification scheme of Miall (1977, 1990), adapted for shallow-marine deposits. (2) Sedimentary geometries: description of the geometry of sedimentary bodies follows the methods of architectural analysis (Miall 1985, 1991; Muñoz *et al.* 1992), used for characterizing the depositional style of a particular body of sedimentary rocks, as well as for determining changes in accommodation space through time (e.g. Wright

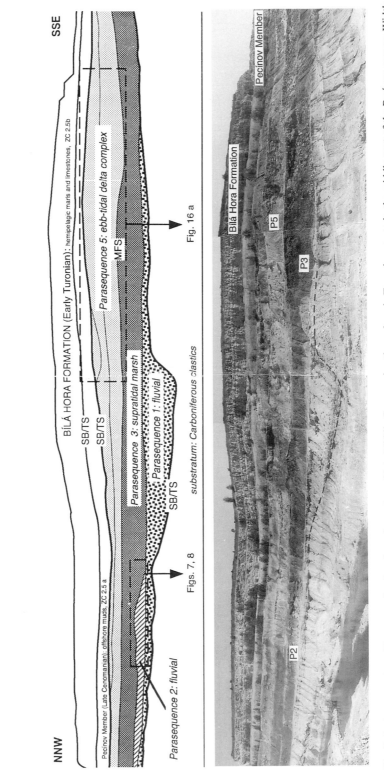

Fig. 5. A photomosaic and interpretive drawing of a section of late Middle Cenomanian through early Turonian strata in the middle part of the Pecínov quarry. Width of view is *c.* 280 m. Dashed frames in the drawing show locations of sections shown in Figs 7, 8 and 16.

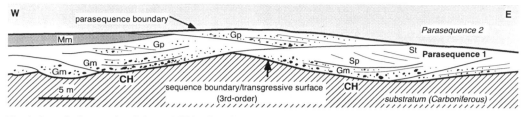

Fig. 6. A typical example of channel-fill bodies of parasequence 1. Note the erosional remnant of an abandoned channel mud plug (Mm) on the upper left hand side. For explanation of lithologic symbols in this and other figures, see Table 1.

& Marriott 1993). We use Miall's symbols for the basic architectural elements (Table 1) when describing large-scale sedimentary geometries within the depositional units of fluvial as well as of shallow-marine origin.

As a consequence of confusion existing in recent literature concerning the relationships between the terms 'sequence' and 'parasequence', especially in the case of high-order (4th-, 5th-order) sequence stratigraphic units (cf. Posamentier & James 1993), we adhere to the concept of a parasequence as expressed by Van Wagoner *et al.* (1988) 'a relatively conformable succession of beds or bedsets bounded by marine flooding surfaces or their correlative conformities', and in agreement with Posamentier & James (1993), we use the term 'parasequence' as a descriptive term, unrelated to the scale of the depositional unit or to the frequency of sea-level change.

Depositional units and bounding surfaces

Parasequence 1

Description. Parasequence 1, up to 6 m thick, rests unconformably on Carboniferous conglomerates and arkoses. The basal surface is erosional and irregular, and locally has up to 5 m of relief. Mainly coarse-grained sediments infill the lowermost parts of the palaeovalley. Parasequence 1 is absent on local topographic highs. The upper bounding surface (base of parasequence 2) is more or less planar, but is slightly undulating in places.

Channels 1–10 m wide and 0.5–1 m deep are the most abundant architectural element of parasequence 1 (Figs 5 and 6). The channels cross-cut one another laterally as well as vertically, in a multistorey pattern. The channel fills are composed of poorly sorted conglomerates with cobbles, coarse- to medium-grained sandstones and mudstones. They are either massively bedded or show lateral accretion architecture. The conglomerates are massive, planar and trough cross-bedded. Overall, the

maximum clast size in the conglomerates of parasequence 1 diminishes upward from 10, exceptionally 20 cm at the bottom, to less than 5 cm at the top. Rarely, massive to horizontally bedded, flat-based conglomerate bodies occur, which could represent gravel bars, formed during peak flow (Miall 1985). Other lithologies include coarse to medium-grained, massive, planar or trough cross-bedded sandstones. They form thin lenses 20–50 cm thick and up to 5 m in lateral extent. Only a few erosional remnants of plant-rich mudstone channel plugs, up to 0.5 m thick, were observed.

Interpretation. Parasequence 1 was deposited in a shallow, gravelly braided river. The lateral shifting of channels in various directions and the size of the channels reflect rapid channel migration due to fluctuations in discharge. The lack of thick planar and trough cross-sets suggests that the channels were relatively shallow. The low angle, planar cross-bedded sandstones and conglomerates could be remnants of linguoid bars modified during decreasing discharge (Blodgett & Stanley 1980). The presence of poorly sorted, matrix supported conglomerates and internal channelling suggests high discharge variations or perhaps an ephemeral river. Slow subsidence and high frequency channel migration permitted only limited accumulation of mudstone in abandoned channels.

The high lateral and vertical connectedness of channels, together with the overall thickness (maximum 6 m) of parasequence 1 suggests that little accommodation space was available. The deposition of parasequence 1 was enabled by a gradual rise in sea-level, which caused a landward shift of the shoreline and backfilling of the river valley.

Parasequence 2

Description. Parasequence 2, up to 5 m thick, rests on a nearly flat, slightly scoured, upper surface of parasequence 1 (Figs 7 and 8). Locally, it sits directly on the Carboniferous

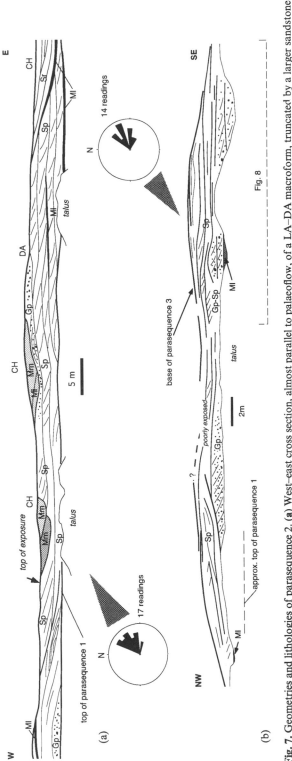
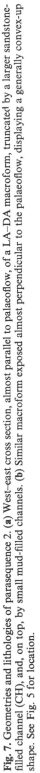

Fig. 7. Geometries and lithologies of parasequence 2. **(a)** West–east cross section, almost parallel to palaeoflow, of a LA–DA macroform, truncated by a larger sandstone-filled channel (CH), and, on top, by small mud-filled channels. **(b)** Similar macroform exposed almost perpendicular to the palaeoflow, displaying a generally convex-up shape. See Fig. 5 for location.

Fig. 8. Deposits of parasequences 1 and 2 in the location shown in Fig. 7b. The arrows in the middle show the approximate position of the base of parasequence 2, hidden beneath the talus. Note the difference in depositional architecture between the two parasequences. The convex-up macroform representing parasequence 2 is conformably overlain by dark muddy sandstones and mudstones of parasequence 3. The person is 1.9 m tall.

Fig. 9. A laterally accreted part of a convex-up (LA–DA) macroform of parasequence 2, directly overlying a small topographic high of basement rock. The origin of the coarse-grain size (coarse conglomerate) is attributed to erosion and redeposition of Carboniferous conglomerates. The basal surface of the parasequence is penetrated by *Arenicolites* burrows (arrows).

Fig. 10. Alternating mud-rich and sand-rich laminasets in the cross-bedded sandstones of parasequence 2, suggesting a tidal influence in the fluvial environment.

bedrock (Fig. 9). The upper bounding surface of parasequence 2 is mainly horizontal, but it can be locally scoured by channels of younger parasequences, or may locally have a convex-up shape. Parasequence 2, like parasequence 1, infills topographic lows and is absent on highs.

Dominant lithologies of parasequence 2 vary from fine-grained conglomerates and pebbly sandstones to fine-grained sandstones. The sandstones are massive, planar and trough cross-bedded with abundant mud drapes on foresets and reactivation surfaces. Laminated mudstone lenses or massive mudstone with pedogenic features, rootlets, burrowing, quartz pebbles and slickenslides, occur in the upper part of the parasequence. Heterolithic lithologies include sigmoidal-shaped bundles of cross-strata which show quasi-rhythmic alternation of mud-rich and sand-rich laminae (Fig. 10). *Thalassinoides* and *Arenicolites* burrows occur locally in the sandstones, and were rarely found penetrating the weathered Carboniferous substratum where it is overlain by parasequence 2 (Fig. 9).

Dominant architectural elements of parasequence 2 are macroforms, 3–4 m thick, which show lateral as well as downstream accretion. These macroforms show similar structures to complex sand flats described by Cant & Walker (1978) from the South Saskatchewan River, or to sand flats documented by Allen (1983). They represent a transition between downstream and lateral-accreting macroforms (cf. Miall 1990, 1994). The planar cross-beds give northeasterly directed palaeocurrents (Fig. 7), whereas the accretion surfaces are orientated in a wide range

of directions from the northwest to the southeast. In cross-sections transverse to palaeoflow, the continuity of convex-up accretion surfaces indicates almost contemporaneous lateral and downstream accretion. Larger (0.7 m thick) sets of planar cross-bedded sandstone may represent linguoid bars, which could have acted as nuclei during the formation of macroforms. Shallow channels infilled with mudstone occur in some macroforms. Almost all the sandstone macroforms fine-upward and are capped by mudstones. The mudstones are massive, with pedogenic features including rootlets and slickensides or with heterolithic laminae.

Interpretation. Parasequence 2 represents the deposits of a large, low-sinuosity river. Dominant elements of fluvial architecture are combined lateral and downstream accretion bars which migrated across the flat bottom of a relatively wide master channel that was hundreds of metres wide. The minor channels at the top of the macroforms indicate that the macroforms were only rarely emergent. The minor channels formed during lower water stage and were passively infilled with mud. The size and convex-up shape of macroforms, together with only minor channel incision suggest generally minor syndepositional erosion reflecting relatively deep channels and regular discharge. The combination of quasi-rhythmic stratification, sigmoidal shape of some cross-bed sets, and trace fossil assemblage suggests the influence of tides (cf. Shanley *et al.* 1992).

Parasequence 2 is composed of finer-grained

Fig. 11. Black, marsh mudstones of parasequence 3, resting directly on a local topographic high of white Carboniferous arkoses and conglomerates. Arrows mark the basal surfaces of parasequences 3 and 4. The strata underlying parasequence 3 are penetrated by a rootlet zone up to 0.8 m thick. The hammer (circled) is 0.4 m long.

Fig. 12. Marginal part of a tidal creek channel, incised in the marsh muds of parasequence 3. The upper part of the heterolithic, lateral accretion channel fill and the mudstone are penetrated by rootlets. Parasequence 3 is overlain by tidal laminites of parasequence 4 (upper half of figure), cf. Fig. 13.

sediments of a large, relatively deep, low-sinuosity river with probable tidal influence, and indicates another step in the flooding of the palaeovalley due to sea-level rise.

Parasequence 3

Description. Parasequence 3, up to 3.5 m thick, rests on a sharp flooding surface which caps the fluvial deposits of parasequence 2. The basal surface is uneven, reflecting local relief on the top of the underlying deposits. In the southern-most part of the quarry, parasequence 3 directly overlies a local basement high. No significant channel incision was observed at the base, and the upper surface is normally flat, but is locally incised by channels belonging to Parasequence 4, which can locally reduce the thickness of parasequence 3 to less than 1 m.

The bulk of the deposits of parasequence 3 consists of dark grey to black mudstones, rich in pyrite concretions, and is capped by a layer of

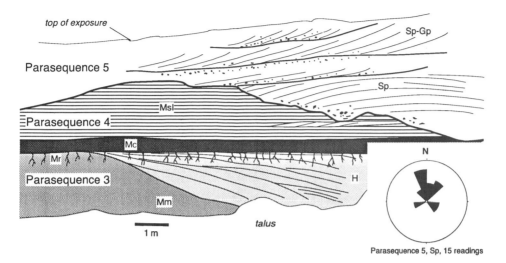

Parasequence 5, Sp, 15 readings

Fig. 13. Marsh mudstones and tidal-creek channel fill sediments of parasequence 3 (see Fig. 12), overlain by intertidal laminites of parasequence 4 (Msl), preserved as an erosional remnant below parasequence 5.

carbonaceous mudstone (5.5 weight % TOC), underlain by a rootlet horizon (Figs 11–13). Dark grey sandstones with muddy interbeds usually occur at the base of the parasequence; locally, fine-grained sand streaks and lenses occur within the mudstones.

Channels, up to 30 m wide and 4 m deep, are encased in dark mudstones at different elevations within the parasequence. The channels are filled by sandstone-dominated heterolithic lateral accretion ('inclined heterolithic stratification' of Thomas *et al.* 1987). Dark, silty laminasets in the channel fills show rhythmic lamination. Small synsedimentary folds indicate soft-sediment deformation due to loading. The channel-fill lithologies are often penetrated from the top by a rootlet zone up to 1 m thick, and overlain by the carbonaceous mudstone layer. In many places, this layer contains compacted fragments of leaves of *Nehvizdya* and *Frenelopsis*, plants attributed by some authors to coastal swamp environments. Svobodová (1995) described marine plankton (dinocysts, acritarchs) in parasequence 3. Pyritized bivalve moulds occur in the mudstones.

Interpretation. Parasequence 3 is interpreted as a shallowing-upward succession of high intertidal mudflat to supratidal marsh deposits. The base of parasequence 3 is a flooding surface created by a relative sea-level rise, which shifted the overall depositional environment from fluvial to high intertidal and supratidal. The flooding surface is marked by relatively abrupt, low

energy drowning of fluvial sedimentary bodies of parasequence 2 because relatively well preserved fluvial macroforms are 'blanketed' by overlying marsh mudstones of parasequence 3 (Figs 7 and 8). We interpret the channel-fill bodies to have been deposited in slightly meandering tidal creeks, incised in marsh muds. The marshes fringed an estuarine complex established in the palaeovalley. During gradual filling of the accommodation space, the topography of the palaeovalley was largely filled by parasequence 3.

Parasequence 4

Description. Parasequence 4, up to 4 m thick, is separated from the underlying deposits by a sharp, erosional surface, with up to 3 m of relief (Fig. 14). The deposits of parasequence 4 are commonly preserved as erosional remnants, due to deep erosion at the base of the overlying parasequence 5, and are totally cut out along approximately 60% of the quarry exposure.

The dominant lithology of the channel fill bodies is cross-bedded, coarse- to medium-grained sandstone with mud-draped foresets and mud couplets forming classical tidal bundles (Fig. 15), organized in tidal bundle sequences (Visser 1980; Nio & Yang 1991). The dominant current direction was towards the northeast. In dip-parallel section, the cross-bedded sandstones show downstream accretion which arose from the migration of superimposed sandwave trains on the channel bottom (Fig. 14). In quarry faces

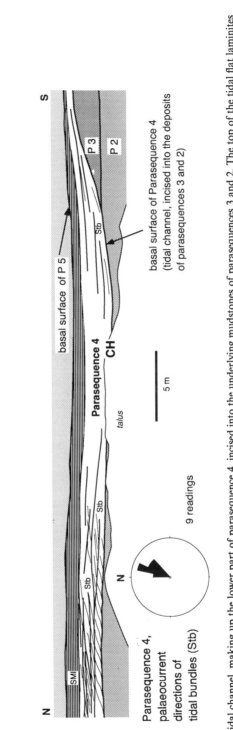

Fig. 14. A tidal channel, making up the lower part of parasequence 4, incised into the underlying mudstones of parasequences 3 and 2. The top of the tidal flat laminites (SM1) is scoured by erosion at the basal surface of parasequence 5.

Fig. 15. The most completely preserved record of parasequence 4, in the left part of the exposure shown in Fig. 14. Note the tidal bundle sequences at the base of the channel (lower arrow).

perpendicular to the channel axis, the lateral accretion geometry of the channel fill is apparent.

The cross-bedded sandstones fine-upwards into a succession of laminites which show continuum of flaser to lenticular-bedded, interlaminated sandstones and mudstones.

Interpretation. Parasequence 4 is interpreted as a fining-upward succession of shallow subtidal to intertidal deposits (cf. Klein 1971; Van Wagoner et al. 1990). Their occurrence above the supratidal marsh mudstones of Parasequence 3 indicates a flooding event and another landward translation of facies belts. The mud couplets and palaeocurrent directions show that the channels at the base of Parasequence 4 were dominated by ebb-directed subtidal currents. The tidal bundle sequences at the base of the channels recorded neap-spring tide cyclicity (Nio & Yang 1991). Lateral-accretion architecture due to meandering of tidal channels is typical of most lower tidal flat deposits (Weimer et al. 1981). The laminites of the upper part of Parasequence 4 are interpreted as intertidal deposits of a mid-tidal flat environment (cf. Terwindt 1988).

Parasequence 5

Description. Parasequence 5, based by a prominent erosional surface (Figs 15 and 16), has a maximum thickness of 11 m. The basal surface of parasequence 5 has a relief of up to 5 m due to deep incision of channels into the underlying deposits of parasequences 4 and even 3. Parasequence 5 is capped by a relatively flat basal surface of the Pecínov Member. The importance of the bounding surfaces of this parasequence for 3rd-order sequence stratigraphy is discussed below.

Parasequence 5 is dominated by yellow, coarse-grained sandstone and locally fine-grained conglomerates. Stratification is dominated by planar and trough cross-bedding, locally with mud drapes on foresets. In a few exposures, thin successions of tidal bundles occur. Mudstone and fusain intraclasts occur near the channel bottoms. *Ophiomorpha* burrows are abundant in the sandstones.

The lower part of the parasequence is dominated by channels with lateral accretion infills. The middle part consists of lateral accretion sandstone lithosomes, showing downstream accretion in dip-parallel section (Fig. 16). Palaeocurrents are dominated by basinward (ebb) flow, with maxima in north-northwestern and northeastern directions.

In the upper portion of parasequence 5, mudstone and heterolithic channel fills occur, with coarse-grained sandstone interbeds in some of the channels. Locally, bivalve shell lags of *Protocardia hillana* (Sowerby), preserved as moulds and casts, occur near the channel bottom (Fig. 17). The dark, organic-rich mudstones range from massive to faintly laminated, with lenticular bedding and abundant horizontal *Thalassinoides* networks. Gently dipping lateral accretion surfaces were locally observed in the laminated mudstones. The mudstones at the top of the parasequence are rich in plant remains and usually densely penetrated by coalified rootlets.

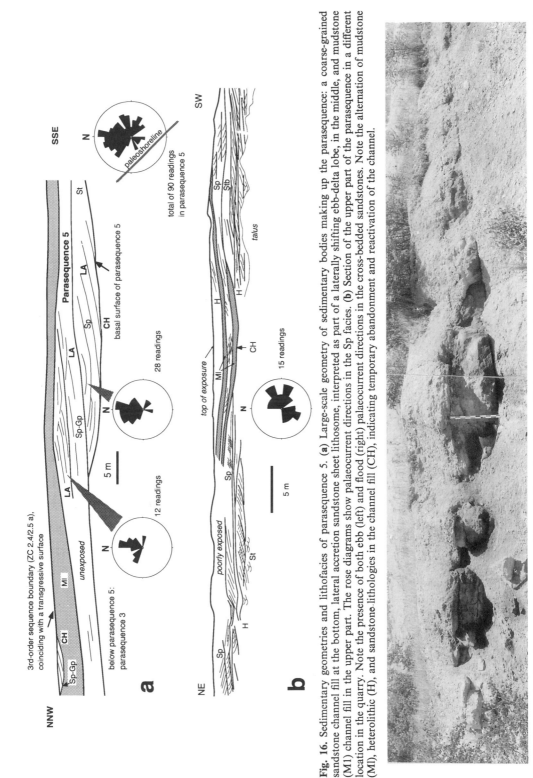

Fig. 16. Sedimentary geometries and lithofacies of parasequence 5. (**a**) Large-scale geometry of sedimentary bodies making up the parasequence: a coarse-grained sandstone channel fill at the bottom, lateral accretion sandstone sheet lithosome, interpreted as part of a laterally shifting ebb-delta lobe, in the middle, and mudstone (MI) channel fill in the upper part. The rose diagrams show palaeocurrent directions in the Sp facies. (**b**) Section of the upper part of the parasequence in a different location in the quarry. Note the presence of both ebb (left) and flood (right) palaeocurrent directions in the cross-bedded sandstones. Note the alternation of mudstone (MI), heterolithic (H), and sandstone lithologies in the channel fill (CH), indicating temporary abandonment and reactivation of the channel.

Fig. 17. Coarse-grained sandstone to conglomerate filling a shallow channel encased in intertidal mudstones in the uppermost part of parasequence 5. The position of this channel, directly overlain by dark, open-shelf mudstones of the Pecínov Member, is shown in Fig. 16a, on the upper left. Rule is 1 m long.

Interpretation. The deposits of parasequence 5 are interpreted as a progradational tide-dominated coastal succession, beginning with subtidal deposits of estuarine ebb-tidal delta, and shallowing upward to muddy and heterolithic intertidal deposits filling abandoned channels. The base of this parasequence is defined as a flooding surface based on the presence of subtidal depositional features appearing above intertidal laminites of Parasequence 4. The interpretation of the estuarine ebb-tidal delta environment is based on comparison to the Eastern Scheldt estuary (Yang & Nio 1989), as well as on other descriptions of modern ebb deltas (Hayes 1980; Imperato *et al.* 1988).

The major accretionary surfaces within the sandstone facies are inclined to the northeast, reflecting the seaward slope of the ebb delta lobe. They also show lateral accretion reflecting lateral migration of the lobe (Fig. 16). Active lobes were covered with mostly ebb-orientated dunes. Because the tidal currents on the ebb delta are less confined than in the tidal channels further landward, mudstone interbeds and mud drapes on cross-bed foresets are less abundant than in Parasequence 4. The coarse grain size of sandstone indicates that the sandstone facies of Parasequence 5 were deposited relatively near the estuarine mouth. The presence of a shore-parallel palaeocurrent component indicates that the sandstone bodies were deposited seaward of the shoreline and influenced by wave-generated longshore currents. This is analogous to the Eastern Scheldt mesotidal estuary, where tidal-current velocity patterns for inshore tidal channels show a confined, bidirectional pattern, at a high angle to the shoreline, whereas seaward, on the ebb-delta lobes, tidal elipses tend to be elongated slightly alongshore (Yang & Nio 1989).

The occurrence of flood-orientated cross-beds reflects migration of flood-orientated dunes on ebb-delta shoals (Hayes 1980), or in marginal flood channels (Imperato *et al.* 1988). During shoreline progradation and shallowing, tidal channels incised into the ebb-delta lobes. The intertidal and supratidal mudstone and heterolithic lithologies filled some of the channels after their abandonment.

Discussion

Parasequences in a fluvial setting

Recent studies show that sequence boundaries and systems tracts can be recognized in fluvial successions (Wright & Marriott 1993; Shanley & McCabe 1993). However, recognition of para-sequences in a fluvial setting is usually considered difficult, if not impossible. Two lines of evidence have led to the conclusion that the fluvial depositional units in the Pecínov quarry represent parasequences.

Firstly the shift in fluvial style between para-sequences 1 and 2 is interpreted to be due to flattening of slope gradient, broadening of the zone of fluvial deposition within the valley, and possible onset of tide-influenced conditions in the fluvial system. This, together with the observed onlap of fluvial deposits of Parasequence 2 onto local basement highs, indicates a base level rise, caused by a marine flooding event.

Secondly regional sequence stratigraphic correlation indicates that fluvial parasequences in Pecínov have downdip equivalents in coastal parasequences (marsh, overlain by tidal flat deposits), at the base of the Cretaceous succession in the Kralupy–Nelahozeves region (Fig. 3; Uličný 1992).

The surface dividing the two fluvial units is treated as a parasequence boundary, as defined by Van Wagoner *et al.* (1990), that is, a correlative conformity to the marine flooding surface occurring farther basinward.

Erosional bounding surfaces: parasequence v. high frequency sequence boundaries

Each of the parasequences present in the Peruc–Korycany Formation at Pecínov is bounded below and above by more or less pronounced erosional surfaces. Judged from only one outcrop, and without regard to autocyclic processes acting within particular depositional environments, such surfaces could be misinterpreted as sequence boundaries of high frequency sequences, i.e. as subaerial exposure surfaces formed in response to base level falls. In the current state of investigation, several lines of evidence lead us to interpret these erosional surfaces as parasequence boundaries, and the units bounded by them as parasequences.

1. There is no basinward shift in facies on any of the bounding surfaces. The facies overlying each particular surface show features of more seaward depositional environment than the underlying deposits. Accordingly, the boundary surfaces are marine flooding surfaces or their correlative equivalents.

2. The units separated by the bounding surfaces show either no marked internal changes in palaeobathymetry in vertical section (para-sequences 1 & 2), or they display simple

shallowing-upward patterns (parasequences 3 through 5). High frequency sequences would show more complex internal organization into systems tracts (cf. Mitchum & Van Wagoner 1991; Posamentier & James 1993), observed neither at Pecínov, nor elsewhere in the study area.

Downdip correlation shows that the parasequence bounding surfaces in Pecínov correlate with marine flooding surfaces (Fig. 3). Moreover, in a basinward direction, no lowstand deposits or any basinward shift in facies were found on surfaces correlative to the bounding surfaces exposed in Pecínov.

Causes of erosion between and within parasequences

In order to understand the causes of channel incision both within and between parasequences, it is necessary to consider the nature of the depositional and erosional processes inherent to particular depositional environments. The bounding surfaces of the parasequences observed are not **conformities** in the strict sense, as a rigorous application of the definition of parasequence (Van Wagoner et al. 1990) would require. Furthermore, the successions of beds within the parasequences in the Pecínov quarry are indeed only **relatively conformable**, showing many internal erosional surfaces and lateral and vertical facies changes.

The parasequence boundaries in fluvial settings show channel incision due to autogenic fluvial processes. The least amount of erosion was observed at the base of the marsh deposits, which probably fringed the estuarine funnel and therefore were situated outside the reach of major current activity.

The parasequences formed in intertidal to subtidal environments show that significant erosion occurred at the flooding surfaces by incision of subtidal channels shifted landward by a relative sea-level rise (cf. Oertel et al. 1991). The channelled flooding surfaces at the base of parasequences 4 and 5 are tidal ravinement surfaces as defined by Allen & Posamentier (1993).

In contrast, the erosional features **within** parasequences 4 and, especially, 5 (Fig. 16b) are attributed to tidal channel migration (Yang & Nio 1989).

3rd-order v. 4th-order sequence stratigraphic signature

Parasequences 1 through 4 are arranged in a clear retrogradational stacking pattern, and

interpreted as a transgressive systems tract of a 3rd-order sequence, correlated to the ZC 2.4 cycle of Haq et al. (1988). The base of parasequence 1 is interpreted as a transgressive surface because it correlates to the base of marsh deposits in the Kralupy region approximately 25 km downdip of Pecínov (Uličný 1992). No lowstand deposits are present beneath the transgressive systems tract at Pecínov, because of the ramp physiography of the basin margin and the updip position of the Pecínov area.

The base of parasequence 5, marked by the appearance of the deepest water facies in the described section, is interpreted as a maximum flooding surface within the Peruc–Korycany sequence. Regional correlation shows onlap of parasequence 5 onto the top of the Unhošt'–Tursko and Jedomělice palaeohighs (Fig. 3b). Parasequence 5, therefore, represents a 3rd-order highstand systems tract, as suggested also by downdip correlation (Fig. 3a). A 3rd-order sequence boundary formed on top of parasequence 5, coinciding with a subsequent transgressive surface at the base of the overlying Pecínov Member (Uličný et al. 1993). This sequence boundary is difficult to prove in the Pecínov quarry alone, and its interpretation is supported by regional data. Occurrence of glauconitic, cross-bedded sandstone between this surface and the transgressive Pecínov mudstone was reported by Uličný (1992) from the central part of the basin c. 50 km downdip of Pecínov and tentatively interpreted as a detached lowstand shoreline deposit. Voigt et al. (1992), on the basis of data from the Saxonian part of the Bohemian Cretaceous Basin, interpreted a relative sea-level fall of c. 20–30 m immediately preceding the *Metoicoceras geslinianum* Zone, i.e. the base of the Pecínov Member.

The shallowing-upward trend within parasequence 5 at Pecínov is the most marked of all the parasequences observed. Fossil plant assemblages in the upper part of this parasequence show a much more xerophytic character than those of the underlying parasequences, which is interpreted as a consequence of marked shoreline progradation and lowering of the groundwater table (J. Kvaček, pers. comm. 1994).

It might be tempting to interpret the channels in the upper part of parasequence 5, filled with intertidal deposits, as incised during a relative sea-level fall and infilled by lowstand deposits. However, channelling of varying vertical extent occurs throughout parasequence 5, and the intertidal heterolithic lithologies and mudstones formed clearly due to channel abandonment. Cross-bedded sandstones, commonly overlying

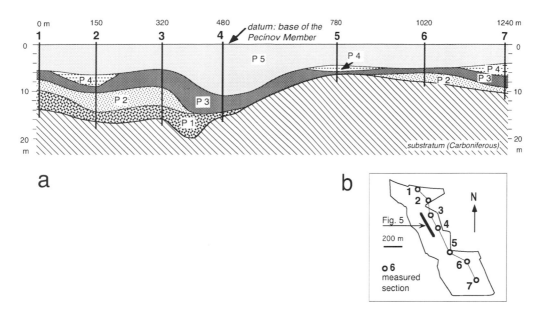

Fig. 18. Correlation of parasequences between vertical measured sections (1–7), situated in a 1.24 km profile across the Pecínov quarry. The correlation was made using outcrop photomosaics. Only the measured section 2 shows the complete number of parasequences. Note large vertical exaggeration. (**b**) Sketch map of the Pecínov quarry, showing the locations of the measured sections 1–7 and of the photomosaic shown in Fig. 5.

the mudstones and laminites in the channels, show that some of the channels were periodically reactivated (Fig. 16b). The channelling within the upper part of parasequence 5 is typical of the depositional environment, which gradually shallowed from subtidal to supratidal conditions.

No fluvial incision was observed at the top of parasequence 5, which may be due to the flat, gently dipping topography of this surface, and due to lack of any physiographic breaks at the basin margin (cf. Posamentier & Allen 1993). Another explanation could be the removal and/or reworking of any lowstand deposits by the subsequent transgression (cf. Van Wagoner et al. 1990).

Frequency and amplitude of the 4th-order sea-level fluctuations

The parasequences in the Peruc and Korycany Members in the Pecínov quarry are interpreted to have formed in response to 4th-order sea-level fluctuations. A comparison of the biostratigraphic constraints available from the study region to chronostratigraphic dating of Juignet & Breton (1992) and Gale (1990) indicates that the transgressive and highstand systems tracts of the ZC 2.4 sequence represent a time span of not more than 0.5 Ma. This gives an average figure

of 100 Ka for the frequency of the 4th-order increments of sea-level rise which formed the five parasequences at Pecínov. This frequency approximates to the duration of orbital eccentricity cycles, raising the possibility of orbitally forced, climate-driven eustatic changes as the main cause of high frequency relative sea-level changes recorded in the Cenomanian deposits at Pecínov. The 4th-order fluctuations in sea-level, as well as the inconspicuous sea-level fall before the *M. geslinianum* Zone, were superimposed on a longer-term trend of sea-level rise. This trend culminated in the highest Phanerozoic sea-level during the early Turonian (cf. Haq et al. 1988).

The palaeobathymetry of particular parasequences in Pecínov allows an estimate of the magnitude of the 4th-order sea-level changes responsible for their formation. The shift between fluvial parasequences 1 and 2 reflects only a moderate base level rise, as confirmed by downdip correlation: a shift from marsh to intertidal facies in the Kralupy region (Uličný 1992). The shifts in facies between parasequences 3, 4 and 5 (generally from supratidal to intertidal to subtidal) also suggest that the amplitude of relative sea-level change probably did not significantly exceed the magnitude of the regional tidal range.

The comparison of the palaeogeography and facies interpretations with recent coasts (cf. Hayes 1980) indicates that the tide-dominated, estuarine coast of west-central Bohemia experienced a high mesotidal, or, even more likely, macrotidal regime during the Cenomanian. Therefore, we estimate the tidal range as greater than 5 m, and the amplitude of the 4th-order relative sea-level change as not more than 10 m.

Preservation of parasequences in fluvial to shallow-marine settings

The relationships of parasequences and their bounding surfaces identified at Pecínov illustrate the complexity of estuarine valley-fill successions. As discussed above, the incision of channels at the base of some parasequences has erased evidence of the previous existence of one or even two underlying parasequences. The depth and lateral extent of erosion, localized at the base of subtidal channels, show the significant potential for loss of stratigraphic record in tide-dominated estuaries during transgression.

The exposures at Pecínov demonstrate that of many closely spaced (150–300 m) vertical sections, only a few may record the actual number and stacking pattern of parasequences in the succession (Fig. 18). This should be taken into account in sequence stratigraphic interpretations of estuarine systems based on well-log data. The complicated pattern of preservation of parasequences in the Pecínov quarry shows that analysis of the geometry of sedimentary bodies (or, architectural-element analysis in the sense of Miall 1985) is of critical importance in sequence stratigraphy of fluvial to estuarine settings.

Conclusions

1. The fluvial to estuarine deposits of late middle to late Cenomanian age, exposed in the Pecínov quarry, form a transgressive and highstand systems tract of a 3rd-order sequence. Within this sequence, five parasequences formed in response to eustatically driven, high frequency (4th-order) steps in sea-level rise, with a periodicity of approximately 100 ka. The parasequences, composed successively of fluvial, tide-influenced fluvial, supratidal marsh, tidal flat and estuarine ebb tidal delta deposits, are separated by erosive bounding surfaces.

2. Erosion at parasequence bounding surfaces, resembling features on sequence boundaries, was not related to subaerial exposure caused by a fall in sea-level. The erosional features at parasequence boundaries are interpreted to have been caused by erosional processes inherent in the coastal depositional environments, shifted landward due to stepwise rise in relative sea-level. The magnitude of steps in relative sea-level rise generating the parasequences is estimated at less than 10 m. Because of the low subsidence rate, the accommodation space created was not sufficient to prevent erosion at parasequence boundaries. The deepest and laterally most extensive erosion occurred at the base of subtidal channels, leading to generation of tidal ravinement surfaces.

3. The complex patterns of mutually erosive surfaces between parasequences, as exposed in the Pecínov quarry, show that the analysis of large-scale geometry of sedimentary bodies is essential for correct sequence stratigraphic interpretation in fluvial to estuarine settings.

The authors express their gratitude to the České lupkové závody, a.s., for kind permission to work in the Pecínov quarry; and to their staff members, especially Mr Sochor and Mr Kubec, for the assistance they provided. Petr Vízdal is thanked for creating the pseudo-3D palaeotopography images. We are indebted to Stanislav Čech for discussing biostratigraphic problems, and to Guy Plint for a fruitful discussion in the field as well as for his detailed and stimulating criticisms of the first version of this paper. We thank the reviewers Pat Brenchley and especially David C. Jennette for important suggestions that helped us to improve the paper. Field work was partly supported by the Czech Republic Grant Agency grant No. 205/94/1744 to DU.

References

ALLEN, G. P. & POSAMENTIER, H. W. 1993. Sequence stratigraphy and facies model of an incised valley fill: the Gironde estuary, France. *Journal of Sedimentary Petrology*, **63**, 378–391.

ALLEN, J. R. L. 1983. Studies in fluviatile sedimentation: bars, bar-complexes and sandstone sheets (low-sinuosity braided streams) in the Brownstones (Lower Devonian), Welsh Borders. *Sedimentary Geology*, **33**, 237–293.

BLODGETT, R. H. & STANLEY, K. O. 1980. Stratification, bedforms, and discharge relations of the Platte braided river system, Nebraska. *Journal of Sedimentary Petrology*, **50**, 139–148.

CANT, D. J. & WALKER, R. G. 1978. Fluvial processes and facies sequences in the sandy braided South Saskatchewan River, Canada. *Sedimentology*, **25**, 625–648.

ČECH, S., KLEIN, V., KŘÍŽ, J. & VALEČKA, J. 1980. Revision of the Upper Cretaceous stratigraphy of the Bohemian Cretaceous Basin. *Věstník Ústředního Ústavu geologického*, **55**, 277–296.

DALRYMPLE, R. W. & ZAITLIN, B. A. 1994. High-resolution sequence stratigraphy of a complex, incised valley succession, Cobequid Bay–Salmon River estuary, Bay of Fundy, Canada. *Sedimentology*, **41**, 1069–1091.

ELDER, W. P., GUSTASON, E. R. & SAGEMAN, B. B. 1994. Correlation of basinal carbonate cycles to nearshore parasequences in the Late Cretaceous Greenhorn seaway, Western Interior U.S.A. *Geological Society of America Bulletin*, **106**, 892–902.

GALE, A. S. 1990. A Milankovitch scale for Cenomanian time. *Terra Nova*, **1**, 420–425.

HADLEY, D. F. & ELLIOTT, T. 1993. The sequence-stratigraphic significance of erosive-based shore-face sequences in the Cretaceous Mesaverde Group of northwestern Colorado. *In:* POSAMENTIER, H. W., SUMMERHAYES, C. P., HAQ, B. U. & ALLEN, G. P. (eds) *Sequence Stratigraphy and Facies Associations.* International Association of Sedimentologists, Special Publication, **18**, 521–536.

HAQ, B. U., HARDENBOL, J. & VAIL, P. R. 1988. Mesozoic and Cenozoic chronostratigraphy and cycles of sea-level change. *In:* WILGUS, C. K., HASTINGS, B. B., KENDALL, C. G. St. C., POSAMENTIER, H. W., ROSS, C. A. & VAN WAGONER, J. C. (eds) *Sea-Level Changes: an Integrated Approach.* SEPM Special Publication, **42**, 71–108.

HARDENBOL, J., CARON, M., AMÉDRO, F., DUPUIS, C. & ROBASZYNSKI, F. 1991. The Cenomanian–Turonian boundary in central Tunisia in the context of a sequence-stratigraphic interpretation. *Geologie Alpine, Mém. h.s.*, **17**, 59.

HAYES, M. O. 1980. General morphology and sediment patterns in tidal inlets. *Sedimentary Geology*, **26**, 139–156.

HILBRECHT, H. 1986. On the correlation of the Upper Cenomanian and Lower Turonian of England and Germany (Boreal and N-Tethys). *Newsletters in Stratigraphy*, **15**, 115–138.

IMPERATO, D. P., SEXTON, W. J. & HAYES, M. O. 1988. Stratigraphy and sediment characteristics of a mesotidal ebb-tidal delta, North Edisto Inlet, South Carolina. *Journal of Sedimentary Petrology*, **58**, 950–958.

JELEN, J. & MALECHA, A. 1988. *Zpráva o vývoji cenomanu na jižním Lounsku (II. etapa), s vyhodnocením prognóz jílovců mezi Jimlínem a Líšt'any.* 1–38. Unpublished Report, Czech Geological Survey, Prague. [In Czech].

JUIGNET, P. & BRETON, G. 1992. Mid-Cretaceous sequence stratigraphy and sedimentary cyclicity in the western Paris Basin. *Palaeogeography, Palaeoclimatology, Palaeoecology*, **91**, 197–218.

KLEIN, G. DE V. 1971. A sedimentary model for determining paleotidal range. *Geological Society of America Bulletin*, **82**, 2585–2592.

McCABE, P. J. 1994. The nature and origin of parasequences. *In:* JOHNSON, S. D. (ed.) *High Resolution Sequence Stratigraphy: Innovations and Applications*, University of Liverpool, Abstract Volume, 40.

MATĚJKA, A. 1936. Kapitola o křídě. *In:* Vysvětlivky ke geologické mapě Československé republiky, list Kladno 3952. *Knihovna Statního geologického ústavu Československé republiky*, **17**, 12–21. [In Czech].

MELLERE, D. 1994. Sequential development of an estuarine valley fill: the Twowells Tongue of the Dakota Sandstone, Acoma Basin, New Mexico. *Journal of Sedimentary Research*, **B64**, 500–515.

MIALL, A. D. 1977. A review of the braided-river depositional environment. *Earth Science Reviews*, **13**, 1–62.

—— 1985. Architectural-element analysis: a new method of facies analysis applied to fluvial deposits. *Earth Science Reviews*, **22**, 261–308.

—— 1990. *Principles of Sedimentary Basin Analysis.* Springer-Verlag.

—— 1991. Hierarchies of architectural units in terrigenous clastic rocks, and their relationship to sedimentation rate. *In:* MIALL, A. D. & TYLER, N. (eds) *The Three-Dimensional Facies Architecture of Terrigenous Clastic Sediments and its Implications for Hydrocarbon Discovery and Recovery.* SEPM, Concepts in Sedimentology and Paleontology, **3**, 6–12.

—— 1994. Reconstructing fluvial macroform architecture from two-dimensional outcrops: examples from the Castlegate Sandstone, Book Cliffs, Utah. *Journal of Sedimentary Research*, **B64**, 146–158.

MITCHUM, R. M. & VAN WAGONER, J. C. 1991. High-frequency sequences and their stacking patterns: sequence-stratigraphic evidence of high-frequency eustatic cycles. *Sedimentary Geology*, **70**, 131–160.

MUÑOZ, A., RAMOS, A., SÁNCHEZ-MOYA, Y. & SOPEÑA, A. 1992. Evolving fluvial architecture during a marine transgression: Upper Buntsandstein, Triassic, central Spain. *Sedimentary Geology*, **75**, 257–281.

NIO, S.-D. & YANG, C. S. 1991. Diagnostic attributes of clastic tidal deposits: a review. *In:* SMITH, D. G. ET AL. (eds) *Clastic Tidal Sedimentology.* Canadian Society of Petroleum Geologists Memoir, **16**, 3–28.

OERTEL, G. F., HENRY, V. J. & FOYLE, A. M. 1991. Implications of tide-dominated lagoonal processes on the preservation of buried channels on a sediment-starved continental shelf. *In:* SWIFT, D. J. P., OERTEL, G. F., TILLMAN, R. W. & THORNE, J. A. (eds) *Shelf Sand and Sandstone Bodies: Geometry, Facies and Sequence Stratigraphy.* International Association of Sedimentologists, Special Publication, **14**, 379–393.

POSAMENTIER, H. W. & ALLEN, G. P. 1993. Variability of the sequence stratigraphic model: effects of local basin factors. *Sedimentary Geology*, **86**, 91–109.

—— & JAMES, D. P. 1993. An overview of sequence-stratigraphic concepts: uses and abuses. *In:* POSAMENTIER, H. W., SUMMERHAYES, C. P., HAQ, B. U. & ALLEN, G. P. (eds) *Sequence Stratigraphy and Facies Associations.* International Association of Sedimentologists, Special Publication, **18**, 3–18.

SHANLEY, K. W. & McCABE, P. J. 1993. Alluvial architecture in a sequence stratigraphic framework: a case history from the Upper Cretaceous of southern Utah, USA. *In:* FLINT, S. S. & BRYANT, I. D. (eds) *The Geological Modelling of Hydrocarbon reservoirs and Outcrop Analogues.* International Association of Sedimentologists,

Special Publication, **15**, 21–56.

——, —— & HETTINGER, R. D. 1992. Tidal influence in Cretaceous fluvial strata from Utah, USA: a key to sequence stratigraphic interpretation. *Sedimentology*, **39**, 905–930.

SVOBODOVÁ, M. 1995. Palynology and stratigraphy of the Peruc-Korycany Formation of Pecínov-Babín, Cenomanian, Czechoslovakia (Preliminary report). *Acta Universitatis Carolinae, Geologica*, in press.

SWIFT, D. J. P., PHILLIPS, S. & THORNE, J. A. 1991. Sedimentation on continental margins, V: parasequences. *In:* SWIFT, D. J. P., OERTEL, G. F., TILLMAN, R. W. & THORNE, J. A. (eds) *Shelf Sand and Sandstone Bodies: Geometry, Facies and Sequence Stratigraphy*. International Association of Sedimentologists, Special Publication, **14**, 153–188.

TERWINDT, J. H. J. 1988. Palaeo-tidal reconstructions of inshore tidal depositional environments. *In:* DEBOER, P. L. *ET. AL.* (eds) *Tide-influenced Sedimentary Environments and Facies*. Reidel Publishing Company, 233–263.

THOMAS, R. G., SMITH, D. G., WOOD, J. M., VISSER, J., CALVERLEY-RANGE, E. A. & KOSTER, E. H. 1987. Inclined heterolithic stratification–terminology, description, interpretation and significance. *Sedimentary Geology*, **53**, 123–179.

ULIČNÝ, D. 1992. *Low and high-frequency sea-level change and related events during the Cenomanian and across the Cenomanian–Turonian boundary, Bohemian Cretaceous Basin*. PhD Thesis, Charles University, Prague.

——, HLADÍKOVÁ, J. & HRADECKÁ, L. 1993. Record of sea-level changes, oxygen depletion and the delta^{13}C anomaly across the Cenomanian–Turonian boundary, Bohemian Cretaceous Basin. *Cretaceous Research*, **14**, 211–234.

VAN WAGONER, J. C., MITCHUM, R. M., CAMPION, K. M., RAHMANIAN, V. D. 1990. *Siliciclastic* sequence stratigraphy in well logs, cores, and outcrops. American Association of Petroleum Geologists, Methods in Exploration Series, **7**, 1–55.

——, POSAMENTIER, H. W., MITCHUM, R. M., VAIL, P. R., SARG, J. F., LOUTIT, T. S. & HARDENBOL, J. 1988. An overview of fundamentals of sequence stratigraphy and key definitions. *In:* WILGUS, C. K. HASTINGS, B. B., KENDALL, C. G. St. C., POSAMENTIER, H. W., ROSS, C. A. & VAN WAGONER, J. C. (eds) *Sea-Level Changes – An Integrated Approach*. SEPM Special Publication, **42**, 39–44.

VISSER, M. J. 1980. Neap-spring cycles reflected in Holocene subtidal large-scale bedform deposits: A preliminary note. *Geology*, **8**, 543–546.

VOIGT, T., POHL, T. & TRÖGER, K.-A. 1992. Geological evidence for the sub-Plenus regression in Saxony. *Abstracts of 4th International Cretaceous Symposium, Hamburg*.

WEIMER, R. J., HOWARD, J. D. & LINDSAY, D. R. 1981. Tidal flats and associated tidal channels. *In:* SCHOLLE, P. A. & SPEARING, D. (eds) *Sandstone Depositional Environments*. American Association of Petroleum Geologists, Memoir, **31**, 191–245.

WRIGHT, V. P. & MARRIOTT, S. B. 1993. The sequence stratigraphy of fluvial depositional systems: the role of floodplain sediment storage. *Sedimentary Geology*, **86**, 203–210.

YANG, C.-S. & NIO, S.-D. 1989. An ebb-tide delta depositional model – a comparison between the modern Eastern Scheldt tidal basin (southwest Netherlands) and the Lower Eocene Roda Sandstone in the southern Pyrenees (Spain). *Sedimentary Geology*, **64**, 175–196.

ZIEGLER, P. A. 1987. Late Cretaceous and Cenozoic intra-plate compressional deformations in the Alpine foreland – a geodynamic model. *Tectonophysics*, **137**, 389–420.

Interbasinal correlation of the Cenomanian Stage; testing the lateral continuity of sequence boundaries

DAVID OWEN

Geology Department, Royal School of Mines, Imperial College London, SW7 2BP, UK

Abstract: Sequence stratigraphic analysis of the Aquitaine, Münster and Paris Basins has shown that many of the major discontinuities found can be traced with good biostratigraphic confidence between the basins, and through a variety of sedimentary, tectonic and palaeobiogeographic regimes. Up to five sequence boundaries appear to be synchronous between the basins when account is taken of the biostratigraphic shortcomings. An extra sequence boundary near the Lower–Middle Cenomanian boundary in the Münster Basin may be a tectonic disruption as all evidence for this relative sea-level fall is from the northern, active margin of the basin, where it is associated with mass flow deposits. Notable features are the absence of lowstand deposits in these epieric, hemipelagic settings and the expression of sequence boundaries as hardgrounds. Multiple transgressive surfaces are common in deeper water settings, and surfaces of maximum flooding are very poorly developed. However, deep water deposits do preserve more accurate evidence of sea-level change, due to increased lithological and biostratigraphical completeness and fewer controls on sedimentation.

The appeal of sequence stratigraphy has been its success in placing sedimentary facies in a basin-wide framework of both spatial and temporal distribution. Basinwide stratigraphic discontinuities are used to divide sedimentary packages into time equivalent units, and facies associations with such discontinuities can be predicted, such as the occurrence of immature clastic deposits in the base of slope area above a sequence boundary. The original conceptual models inferred eustasy as the prime control on the development of passive margin sequences and therefore related the development of sequence boundaries and other discontinuities directly to changes in sea-level. The potential to date sequence boundaries, their assumed eustatic origin and the predictable spatial distribution of associated facies ambitiously promised a technique that would allow the correlation of sedimentary packages independently of a detailed biostratigraphic scheme.

Thus Haq *et al.* (1987, 1988) of the Exxon Group produced their Global Sea-level Cycle Chart, a detailed interpretation of sea-level changes during the Mesozoic and Cenozoic in a sequence stratigraphic framework tied to regional biostratigraphic, magnetostratigraphic and geochronologic data. Although widely used (e.g. Plint 1988; Juignet & Breton 1992) this curve has been the subject of derision for many reasons (Hallam 1992*a*; Hancock 1989, 1993). Miall (1992) and others criticize the basic assumption of the Exxon Group that sequence formation is related directly to eustatic sea-level changes and thus globally correlatable. Miall (1992) also points out that the precision of much of the Cycle Chart is implied to be greater than that of any existing chronostratigraphic method, making it impossible to test and unrealistic owing to the large possible errors that can occur when calculating biostratigraphic and absolute ages. Miall notes that the 'critical test of the Exxon Chart is to demonstrate that successions of cycles of precisely similar age do indeed exist in many tectonically independent basins around the world'.

There are other arguments against the concept of sedimentary packages being generated purely by eustatic sea-level changes, e.g. Parkinson & Summerhayes (1985), Watts (1982), Thorne & Watts (1984), and thus it is important to establish whether it is possible to correlate stratigraphically significant surfaces over a substantial area with sufficient accuracy to test the concepts. Many detailed local studies show significant local variations (Hubbard 1988; Plint 1988), which can in some cases be related directly to tectonic activity. Amongst some geologists there has been a complete shift in dogma away from any eustatic influence at all.

Cartwright *et al.* (1993) criticized the low resolution, large-scale seismic origin of sequence stratigraphic concepts which may have led to oversimplifications in the understanding of true stratigraphic relationships when dealing with high resolution outcrops. They question the lateral continuity of sequence boundaries on a regional scale, pointing out that many local

From Howell, J. A. & Aitken, J. F. (eds), *High Resolution Sequence Stratigraphy: Innovations and Applications,*
Geological Society Special Publication No. 104, pp. 269–293.

disconformity surfaces show the diagnostic relationships of onlap, toplap, downlap and erosional truncation used to define sequence boundaries. They postulate that seismically defined sequence boundaries are likely to cross time lines due to possible contemporaneous upslope deposition of, for instance, hemipelagic deposits. Sequence boundaries may thus be composed of groups of surfaces crossing time lines, and classing them as regionally continuous surfaces would violate the principles of super-position. Low resolution seismic data cannot image such departures and the limited three dimensional extent of outcrops may also miss such relationships. Although valid on a seismic scale this argument is less convincing for areas where there is good regional outcrop and a detailed biostratigraphic scheme. Diagnostic lithostratigraphic relationships, such as major facies changes or hardgrounds with distinctive mineralized surfaces or characteristic trace fossil assemblages can also be combined with bio-stratigraphic information to constrain the lateral stratigraphic continuity of a surface.

Aims

Three basin remnants have been examined at outcrop; the Münster Basin, Anglo-Paris Basin and Aquitaine Platform. The three water masses were connected during the Cenomanian, though a variety of tectonic and sedimentary, climatic and oceanographic regimes prevailed. This paper aims to show how detailed lithological logging and sequence stratigraphic interpreta-tion, superimposed on a detailed biostrati-graphic framework can be combined (1) to give a coherent and detailed sequence stratigraphic model for each basin studied; and (2) to compare accurately the synchroneity and lateral extent of major surfaces and sediment packages within and between basins, to derive conclusions about their driving mechanisms.

The study is complicated by the epeiric, continental interior setting of some of the basins (Hallam 1992*b*) and the deep water, hemipelagic nature of much of the sediment studied (Robas-zynski *et al.* 1993, 1995; Juignet & Breton 1992). When dealing with biogenically derived pelagic sediments, sea-level may be secondary to cli-matic or oceanographic factors (changes in the carbonate compensation depth/upwelling/cur-rent patterns) in terms of sedimentary package generation. Sequence stratigraphic models are largely based upon clastic and shelf carbonate sediments, not pelagic sediments and the sig-nificance of laterally persistent hardground surfaces, for instance, Bromley & Gale (1982),

has yet to be integrated fully into sequence stratigraphic theory.

Hancock & Kauffman (1979) suggest that deeper water sediments provide better facies for sea-level assessment. The lack of extreme topo-graphy, more constant subsidence and the increased completeness of sections, remove the problem of a lack of regressive facies encoun-tered in more marginal areas. Also higher stratigraphic resolution is achieved in thicker, more complete and generally more fossiliferous deep water sediments.

The Cenomanian

The Cenomanian Stage has been chosen for this study as it possesses an excellent biostratigraphic timescale based upon ammonites and inocera-mid bivalve species, as well as other short lived faunal events, many of which can be traced over wide areas of Europe. Subdividing by ammo-nites and inoceramids alone it is possible to form up to 12 unequal subdivisions of the Cenoma-nian. The stage's duration has been estimated at 4–5 Ma (Hallam *et al.* 1985) thus giving a potential resolution of 0.3 Ma by simple divi-sion. When this is augmented by other fossil taxa an excellent biostratigraphic scale is achieved.

The current ammonite zonation for southern England was constructed by Wright & Kennedy (1984), also discussed by Hancock *et al.* (1975, 1993). With local modifications this scale is applicable further afield: Germany (Klinger & Wiedmann 1983; Kaplan 1992), Romania (Ion & Szasz 1994), Tunisia (Robaszynski *et al.* 1993), Poland and the Soviet Union (Marci-nowski 1980), (the Northwest European Boreal Realm). This standard European zonation can thus be used with local modifications over much of the area studied. However, the Aquitaine Basin has greater faunal affinities with the Southern European Tethyan Realm and diag-nostic Boreal fauna are rare, causing problems with correlations.

Inoceramid data are also of excellent use thanks to their wide geographic range (Europe and beyond) and general facies independence. The zonations follow the work of Tröger (1981) and Wiedmann *et al.* (1989).

Cenomanian deposits are widespread in Eur-ope and much of the world. A wide range of facies occurs from basinal hemipelagic argillace-ous micrites, i.e. chalks and marls to fluvial and shallow marine deposits of both clastic and carbonate origin. In many parts of the world (e.g. Nigeria and SE India), the first marine sediments of the Mesozoic belong to the *M. inflatum Zone* (U. Albian), thus marking the beginning of the Cenomanian Transgression (Suess 1906; Hancock & Kauffman 1979). The Cenomanian may therefore be considered an anomalous stage in terms of sea-level because of its position early in this great transgression which led to the encroachment of deep water hemipelagics onto the world's continental shelves, likely to be a true large-scale eustatic rise culminating in the Middle Turonian. Hays & Pitman (1983) ascribed this great transgression to an increase in ocean ridge volume due to increased spreading rates. Thus when studying the relative effects of eustatic versus tectonic signatures a somewhat biased result may occur.

Methodology

Recently, Robaszynski et al. (1993) have published a set of criteria for sequence analysis in distal platform and basinal settings; some of their concepts have been adapted here. For each section, the following grounds have been adopted as the best method of classifying such subtle sediments. A large-scale approach has been taken to the classification of sedimentary units between major surfaces to accommodate the large number of sections studied.

Firstly the position and style of discontinuity surfaces are noted:

Erosive surfaces/non deposition/hardgrounds
Biostratigraphic gap/condensation
Facies change–clay minerals increase
Faunal changes/markers–influx/acme

These factors will vary between different types of surface and along a surface depending upon position within the basin morphology.

Sequence boundaries (SB). Erosive surfaces and biostratigraphic gaps will pass into deep basin facies changes (basinward shift in sedimentation, usually clay influx) and a faunal influx, shallow water forms, oysters, etc. A change in ammonite or inoceramid zone is often associated with a sequence boundary hiatus. Hardgrounds can form in pelagic settings where there is no clastic sediment influx.

Transgressive surfaces (T.surf.). Also associated with erosion and biostratigraphic gaps. Characterized by marine onlap and vertical evidence of deepening in logged sections. In marginal areas they are often combined with the sequence boundaries.

Maximum flooding (MFS)–downlap surface. The most subtle and problematic transition to

recognize in deep water sediments. There is often no obvious expression, although they may be associated with increased faunal preservation. After the maximum rate of marine flooding a gradual increase in the detrital clay content associated with increased terrigenous influence may be seen.

Results

Münster Basin (Westphalia, Germany)

Geological setting. The deposits occur as an asymmetric wedge of sediment in an east–west trending trough, bounded to the south by the Hercynian Saurland Massifs, to the east and northeast by the Teutoburger Wald and Egge Wald. The sediments of the southern margin form a sheet inclined from southwest to northeast; consequently the deep shelf sediments occur in the north and northeast, grading to shallower coastal sediments as a series of facies belts, with increasing stratigraphic gap in a southwesterly direction. The deepest part of the asymmetrical basin occurs near the northern margin, with a total Upper Cretaceous sediment thickness of over 2000 m.

From the end of the Variscan Orogeny to early Cretaceous times the Münsterland was largely a landmass. Marine sedimentation was initiated in the late Albian–Cenomanian, depositing a clay–ironstone conglomerate and greensands which transgressed the Carboniferous of the Rhenish Massif. A record of this early transgressive phase is preserved along the southern margin of the basin. The Cenomanian in these marginal areas can vary from 0–25 m in thickness and contains major gaps.

Biostratigraphy. The base of the Cenomanian in Germany is taken at the first appearance of the genera *Mantelliceras*, *Sharpeiceras*, *Acompsoceras* together with an abundance of *Schloenbachia* and *Neostlingoceras carcitanense*. The actual boundary is represented within an omission surface and a level of reworking in most areas, followed by a centimetre-scale conglomerate and then the Essen Greensands or Cenomanian Marls (Hiss 1983).

Eucalycoceras pentagonum is a more suitable zonal fossil for the Upper Cenomanian boundary as *Calycoceras naviculare* and *Calycoceras guerangeri* are rare in the Münster Basin. The top *M. dixoni* to low *Acanthoceras rhotomagense* zones are poorly fossiliferous in the basin because of condensation and preservation effects (Kaplan 1992). Thus in Lengerich the beginning of the Mid Cenomanian is taken at the entry of

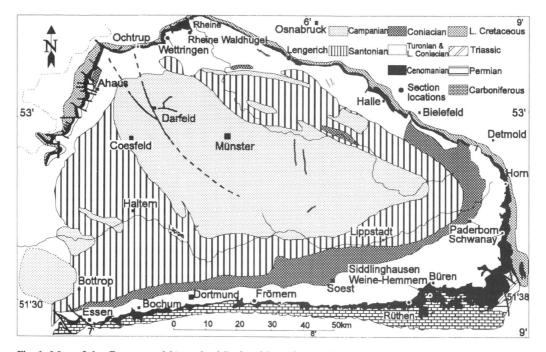

Fig. 1. Map of the Cretaceous Münsterland Basin with section locations. Modified after Ernst *et al.* (1992).

Turrilites costatus in an ammonite bed, rather than at the first appearance of *A. rhotomagense*, which is rare in this part of the section. This uncertainty over the position of the Lower–Mid Cenomanian boundary is a problem which will be discussed later.

German authors Ernst *et al.* (1983) proposed a classification of various 'events' in the Münster Basin. These 'events' consist of various types of short lasting isochronous events, classified as eustato, eco, tephro, litho, phylo and oxic/ anoxic events. As noted by Meyer (1990) and Kaplan (1992), however, these terms are largely redundant as it is generally impossible to distinguish the causes of events as either eco-, eustato- and litho- events in practice. However, flood occurrences of distinctive species in distinctive lithologies do occur (e.g. the *Pycnodonte* and *Amphidonte* oyster events and the *Actinoca-max plenus/Pachydesmoceras denisonianum* Event), which provide an excellent means of correlation across the basin and beyond, regardless of their cause.

As noted by Kaplan (1992), the lower Cenomanian up to some way into the *Inoceramus crippsi* or *I. virgatus* Zone is rarely exposed in the northern parts of the basin and they were not examined in this study. Haack (1935) discussed the Albian–Cenomanian transition in

the region; this work was summarized by Kaplan (1992).

Section descriptions

Logged sections were taken from ten districts around the basin remnant (Fig. 1), showing a range of facies from the highly condensed southern marginal deposits to the thick (140 m) hemipelagic sections from the deeper parts of the basin (offset to the northern margin). Exposures are confined to quarry sections. For discussion see Figs 2 and 3.

Sequence MB.1. The beginning of Cenomanian sedimentation throughout the Münsterland Basin is marked by an omission surface constrained between the Mid–Upper Albian (*Neohibolites ultimus*) and the lowest Lower Cenomanian *M. mantelli* Zone (*I. crippsi* occurrences and the *N. ultimus/Aucellina* Event; Kaplan 1992; Ernst *et al.* 1983; Morter & Woods 1983). The stratigraphic gap increases from the deep northern areas to the shallower southern margin. The *N. ultimus/Aucellina* Event is a flood occurrence of bivalves and belemnopseids which occurs in the lowest Cenomanian *M. mantelli* Zone in a number of areas including, Soest to Paderborn, Kassenberg and also in the Lower Saxony Basin.

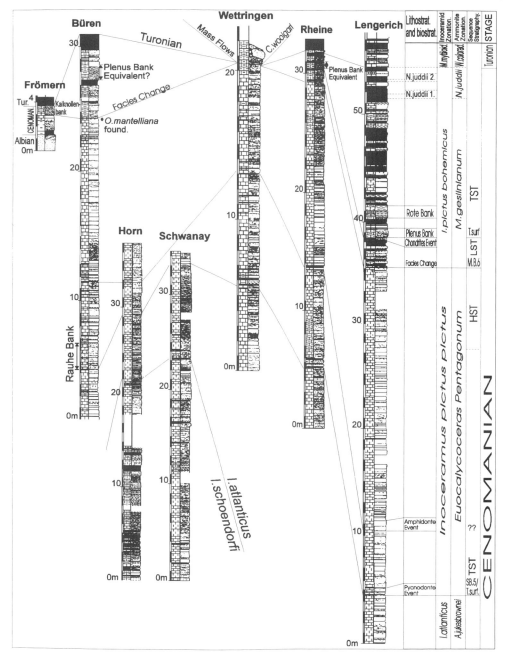

Fig. 2. Correlation of the Upper Cenomanian of the Münster Basin. MB, sequence boundary; LST, lowstand; T.surf, transgressive surface; TST, Transgressive tract; HST, Highstand.

It lies on the erosional surface with phosphorite & glauconitic material, passing into marly limestones in the basin. In the more marginal sections of the southern Münsterland the hiatus at the Albian–Cenomanian boundary is of a longer duration (e.g. Rüthen, where the 5 m *Inoceramus crippsi* Zone section would suggest a larger hiatus/reworking at the base, with the absence of the *N. ultimus/Aucellina* Event. The Albian–Cenomanian omission surface thus

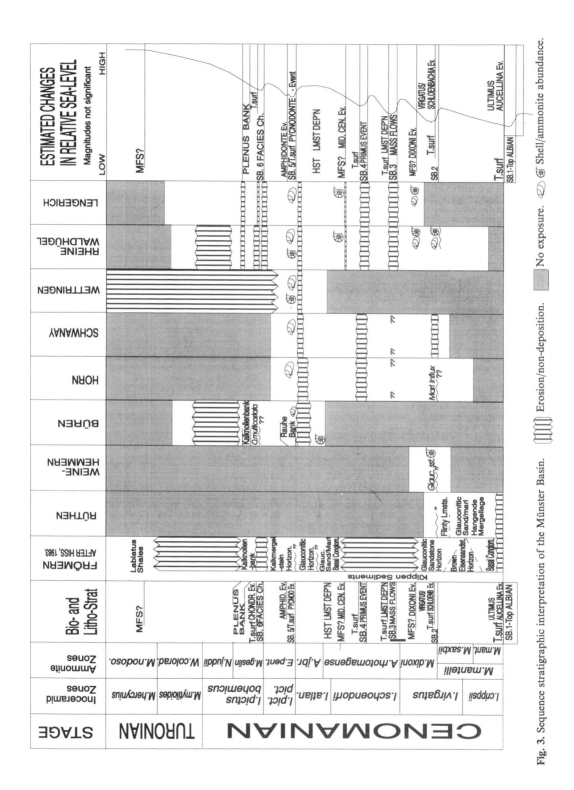

Fig. 3. Sequence stratigraphic interpretation of the Münster Basin.

marks the sequence boundary of the first Cenomanian sequence. In marginal areas it is combined with the transgressive surface, which is overlain by the *N. ultimus/Aucellina* Event. The lack of basinal outcrop precludes a definitive timing of the late Albian fall. The thick, marl-rich Lower Cenomanian deposits of Lengerich, Rheine and Horn onlap this transgressive surface, gradually thinning onto the southern margin and becoming more detrital and glauconitic in places.

Sequence MB.2. At Rheine, Rüthen, Horn and formerly in Lengerich (Haack 1935), as well as other areas within and outside the basin (Ernst *et al.* 1983), a faunal influx, known as the *Schloenbachia/virgatus* Event occurs early in the *M. dixoni* Zone. This faunal event is associated with glauconitic greensands in some areas; at the Weine–Hemmern section a glauconitic sandy unit occurs at this level. Owing to a lack of examinable sections, interpretation of this event is difficult, though it has similarities with the *N. ultimus/Aucellina* Event and is thus here interpreted as marking a transgressive surface (though the presence of an underlying sequence boundary can only be inferred).

The lower part of the *dixoni* Zone, when exposed in the northern basin is marl rich, progressing up into white flasery limestones during the continuing transgressive systems tract. At Lengerich and Rheine a fossil occurrence called the *dixoni* Event occurs. The ammonites occur in relatively large numbers in a prominent white limestone, indicating deep water conditions with little clastic input, good preservation conditions and probably some condensation during continued transgression. This is succeeded by further massive white limestones into the highstand.

Sequence MB.3. At Lengerich and Rheine mass flow deposits occur, consisting of redeposited beds of limestone material in a marly matrix, with interdigitating highly burrowed beds. These beds occur after the proposed Lower–Middle Cenomanian Boundary. Hilbrecht (1988) proposed a mechanism by which such mass flow deposits can be related to sea-level falls. They are interpreted as evidence of a relative sea-level fall, with the sequence boundary at the base of the flow deposits in the basin. The flow deposits are capped by the transgressive surface and overlying transgressive systems tract limestones indicating deepening conditions.

The beginning of the Mid Cenomanian in the Münster Basin has been taken at an ammonite bed (18 m above the *dixoni* Event at Lengerich), which contains *I. schoendorfi, T. costatus, Hypoturrilites* sp. and *Mesoturrilites boersumensis*. This is problematic in the northern basin because of the lack of occurrences of *A. rhotomagense* until its acme at the top of the *Actinocamax primus* Event (see later). Kaplan (1992) proposed the present ammonite bed as the beginning of the Middle Cenomanian; however, it is considerably too low biostratigraphically as suggested by the presence of *Hypoturrilites* (typically low Cenomanian) and *T. costatus* which considerably precedes *A. rhotomagense* in England (Hancock *et al.* 1993; Gale 1995). It is very likely that the *dixoni–rhotomagense* boundary is higher than it is placed here and the third sequence boundary (i.e. the base of the mass flow deposits) is late *dixoni* rather than early *rhotomagense* age.

Sequence MB.4. The *A. primus* Event is the next major event in the basin. It is marked by a thick marly influx in the Lengerich and Rheine sections which can be traced as far south as Schwanay, but becomes more obscure south of this section. Christensen *et al.* (1992) noted occurrences of *Belemnocamax boweri* from this marly unit at Dörenthe (southwest of Osnabrück) and *Actinocamax primus* from a correlative horizon at Wünsdorf near Hannover, Halle Hesseltal and Halle Künesbeck near Bielefeld in the Münster Basin. Due to the lack of sections, the *A. primus* Event has not been recognized near the southern margin, though Hiss (1989) correlated the *Primus* Event with a silty, glauconite and organic rich bed known as the Rauhe Bank at Büren, taking the bed as the base of the Mid Cenomanian. This correlation has been revised during the course of the present study, however, based upon a newly discovered Upper Cenomanian fauna (see later).

The *A. primus* Event is here interpreted as a fourth sequence boundary due to a major sea-level fall, causing a marl shift in the basin centre and an influx of shelf derived fauna including belemnites, brachiopods, etc.

The *A. primus* Event is succeeded by a return to white limestone deposition (8 m) in the central parts of the basin which have been assigned to late transgressive systems tract deposition. Above these there is an indurated, marly bed with a rich fauna of *Sciponoceras baculoide, Cunningtoniceras inerme, Holaster subglobosus, Inoceramus tenustriatus* and a change from a predominantly benthic to a planktonic foraminiferal fauna (Ernst *et al.* 1983). This thus correlates biostratigraphically with the Mid-Cenomanian non-sequence of Carter & Hart (1977) and is known as the Mid-Cenomanian

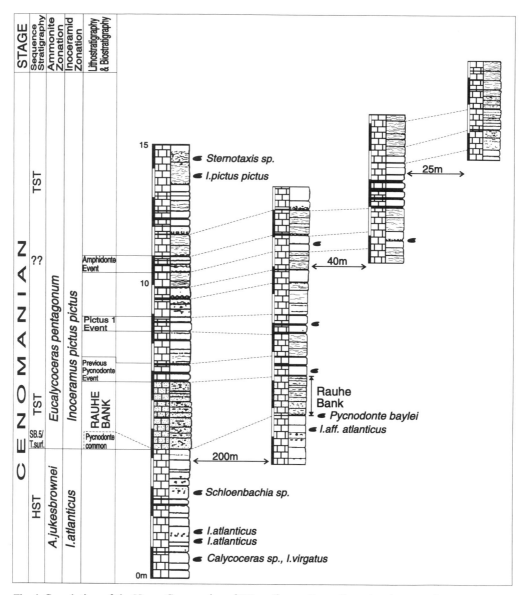

Fig. 4. Correlation of the Upper Cenomanian of Büren (former Evers Cementworks quarry).

Event by German authors. Its stratigraphic significance is dubious, however. In the Saxony Basin the Mid-Cenomanian Event is represented as a hardground with a pebble lag on top, suggesting regression (Hilbrecht 1986). The event is less well developed in the Münster Basin and shows little evidence of regression (the increase in planktonics suggests deepening). It is provisionally suggested to be a possible maximum flooding surface, although it is recognized that there is little evidence for this

and other interpretations unrelated to sea-level change are equally possible.

The Mid-Cenomanian Event is followed by further limestone deposition in the basin, with little faunal data, the thick, carbonate rich limestones being suggestive of deep water highstand sedimentation.

Sequence MB.5. The *I. schoendorfi/I. atlanticus* boundary is not currently exposed at Lengerich and was not examined in the northern parts of

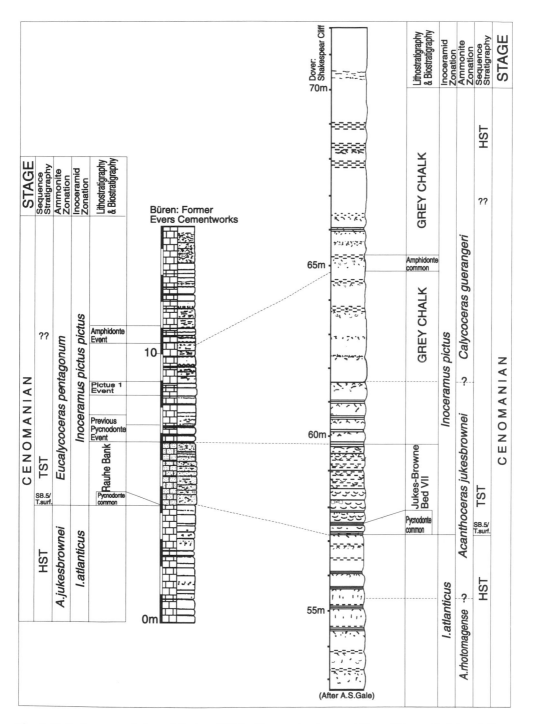

Fig. 5. Preliminary correlation of the top Middle Cenomanian between Büren (Münster Basin) and Dover–Folkestone. NB. Apparent discrepancy in ammonite zonation is due to the German usage of *A. jukesbrownei* as a partial range zone, terminated by the first appearance of *E. pentagonum*. In England the *A. jukesbrownei* Zone is a total range zone.

the basin during this study. The top of the *atlanticus* Zone is marked by the minor hiatus (picked out by the change in ammonite and inoceramid zones) and overlying flood of oysters known as the *Pycnodonte* Event. Occurrences of *Acanthoceras jukesbrownei* at Lengerich and *Calycoceras asiaticum*, Eucalycoceras sp., *Sciponoceras baculoide*, Scaphites obliquus and *Inoceramus pictus* are noted from the *Pycnodonte* Event and this is taken as the base of the Upper Cenomanian (after Kaplan 1992). The *Pycnodonte* Event can be traced throughout the basin as an obvious marl influx, leading to the deposition of three marl beds rich in *Pycnodonte baylei*, as well as a good ammonite fauna in some areas. At Wettringen, phosphatized ammonites occur, indicating a long residence time on the sea-floor, suggesting the *Pycnodonte* Event is a highly condensed deposit.

At Büren (Figs 4, 5) the lowest deposits exposed are of the *atlanticus* Zone. These deposits were originally considered to be of Lower to Middle Cenomanian age. In the Büren Geological Map Guide (Hiss 1989), the Rauhe Bank, a reworked, glauconitic sandy body was correlated with the *A. primus* Event of the basin (in Sequence MB.4). This was due to a supposed find of *Mantelliceras* sp. in the beds below the Rauhe Bank. The Rauhe Bank has now been re-evaluated during the present study and is re-assigned to an Upper Cenomanian age on occurrences of *Inoceramus atlanticus*, *Calycoceras* sp. and *Rotalipora cushmani* (S. Mitchell pers. comm.), which have been noted during this study. The Rauhe Bank is a winnowed, transgressive deposit with the *Pycnodonte* Event equivalent at its base. The *Pycnodonte* Event, by analogy with the *N. ultimus* and *Schloenbachia/virgatus* Events is also indicative of a transgressive surface. The position of the corresponding sequence boundary is difficult to assess. There is no obvious marl-shift in the basinal sections before the *Pycnodonte* Event and no other deposits that could be assigned as lowstand deposits below the *Pycnodonte* Event.

Provisionally the sequence boundary and transgressive surface are interpreted to concur at the base of the *Pycnodonte* Event with lowstand and initial transgressive deposits being confined to the condensed, marl rich beds of the *Pycnodonte* Event in the exposed sections of the basin. A lack of lowstand deposits may indicate that this was a sea-level fall of relatively rapid duration, or of a low magnitude, causing little exposure and erosion. This highlights the problem of applying sequence stratigraphic concepts to epicontinental sea deposits as it seems likely the shallow gradient and lack of

shelf edge leads to little clastic generation during sea-level falls, as noted for the Jurassic (Hallam 1992*b*).

The transgressive systems tract is shown in marginal areas (Büren section) by the gradation into the winnowed Rauhe Bank and in the deeper basin by pelagic limestone deposition which overlies the *Pycnodonte* Event.

The Upper Cenomanian is much more clearly exposed throughout the basin. The Lengerich section shows little evidence of condensation, though there is diagenetic alteration (pressure solution) and tectonic disruption. Towards the southern margin (e.g. Rheine to Schwanay to Horn) there is increasing evidence of induration and condensation as the units become more marl-rich and flasered, leading to amalgamation of rhythmic bedded marl–limestone pairs into flasered beds showing pressure solution as well.

The *Amphidonte* Event is an obvious lithological and faunal marker consisting of three marl beds with occasional *Amphidonte* sp. and subordinate *Pycnodonte* sp. oysters, which is therefore provisionally interpreted as a shallowing event.

The rest of the *E. pentagonum/I. pictus* Zone is expressed in white limestone deposition, progressively more marly away from central parts of the basin. These white, poorly fossiliferous limestones are referred to as the Arme Rhotomagense Schichten in the literature. However, these beds constitute the *E. pentagonum/C. naviculare* Zone (Ernst *et al.* 1983) and *A. rhotomagense* is absent. Near the top of the Arme Rhotomagense Schichten a second occurrence of *Inoceramus pictus* occurs; the *I. pictus* 2 Event which is again rare in specimens but is a useful marker for the top few metres of the Rhotomagense Schichten.

The *I. pictus* 1 & 2, and *Amphidonte* Events (Ernst *et al.* 1983) are all relatively minor and problematic events in terms of sequence stratigraphic interpretation. The *I. pictus* Events in particular show no major lithological change above or below and can equally be inferred to be the result of productivity events, preservation events, climate effects, imagination, as well as sea-level fluctuations. They are not, however, considered to represent major changes in the overall depositional environment.

Sequence MB.6. The final regressive break in sedimentation during the Cenomanian occurs throughout the basin at the top of the *E. pentagonum/I. pictus pictus* Zone. This is known as the Facies Change, marked in basinal areas by a major influx of dark, carbonate poor, marl/black shale lithologies. At Lengerich, the Facies

Change occurs *c.* 1 m above the *I. pictus* 2 Event and a strong surface marks the change in clay composition from white limestones with a small kaolinite content below, to a black, montmorillionite rich clay facies above (acid insoluble residue change from 0–65% montmorillionite (Heim 1957; Schöner 1960). More marginal areas such as Rheine show major hiatal surfaces related to this event and are overlain by well oxygenated brightly coloured shallow water equivalent of the anoxic marl facies, known as the Rotpläner. The Rotpläner shows features of shallow water deposition such as rapid lateral variations, intraformational scours and shell lags (Ernst *et al.* 1983, 1984). Hilbrecht (1986) notes 1 m deep *Bathichnus* sp. at Rheine and local pebble concentrations on the surface (see Heinz 1928; Wegner 1925).

At Lengerich, *c.* 2 m above the Facies Change is an interval of *c.* 0.5 m in which *Chondrites* sp. made of a white matrix are piped down into the black marls from an overlying omission surface, known as the *Chondrites* Event. The extinction of the planktonic foraminiferan *R. cushmani* occurs at the level of the *Chondrites* Event at Lengerich (Ernst *et al.* 1992).

Ernst *et al.* (1984) recognize two closely spaced *Chondrites* horizons (the *Chondrites* sp. being more pronounced in the upper bed) which can be traced right across Germany from eastern Lower Saxony to Wüllen on the western margin of the Münsterland. A comparable *Chondrites* Event occurs near the base of the Eibrunn Marls in the Regensburg Basin (Bavaria) (Ernst *et al.* 1984). According to Ernst *et al.* (1984) the extinction of the planktonic foraminiferan *Rotalipora greenhornensis* lies at the top of the lower and the extinction of *R. cushmani* at the top of the upper *Chondrites* Event in the Rotpläner successions of Lower Saxony.

The *Chondrites* Events are inferred to mark a period of omission, preceding the onset of transgression. They are overlain by an indurated and condensed nodular limestone unit known as the Plenus Bank. *A. plenus* is rare in the Münster Basin, though it has been noted from the Plenus Bank (Kaplan 1992; Schmid 1965). *Pachydesmosceras denisonianum*, *Inoceramus pictus bohemicus* and *Metoicoceras geslinianum* have also been noted (Ernst *et al.* 1984). The Plenus Bank formed over a considerable period of time and a wide area during transgression. Equivalent (though probably slightly diachrounous) facies occur in marginal areas of the basin (e.g. Fromern). Kaever (1985) noted a hiatus at the top of the Plenus Bank on glauconite content and referred to it as an incipient to true hardground *sensu* Kennedy & Garrison (1975).

This marks a second flooding surface. The Plenus Bank also corresponds to a ^{13}C isotope peak (Scholle & Arthur 1980; Hilbrecht & Hoefs 1986). The Plenus Bank throughout the basin is overlain by a further series of dark marls/shales, in some places showing purple/orange colourations, as the transgression continues into the Turonian.

The final bioevents of the Cenomanian at Lengerich are the *Neocardioceras juddii* Events (Hilbrecht 1986; Ernst *et al.* 1983, 1984; Kaplan 1992). *N. juddii* has been recorded from Hannover (the Misburg section) as well as *Sciponoceras* sp. and *Thomelites serotinus*, in a 5 cm interval of a 60 cm black shale bed (Ernst *et al.* 1983). Kaplan (1992) recognized *N. juddii* and *Sciponoceras* sp. from two horizons in the black shales at Bielefeld near the Teutoburger Wald. The top part of the Cenomanian, above the Plenus Bank is very variable, due to local variations in CaCO$_3$ content and thus development of limestones, as well as the numerous hiati that occur in this transgressive interval that continued through the earliest *Watinoceras coloradoense* Zone of the Turonian.

The top of the Cenomanian-base Turonian at Lengerich is marked by a CaCO$_3$ increase and influx of bivalves of the genus *Mytiloides* spp. as well as *I. ex. gr. pictus*. The genus *Mytiloides* is accepted as defining the base of the Turonian in boreal inoceramid stratigraphy. At Rheine the Upper Cenomanian section is much reduced, with significant erosion on SB6, reflecting the more marginal position of the section.

Anglo-Paris Basin

Geological setting. The Anglo-Paris Basin (Fig. 6) is a classic area for the study of the Cenomanian. The type area of the stage occurs within the basin in the marginal platform deposits exposed around Le Mans. The basin remnant is characterized by strong clastic inputs in marginal areas next to major massifs (Devon, SW England, La Sarthe, N. France) which decrease into the basin, being succeeded by hemipelagic marl rhythms in the basin centre (Boulonnais). Evidence of fault sedimentation occurs (Mortimore & Pomerol 1991), particularly on the Normandy coast, strongly modifying any response to sea-level change. The central parts of the basin can be regarded as more tectonically quiescent and hemipelagic deposits onlap onto earlier structures e.g. Isle of Wight.

The Anglo-Paris Basin occupies a position forming the meeting point of the Atlantic, Boreal and Tethyan Realms. The main marine inundation began as a result of basin subsidence

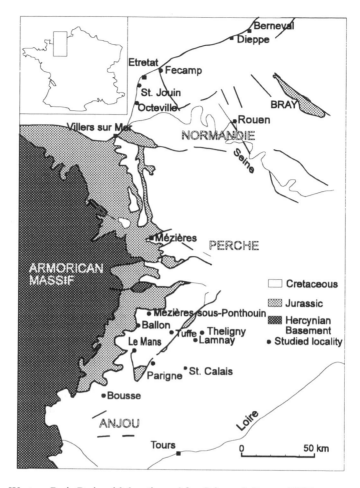

Fig. 6. Map of the Western Paris Basin with locations. After Juignet & Breton (1992).

in the Aptian, with the Lower Greensand deposits (Hancock 1969; Rawson *et al.* 1978) transgressing the Wealden facies of southern England. Lower Greensand is overlapped westward by Gault and Upper Greensand (Albian), which overstep older rocks westward, resting on Lias in west Dorset, Trias in east Devon and Devonian near Dartmoor. A similar onlap relationship is seen within the Upper Cretaceous Chalk basement beds which young westward. More shoreward facies of the early transgression (Cenomanian) are seen in west Dorset as the glauconitic Eggardon Grit and to the west of Lyme Regis as the Cenomanian Beer Head Limestone. Cenomanian successions range from a maximum of 80 m in the more basinal chalk facies of the Boulonnais to as low as 0.5 m in the marginal Devon sections.

The basin contains a classic Boreal European

Realm succession, studied by Haq *et al.* (1988) for their Mesozoic sea-level curve. Good sequence stratigraphic models therefore already exist (Juignet and Breton 1992; Robaszynski *et al.* 1995). Haq *et al.* (1988) used the Cenomanian outcrop of the basin as a showcase for their methods, though as detailed by Hancock (1989, 1993) they made a number of errors. Thus this basin differs substantially from the others studied due to the high proportion of sequence stratigraphic literature derived from its sediments and this paper is influenced by these previous studies, especially for the Lower Cenomanian where outcrops are poor.

Biostratigraphy. The biostratigraphy of the basin for the Cenomanian is well constrained from the works of Kennedy (1969, 1971) and Wright & Kennedy (1981, 1984, 1987, 1990).

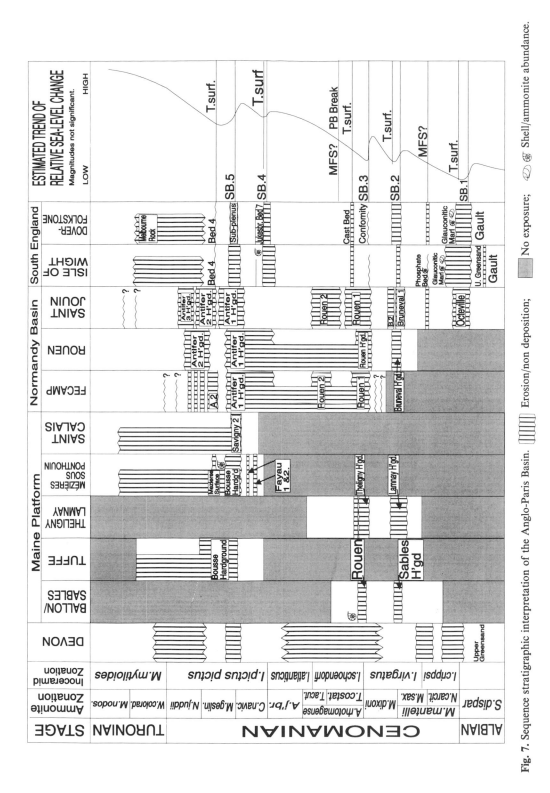

Fig. 7. Sequence stratigraphic interpretation of the Anglo-Paris Basin. ||||| Erosion/non deposition; ▒ No exposure; ◇ ⊛ Shell/ammonite abundance.

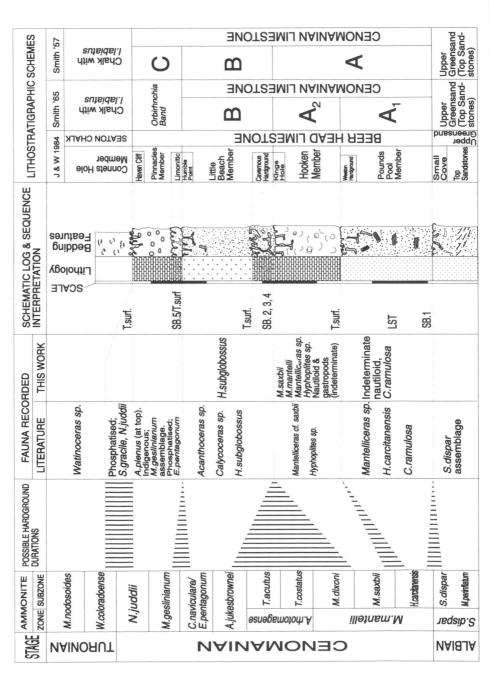

Fig. 8. Sequence stratigraphic interpretation of the Beer Head Limestone Formation. Lithostratigraphic information after Smith (1957, 1965); Kennedy (1970); Rawson *et al.* (1978); Kennedy & Juignet (1973); Jarvis & Woodroof (1984).

Section descriptions

Over 26 sections were studied from areas around the basin (Fig. 8) showing a range of facies from the highly condensed western marginal sections (Devon) through thick clastic platform deposits (La Sarthe) to basinal hemipelagics (Normandy, Boulonnais/Kent coast).

A number of discontinuities can be traced from these clastic-rich, marginal deposits into the basinal areas of Normandy and east Kent. Five sequence boundaries have been located in the basinal section from Dover–Folkestone (see main correlation panel) and these can be traced around the basin, becoming amalgamated in marginal areas. At Devon a highly condensed succession of calcarenitic limestones occurs. Major hiatuses occur as hardgrounds. Hardground surfaces are common compared to the other basins studied. They often have distinctive styles of mineralization and encrustation making them laterally traceable for 10–15 km (e.g. the St Jouin Hardground). For discussion see Figs 7 and 8.

Sequence APB1. Cenomanian sedimentation was initiated in the basin during the *M. mantelli* Zone (*N. carcitanense* Subzone). In basinal areas such as the Boulonnais (Dover–Folkestone), a burrowed omission surface occurs between the Albian (*Stoliczkaia dispar* age) Gault and the transgressive Cenomanian Glauconitic Marl (Gale 1989). The Glauconitic Marl, a grey–green marly silt, rich in phosphatic clasts and *Spongeliomorpha*, is laterally variable with occasional poorly defined omission surfaces. Some parts contain abundant *Aucellina* sp. (Gale 1989; Morter & Woods 1983).

The Albian–Cenomanian omission surface can be traced a considerable distance and is here interpreted as a combined sequence boundary and transgressive surface to a relative sea-level fall initiated in the Upper Albian. At St Jouin the underling silty, pyritic marls of the Gaize d'Octeville are overlain by glauconitic, silty chalk containing bored intraclasts of grey silt, often with glauconitic coatings. These sediments are transgressive deposits of Lower Cenomanian *M. mantelli* age.

At Beer in Devon, the Albian–Cenomanian surface occurs as the Small Cove Hardground, which separates the underlying Upper Greensand from the marginal, laterally variable deposits of the Beer Head Limestone Formation.

The rest of the *N. carcitanensis* Zone consists of transgressive sediments with high glauconite contents which can be seen to onlap topographic structures, as in the Isle of Wight (Kennedy 1969).

The significance of the *N. carcitanensis–M. saxbii* transition is difficult to assess. Other authors have interpreted it as a sequence boundary (Robaszynski *et al.* 1995). This interpretation is rejected here. In the marginal deposits of the Beer Region, the Pounds Pool Member shows lateral thickness variations capped by a hardground. The overlying Hooken Member onlaps onto the Upper Greensand, indicating continued relative transgression in the *saxbii* Zone at the Devon margin.

In other areas the zonal transition is marked by a hardground (e.g. St. Jouin; see Juignet & Breton 1992) or, for instance, the Ballon erosional surface (Juignet 1974). The presence of these laterally continuous surfaces suggests a major oceanographic change of some kind, here interpreted to be connected with the transgressive systems tract or maximum flooding.

Sequence APB2. The second major break in sedimentation in the basin is at the base of the Zone of *M. dixoni*. This regressive break occurs throughout the basin. In the more marginal areas of La Sarthe a surface occurs, known as the Lamnay Hardground (also the Perronerie, Sables and Longuville Hardgrounds). This is overlain by the coarse-grained, regressive, tidally influenced Sables de Lamnay which contain an excellent *M. dixoni* assemblage and also abundant bivalve and crustacean material. These represent a shelf margin wedge as described by Juignet & Breton (1992).

In the Pays de Caux region around St Jouin the sequence boundary is exposed as a nodular, glauconitized calcarenitic chalk, penetrated by *Thalassinoides* sp. known as the Bruneval 1 Hardground. Overlying the hardground is intensely glauconitic, marly chalk with oyster, inoceramid, sponge and serpulid debris as well as occasional glauconitic clasts forming a transgressive lag. At Dover, a glauconitic, silty chalk bed rests upon this surface, marking the transgressive surface. (For a detailed log of the Cenomanian at Dover see Jenkyns *et al.* 1994).

The upper *M. dixoni* Zone records a major transgressive period, onlapping structures on the Isle of Wight and parts of the Maine Platform. The basal parts often contain reworked fossils from the *M. mantelli* Zone in an *M. dixoni* Zone matrix rich in winnowed glauconitic sand and *I. virgatus*.

Sequence APB3. This sequence is initiated by a major break throughout the basin, shortly before the end of the *M. dixoni* Zone (as seen in Dover). It led to extensive erosion in marginal areas (Devon) and the Chilterns, followed by the debris flow deposits of the *A. rhotomagense* Zone; the Totternhoe Stone and the Grey Bed of Yorkshire (Hancock 1989). This break marks the Lower–Middle Cenomanian boundary. In the Dover–Folkestone succession, there is a subtle and gradual increase in carbonate content, rather than an obvious break in sedimentation shortly before the end of the *M. dixoni* Zone. The preceding high sea-levels desensitized the basin centre to even the effects of this major fall in sea-level, and the sequence boundary is here represented as a correlative conformity.

In the St Jouin region a strong hardground surface is developed; the Rouen 1 Hardground. This is overlain by a lag of shell debris, calcarenitic, glauconitic chalk with phosphatized fossil fragments. There is reworking of the lower surface and a *T. costatus–T. acutus* Zone assemblage has been assigned to this 3.5 m interval by Juignet (1974). At Rouen and Fecamp a strong hardground occurs with a hiatus spanning much of the Middle Cenomanian. This major break is related to condensation on the hangingwall of the Fecamp–Lillebonne–Villequier tectonic axis; one of a number of structures controlling sedimentation in the area.

On the Maine platform, at Theligny-Lamnay, a strong, highly glauconitized hardground occurs in a calcarenitic sediment at the top of the *M. dixoni* Zone; the Theligny Hardground. This hardground, developed at the top of the Sables de Lamnay is onlapped by the transgressive, glauconitic, chalk deposits of the Craie de Theligny.

In Devon, sediments of *M. dixoni–T. costatus–T.* acutus age are thin or absent, indicating the scale of the top *dixoni* regression and probable amalgamation of the third and fourth sequence boundaries in this marginal position.

The transgressive surface is more defined in the basin at Dover, marked by a prominent trace fossil horizon known as the Cast Bed which marks an omission surface and a decrease in carbonate content. The Rouen 1 Hardground represents a combined sequence boundary and transgressive surface, the sediments overlying the hardground consisting of a winnowed and condensed, shelly lag with phosphatized ammonites and intraclasts. On the Maine platform at Theligny and Lamnay a second, minor hardground marks the transgressive surface. It is overlain by a glauconitic, chalky marl which grades into massive white chalk containing ammonites which include *Acanthoceras* sp., marking the onset of Middle Cenomanian

sedimentation.

The Planktonic–Benthonic (PB) Break is a conspicuous erosive surface which occurs in the basin at the top of a marl–chalk rhythm. It marks a change from a dominantly benthic to a planktonic microfauna (Carter & Hart 1977). It marks continued deepening, as shown by the faunal changes across the omission surface, and also facies relationships in more marginal areas (Carter & Hart 1977, p. 66). This deepening appears to continue for a considerable length of time and thus it seems unlikely to be the product of maximum flooding, and may represent a secondary transgressive surface, or be related to other oceanographic changes. However, it is a major surface, which can be picked out throughout the basin and beyond. The remainder of the *rhotomagense* Zone in the basin consists of a thick package of sediment, clay content within the marl–chalk rhythms gradually decreasing upward as water depth increases, leading to dilution of the detrital clay component.

Sequence APB 4. At Dover, the next major break in the succession is in the *jukesbrownei* Zone, at the base of Jukes-Browne's Bed 7 (Jukes-Browne & Hill 1903). This winnowed, calcarenitic chalk unit overlies an omission surface containing structures variously described as scours or burrows, which mark an amalgamated sequence boundary and transgressive surface. The winnowed deposits of Jukes-Browne's Bed 7 constitute a transgressive lag overlying the sequence boundary. The remainder of the *jukesbrownei* Zone and overlying *naviculare* Zone are strongly transgressive, depositing thick, massive chalk deposits in the basinal areas. In the less well dated clastic deposits of the Maine Platform, a number of laterally variable discontinuities occur at the levels described. At Allones, the Jalais Hardground, dated to near the *rhotomagense–jukesbrownei* transition, is likely to correspond to the sequence boundary, overlain by the regressive deposits of the Perche Formation. The top of the Perche Formation is marked by the Fayau 1 & 2 hardgrounds which mark the transgressive surface, dated here to the *jukesbrownei–naviculare* transition and overlain by the Marnes a *Ostrea biauriculata*. In such marginal deposits (which are influenced by a large number of controls) the validity of any interpretation placed upon the laterally variable surfaces which occur is not assured. The Normandy coast area was undergoing tectonic disturbance during much of this period and there is a high degree of lateral variation and condensation associated with fault uplift in these areas.

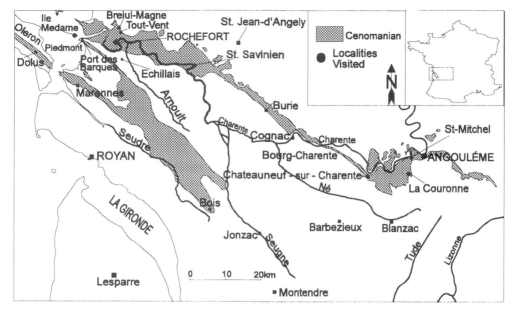

Fig. 9. Locality map for the Aquitaine Basin.

Sequence APB5. The final most spectacular break during the Cenomanian occurs at the base of the *Metoicoceras geslinianum* Zone. This break and the overlying sediments have been studied extensively due to the hypothesis of the Global Oceanic Anoxic Event II which is said to occur at this level. At Dover the break is expressed as an erosive surface; the Sub-plenus Erosion surface of Jefferies (1963). It occurs in the Normandy region as the Antifer Hardground, though here tectonic modification is apparent, and up to four strong hardgrounds can be recognized in the *M. geslinianum* Zone. The transgressive surface in the Dover section occurs at the top of Jefferies' (1962, 1963) Bed 3. This slightly erosive horizon marks the extinction of the planktonic foram *Rotalipora cushmani* and the influx of *A. plenus* in Bed 4. Considerable lateral variation in bed thickness is confined to the thin packet of lowstand sediment in beds 1–3.

On the Maine Platform the *C. naviculare–M. geslinianum* break is marked by a series of strong hardgrounds; the Savigny hardground at Saint Calais or the Bousse Hardground. In these areas the overlying sediments are generally highly condensed, glauconitic transgressive deposits which persist until the onset of chalk deposition in the Lower Turonian. The *N. juddii* Zone is typically highly condensed within this transgressive systems tract. In the Devon sections a strong hardground occurs at the *C. naviculare–M.*

geslinianum boundary which is overlain by a unit containing a phosphatized *E. pentagonum* fauna with an indigenous *M. geslinianum* Zone fauna including *Actinocamax plenus*. This is the combined sequence boundary and transgressive surface as shown by the change in facies from the clastic Little Beach Member to the carbonate rich Pinnacles Member. The Pinnacles Member is capped by the Haven Cliff Hardground which spans the *N. juddii* Zone and part of the *Watinoceras coloradoense* Zone. A phosphatized *N. juddii* fauna occurs above this hardground. The hardground surface is here interpreted in consistency with the sections on the Maine Platform as a secondary transgressive surface probably related to the hiatus associated with the onset of chalk deposition. At Mézières sous Ponthoiun an erosional surface occurs at the same level which is overlain by deeper water transgressive sediments confirming this hypothesis. However, a number of authors (Juignet & Breton 1992; Haq *et al.* 1988) have interpreted this surface as a sequence boundary mainly on the evidence of hiatus, which by itself is not a sufficient criterion. This transgressive period continues, marked by significant onlap and chalk deposition into the low Middle Turonian.

Aquitaine basin

Geological setting. The Aquitaine Basin (Fig. 9) differs substantially from the two previous

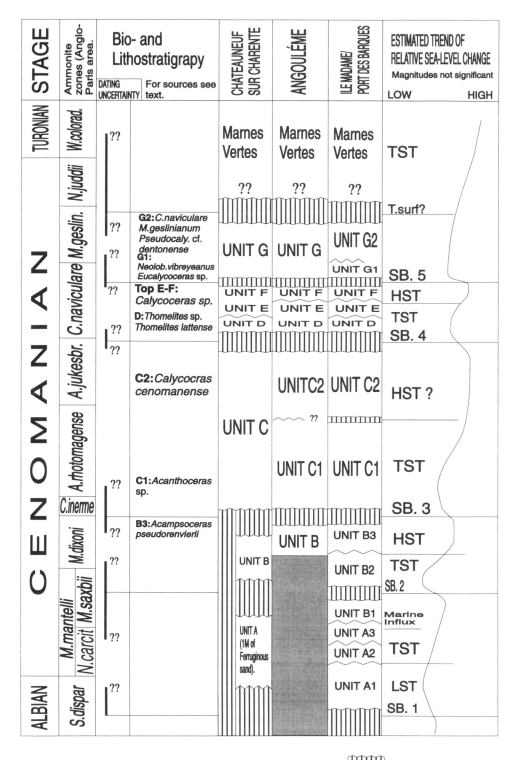

Fig. 10. Sequence stratigraphic interpretation of the Aquitaine Platform. 〔illillill〕 Erosion/non-deposition.

basins in terms of palaeobiogeography and sedimentary development. The basin is situated on the margin of the Atlantic with open marine connections. During the Cenomanian it was in the Tethyan Faunal Realm possessing a fauna poor in ammonites, though rich in rudists and orbitolinids. Juignet (1974) and Francis (1981) have shown that by the late Lower Cenomanian, connections had been established with the Paris Basin, leading to northward incursions of Tethyan fauna into the Paris Basin, most obviously late Lower Cenomanian northward migrations of *Orbitolina* species through the Straits of Poitou. Biostratigraphic correlation with the Anglo-Paris Basin has been attempted previously (e.g. Francis 1984), though the lack of unequivocal faunas has hindered a full correlation. As elsewhere, the early Cenomanian marks the Upper Cretaceous transgression onto the Jurassic limestones and continental deposits of the Lower Cretaceous.

Section descriptions

Sections were taken from a number of positions around the basin, looking at many of the outcrops described by Platel (1989). However, the lack of biostratigraphic data and poor quality of the sections have limited the number of areas used. The sections are in general much more condensed and of more marginal facies than seen in the other two basins studied, being deposited in a carbonate platform setting with a relatively high clastic input. No deep water, Cenomanian pelagic successions crop out on land in this basin. The unit subdivisions used correspond to those used by Francis (1984) and Platel (1989) which were adapted from earlier works detailed therein. Detailed facies descriptions and faunal lists can be found in Platel (1989).

Sequence AQB.1. Sedimentation begins with the fluviatile, coarse-grained, ferruginous sands and gravels of Units A1, A2 resting on Kimmeridgian marking the beginning of transgression. These are best exposed in the Tout-Vente, Breuil Magne area. These sediments have not yielded any stratigraphically useful fossils and their age is poorly constrained between the Upper Albian and early Cenomanian. These are subsequently drowned by lacustrine sediments of A3 which have yielded a potentially Cenomanian microfauna. This transgressive period continues with Unit B1 (exposed in coastal sections), which marks the major marine influx in the form of a bioclastic limestone unit containing a rich marine fauna including rudists, orbitolinids

and echinoids. The presence of orbitolina faunas in Unit B1, which can be correlated with a similar fauna in the Anglo-Paris Basin, confirms a Lower Cenomanian age.

Sequence AQB.2. The transgressive sediments of sequence AQB1 are truncated by an erosional surface and overlain by Unit B2, consisting of glauconitic sands and shales containing *Rhynchostreon suborbiculatum minor* rudists and a microfauna dominated by *O. conica* and *Simpalveolina*, once again marking a transgressive systems tract, with an absence of intervening highstand and lowstand deposits. This unit passes up into Unit B3, a calcarenitic unit with rudists, oysters and small orbitolinids, marking the onset of highstand sedimentation with increased clastic input onto the platform.

This lower Cenomanian succession is very laterally variable with, for instance, the three B units being amalgamated into one, of unknown duration farther inland. On an outcrop scale the deposits are often thin, show lateral variation and can be difficult to correlate. Dating of the units is also poor (Fig. 10), and so only a preliminary interpretation is given.

Sequence AQB.3. The top of Unit B3 (Unit B up dip), is the first well-dated break in the basin, corresponding to the Lower–Middle Cenomanian Boundary. This is confirmed by the presence of *Acanthoceras* sp. in Unit C1 (Moreau *et al.* 1983). The top of Unit B3 shows some dolomitization at the coastal outcrops of Piedmont. The overlying transgressive unit, C1, is composed of grey–white nodular, amygdaloidal limestones with a diverse fauna including rudists and prealveolinids. The orbitolinids disappear at this level, though new forms of benthic forams and alveolinids appear. Unit C1 is succeeded by Unit C2, with an erosional hiatus in some areas, marking the transition to highstand deposition. C2 is a more clastic dominated limestone. Moreau (1976), noted its excellent internal platform fauna.

Sequence AQB.4. The carbonate dominated units of the Middle Cenomanian; C1 and C2 are capped by a strong erosional surface and overlain by *c.* 1.5 m of blue–grey silty marl and limestone at the coastal outcrops. Farther inland at Angoulême the underlying, condensed C2 deposit is capped by a hardground which is overlain by 5 m of blue clay comprising Unit D. This unit marks the start of Upper Cenomanian sedimentation as shown by *Thomelites lattense*, probably indicating a position low in the *C. naviculare* Zone. The hardground between C2 and D marks a sequence boundary, overlain by the terrigenous influenced clay of Unit D. Unit E marks a clastic input onto the platform, consisting of fine-grained, indurated sands with abundant *Ostrea biauriculata*, *Rhynchostreon suborbiculatum minor* and other fauna concentrated in storm lags formed in a medio-littoral environment (Platel 1989). Unit E represents a shallowing relative to Unit D, though they are relatively conformable with no major shift in environment. The transgressive systems tract then continues with Unit F, which marks a full return to the marine domain, with abundant rudists, gastropods, nautiloids and occasional *Calycoceras* sp. at the top. It consists of a complex group of detrital limestone units separated by five hardground surfaces forming the tops of shallowing up parasequence cycles.

Sequence AQB.5. The final hardground of Unit F possesses a very complex and reworked surface. The overlying Unit G transgresses this surface with a shelly lag containing lumachelles of *Rhynchostreon suborbiculatum minor* and lesser *Rastellum carinatum*, as well as pectinids, bryozoans and annelids in a 60–90 cm thick orange stained, highly biotubated, glauconitic marl which passes up into fine-grained sandy deposits with layered accumulations of *R. suborbiculatum minor* and other oysters. The transition to G2 marks further deepening, passing up into nodular sandy limestone with the various oysters still present, though in smaller numbers. Also present here are *Terebratella carantonensis*, *T. phaseolina* and the bivalves *Arca tailleburgensis*, *Neithea* sp. and abundant echinoids. These fully marine units provide much better ammonite controls; for instance, Moreau (1976) discovered *C. naviculare* in G2 and later, *M. geslinianum* (Moreau *et al.* 1983), indicating a high Upper Cenomanian age. The top contact of G2 is poorly exposed throughout the basin. It is overlain by a green marly deposit (Les Marnes Vertes), rich in *Exogyra columba gigas* indicating continued transgression and eventually passing up into true chalk deposition in the early Turonian as the transgression continues, with fluctuations, until the Mid Turonian.

Summary and conclusions

Figures 3, 7 and 10 are designed only to show the general trend of sea-level with respect to the major surfaces and sediments. Magnitudes of sea-level change are difficult to quantify from outcrop data, and the graphs are not quantitative, but based upon knowledge of regional

distribution of any particular sediment unit, and also vertical facies contrasts at outcrop, which give some evidence of rates of change. The timescale on the graphs is also misleading, being plotted against equally divided biozones, when sediment thickness comparisons would suggest that, for instance, the *A. rhotomagense* Zone is considerably longer than the *N. juddii* Zone (Gale 1990).

Many of the surfaces noted are synchronous between the three basins studied at the resolution of the biostratigraphy and it is likely that they are responses to the same environmental event, i.e. changes in sea-level affecting large parts of Europe (Fig. 11).

Apparent instances of poor correlation occur due to the integration of differently applied biostratigraphic schemes in the different regions. The base Middle Cenomanian boundary appears too low in the Münster Basin succession due to the use of *T. costatus* as the Middle Cenomanian marker, rather than the diagnostic *A. rhotomagense*. The boundary is therefore difficult to correlate between the Münster and Paris Basins. Because the Münster Basin sections are ambiguous on biostratigraphic grounds, it is unclear whether the third or fourth sequence boundary in the Münster Basin marks the Middle Cenomanian boundary, or whether there is any correlation with the Anglo-Paris Basin. The *A. primus* Event on strong biostratigraphic grounds is a direct correlative of the Grey Bed of Yorkshire and the Totternhoe Stone of Lincolnshire. Hardgrounds occur around Normandy and on the Maine Platform at this level and in the Aquitaine Basin the break is developed as a series of hardground surfaces. Thus this period has proved difficult to correlate. It is likely that sequence MB3, evidence for which is confined to the mass flow deposits of the northern Münster Basin, is in fact a tectonic disruption not a major sea-level change, and as suggested earlier this sequence may in fact be of late *dixoni* Zone age, with sequence boundary MB4 marking the Lower–Middle Cenomanian Boundary as defined in the Anglo-Paris Basin.

There is also confusion in the *A. jukesbrownei–C. naviculare* zone due to the different usage of the *A. jukesbrownei* Zone as a total range zone in the Paris Basin and a partial range zone in the Münster Basin. This has the effect of lowering the apparent level of sequence boundary APB.4, compared to sequence boundary MB.5, which on other criteria (Fig. 5) appear to be synchronous. More significant biostratigraphic error also exists when correlating between faunal realms, as noted for the Aquitaine Basin and large error bars have been assigned to sedimen-

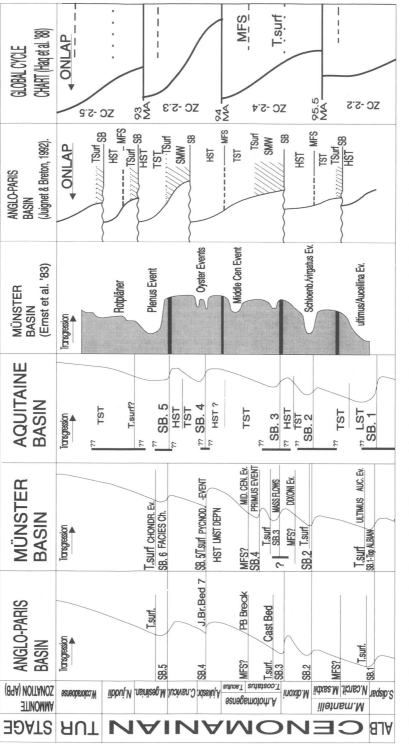

Fig. 11. Correlation panel comparing the results of this study with other works. Data plotted against Anglo-Paris Basin Zonation.

tary breaks when compared to the other basins, purely due to the lack of significant ammonite finds.

Lithologically the sequence boundaries are very varied, but in most facies tend to be marked much more by omission than any major erosion, especially as hardground surfaces. In the deeper sections from the Anglo-Paris and Münster Basins, the sequence boundaries can be subtle, or occur as relatively conformable surfaces, picked out by increases in clay content.

The distribution of sequence boundaries and other surfaces within the basins shows a strong correlation, and the coincidence of faunal changes (for instance ammonite zone boundaries) with the major stratigraphic surfaces suggests an extrabasinal control on their formation. As mentioned in the introduction, many of these faunal changes (and indeed, many of the stratigraphic breaks) have been described from other continents. In effect they are similar responses to the same environmental changes, rather than discrete laterally continuous surfaces, each marked by different features depending on where they form in the basin palaeogeography.

The successions in the three basins can be strikingly similar when similar facies are compared as can be seen with the *jukesbrownei–naviculare* break (Fig. 5). This is also true of the *M. geslinianum* Zone Break, the strongest break in the Cenomanian. The Facies Change in Westphalia and the beginning of the Plenus Marls succession in England both mark major clay inputs into the basins, associated with the Global Oceanic Anoxic Event II hypothesis, which has long been used in long range intrabasinal correlations. In more marginal areas, the anoxic, black shale facies are not developed, though the break is still pronounced as, for instance, the Antifer Hardground in the Normandy Region. On the Aquitaine Platform there is a lack of the diagnostic faunas and no development of the anoxic facies which characterize the deep water successions of the other basins. This is largely a reflection of the much shallower setting of the Aquitaine Platform.

Transgressive surfaces are very well developed in all the successions studied. They are generally marked by a strong marl shift, evidence of condensation (glauconitization, phosphatization) and associated winnowing and scouring. Faunal preservation is usually good in these transgressive deposits, such as the *Pycnodonte* Event of the Münster Basin or the lag overlying the Rouen Hardground in the Paris Basin. Many of the preservation events associated with transgression have their equivalents at the same

stratigraphic levels in other basins, e.g. the *N. ultimus/Aucellina* Event of the Münster Basin has a direct equivalent in the Glauconitic Marl (Gale 1989; Morter & Woods 1983). As noted earlier, the *Chondrites* Event in the Münster Basin and the top of Bed 3 of the English Plenus Marls succession both mark the extinction of *R. cushmani* and the influx of *A. plenus*, representing the beginning of the transgressive systems tract and transgressive surface, hence the wide occurrence of the Plenus Bank.

Of interest is the fact that many of the transgressive surfaces are combined with the underlying sequence boundaries in the successions studied. This lack of lowstand deposits in epeiric seas has been noted previously (Hallam 1992*b*; Robaszynski *et al.* 1993, 1995; Juignet & Breton 1992). Lowstand deposits are also thin during much of the Cenomanian due to the overall rising sea-level trend and the lack of clastic sedimentation away from major massifs.

Major biostratigraphic gaps in the succession can be demonstrated during the transgressive systems tracts of some of the thicker sequences. In sequence MB.6 of the Münster Basin, there are large hiatuses at Büren and Rheine during continued transgression in the *N. juddii* Zone (Fig. 3). This transgressive level is also highly condensed throughout the Anglo-Paris Basin. These breaks are presumably due to local differences in sediment supply and accommodation space, the gaps in some areas are associated with scour surfaces representing multiple transgressive surfaces. This may be evidence for the ideas of Cartwright *et al.* (1993), i.e. multiple transgressive surfaces exist, separated by laterally variable packets of hemipelagic sediments and it is not necessarily the same surface being traced between outcrops. In some sections there may be major gaps only visible on biostratigraphic evidence (e.g. *N. juddii* Zone at Büren).

Surfaces indicating the period of maximum flooding have proved extremely difficult to locate in any of the sections. In the hemipelagic deposits, the point of maximum flooding is only discernible on broad characteristics such as an overall decrease in clay content. This is mainly because of the pelagic, biogenic nature of the sediment, which renders it less sensitive to recording sea-level maxima.

The thicker, basinal deposits do preserve a much better record of the changes in sea-level which have taken place. In more marginal areas, many of the surfaces have long-ranging and complex histories, associated with levels of condensation and reworking. The histories of the sediments in these sections cannot be fully ascertained without reference to the more complete basinal successions, which also, by their nature usually contain better biostratigraphic data. The Aquitaine Basin has more in common with the marginal deposits of Devon than any of the main sections from the other basins i.e. large biostratigraphic gaps and an impoverished fauna compared to the deep water successions of other basins.

This paper has shown that it is possible to demonstrate the synchronous effect of changes in sea-level with a high stratigraphic precision over a large part of the European continent. Records of relative sea-level change can be acquired for basins in which strong tectonic activity is present if a large number of sections are examined from around the basin (e.g. the Anglo-Paris Basin). Although confinement of the study area to the European plate limits any conclusions on eustatic controls, it is very likely that these changes are of eustatic origin, though further integrated studies on sections outside the European region, at the same level of accuracy will be necessary to confirm this hypothesis.

This work has been carried out during the course of a NERC PhD studentship. Many thanks to Jake Hancock and Andy Gale for reading the draft versions, to Ulrich Kaplan and Pierre Juignet for help with fieldwork, and to Chris Wood and Simon Mitchell for general discussion.

References

BROMLEY, R. G. & GALE, A. S. 1982. The Lithostratigraphy of the English Chalk Rock. *Cretaceous Research*, 3, 273–306.

CARTER, D. J. & HART, M. D. 1977. Aspects of Mid-Cretaceous Stratigraphical Micropalaeontology. *Bulletin of the British Museum (Natural History), London*, 29.

CARTWRIGHT, J. A., HADDOCK, R. C. & PINHEIRO, L. M. 1993. The Lateral Extent of Sequence Boundaries. *In:* WILLIAMS, G. D. & DOBB, A. (eds) *Tectonics and Seismic Sequence Stratigraphy.* Geological Society, London, Special Publication, 71, 15–34.

CHRISTENSEN, W. K., DIEDRICH, C. & KAPLAN, U. 1992. Cenomanian Belemnites from the Teutoburger Wald, N.W. Germany. *Paläontologie Zeitschrift*, 66, 265–275.

ERNST, G., SCHMID, F. & SIEBERTZ, E. 1983 Event Stratigraphie im Cenoman und Turon von N.W. Deutschland. *Zitteliana*, 10, 531–554.

——, WOOD, C.J. & HILBRECHT, H. 1984. The Cenomanian–Turonian Boundary Problem in N.W. Germany with comments on the north–south correlation to the Regensberg Area. *Bulletin of the Geological Society of Denmark*, 33, 103–133.

——, HARRIES, P., HISS, M., KAEVER, M., KAPLAN, U. *ET AL.* 1992. *The Middle and Upper Cretaceous*

of Münsterland, Westphalia. Field Guide, Fourth International Cretaceous Symposium. Hamburg.

FRANCIS, I. H. 1981. *Palaeoecological observations in the Cenomanian of Northern Aquitaine, Touraine and Sarthe, Western France.* PhD Thesis, University of Oxford.

—— 1984. Correlation between the North Temperate and Tethyan Realms in the Cenomanian of Western France and the Significance of Hardground Horizons. *Cretaceous Research*, 5, 259–269.

GALE, A. S. 1989. Field Meeting at Folkestone Warren, 29th November, 1987. *Proceedings of the Geologists Association*, 100, 73–82.

—— 1990. A Milankovitch scale for Cenomanian Time. *Terra Nova*, 1, 420–425.

—— 1995. Cyclostratigraphy and Correlation of the Cenomanian Stage in Western Europe. *In:* GALE, A. S. & HOUSE, M. (eds) *Orbital Forcing Timescales and Cyclostratigraphy.* Geological Society, London, Special Publication, 85, 177–197.

HAACK, W. 1935. *Erläuterungen zur Geologischen Karte von Preussen und benachbarten deutchen Ländern* 1:25,000, Blatt Lengerich. 48 S., Berlin.

HALLAM, A. 1992a. *Phanerozoic Sea-level Changes.* Columbia University Press.

—— 1992b. Problems in Applying the Concepts of Sequence Stratigraphy to deposits laid down in Epicontinental Seas. *In: Sequence Stratigraphy of European Basins.* Abstracts. CNRS-IFP, Dijon, France. 79–81.

——, HANCOCK, J. M., LABREQUE, J. L., LOWRIE, W. & CHANNEL, J. E. T. 1985. Jurassic to Palaeogene: Part 2, Jurassic and Cretaceous geochronology and Jurassic to Palaeogene Magnetostratigraphy. *In:* SNELLING, N. J. (ed.) *The Chronology of the Geological Record.* Geological Society, London, Memoir, 10, 118–140.

HANCOCK, J. M. 1969. Transgression of the Cretaceous Sea in south-west England. *Proceedings of the Ussher Society*, 2, 61–83.

—— 1975. The Sequence of Facies in the Upper Cretaceous of northern Europe compared with that in the Western Interior. *In:* CALDWELL, W. G. E. (ed.) *The Cretaceous System in the Western Interior of North America.* Geological Association of Canada, Special Paper 13, 83–118.

—— 1989. Sea-level changes in the British region during the Late Cretaceous. *Proceedings of the Geologists Association*, 100, 565–594.

—— 1993. Comments on the Exxon cycle chart for the Cretaceous System. *Cuadernos de Geología Ibérica*, 17, 57–78.

—— & KAUFFMAN, E. G. 1979. The great transgressions of the Late Cretaceous. *Journal of the Geological Society, London*, 136, 175–186.

——, KENNEDY, W. J. & COBBAN, W. A. 1993. A Correlation of the Upper Albian to Basal Coniacian Sequences of Northwest Europe, Texas and the United States Western Interior. *In:* CALDWELL, W. G. E. & KAUFFMAN, E. G. (eds) *Evolution of the Western Interior Basin.* Geological Association of Canada, Special Paper, 39, 453–476.

HAQ, B. U., HARDENBOL, J. & VAIL, P. R. 1987.

Chronology of Fluctuating Sea Levels Since the Triassic. *Science*, **235**, 1156–1167.

——, —— & —— 1988. Mesozoic and Cenozoic chronostratigraphy and cycles of sea-level change. *In:* WILGUS, C. K., HASTINGS, B. S., KENDALL, C. G. St. C., POSAMENTIER, C. A., ROSS, C. A. & VAN WAGONER, J. C. (eds) *Sea-level Changes: An Integrated Approach.* Society of Economic Paleontologists and Mineralogists, Special Publication, **42**, 71–108.

HAYS, J. D., PITMAN, W. C. 1983. Lithospheric plate motion, sea-level changes and climatic and ecological consequences. *Nature*, **246**, 16–22.

HEIM, D. 1957. Über die mineralischen, nichtcarbonatischen Bestandteile des Cenoman und Turon der mitteldeutschen Kriedemulden und ihre Verteilung. *Heidelberger Beiträge zur Mineralogie und Petrographie*, **5**, 302–330.

HEINZ, R. 1928. Über Cenoman und Turon bei Wunsdorf westlich von Hannover. (Zugleich Beiträge zur Kenntnis der oberkretazischen Inoceramen II). *Jahresbericht niedersachsen geologischen Veröffentlichung*, **21**, 18–28.

HILBRECHT, H. 1986. On the Correlation of the Upper Cenomanian and Lower Turonian of England & Germany (Boreal and Northern Tethys). *Newsletters on Stratigraphy*, **15**, 115–138.

—— 1988. Hangfazies in pelagiscshen Kalken und synsedimentäre Tektonik in Beispielen aus dem Mittel-Turon (Oberkriede) von N.W. Deutschland. *Zeitschrift Deutscher Geologischen Gesellschaft*, **139**, 83–109.

——, HOEFS, J. 1986. Geochemical and Palaeontological Studies of the ^{613}C anomaly in Boreal and North Tethyan Cenomanian–Turonian sediments in Germany and adjacent areas. *Palaeogeography, Palaeoclimatology, Palaeoecology*, **53**, 169–189.

HISS, M. 1983. Biostratigraphische Der Kriede–Basisschichten am Haarstrang (S.E. Westfalen) zwischen Unna und Möhnesee. *Zitteliana*, **10**, 43–54.

—— 1989. *Geologische Karte von Nordrhein-Westfalen 1:25,000. Erläuterungen zu Blatt 4417 Büren.* Geologisches Landesamt Nordrhein-Westfalen. Krefeld.

HUBBARD, R. J. 1988. Age and Significance of Sequence Boundaries on Jurassic and Early Cretaceous Rifted Continental Margins. *American Association of Petroleum Geologists Bulletin*, **72**, 49–72.

ION, J., SZASZ, L. 1994. Biostratigraphy of the Upper Cretaceous of Romania. *Cretaceous Research*, **15**, 59–87.

JARVIS, I. & WOODROOF, P. B. 1984. Stratigraphy of the Cenomanian and basal Turonian (Upper Cretaceous) between Branscombe and Seaton, SE Devon, England. *Proceedings of the Geologists Association*, **95**, 193–215.

JEFFERIES, R. P. S. 1962. The palaeoecology of the *Actinocamax plenus* subzone (lowest Turonian) in the Anglo-Paris Basin. *Palaeontology*, **4**, 609–647.

—— 1963. The Stratigraphy of the *Actinocamax plenus* Subzone (Turonian) in the Anglo-Paris

Basin. *Proceedings of the Geologists Association*, **74**, 1–33.

JENKYNS, H. C., GALE, A. S. & CORFIELD, R. M. 1994. Carbon- and Oxygen-Isotope stratigraphy of the English Chalk and Italian Scaglia and its palaeoclimatic significance. *Geological Magazine.* **131**, 1–34.

JUIGNET, P. 1974. *La transgression crétacée sur la bordure orientale du Massif armoricain. Aptien, Albien, Cénomanien de Normandie et du Maine. La stratotype du Cénomanien.* PhD Thesis, Université de Caen.

—— & BRETON, G. 1992. Mid-Cretaceous sequence stratigraphy and sedimentary cyclicity in the western Paris Basin. *Palaeogeography, Palaeoclimatology, Palaeoecology*, **91**, 197–218.

JUKES-BROWNE, A. J. & HILL, W. 1903. *The Cretaceous Rocks of Great Britain. Volume 2: The Lower and Middle Chalk of England.* Memoir of the Geological Survey of Great Britain.

KAEVER, M. 1985. Beiträge zur Stratigraphie, Fazies, und Paläogeographie der Mittleren und Oberen Kriede Westfalens (N.W. Deutschland). *Münster Forschrift Geologie, Paläontologie*, **63**.

KAPLAN, U. 1992. Die Oberkriede-Aufschlüsse im Raum Lengerich/Westfalen. *Geologische, Paläontologische, Westfalen*, **21**, 7–37.

KENNEDY, W. J. 1969. The Correlation of the Lower Chalk of South East England. *Proceedings of the Geologists Association*, **80**, 460–547.

—— 1970. A Correlation of the Uppermost Albian and the Cenomanian of south-west England. *Proceedings of the Geologists Association*, **81**, 613–677.

—— 1971. *Cenomanian Ammonites from Southern England.* Special Papers in Palaeontology, **8**.

—— & GARRISON, J. 1975. Morphology and Genesis of nodular chalks and hardgrounds in the Upper Cretaceous of Southern England. *Sedimentology*, **22**, 311–416.

—— & JUIGNET, P. 1973. Observations on the Lithostratigraphy and Ammonite succession across the Cenomanian–Turonian Boundary in the environs of Le Mans (Sarthe, N.W. France). *Newsletters on Stratigraphy*, **2**, 189–202.

KLINGER, H. C. & WIEDMANN, J. 1983 Palaeobiogeographic affinities of Upper Cenomanian Ammonites of Northern Germany. *Zitteliana*, **10**, 413–425.

MARCINOWSKI, R. 1980. Cenomanian Ammonites from the German Democratic Republic, Poland and the Soviet Union. *Acta Geologica Polonica*, **30**.

MEYER, T. 1990. Biostratigraphische und sedimentologische Untersuchungen in der Planerfazies des Cenoman von Nordwestdeutschland. *Mitteilungen aus dem Geologischem Institut der Universität Hannover*, **IV S**.

MIALL, A. D. 1992. Exxon global cycle chart: An event for every occasion? *Geology*, **20**, 787–790.

MOREAU, P. 1976. Cadre stratigraphique et rythmes sédimentaires du Cénomanien nord-aquitain (région de Rochefort). *Bulletin de la Société géologique de France*, **7**, 747–755.

——, FRANCIS, I. H. & KENNEDY, W. J. 1983.

Cenomanian Ammonites from Northern Aquitaine. *Cretaceous Research*, **4**, 317–339.

MORTER, A. A. & WOOD, C. J. 1983. The biostratigraphy of Upper Albian – Lower Cenomanian *Aucellina* in Europe. *Zittelania*, **10**, 515–529.

MORTIMORE, R. N. & POMEROL, B. 1991. Upper Cretaceous tectonic disruptions in a placid Chalk sequence in the Anglo-Paris Basin. *Journal of the Geological Society, London*, **148**, 398–404.

PARKINSON, N. & SUMMERHAYES, C. 1985. Synchronous global sequence boundaries. *American Association of Petroleum Geologists Bulletin*, **69**, 658–687.

PLATEL, J-P. 1989. Le Cretace Supérieur de la Plate-Forme Septentrionale du Bassin d'Aquitaine. Stratigraphie et Évolution Géodynamique. *Bureau de Recherches Géologiques et Minières*, **164**.

PLINT, A. G. 1988. Global eustasy and the Eocene sequence in the Hampshire Basin, England. *Basin Research*, **1**, 11–22.

RAWSON, P. F., CURRY, D., DILLEY, F. C., HANCOCK, J. M., KENNEDY, W. J. ET AL. 1978. *A correlation of Cretaceous rocks in the British Isles*. Geological Society, London, Special Report, **9**.

ROBASZYNSKI, F., JUIGNET, P., GALE, A. S., AMEDRO, F. & HARDENBOL, J. 1995. Sequence stratigraphy in the European Upper Cretaceous of the Anglo-Paris Basin, as exemplified by the Cenomanian Stage. *In: Sequence Stratigraphy of European Basins*. Conference Volume CNRS–IFP, Dijon, France, in press.

——, HARDENBOL, B. U., CARON, M., AMEDRO, F., DUPUIS, C. ET AL. 1993. Sequence stratigraphy in a distal environment: the Cenomanian of the Kalaat Senan Region (Central Tunisia). *Bulletin des Centres de Recherche, Exploration–Production. Elf-Aquitaine*, **17**, 395–433.

SCHMID, F. 1965. *Actinocamax plenus* (Blainville) ein seltener Belemnitenfund im Rötplaner (Oberkreide, Niedersachsen). *Geologisches Jahrbuch*, **83**, 517–532.

SCHOLLE, P. A. & ARTHUR, M. A. 1980. Carbon Isotope Fluctuations in Cretaceous Pelagic Limestones: Potential Stratigraphic and Petroleum Exploration Tools. *American Association of Petroleum Geologists Bulletin*, **64**, 67–87.

SCHÖNER, H. 1960. Über die Verteilung und Neubildung der Nichtcarbonatischen Mineralkomponenten der Oberkriede aus der Umgebung von Hannover. *Heidelberger Betreit Mineralogie, Petrologie*, **7**, 76–103.

SMITH, W. E. 1957. The Cenomanian Limestone of the Beer District, South Devon. *Proceedings of the Geologists Association*, **68**, 115–133.

—— 1965. The Cenomanian deposits of southeast Devonshire: The Cenomanian Limestone east of Seaton. *Proceedings of the Geologists Association*, **76**, 121–136.

SUESS, E. 1906. *The Face of the Earth* **2**. (Das Antlitz der Erde), Oxford, Clarendon Press.

THORNE, J. & WATTS, A. B. 1984. Seismic reflections and conformities at passive continental margins. *Nature*, **311**, 365–367.

TRÖGER, K. A. 1981. Zu Problemen der Biostratigraphie der Inoceramen und der Untergliederung Des Cenomans und Turons in Mittel und Österuropa. *Newsletters on Stratigraphy*, **9**, 139–156.

WATTS, A. B. 1982. Tectonic subsidence, flexure and global changes of sea level. *Nature*, **297**, 469–474.

WEGNER, T. 1925. Cenoman und Turon bei Lengerich. *Führer zur Tagung der Deutschen Geologie in Münster*, 66–70.

WIEDMANN, J., KAPLAN, U., LEHMANN, J. & MARCINOWSKI, R. 1989. Biostratigraphy of the Cenomanian of NW Germany. *In:* WIEDMANN, J. (ed.) *Cretaceous of the Western Tethys. Proceedings of the Third International Cretaceous Symposium, Tübingen 1987*. 931–948.

WRIGHT, C. W. & KENNEDY, W. J. 1981. *The Ammonoidea of the Plenus Marls and the Middle Chalk*. Palaeontographical Society, London, Monograph.

—— & —— 1984. *The Ammonoidea of the Lower Chalk. Part 1*. Palaeontographical Society, London, Monograph, 1–126.

—— & —— 1987. *The Ammonoidea of the Lower Chalk. Part 2*. Palaeontographical Society, London, Monograph, 127–218.

—— & —— 1990. *The Ammonoidea of the Lower Chalk. Part 3*. Palaeontographical Society, London, Monograph, 219–294.

Subaerial exposure unconformities on the Vercors carbonate platform (SE France) and their sequence stratigraphic significance

BRUCE W. FOUKE,[1] ARNOUT-JAN W. EVERTS,[2] ERIK W. ZWART,[1]
WOLFGANG SCHLAGER, P. C. SMALLEY[3] & HELMUT WEISSERT[4]

[1] *Institute for Earth Sciences, Vrije Universiteit, De Boelelaan 1085, 1081 HV Amsterdam, The Netherlands*
[2] *Koninklijke Shell Exploratie en Productie Laboratorium, Volmerlaan 8, 2280 AB Rijswijk, The Netherlands*
[3] *BP Research Centre, Chertsey Road, Sunbury-on-Thames, TW16 7LN, UK*
[4] *Geological Institute, Swiss Federal Institute of Technology, CH-8092 Zurich, Switzerland*

Abstract: The integration of data on diagenesis and stratal geometry at the margin of the Vercors carbonate platform (SE France) shows that the most prominent break in depositional style does not coincide with the platform-top horizon exhibiting the most extensive meteoric alteration. This observation again illustrates the ambiguity of geometrical criteria to define sequence boundaries related to subaerial exposure.

Outcrops at the margin of the Cretaceous Vercors platform expose prograding to aggrading tongues of platform grainstones. Growth and lateral progradation of these platform tongues was frequently interrupted, as evidenced by the deposition of wedges of fine-grained deeper water sediments that encroached the clinoform slopes. Petrographic and geochemical analyses have been carried out at strategic bedding surfaces in order to evaluate the extent to which these breaks relate to sea-level falls and subaerial exposure. The analyses reveal evidence for minor meteoric alteration at all four of the bedding surfaces on the platform top that were studied. However, the Surface 3 bedding plane is unique in that it shows the overprinting of several events of meteoric diagenesis. Petrographic and geochemical analyses suggest that the rudist floatstones at this particular surface were diagenetically overprinted at least three times by meteoric groundwaters that dissolved skeletal grains and precipitated bladed and blocky calcite cements that exhibit bladed to blocky morphologies, low Mg, Mn and Fe abundances, depleted $\delta^{18}O$ and $\delta^{13}C$ signatures, and freshwater fluid inclusions. Hardground borings then cross-cut the meteoric calcite cements and biomolds, indicating that the subaerial exposure and meteoric overprinting took place prior to deposition of the overlying marine grainstones. This ensuing period of marine inundation was also accompanied by the deposition of red argillaceous internal sediments and dolomitization. The marine grainstones overlying Surface 3 contain lithoclasts with truncated dolomite rhombs at their margins, suggesting that the dolomitization at Surface 3 also relates to early stage diagenesis.

The offbank continuation of the Surface 3 bedding plane exhibits a distinct wedge of dolomitized lithoclastic debris. No such debris was found along the slope continuation of the other studied surfaces, confirming that the Surface 3 bedding plane relates to a particularly significant episode of subaerial exposure and erosion. However, within the framework of changing stratal geometries at the margin of the Vercors platform, the exposure Surface 3 represents a rather insignificant event. In contrast, the most prominent break in depositional style at the Vercors platform-margin exhibits only minor meteoric alteration along its platform-top continuation. This 'stratigraphic mismatch' suggests that sea-level falls exposing the platform-top exerted only a minor influence on the resulting platform stratal geometries.

In any marine depositional setting where sediment input varies with time, classically defined systems tracts (Van Wagoner *et al.* 1988) represent a dynamic equilibrium between accommodation and sediment supply rather than changes in relative sea-level alone. As a result, the accurate reconstruction of sea-level history from the sedimentary record requires that the effects of eustatic and tectonically controlled sea-level fluctuations be separated from environmental changes (Kendall & Schlager 1981; Schlager, 1991, 1993).

However, two stratigraphic features are exempt from the ambiguities associated with distinguishing changes in accommodation versus changes in sediment supply. These include

From Howell, J. A. & Aitken, J. F. (eds), *High Resolution Sequence Stratigraphy: Innovations and Applications*, Geological Society Special Publication No. 104, pp. 295–320.

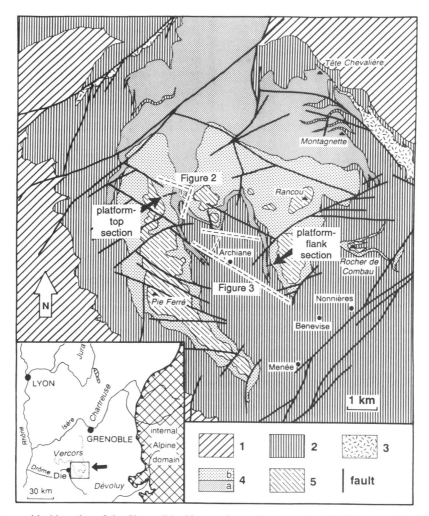

Fig. 1. Geographical location of the Cirque d'Archiane at the southern margin of the Vercors carbonate platform in southeastern France (modified after Arnaud 1981). The perspectives from which the outcrop photographs in Figs 2 and 3 are taken are indicated by the dashed lines adjacent to each figure number. Legend: 1. Neocomian; 2. Barremian–Aptian fine-grained slope and basinal sediments; 3. Lower Barremian allodapic limestones (Borne Formation); 4. Barremian platform and foreslope sediments (Glandasse Formation); 4a. members Bi 1 to Bi 5; 4b. members Bi 6 to Bs 1; 5. Aptian platform sediments (Urgonian Formation).

subaerial exposure unconformities and abrupt basinward shifts of shallow water facies (*forced regressions* of Posamentier *et al.* 1992; Schlager 1993; Franseen *et al.* 1993; Pomar, 1993; Goldstein & Franseen 1995), both of which are indicative of relative falls in sea-level. The diagenetic susceptibility and flat-top to steep-flank morphology of carbonate platforms make the preservation, distribution and resulting recognition of subaerial exposure unconformities in these deposits greater than in siliciclastic systems (e.g. James & Choquette 1984; Weimer

1992). Moreover, several diagenetic criteria have been established with which to recognize subaerial exposure surfaces in carbonate platforms (i.e. Allen & Matthews 1982; Meyers & Lohmann 1985; Saller & Moore 1989; Goldstein *et al.* 1991). Diagenetic analyses provide a means to verify the subaerial character of presumed lowstand unconformities, which is essential for an accurate reconstruction of sea-level history from depositional sequences.

The spectacular outcrops of the Barremian Glandasse Formation limestones comprising the

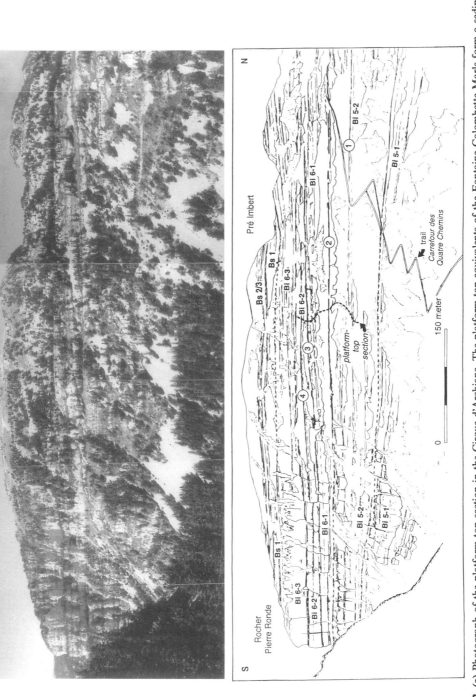

(a)

(b)

Fig. 2. (a) Photograph of the platform-top section in the Cirque d'Archiane. The platform-top equivalents of the Fontaine Colombette Marls form a sediment drape immediately overlying Surface 2. **(b)** Line tracing of photograph in (a). The position of the platform-top section is indicated. Numbered circles indicate the different bedding surfaces on which diagenetic analyses were completed. Note that the stratigraphic relationships of the different platform tongues (Bi 5-1 to Bs 2/3) are plane-parallel all along this domain.

(a)

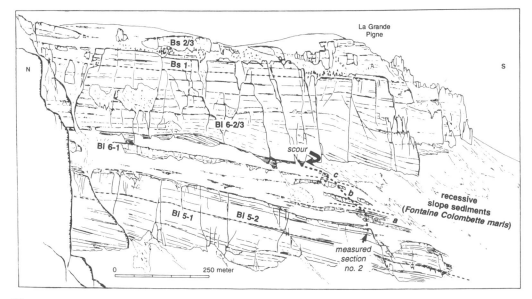

(b)

Fig. 3. (a) Photograph of the platform-flank section on the eastern side of the Cirque d'Archiane. **(b)** Line tracing of photograph in (a). The position of the platform-flank section is indicated. Visible are the different platform tongues (Bi 5-1 to Bs 2/3) as well as their continuation downslope. The labels *a* to *c* refer to the subunits of the Fontaine Colombette Marls as identified in the platform-flank section (Fig. 5; Everts *et al.* 1995)

Vercors carbonate platform in southeastern France (Fig. 1) provide an excellent natural laboratory in which to integrate diagenetic studies into a sequence stratigraphic framework. The platform-to-basin transition exposed in the Cirque d'Archiane exhibits two prograding grainstone tongues (called Bi 5 and Bi 6 respectively) separated by a basinward wedge of finer-grained sediments (called the Fontaine Colombette Marls; Figs 2a, b and 3a, b; Arnaud 1981). Significant controversy has developed over the extent to which these stratal geometries were created by third-order sea-level cycles (Arnaud 1981; Jacquin et al. 1991; Hunt & Tucker 1993; Stafleu et al. 1994; Everts et al. 1995). However, to date no detailed diagenetic analyses have been completed at the stratigraphic intervals where the major changes in depositional geometries occur.

In the present study, petrographic, geochemical, and fluid inclusion analyses have been completed on strategic Bi 5 and Bi 6 platform-top and platform-flank bedding surfaces to identify subaerial exposure unconformities in the Vercors carbonate platform. The stratigraphic position of these exposure unconformities are then evaluated with respect to the large-scale stratal geometries and sedimentological framework of the Vercors carbonate platform to evaluate reconstructions of sea-level history.

Methods

Polished thin-sections were examined with cathodoluminescence on a Technosyn 800 Cold Luminoscope operating at 12 Kv. Carbon and oxygen isotope analyses of calcite and dolomite were measured using two different techniques, the first of which was by the analysis of CO_2 released during digestion of drilled powders in 100% phosphoric acid at 50°C on a V.G. Micromass-903 mass spectrometer. Acid digestion data are reported as $\delta^{18}O$ and $\delta^{13}C$-values for CO_2 gas relative to PDB in the standard per mil notation, and yield an analytical precision of 0.1 per mil for O and 0.2 per mil for C. Isotopic analyses were also completed on CO_2 liberated by laser ablation on a Finnegan MAT 251 mass spectrometer (Smalley et al. 1989). The laser-microprobe operated with a spot diameter of approximately 30 mm, and required empirical corrections of –0.8‰ $\delta^{13}C$ and 1.2‰ $\delta^{18}O$ for calcite and –0.8‰ $\delta^{13}C$ and 0.4‰ $\delta^{18}O$ for dolomite. Reproducibility (σ) of the laser-microprobe data was approximately 0.4 per mil for O and 0.2 per mil for C.

In situ elemental abundances were obtained for Ca, Mg, Fe and Mn by electron probe

microanalysis (EPMA) operating at 20 kV and 0.015 uA with a 13 μm beam diameter. Estimated precision is expressed as 95% confidence limits and are typically less than 2 wt% for Ca and Mg and 4 wt% for Mn and Fe as calculated from replicate analyses. Doubly polished 150 mm-thick sections were analysed for fluid inclusion microthermometry on a Linkam TP/91-THMS 600 gas flow heating and freezing stage. Melting temperatures in all-liquid inclusions are unreliable due to metastability (Roedder, 1984), therefore the Event A7 calcites and Event A12 calcites were artificially stretched by heating them to 245°C and 275°C respectively to induce a vapour bubble (Bodnar & Bethke 1987; Goldstein 1990, 1993). Inclusions that did not develop a vapour-phase when returned to room temperature were not used for further analyses. Inclusions were cooled until all of the liquid in the inclusion was frozen, and then slowly heated at a rate of 0.5–1°C per minute until all of the ice was melted, yielding a final melting temperature (T_m). The techniques used to determine the primary versus secondary origin of inclusions, liquid/vapour ratios, and gas phase pressures follow Roedder (1984).

Geological and stratigraphic setting

The Vercors carbonate platform crops out in the External Alpine units (Dauphiné Region) of the Western Alps (Fig. 1). Late Cretaceous to Tertiary Alpine tectonics have been relatively mild in this area, and therefore the strata are largely undeformed except for some local faults. During the Early Cretaceous, the Vercors platform formed the southern extension of the Jura platform and rimmed the northern margin of the Vocontian Basin (Fig. 1; Arnaud 1981; Arnaud-Vanneau & Arnaud 1990). The Late Hauterivian and Early Barremian showed a rapid southward progradation of the platform edge, while the platform gradually evolved from progradation to aggradation and eventually retrogradation during the Late Barremian to Early Aptian. In mid-Aptian times the Vercors platform finally drowned (Arnaud-Vanneau & Arnaud 1990; Jacquin et al. 1991). Basinward progradation of the southern margin of the Vercors platfom during the Barremian was not continuous. It was interrupted by several episodes of platform demise when deeper water limestones and marls encroached the platform flanks. The resulting stratal geometries are best exposed in the Cirque d'Archiane, where prograding tongues of shallow water grainstones alternate with wedges of deeper water sediments (Figs 2a, b and 3a, b; Arnaud 1981; Everts et al. 1995). These deposi-

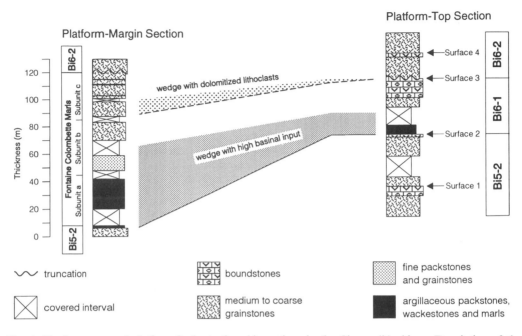

Fig. 4. Platform-top and platform-flank stratigraphic sections in the Cirque d'Archiane. Description of the lithologies in each section are presented in the text. Correlations between sections are based on the presence of red dolomitized clasts and limestones containing basinal sediments.

tional geometries of the Vercors platform have avoided being distorted during ensuing compaction and Alpine tectonism (i.e. horizontal top-set beds, consistent clinoform angles).

A complete sedimentological and stratigraphic description of the exposures in the Cirque d'Archiane is presented in Everts *et al.* (1995). Based on integrated studies of stratal geometries and microfacies, Everts *et al.* (1995) propose that the Bi 5 and Bi 6 grainstone tongues originally described by Arnaud (1981) be subdivided into units Bi 5-1, 2 and Bi 6-1, 2, 3 (Figs 2a, b and 3a, b). One section from the platform-top and one section from the platform-flank have been analysed in the present study (Figs 2b, 3b and 4). The platform-top lithologies are composed (Fig. 4) of: (1) Bi 5-2, a 60 m-thick sequence of coarse-grained grainstones alternating with coral–stromatoporoid boundstones, which is capped by a 1 m-thick layer of coral rudist boundstones; (2) a *c.* 15 m-thick recessive unit of fine-grained packstones and grainstones that separates Bi 5-2 and Bi 6-1; (3) Bi 6-1, a 25 m-thick sequence of cross-bedded, coarse-grained grainstones overlain by coral floatstones and a 3 m-thick cap of rudist pack-floatstones; and (4) Bi 6-2, a 30 m-thick sequence of cross-

bedded algal foraminifer and oolitic grainstones, with a coral floatstone bed at 15 m from the base.

Four prominent bedding surfaces have been observed in the platform-top sequence of the Cirque d'Archiane (Figs 2b and 4). Surface 1 is the bedding plane at the top of the coral-stromatoporoid boundstones 40 m below the top of the Bi 5-2 sequence. The Surface 2 bedding plane is at the top of coral rudist boundstones that cap the Bi 5-2 sequence. Surface 3 is the bedding surface at the top of rudist pack-floatstones capping the Bi 6-1 sequence. And finally, the Surface 4 bedding plane occurs within the lower portion of the Bi 6-2 limestones, approximately 15 m above Surface 3 at the top of a coral floatstone bed. Some of these surfaces have been assigned sequence stratigraphic significance by previous authors. Jaquin *et al.* (1991) proposed that a sequence boundary occurs at approximately the level of Surface 2. Conversely, Hunt & Tucker (1993) interpreted the rudist bed at the top of Bi 6-1 as a 'forced regressive wedge' topped by a sequence boundary in a position equivalent to Surface 3.

The platform-flank lithologies are composed (Figs 3b and 4) of: (1) the uppermost 10 m of Bi

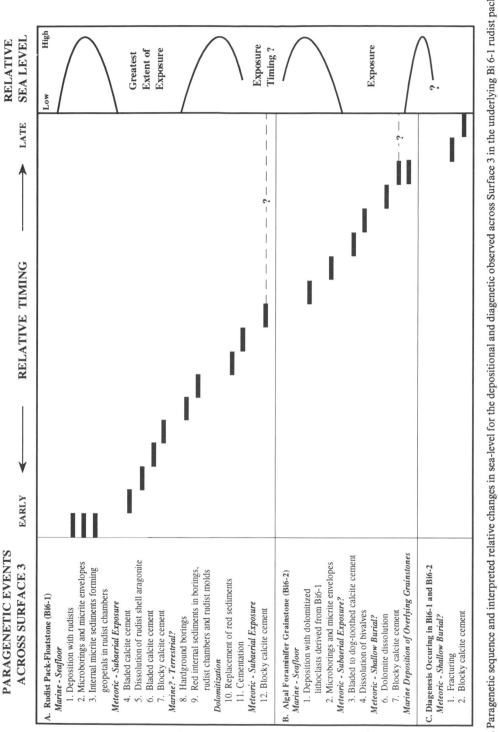

Fig. 5. Paragenetic sequence and interpreted relative changes in sea-level for the depositional and diagenetic observed across Surface 3 in the underlying Bi 6-1 rudist pack-floatstones and the overlying Bi 6-2 grainstones.

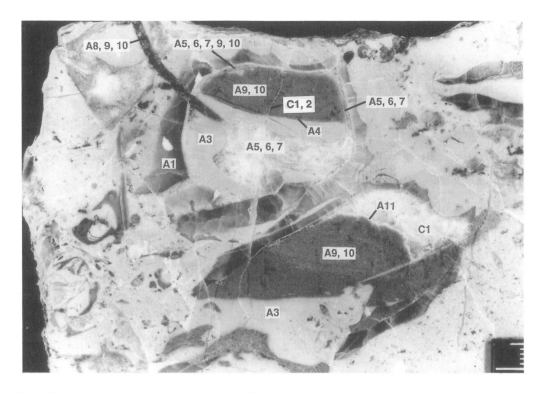

Fig. 6. Handsample photograph of the Bi 6-1 rudist pack-floatstones. Numbered features refer to paragenetic events presented in Fig. 5. The Surface 3 bedding plane is the surface at the top of the photograph. Scale bar is 1 cm.

5-2, a thick sequence of coarse-grained poorly sorted skeletal packstones; (2) the *Fontaine Colombette Marls Subunit A*, a 40 m-thick sequence of bioturbated fine-grained skeletal grainstones, packstones and wackestones that rhythmically alternate with mudstones and marls; (3) the *Fontaine Colombette Marls Subunit B*, a 45 m-thick sequence of fine-grained peloidal packstones grainstones; (4) the *Fontaine Colombette Marls Subunit C*, a 35 m-thick sequence of bioturbated coarse-grained skeletal/lithoclastic grainstones and packstones with interlayered clay partings; and (5) Bi 6-2/3, an over 120 m-thick sequence of massive clinostratified medium- to coarse-grained skeletal grainstones. Tracing of bedding surfaces (Fig. 4) shows that the boundary between *Subunits B* and *C* of the *Fontaine Colombette Marls* is the downslope continuation of Surface 3 (i.e the Bi 6-1 top). A conspicuous scour at the base of the Bi 6-2/3 clinoforms is likely to represent the downward continuation of Surface 4 (Everts *et al.* 1995).

Paragenetic sequence

Platform-top paragenesis

A total of 19 paragenetic events have been indentified at the Surface 3 bedding plane in the uppermost Bi 6-1 rudist pack-floatstones and the lowermost Bi 6-2 algal foraminifer grainstones. Conversely, a total of only 4 paragenetic events have been observed at Surfaces 1, 2 and 4. Because Surface 3 exhibits more diagenetic events than Surfaces 1, 2 and 4, it will serve as a comparative standard with which to evaluate diagenesis on Surfaces 1, 2 and 4. The stratigraphic distribution of each of the paragenetic events with respect to the Surface 3 bedding plane has been used to establish the relative timing of the diagenetic alteration.

Paragenetic sequence across Surface 3. The Bi 6-1 pack-floatstones are composed of large rudist shells, corals, bivalves, algae, foraminifers, microborings, micritic envelopes, and horizon-

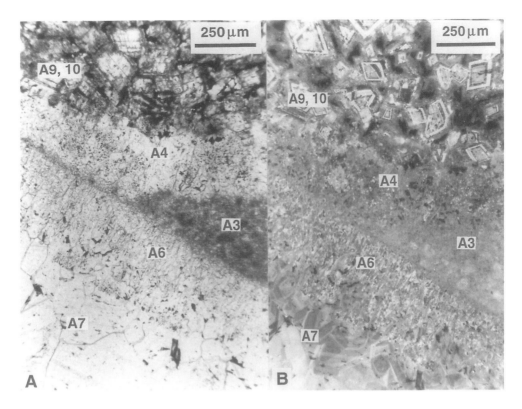

Fig. 7. Paired plane-light and cathodoluminescence photomicrographs of the Bi 6-1 rudist pack floatstones immediately underlying Surface 3. Numbered features refer to the paragenetic sequence presented in Fig. 5. The dissolved aragonitic layer of the rudist shell, which is now filled with the A6 and A7 cement generations, is in the lower left-hand corner of each photograph. This photograph is an enlargement of the upper left-hand side of the uppermost rudist shell in Fig. 6. The photograph is taken where the contact between the A3 and A9 internal sediments meets the inner rudist shell wall, a few millimetres above the position where the A8 boring penetrates the shell.

tally-orientated internal sediment geopetal fabrics (*Events A1, A2 and A3*; Figs 5 and 6). A succession of calcite cementation and dissolution then occurs, that is initiated with the precipitation of bladed calcite cements on the inside wall of rudist chambers and the upper surface of micrite geopetals (*Event A4*, Figs 5 and 6). The Event A4 bladed calcites are up to 100 μm in length and exhibit noncathodoluminescent patches within an overall mottled cathodoluminescence fabric, while the A2 and A3 micrites are brightly cathodoluminescent (Fig. 7). This was followed by dissolution of the 200–300 μm-thick inner aragonitic layer of many of the Bi 6-1 rudist shells (*Event A5*; Figs 5, 6 and 7), with the remaining low-Mg calcite layers being texturally unaltered and noncathodoluminescent (A1, Fig. 8). A second generation of bladed calcite cement rims was then precipitated exclusively within the rudist shell molds (*Event A6*; Figs 5, 6 and 7).

The Event A6 bladed calcites are distinct from the previous Event A4 cements not only in their distribution, but also in that they are slightly larger (150–200 μm in length) and exhibit a brightly zoned cathodoluminescence (Fig. 7). And finally, a blocky calcite cement was precipitated within the rudist shell molds (*Event A7*; Figs 5 and 6) that encrusts the Event A6 bladed calcites and occluded most of the remaining moldic porosity (Fig. 7). The Event A7 blocky calcites are up to 300 μm in diameter and exhibit concentric cathodoluminescence zonations as well as sectoral zoning (Fig. 9). Elongate cylindrical borings were then formed (*Event A8*; Figs 5 and 6) that cross-cut all previous paragenetic events (Fig. 10). The Event A8 borings are up to 2 cm in length with smooth walls that taper toward a rounded termination (Fig. 10). Red internal argillaceous sediments (*Event A9*; Figs 5 and 6) then partially to

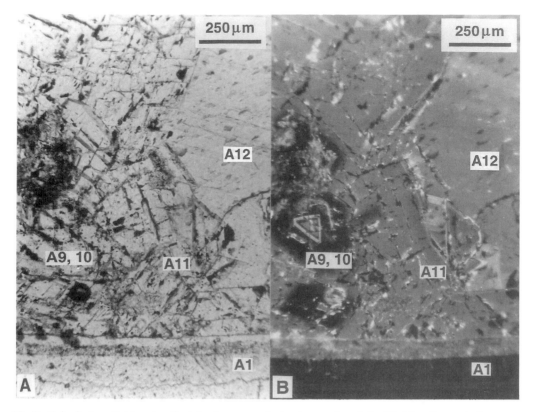

Fig. 8. Paired plane-light and cathodoluminescence photomicrographs of the Bi 6-1 rudist pack floatstones immediately underlying Surface 3. Numbered features refer to the paragenetic sequence presented in Fig. 5.

completely filled rudist shell chambers, rudist shell molds, shelter porosity and Event A8 burrows (Fig. 10). The internal sediments have filtered into primary and secondary porosity within the Bi 6-1 limestones down to 7 m below the Surface 3 bedding plane.

Two separate dolomitization episodes then took place on Surface 3. The first is composed of dolomite that preferentially replaces the Event A9 red internal sediments (*Event A10*; Figs 5 and 6). The Event A10 replacement dolomites are 75–125 μm-diameter euhedral crystals that exhibit bright concentric cathodoluminescent zonations (Figs 7 and 8). The second generation of dolomite (*Event A11*; Figs 5 and 6) is a cement that encrusts the upper surface of the Event A9/A10 dolomitized red internal sediment geopetals. The Event A11 cement dolomites are 100–300 μm-diameter rhombohedra with a bright homogeneous cathodoluminescence (Fig. 8), and are therefore distinct in distribution, size and cathodoluminescence fabric from the Event A10 dolomites (Fig. 8). A blocky calcite cement

(*Event A12*; Figs 5 and 6) then encrusted the Event A11 dolomite cements to occlude the remaining primary porosity. The Event A12 blocky calcites are 100–200 μm-diameter crystals that exhibit concentric zonations under cathodoluminescence in addition to sectoral and intrasectoral zoning (Fig. 8).

The Bi 6-2 grainstones overlying Surface 3 are composed of well rounded fragments of algae, rudists, oysters, foraminifers and red lithoclasts (*Events B1 and B2*, Fig. 5). Several of the Bi 6-2 lithoclasts are dolomitized, and contain truncated dolomite rhombohedra at their margins that are encrusted by calcite cements (Fig. 11). The initial phase of intergranular Bi 6-2 calcite cementation is composed of rims of bladed to dog-toothed calcite crystals 25–75 μm in length that have concentric cathodoluminescent zonations (*Event B3*; Figs 5 and 11). This was followed by the partial dissolution of bivalve shells (*Event B4*; Figs 5 and 11), the timing of which is indicated by the absence of Event B3 bladed calcites within bivalve molds. And finally

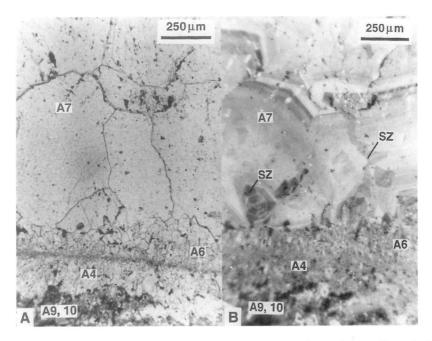

Fig. 9. Paired plane-light and cathodoluminescence photomicrographs of the Bi 6-1 rudist pack floatstones immediately underlying Surface 3. Numbered features refer to the paragenetic sequence presented in Fig. 5. Event A7 blocky calcite cements exhibit sectoral zoning (SZ) under cathodoluminescence.

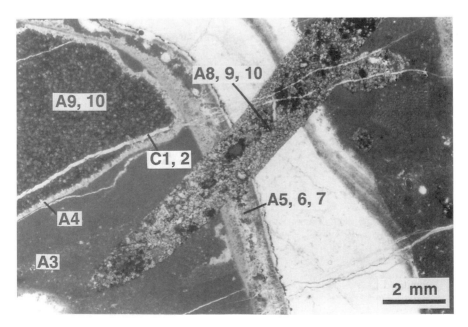

Fig. 10. Plane-light photomicrograph of a Surface 3 boring cross-cutting Bi 6-1 rudists, internal sediments and calcite cements. Numbered features refer to the paragenetic sequence presented in Fig. 5. This photograph is a mirror image enlargement of the upper left-hand side of the upper rudist shell in Fig. 6, at the position where the A8 boring cuts through the shell wall.

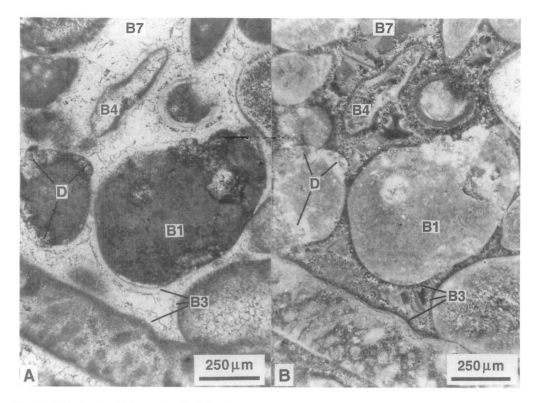

Fig. 11. Paired plane-light and cathodoluminescence photomicrographs of the Bi 6-2 grainstones directly overlying Surface 3. Truncated dolomite rhombohedra at the edge of rounded lithoclasts are labelled 'D'. Numbered features refer to the paragenetic sequence presented in Fig. 5.

a blocky calcite cement (*Event B7*; Figs 5 and 11) was precipitated. The Event B7 blocky calcites are approximately equivalent in size to the A12 calcites (75–150 μm), but exhibit distinctly thinner crystal zonations under cathodoluminescence (Figs 8 and 11). In addition, the A12 calcite cements occur exclusively within the underlying Bi 6-1 limestones, while the B7 calcites are confined to the overlying Bi 6-2 limestones.

The late-stage Surface 3 paragenetic events occur in both the underlying Bi 6-1 pack-floatstones and the overlying Bi 6-2 grainstones. The first is extensive fracturing (*Event C1*; Figs 5, 6 and 12) that cross-cuts all previous depositional and diagenetic events. This was followed by a blocky calcite cement (*Event C2*; Figs 5, 6 and 12) that was precipitated within the Event C1 fractures. The Event C2 blocky calcites are up to 200 μm in diameter and exhibit concentrically zoned cathodoluminescence (Fig. 12).

Paragenetic sequence across Surfaces 1, 2 and 4. Surfaces 1, 2 and 4 exhibit significantly less diagenetic alteration than Surface 3, with each surface containing only four paragenetic events. Following deposition, bivalves and other skeletal grains were partially to completely dissolved (Fig. 13). This was followed by the precipitation of patchy rims of bladed to dog-toothed calcite cements within intergranular and biomoldic porosity. These 35–50 μm-long crystals have scalenohedral terminations and exhibit concentric cathodoluminescent zonations (Fig. 13). Blocky calcite cements were then precipitated that occluded the remaining porosity. These blocky calcite crystals are 100–150 μm in diameter and exhibit a concentrically-zoned cathodoluminescence (Fig. 13). And finally, fractures cross-cutting all previous events are filled with blocky calcites exhibiting concentric cathodoluminescent zonations identical to those of the Event C2 calcites. As observed on Surface 3, some internal sediments composed of red,

Fig. 12. Paired plane-light and cathodoluminescence photomicrographs of the Bi 6-1 rudist pack floatstones immediately underlying Surface 3. Numbered features refer to the paragenetic sequence presented in Fig. 5.

argillaceous materials have infilled primary and secondary (intraskeletal) porosity on Surfaces 1, 2 and 4. However, unlike Surface 3, the Surface 1, 2 and 4 internal sediments did not penetrate more than 20–30 cm into the underlying limestones. In addition, the rocks overlying platform-top Surfaces 1, 2 and 4 do not contain red dolomitized lithoclasts similar to those found above Surface 3.

Platform-flank paragenesis

The paragenesis of the packstones and grainstones comprising the platform flank succession has not been studied in detail except for an 8 m-thick interval at the base of *Subunit C* of the *Fontaine Colombette Marls* (i.e. at the downslope continuation of platform top Surface 3; Fig. 4). The rocks in this interval are grainstones composed of coarse-grained, well rounded fragments of algae, rudists, oysters and foraminifers, as well as significant quantities of red lithoclasts. The red lithoclasts are composed of a red micrite similar to the Event A9 internal

sediments at the base of the Bi 6-2 grainstones on the platform-top section, and have not been observed in any other stratigraphic interval within the platform-flank section. Each of these red lithoclasts is composed of a complex mosaic of original grain fabrics and secondary calcitization. The lithoclasts commonly contain rhomb-shaped molds filled with small equant calcite crystals with concentric cathodoluminescent zonations (Fig. 14). These calcite filled molds occurring at the margin of rounded clasts are truncated (Fig. 14). In addition, host grains and rhombic molds are cross-cut by fractures filled with blocky calcites. These calcites are less than $50\,\mu m$ in diameter and have concentric cathodoluminescent zonations.

Geochemical analyses

The isotopic and trace element composition of the micrites, skeletal components, and calcite and dolomite cements comprising the Surface 3 paragenetic sequence are presented in Table 1 and Fig. 15. In addition, bulk rock O and C

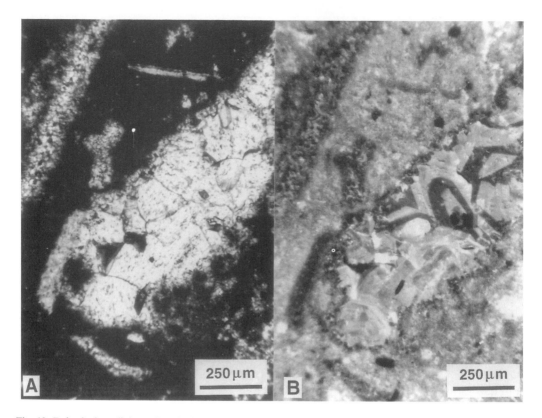

Fig. 13. Paired plane-light and cathodoluminescence photomicrographs of the Bi 5-2 grainstones immediately underlying Surface 2. Photographs exhibit dog-toothed and blocky calcite cements with cathodoluminescent zonations precipitated in biomoldic porosity.

isotopic analyses taken throughout the platform-top section are presented in Figs 16 and 17. The most distinct geochemical characteristics of the Surface 3 paragenetic sequence are: (1) positive and negative excursions in $\delta^{18}O$ and $\delta^{13}C$ that are not sequentially consistent with the paragenetic sequence (Fig. 15); and (2) a negative peak in bulk rock $\delta^{13}C$ at Surface 3 that decreases in the underlying limestones from the average baseline ratios exhibited in the rest of the platform-top section (Figs 16 and 17). The depositional and diagenetic components comprising Surface 3 generally exhibit low abundances of Mg, Mn and Fe (Table 1). The only exceptions to this are: (1) the high Mn abundances (104 to 156 ppm) in the Event A11 dolomite cements; and (2) the high Mn (< 47–204 ppm) and Fe (900–1600 ppm) abundances in the late stage fracture filling C2 blocky calcite cements. X-ray diffraction analyses of the dolomitized Event A9 red argillaceous internal sediments indicate that they are composed of

smectite, kaolinite, illite, mica, goethite, hematite, quartz, feldspar and talc.

Fluid inclusion analyses

Two of the generations of blocky calcite cement observed across Surface 3 (Events A7 and A12; Figs 5 and 6) contain abundant one and two-phase fluid inclusions. Event A7 blocky calcite cements contain mainly single-phase fluid inclusions that are irregularly shaped and range from 2–30 μm in diameter (Fig. 18). Conversely, Event A12 blocky calcite cements contain abundant two-phase liquid–vapour fluid inclusions that are 2–65 μm in diameter (Fig. 18). The majority of the inclusions in both Event A7 and A12 calcites are necked and irregular in shape. While most of the inclusions tend to be flat, especially solitary inclusions exhibit a clear third dimension. The flat inclusions generally occur within short inclusion trails that terminate at cleavage plains and crystal growth boundaries

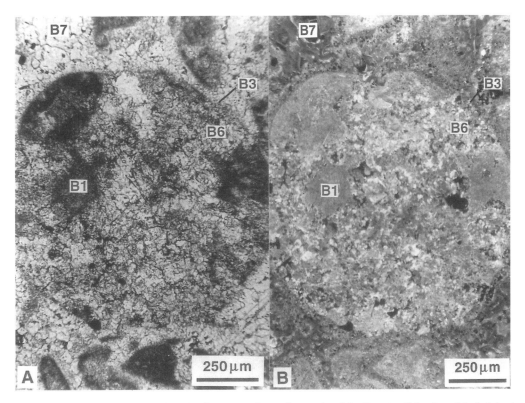

Fig. 14. Paired plane-light and cathodoluminescence photomicrographs of the *Fontaine Colombette Marls Subunit C* grainstones in the platform-flank section (Fig. 4). Note the presence of dolomitized lithoclasts. Numbered features refer to the paragenetic sequence presented in Fig. 5.

defined by cathodoluminescence, implying that they are pseudosecondary (Roedder 1984). Although most of the three-dimensional inclusions are not grouped, some are part of widely spaced clusters that also contain flat inclusions. Close examination under cathodoluminescence reveals that these inclusion clusters occur at crystal growth triple boundaries. This restriction to crystal growth-confined zones combined with the occurrence of random single inclusions, suggests that these inclusions are primary in origin (Roedder 1984). An additional phenomenon is the significantly different liquid-to-vapour ratios observed in all originally two-phase inclusions, which can range from all-liquid to all-vapour within a single inclusion cluster. However, no consistent relationship was observed between the relative position of inclusions and their liquid-to-vapour ratio.

Microthermometric analyses of these inclusions yield ice melting point temperatures as follows: (1) 37 inclusions in Event A7 calcites gave an overall range of $-17.2°$ to $0°C$, with 55% of the analyses falling between $-1°$ to $0°C$ ($0-1.7 \text{wt}\%$ NaCl, conversions after Potter *et al.* 1978) suggesting that they are primarily freshwater inclusions (Fig. 19); and (2) 49 inclusions in Event A12 calcites yield an overall range of $-3.7°$ to $0°C$, with 83% of the analyses falling between $-1°$ and $0°C$ ($0-1.7 \text{wt}\%$ NaCl) suggesting that they are also primarily freshwater inclusions (Fig. 19). Crushing experiments suggest that the vapour pressure of these inclusions is close to 1 atmosphere, as no significant change in the volume of the vapour phase was observed at the moment the inclusion was broken.

Discussion

No evidence was observed to suggest that the larger number of diagenetic events observed on Surface 3 is a preservational bias caused by the preferential erosion of Surfaces 1, 2 and 4. On the contrary, Surface 3 locally exhibits up to 1 m of erosional relief (Hunt & Tucker 1993; Everts *et al.* 1995), whereas no evidence for erosion was

Table 1. *Trace element and isotopic compositions of micrite, skeletal components, calcites and dolomites in the Bi 6-1 rudist pack-floatstones directly underlying Surface 3*

			Standard Deviation	0.04	21	51	0.10	0.20
			Detection Limit	0.03	47	78		
Paragenetic sequence	Mineralogy	Component	Mg mol%	Mn ppm	Fe ppm	$\delta^{18}O$ PDB	$\delta^{13}C$ PDB
A1	Calcite	Rudist shell	1.3	bd	211	−5.37	−1.82
A1	Calcite	Rudist shell	1.0	bd	478	−4.55	−1.34
A1	Calcite	Rudist shell	0.8	bd	139		
A1	Calcite	Rudist shell	0.6	bd	111		
A1	Calcite	Rudist shell	1.4	bd	201		
A1	Calcite	Rudist shell	1.1	bd	134		
Average composition of A1			*1.0*	*bd*	*212*	*−4.96*	*−1.58*
A3	Calcite	Micrite	0.7	203	bd	−3.96	1.49
A3	Calcite	Micrite	1.1	159	bd		
A3	Calcite	Micrite	1.4	bd	bd		
A3	Calcite	Micrite	1.0	132	bd		
A3	Calcite	Micrite	0.8	199	bd		
A3	Calcite	Micrite	0.9	200	bd		
Average composition of A3			*1.0*	*179*	*bd*	*−3.96*	*1.49*
A4	Calcite	Bladed cement	1.5	bd	132	−4.58	2.07
A4	Calcite	Bladed cement	1.7	bd	401	−3.53	2.32
A4	Calcite	Bladed cement	1.1	bd	234		
A4	Calcite	Bladed cement	1.7	bd	bd		
A4	Calcite	Bladed cement	1.0	bd	125		
A4	Calcite	Bladed cement	0.9	bd	111		
Average composition of phase A4			*1.3*	*bd*	*201*	*−4.06*	*2.20*
A6	Calcite	Bladed cement	0.7	bd	101	−4.11	3.06
A6	Calcite	Bladed cement	0.5	bd	bd	−2.64	3.47
A6	Calcite	Bladed cement	0.5	bd	bd		
A6	Calcite	Bladed cement	1.1	bd	123		
A6	Calcite	Bladed cement	1.1	bd	189		
A6	Calcite	Bladed cement	0.8	bd	169		
Average composition of phase A6			*0.8*	*bd*	*146*	*−3.38*	*3.27*
A7	Calcite	Blocky cement	0.5	bd	101	−8.29	2.22
A7	Calcite	Blocky cement	0.2	bd	bd	−6.20	3.40
A7	Calcite	Blocky cement	1.1	bd	bd	−6.82	2.05
A7	Calcite	Blocky cement	1.1	bd	bd		
A7	Calcite	Blocky cement	1.2	bd	399		
A7	Calcite	Blocky cement	1.2	bd	bd		
Average composition of phase A7			*0.9*	*bd*	*250*	*−7.10*	*2.56*
A10	Dolomite	Replacement	45.5	bd	902	−4.91	0.50
A10	Dolomite	Replacement	46.9	bd	458	−5.63	0.84
A10	Dolomite	Replacement	47.1	bd	421		
A10	Dolomite	Replacement	45.7	bd	590		
A10	Dolomite	Replacement	48.2	bd	1124		
A10	Dolomite	Replacement	45.8	bd	1042		
Average composition of phase A10			*46.5*	*bd*	*756*	*−5.27*	*0.67*
A11	Dolomite	Rhomb cement	47.3	121	609	−5.04	2.15
A11	Dolomite	Rhomb cement	47.5	104	615	−5.05	2.78
A11	Dolomite	Rhomb cement	46.7	125	498		
A11	Dolomite	Rhomb cement	46.8	156	499		
A11	Dolomite	Rhomb cement	46.6	125	379		
A11	Dolomite	Rhomb cement	46.6	116	401		
Average composition of phase A11			*46.9*	*125*	*500*	*−5.05*	*2.47*
A12	Calcite	Blocky cement	0.3	bd	bd	−3.61	3.60
A12	Calcite	Blocky cement	0.3	bd	bd	−1.34	3.78
A12	Calcite	Blocky cement	0.6	bd	bd		

A12	Calcite	Blocky cement	0.3	bd	bd		
A12	Calcite	Blocky cement	0.4	bd	bd		
A12	Calcite	Blocky cement	0.3	bd	bd		
Average composition of phase A12			*0.4*	*bd*	*bd*	*−2.63*	*3.69*
C2	Calcite	Blocky cement	1.2	bd	1100	−6.51	0.75
C2	Calcite	Blocky cement	0.9	101	954	−6.67	0.65
C2	Calcite	Blocky cement	0.8	107	1267		
C2	Calcite	Blocky cement	1.6	204	1600		
C2	Calcite	Blocky cement	1.4	bd	903		
C2	Calcite	Blocky cement	1.7	126	900		
Average composition of phase C2			1.3	135	1121	−6.59	0.70

bd, below detection. All data from laser ablation, except A3 micrites from acid digestion.

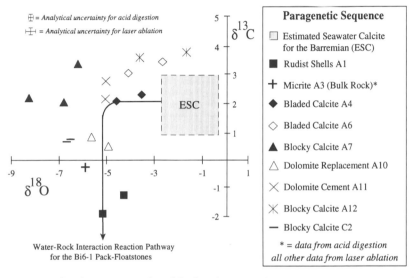

Fig. 15. Oxygen versus carbon isotope cross-plot of Surface 3 paragenetic sequence components presented in Fig. 6. Data are present in Table 1. The Barremian estimated sea-water calcite (ESC) composition is from Weissert *et al.* (1985), Arthur *et al.* (1990) and Föllmi *et al.* (1994). A water–rock interaction pathway was calculated using equations in Banner & Hanson (1990). The model predicts isotopic changes of the ESC during equilibration with meteoric waters that have a composition of −1.0 to −3.0‰ SMOW $\delta^{18}O$ and −9.0 to −11.0‰ PDB $\delta^{13}C$, which are comparable to modern meteoric groundwaters in carbonate terrains of the Yucatan Peninsula, Mexico (Stoessel *et al.* 1989).

found along the other platform-top bedding surfaces. In addition, similarity in mineralogy, grain size, permeability and porosity of the lithologies comprising Surfaces 1, 2, 3 and 4 (Everts *et al.* 1995) suggests that the differences in diagenesis are not due to facies control. Therefore, the multiple well developed diagenetic phases observed on Surface 3 suggest that this bedding horizon experienced a greater extent of diagenetic alteration than Surfaces 1, 2 and 4. Only the difference in the intensity of diagenesis as represented by the greater number of events on Surface 3 is acknowledged, and no direct correlation regarding differences in the

absolute length of time over which diagenesis occurred on Surfaces 1, 2, 3 and 4 is implied.

Diagenetic alteration along Surface 3

Based on the petrographic and geochemical evidence discussed below, we concluded that diagenetic alteration at Surface 3 occurred in four major stages. These are early stage meteoric alteration, marine hardground formation, dolomitization and late stage cementation.

Early meteoric alteration (Events A4 to A7; Fig. 5). The A4 calcite cements consistently show a

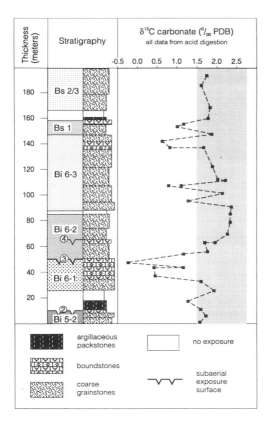

Fig. 16. Carbon isotopic record of the Bi 5 to Bs 2/3 platform-top section of the Vercors carbonate platform exposed in the Cirque d'Archiane. The stratigraphic position of Surfaces 2, 3 and 4 are indicated. A strong negative excursion in $\delta^{13}C$ relative to the baseline values for the section of +1.5 to +2.5‰, occurs at Surface 3.

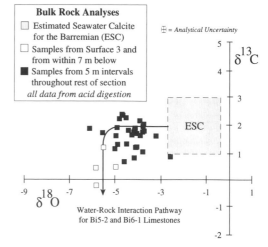

Fig. 17. Oxygen versus carbon isotope cross-plot of bulkrock analyses from the Bi 5 and Bi 6 platform-top section in the Cirque d'Archiane. See Fig. 18 for a description of water–rock interaction modelling.

bladed crystal morphology, low Mg (≤1.7 mol%), Mn (≤47 ppm) and Fe (≤401 ppm) abundances, and low $\delta^{18}O$ (−3.53 to −4.58‰) signatures relative to a Barremian Estimated Seawater Calcite (ESC) of *c.* −1.5 to −2.5‰ $\delta^{18}O$ and +1 to +3‰ $\delta^{13}C$ (Fig. 18; Weissert *et al.* 1985; Arthur *et al.* 1990; Föllmi *et al.* 1994). These features are consistent with precipitation from meteoric or mixing zone fluids (Meyers & Lohmann 1985; Saller & Moore 1989; Goldstein *et al.* 1991; Fouke 1994), except for the lack of depletion in ^{13}C, which is discussed below. Dissolution of the aragonitic rudist shell layers (Event A5; Fig. 5) is also interpreted to have taken place in undersaturated meteoric waters. For the Event A6 calcite cements (Fig. 5), bladed crystal morphology and stoichiometry

(low Mg (≤1.1 mol%), Mn (≤47 ppm) and Fe (≤189 ppm) abundances and a low $\delta^{18}O$ (−2.64 to −4.11‰) relative to the ESC are all suggestive of precipitation in freshwater (Meyers & Lohmann 1985; Saller & Moore 1989; Goldstein *et al.* 1991; Fouke 1994). Similarly, the Event A7 calcite cements are suggestive of meteoric precipitation due to their blocky crystal morphology, low Mg (≤1.2 mol%), Mn (≤47 ppm), and Fe (≤399 ppm), and low $\delta^{18}O$ (−6.20 to −8.29‰) relative to the ESC (Meyers & Lohmann 1985; Saller & Moore 1989; Goldstein *et al.* 1991). In addition, a meteoric origin for the Event A7 calcites is suggested by their predominantly freshwater single-phase fluid inclusions (Fig. 19). Although the origin of the higher salinity fluid inclusions is uncertain, similar saline inclusions have been observed in other meteoric blocky calcites and may suggest that some re-equilibration has taken place (Goldstein & Reynolds 1994). However, the presence of distinct concentric, sectoral and intrasectoral cathodoluminescence zoning suggest that the crystals have not been recrystallized in saline fluids since their original precipitation at disequilibrium (Reeder & Grams 1987; Paquette & Reeder 1990; Reeder 1991; Fouke & Reeder 1992; Ward & Reeder 1993). Another possibility is that these were meteoric waters that had been mixed with sea water and/or hypersaline sea water-derived brines, thus creating elevated salinities in primary inclusions and relatively

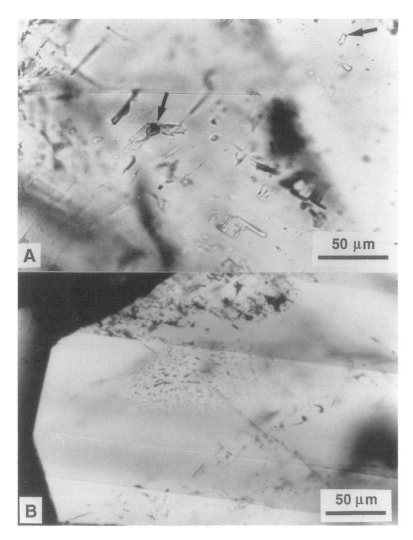

Fig. 18. Plane-light photomicrographs of fluid inclusions observed in Surface 3 blocky calcite cements. (**A**) Single-phase fluid inclusions in event A7 blocky calcite cements, in which vapour bubbles (marked by arrows) have been thermally induced. (**B**) Two-phase fluid inclusions in event A12 blocky calcite cements.

small shifts in the oxygen isotopes.

Although initially deposited in a shallow marine environment, the petrographic and geochemical composition of the Event A1 rudist shells and Event A3 micrites suggests that they were also diagenetically equilibrated with meteoric waters during subaerial exposure. The rudists contain significantly lower $\delta^{18}O$ (−4.55 to −5.37‰) and $\delta^{13}C$ (−1.34 to −1.82‰) compositions than a Barremian ESC. This indicates that the rudists are geochemically altered despite their noncathodoluminescent texturally unal-

tered fabrics (Fig. 9; e.g. Popp *et al.* 1986; Fouke 1994). Similarly, the Event A3 micrites are altered based on their bright cathodoluminescence (Fig. 7) and low $\delta^{18}O$ (−5.84‰) and $\delta^{13}C$ (−0.23‰) compositions relative to the ESC (Fig. 15). We have no direct evidence for the timing of alteration of the shells and micrite, but at least for the shell material it is likely that alteration occurred in association with leaching of the aragonite layers.

During meteoric alteration by $\delta^{13}C$-depleted meteoric waters, the Bi 6-1 sediments would be

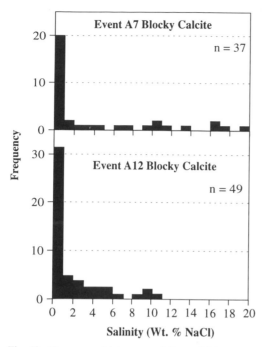

Fig. 19. Frequency histogram of the weight percent NaCl composition of fluid inclusions in the event A7 and A12 blocky calcite cements.

expected to evolve along an inverted J-shaped pathway originating with the ESC (Fig. 15; Deines *et al.* 1974; Carling 1984; Meyers & Lohmann 1985; Lohmann 1987; Banner & Hanson 1990). Bulk rock samples of the Bi 6-1 limestones at and immediately below Surface 3 exhibit a strong depletion in both $\delta^{18}O$ and $\delta^{13}C$, relative to the ESC as well as to most of the other platform sediments of the Cirque d'Archiane (Figs 16 and 17). However, isotope analyses of the individual components (grains, matrix, cements) within the Bi 6-1 rocks revealed a strongly depleted O and C signal for the rudists and micrites only. The cements A4, A6 and A7 do exhibit a depletion in $\delta^{18}O$ relative to the ESC, but their $\delta^{13}C$ composition is equivalent or slightly higher to that of the ESC (Fig. 15). This suggests that some of the cements may have precipitated from multiple meteoric and/or mixed sea water–fresh water fluids that subsequently overprinted Surface 3 after its initial meteoric stabilization. Under these circumstances, the composition of such waters would have been controlled by previous differences in orographic and/or water–rock interaction histories of the fluids themselves (Lohmann 1987), and not by the water–rock buffering of the Bi 6-2 limestones.

Marine hardground stage (Events A8 and A9; Fig. 5). The cylindrical borings of Event A8 most likely formed during a period when Surface 3 was a marine hardground. This is suggested by the shape and lateral distribution of the borings (Ekdale *et al.* 1984). In addition, kaolinite–smectite–illite clay assemblages of the Event A9 red internal sediments filling the borings and biomolds are suggestive of the marine deposition of terrestrial-derived clays (Biscaye 1965). Distribution of the borings of Event A8 (Fig. 6) provides an unequivocal constraint on the timing of meteoric alteration at the Surface 3 bedding plane. The borings cross-cut the cements of A4 through A7 (Fig. 10), indicating that the Bi 6-1 sediments had been lithified by meteoric alteration prior to this time.

Dolomitization (Events A10 and A11; Fig. 5). Application of a 3±1‰ dolomite–calcite fraction factor (Land 1980) to the narrow range of $\delta^{18}O$ (–4.91 to –5.63‰) observed in the Event A10 and A11 dolomites suggest a coeval calcite O isotope composition equivalent to that of the ESC (Fig. 15). When combined with the non-burial platform-top setting of Surface 3 and the requirements for Mg supply during dolomitization (Land 1985), this implies that sea water was an important component of the Event A10 and A11 dolomitizing fluids. However, differences in the Mn and $\delta^{13}C$ compositions suggest that the replacement dolomites may have precipitated from different fluids than those that precipitated the dolomite cements. The Ca-rich Event A10 replacement dolomites contain low Mn abundances (<47 ppm), a range of Fe abundances (421–1124 ppm), and a $\delta^{13}C$ signature (+0.05 to +0.84‰) that is lower than the ESC by c. 1‰. This suggests that precipitation may have taken place in a mixing zone environment, where well-oxygenated ^{13}C-depleted freshwater mixed with seawater to draw down the dolomite $\delta^{13}C$ and Mn signatures (e.g. Land 1973; Plummer 1975; Humphrey 1988; Fouke 1994; Cander 1994). Alternatively, the Ca-rich Event A11 cement dolomites have much higher Mn abundances (104–156 ppm), similar Fe abundances (379–615 ppm) and $\delta^{13}C$ compositions (+2.15 to +2.78‰) that range from being equivalent to +0.28‰ higher than the ESC. This suggests that dolomitizing fluids may have been composed of reduced sea water, in which Mn was mobilized and $\delta^{13}C$ was driven up during the anaerobic degradation of organic matter (Irwin *et al.* 1977).

The distribution of dolomite in the Bi 6-2 grainstones overlying Surface 3 provides some constraints on the timing of dolomitization at

Surface 3. Within the Bi 6-2 grainstones, dolomite only occurs within rounded lithoclasts that commonly feature truncated dolomite rhombohedra at their margins (Fig. 11). This indicates that at least some dolomitization of the Bi 6-1 limestones must have occurred prior to the erosion of Surface 3 and subsequent deposition of the overlying grainstones. Lithoclasts with truncated calcite-filled rhombohedral molds at their margins also occur in an 8 m-thick interval of platform-flank grainstones at the downslope continuation of Surface 3 (Figs 4 and 14). The rhombic shape of these molds suggest that they are dissolved replacement dolomite crystals that were filled with an equant calcite cement. This dolomitization fabric is similar to that observed in lithoclasts within the Bi 6-2 grainstones on the platform top. Therefore we assume that, like the Bi 6-2 lithoclasts on the platform top, the platform-flank lithoclasts were also derived from erosion of the dolomitized Bi 6-1 pack-floatstones at Surface 3.

Late stage meteoric and shallow burial cements (Events A12 and C2; Fig. 5). The Event A12 calcite cements are suggestive of a renewed phase of meteoric and/or mixing zone alteration. Blocky crystal morphologies, low Mg (0.3–0.6 mol%), Mn (< 47 ppm) and Fe (< 78 ppm) abundances, and freshwater two-phase fluid inclusions are consistent with precipitation from meteoric waters (e.g. Meyers & Lohmann 1985; Saller & Moore 1989; Goldstein et al. 1991). The variety of liquid-to-vapour ratios observed in the A12 calcites may result from any of three processes: (1) necking down of originally homogeneous (all-liquid) inclusions trapped at high temperatures (> 50°C; Goldstein 1993) that have developed a vapour bubble upon returning to surface conditions; (2) entrapment under 'boiling' conditions (i.e. co-existing liquid + vapour), which is 100°C at 1 bar pressure and increases with higher pressures (Roedder 1984); and (3) entrapment in a meteoric vadose environment where the liquid-to-vapour ratios are heterogeneous as a result of the mixed liquid and vapour environment (Goldstein 1990). The first two options are highly unlikely because the Vercors carbonate platform has experienced only shallow burial to depths not exceeding 500 m (Arnaud 1981). With a normal palaeo-geothermal gradient of 3°C/100 m and a surface depositional temperature of 25°C, the maximum burial temperature could not have exeeded 40°C, which also excludes thermal re-equilibration. Trapping temperatures of less than 50°C are

further supported by the observation that the vapour bubble in each inclusion does not implode during crushing. Therefore, it is most probable that the Event A12 inclusions were trapped during precipitation in a near-surface vadose environment.

The origin of the higher salinity inclusions is uncertain. Goldstein (1990, 1993) and Goldstein & Reynolds (1994) have shown how primary inclusions can be re-equilibrated and filled with later burial brines. However, in the Event A12 blocky calcites, highly saline inclusions are interspersed with freshwater inclusions within crystals that have cathodoluminescence banding, suggesting that recrystallization has not occurred. Furthermore, the shallow burial history of the platform implies that re-equilibration with burial brines was unlikely. A further possible explanation is that the near sea-level diagenetic environments affecting Surface 3 at this time included mixtures between fresh water and sea water and hypersaline brines derived from the evaporation of sea water. Therefore, periodic infiltration of these mixed fresh water–brine waters could have created anomalously saline primary fluid inclusions.

As was the case for Events A6 and A7, the Event A12 calcite cements exhibit isotopic excursions that are not consistent with the previous Event A1 through A11 compositions (Fig. 15). This suggests that the composition of the fluids that precipitated the Event A12 calcites may have been controlled by previous orographic and water–rock interaction fluid histories (Lohmann 1987), and not buffered by the Bi 6-1 limestones. Unfortunately we have no evidence to determine whether the A12 meteoric cements formed before or after deposition of the overlying Bi 6-2 grainstones. However, the complete absence of A12 in the overlying Bi 6-2 limestones does imply that the A12 calcites were precipitated prior to the deposition of Bi 6-2 sediments (Fig. 5).

Event C2 blocky calcites fill fractures that cross-cut all previous diagenetic events within the Bi 6-1 limestones (Fig. 12). Low Mg abundances (0.8–1.7 mol%), low $\delta^{18}O$ (–6.51 to –6.67‰) and $\delta^{13}C$ (0.65–0.75‰) compositions with respect to the ESC, and relatively high Mn (< 47 ppm) and Fe (< 78 ppm) abundances imply that the Event C2 calcites were precipitated from reducing shallow-burial meteoric waters (Meyers & Lohmann 1985; Saller & Moore 1989). Similar cements also occur within the Bi 6-2 grainstones, suggesting they represent late stage diagenetic precipitation that was probably associated with burial and Alpine tectonics.

Diagenetic alteration along Surfaces 1, 2 and 4

The grainstones immediately below Surfaces 1, 2 and 4 show some dissolved shell molds filled with bladed to blocky calcite cements (Fig. 13). These are also diagenetic fabrics consistent with meteoric diagenesis (Meyers & Lohmann 1985; Saller & Moore 1989). However, there is no evidence for multiple phase meteoric overprinting such as was observed at Surface 3. In addition, platform-top bulkrock $\delta^{13}C$ analyses do not show the type of strong negative excursions at Surfaces 2 and 4 that they do at Surface 3 (Figs 16 and 17). The presence of some red internal sediments at Surfaces 1, 2 and 4 is suggestive of a marine hardground stage. However, unlike Surface 3, the internal sediments were unable to penetrate more deeply than 20–30 cm. This is probably because the creation of secondary porosity during the previous exposure event was less extensive than that on Surface 3. Finally, the absence of red dolomitized lithoclasts in the rock units overlying Surfaces 1, 2 and 4 and their downslope equivalents confirms that erosion of the early lithified material on these surfaces was relatively insignificant.

Sequence stratigraphic implications

Of the different platform-top bedding surfaces that were analyzed in this study, the top of the Bi 5-2 platform tongue (Surface 2) represents by far the most prominent change in depositional style. Basinward progradation of the platform suddenly ceased and fine-grained deeper water sediments (Fontaine Colombette Marls) encroached on the clinoform slopes. The stratal pattern created by this shift was the main criteria for Jacquin *et al.* (1991) to interpret the Bi 5-2 top as a Type 1 sequence boundary (i.e. as a major lowstand unconformity).

However, our diagenetic studies revealed only minor meteoric alteration at the top of the Bi 5-2 limestones (Surface 2). The alteration at this surface is equally intense to that observed on Surfaces 1 and 4, where changes in stratal geometry are insignificant. In contrast, petrographic, geochemical, and sedimentological evidence clearly suggests that the Surface 3 bedding plane underwent significantly more early stage meteoric alteration then the other platform-top surfaces. Although the Surface 3 bedding plane does exhibit minor onlap (Everts *et al.* 1995) its geometric expression is again significantly less prominent that of Surface 2 (Figs 2b and 3b).

Consequently, the Cirque d'Archiane exposures illustrate a stratigraphic mismatch, wherein the exposure surface exhibiting the most subaerial alteration does not coincide with the most prominent change in stratal geometry. This suggests that sea-level lowstands exposing the platform top were not the only cause for the changes in stratal geometry at the Vercors platform margin (Hunt & Tucker 1993; Everts *et al.* 1995). Apparently, processes other than subaerial exposure were capable of terminating platform growth for extended periods of time and creating major lateral shifts in deposition. For example, a drowning event may have caused the significant interruption in platform growth at the top of the Bi 5-2 tongue and the encroachment of deeper water sediments on the clinoforms (Hunt and Tucker, 1993; Everts *et al.* 1995).

One may speculate that the extent of meteoric alteration observed at exposure surfaces provides a relative measure of the duration of the exposure event, the amplitude of the sea-level drop or some combination of the two. In the case of the Cirque d'Archiane, this would imply that the extensive alteration affecting the top of the Bi 6-1 platform tongue (Surface 3) represents the most significant emersion. This hypothesis agrees with the results of detailed petrographical analyses of sections on the flank of the Vercors platform (Everts & Reijmer 1995; Everts *et al.* 1995). The grain composition of the slope sediments equivalent to platform top Surface 3 yielded significant quantities of lithoclastic debris, presumably derived from erosion of the platform top, whereas no such debris was found along the downslope continuation of platform-top Surfaces 1, 2 and 4. These observations are again suggestive of a long period of subaerial exposure, meteoric lithification and erosion at Surface 3, whereas the other Surfaces 1, 2, and 4 were exposed for only a short time and were only moderately affected by erosion.

Biostratigraphic data (Arnaud-Vanneau *et al.* 1976; Arnaud 1981; Arnaud-Vanneau & Arnaud 1991; Jacquin 1993) consistently indicate a Barremian age for all of the platform sediments in the southern Vercors. The relatively low baseline of $\delta^{13}C$ values (between 1.5‰ and 2.25‰) that we observed throughout the entire platform-top sequence in the Cirque d'Archiane (Fig. 16) also compares well with the average carbon-isotope signature for Barremian sediments (e.g. Arthur *et al.* 1990; Föllmi *et al.* 1994). However, our exposure Surfaces 1, 2, 3 and 4 are spaced only tens of metres apart, while the entire platform sequence in the southern Vercors is about 700 m thick. This would imply that Surfaces 1 to 4 relate to exposure events in the 10^4 to 10^5 year domain, given a duration for the Barremian that ranges between 3.5 Ma (Haq *et al.* 1988) and 7.3 Ma (Harland *et al.* 1989). The more extensive alteration observed along Surface 3 may also result from the amalgamation of several high frequency exposure events, which are a common expression of low frequency cycle boundaries in other platform carbonates (e.g. Goldhammer *et al.* 1990, 1991, 1993; Montañez & Osleger 1993).

Conclusions

Sedimentological and diagenetic evidence for subaerial exposure has been evaluated with respect to the stratal geometric framework at the southern margin of the Vercors carbonate platform in southeastern France. At several bedding surfaces on the platfom top evidence for early stage meteoric alteration was found, including bladed and blocky calcite cement generations cut by hardground borings, leaching, depleted $\delta^{18}O$ and $\delta^{13}C$ signatures, and fresh water fluid inclusions. Sedimentologic, petrographic and geochemical data indicate that one of these bedding planes (Surface 3 at the top of the Bi 6-1 platform tongue) has been significantly more affected by meteoric diagenesis and subaerial erosion than the other surfaces. Although this particular bedding surface does exhibit the most prominent subaerial exposure features, it represents a rather insignificant change in stratal geometry. This observation illustrates the ambiguity of geometric criteria to identify lowstand unconformities. Processes like platform drowning may create major lateral shifts in deposition that have a similar geometric expression as lowstand unconformities. For a correct interpretation of the sea-level history of depositional systems, studies of stratal geometry need to be accompanied by sedimentological and diagenetic evidence.

This research was supported by the Vrije Universiteit Industrial Associates in Sedimentology research grants held by W. Schlager. AE acknowledges support by the Netherlands Foundation for Earth Science Research (AWON/GOA, Project No. 751.356.025). We are grateful to S. van de Gaast for conducting X-ray diffraction analyses at the Netherlands Institute for Sea Research in Texel. Trace element analyses were completed on the electron probe at the State University in Utrecht. W. Rave-Koot, B. Lacet and V. Wiederhold are thanked for their excellent preparation of customized thin-sections for the analyses. Reviews by R. Goldstein, D. Hunt and V. Vahrenkamp added significantly to the manuscript. Discussions with J. Kenter, J. Stafleu and S. Davey were also quite helpful in refining our ideas.

References

ALLEN, J. R. & MATTHEWS, R. K. 1982. Isotope signatures associated with early meteoric diagenesis. *Sedimentology*, **29**, 797–818.

ARNAUD, H. 1981. De la plate-forme urgonienne au bassin voconien: le Barrémo–Bédoulien des Alpes occidentales entre Isère et Buëch (Vercors méridional, Diois oriental et Dévoluy). *Géologie Alpine*, Mémoir, **12**.

ARNAUD-VANNEAU, A. & ARNAUD, H. 1990. Hauterivian to Lower Aptian carbonate shelf sedimentation and sequence stratigraphy in the Jura and northern Subalpine chains (SE France and Swiss Jura). *In:* TUCKER, M. E., WILSON, J. L., CREVELLO, P. D., SARG, J. R. & READ, J. F. (eds) *Carbonate Platforms, Facies, Sequences and Evolution.* International Association of Sedimentologists, Special Publication, **9**, 203–234.

—— & —— 1991. Sédimentation et variations relatives du niveau de la mer sur les plate-formes carbonatées du Berriasian–Valanginien et du Barrémien dans les massifs subalpins septentrioneaux et la Jura (SE de la France). *Bulletin of the Geological Society of France*, **162**, 535–545.

——, —— & THIEULOY, J. P. 1976. Bases nouvelles pour la stratigraphie des calcaires urgoniens du Vercors (Massifs subalpins septentrioneaux, France). *Newsletter of Stratigraphy*, **5**, 143–159.

ARTHUR, M. A., JENKYNS, H. C., BRUMSACK, H. J. & SCHLANGER, S. O. 1990. Stratigraphy, geochemistry and paleooceanography of organic-rich Cretaceous sequences. *In:* GINSBURG, R. N. & BEAUDOIN, B. (eds) *Cretaceous Resources, Events and Rythms.* Kluwer, Dordrecht, 75–119.

BANNER, J. L. & HANSON, G. N. 1990. Calculation of simultaneous isotopic and trace element variations during water–rock interaction with applications to carbonate diagenesis. *Geochimica et Cosmochimica Acta*, **54**, 3123–3127.

BISCAYNE, P. E. 1965. Mineralogy and sedimentation of recent deep-sea clay in the Atlantic Ocean and adjacent seas and oceans. *Geological Society of America Bulletin*, **76**, 803–832.

BODNAR, R. J. & BETHKE, P. M. 1984. Systematics of stretching of fluid inclusions I: Fluorite and calcite at 1 atmosphere confining pressure. *Economic Geology*, **79**, 141–161.

CANDER, H. S. 1994. An example of mixing zone dolomite, Middle Eocene Avon Park Formation, Floridan aquifer system. *Journal of Sedimentary Research*, **A64**, 615–629.

CERLING, T. E. 1984. The stable isotopic composition of modern soil carbonate and its relationship to climate. *Earth and Planetary Science Letters*, **71**, 229–240.

DEINES, P., LANGMUIR, D. & HARMON, R. S. 1974. Stable carbon isotope ratios and the existence of a gas phase in the evolution of carbonate groundwaters. *Geochimica et Cosmochimica Acta*, **38**, 1147–1164.

EKDALE, A. A., BROMLEY, R. G. & PEMBERTON, S. G. 1984. *Ichnology. The use of trace fossils in sedimentology and stratigraphy.* Society of Economic Paleontologists and Mineralogists, Short Course, **15**.

EVERTS, A. J. W. & REIJMER, J. J. G. (1995). Clinoform composition and margin geometries of a Lower Cretaceous carbonate platform (Vercors, SE France). *Palaeogeography Palaeoclimatology Palaeoecology*, in press.

——, STAFLEAU, J., SCHLAGER, W., FOUKE, B. W. & ZWART, E. W. 1995. Stratal patterns and sequence stratigraphy at the margin of the Vercors carbonate platform (Lower Cretaceous, SE France). *Journal of Sedimentary Research*, **B65**, 119–131.

FÖLLMI, K. B., WEISSERT, H., BISPING, M. & FUNK, H. 1994. Phosphogenesis, carbon isotope stratigraphy, and carbonate platform evolution along the Lower Cretaceous northern Tethyan margin. *Geological Society of America Bulletin*, **106**, 729–746.

FOUKE, B. W. 1994. *Deposition, Diagenesis and Dolomitization of Neogene Reef-Derived Limestones on Curacao, Netherlands Antilles.* Publications Foundation for Scientific Research in the Caribbean Region, **133**, The Hague, Netherlands.

—— & REEDER, R. J. 1992. Surface structural controls on dolomite composition: Evidence from sectoral zoning. *Geochimica et Cosmochimica Acta*, **56**, 4015–4024.

FRANSEEN, E. K., GOLDSTEIN, R. H. & WHITESELL, T. E. 1993. Sequence stratigraphy of Miocene carbonate complexes, Las Negras area, Southeastern Spain: implications for quantification of changes in sea-level. *In:* LOUCKS, R. G. & SARG, G. F. (eds) *Carbonate Sequence Stratigraphy – Recent Developments and Applications.* American Association of Petroleum Geologists, Memoir, **57**, 409–434.

GOLDHAMMER, R. K., DUNN, P. A. & HARDIE, L. A. 1990. Depositional cycles, composite sea-level changes, cycle stacking patterns, and the hierarchy of stratigraphic forcing: examples from platform carbonates of the Alpine Triassic. *Geological Society of America Bulletin*, **102**, 535–562.

——, LEHMANN, P. J. & DUNN, P. A. 1993. The origin of high-frequency platform carbonate cycles and third-order sequences (Lower ordivician El Paso Group, West Texas): constraints from outcrop data and stratigraphic modeling. *Journal of Sedimentary Petrology*, **63**, 318–359.

——, ——, TODD, R. G., WILSON, J. L., WARD, W. C. & JOHNSON, C. R. 1991. *Sequence Stratigraphy and Cyclostratigraphy of the Mesozoic of the Sierra Madre Oriental, Northeast Mexico, A Field Guide Book.* Society of Economic Paleontologists and Mineralogists, Gulf Coast Section.

GOLDSTEIN, R. H. 1990. Petrographic and geochemical evidence for origin of paleospeleotherms, New Mexico: Implications for the application of fluid inclusions to studies of diagenesis. *Journal of Sedimentary Petrology*, **60**, 282–292.

—— 1993. Fluid inclusions as carbonate micro-

fabrics: A petrographic method to determine diagenetic history. *In:* REZAK, R. & LAVOIE, D. L. (eds) *Carbonate Microfabrics: Frontiers in Sedimentary Geology.* Springer-Verlag, 279–290.

—— & FRANSEEN, E. K. 1995. Pinning points: a method providing quantitative constraints on relative sea-level history. *Sedimentary Geology*, **95**, 1–10.

—— & REYNOLDS, T. J. 1994. *Systematics of Fluid Inclusions in Diagenetic Minerals.* Society of Sedimentary Geology, Short Course, **31**.

——, ANDERSON, J. E. & BOWMAN, M. W. 1991. Diagenetic responses to sea-level change: Integration of field, stable isotope, paleosol, paleokarst, fluid inclusion and cement stratigraphy research to determine history and magnitude of sea-level fluctuation. *In:* Sedimentary modelling: Computer Simulations and Methods for Improved Parameter Definition. *In:* FRANSEEN, E. K., WATNEY, W. L., KENDALL, C. G. St. C. & ROSS, W. *Sedimentary Modelling: Computer simulations and methods for improved parameter definition.* Bulletin of the Kansas Geological Survey, **223**, 139–162.

HAQ, B. U., HARDENBOL, J. & VAIL, P. R. 1988. Mesozoic and Cenozoic chronostratigraphy and cycles of sea-level change. *In:* WILGUS, C. K., HASTINGS, B. S., KENDALL C. G. St. C., POSAMENTIER, H. W., ROSS, C. A. & VAN WAGONER, J. C. (eds) *Sea-Level Changes: An Integrated Approach.* Society of Economic Paleontogists and Mineralogists, Special Publication, **42**, 71–108.

HARLAND, W. B., ARMSTRONG, R. L., COX, A. V., CRAIG, I. E., SMITH, A. G. & SMITH, D. G. 1989. *A Geologic Time Scale.* Cambridge University Press.

HUMPHREY, J. D. 1988. Late Pleistocene mixing zone dolomitization, southeastern Barbados, West Indies. *Sedimentology*, **35**, 327–348.

HUNT, D. & TUCKER, M. E. 1993. The Mid-Cretaceous Urgonian Platform of SE France. *In:* SIMO, J. A., SCOTT, R. W. & MASSE, J. P. (eds) *Cretaceous Carbonate Platforms.* American Association of Petroleum Geologists, Special Publication, **56**, 409–453.

IRWIN, H. C., CURTIS, C. & COLEMAN, M. L. 1977. Isotopic evidence for the source of diagenetic carbonates formed during burial of organic-rich sediments. *Nature*, **269**, 209–213.

JACQUIN, TH. 1993. Sequence-stratigraphic framework of the Urgonian platform. *In:* MULOCK-HOUWER, J. A., PILAAR, W. F. & V. D. GRAAF-TROUWBORST, T. (eds) *Vercors, France: Sequence Stratigraphy of an Early Cretaceous Carbonate Platform.* American Association of Petroleum Geologists, International Conference 1993, Field Guidebook, **6**, The Hague, Netherlands, 59–75.

——, ARNAUD-VANNEAU, A., ARNAUD, H., RAVENNE, C. & VAIL, P. R. 1991. System tracts and depositional sequences in a carbonate setting: a study of continuous outcrops from platform to basin at the scale of seismic lines. *Marine Geology*, **8**, 122–139.

JAMES, N. P. & CHOQUETTE, P. W. 1984. Diagenesis 9. Limestones – the meteoric diagenetic environment. *Geosciences Canada*, **11**, 161–194.

KENDALL, C. G. St. C. & SCHLAGER, W. 1981. Carbonates and relative changes in sea-level. *Marine Geology*, **44**, 181–212.

LAND, L. S. 1973. Contemporaneous dolomitization of Middle Pleistocene reefs by meteoric water, North Jamaica. *Bulletin of Marine Science*, **23**, 64–92.

—— 1980. The isotopic and trace element geochemistry of dolomite: The state of the art. *In:* ZENGER, D. H., DUNHAM, J. B. & ETHINGTON, R. L. (eds) *Concepts and Models of Dolomitization*. Society of Economic Paleontologists and Mineralogists, Special Publication, **28**, 87–110.

—— 1985. The origin of massive dolomite. *Journal of Geological Education*, **33**, 112–125.

LOHMANN, K. C. 1987. Geochemical patterns of meteoric diagenetic systems and their application to studies of paleokarst. *In:* CHOQUETTE, P. W. & JAMES, N. P. (eds) *Paleokarst*. Springer-Verlag.

MEYERS, W. J. & LOHMANN, K. C. 1985. Isotope geochemistry of regionally extensive calcite cement zones and marine components in Mississippian limestones. *In:* SCHNEIDERMAN, N. & HARRIS, P. M. (eds) *Carbonate Cements*. Society of Economic Paleontologists and Mineralogists, Special Publication, **36**, 223–239.

MONTAÑEZ, I. P. & OSLEGER, D. A. 1993. Parasequence stacking patterns, third-order accommodation events, and sequence stratigraphy of Middle to Upper Cambrian platform carbonates, Bonanza King Formation, Southern Great Basin. *In:* LOUCKS, R. G. & SARG, J. F. (eds) *Carbonate Sequence Stratigraphy: Recent Developments and Applications*. American Association of Petroleum Geologists, Memoir, **57**, 305–326.

PAQUETTE, J. & REEDER, R. J. 1990. New type of compositional zoning in calcites: Insights into crystals-growth mechanisms. *Geology*, **18**, 1244–1247.

PLUMMER, L. N. 1975. Mixing of seawater with calcium carbonate groundwater. *In:* WHITTEN, E. H. T. (ed.) *Quantitative Studies in Geological Sciences*. Geological Society of America Memoir, **142**, 219–236.

POMAR, L. 1993. High-resolution sequence stratigraphy in prograding Miocene carbonates: Application to seismic interpretation. *In:* LOUCKS, R. G. & SARG, G. F. (eds) *Carbonate Sequence Stratigraphy – Recent Developments and Applications*. American Association of Petroleum Geologists, Memoir, **57**, 389–407.

POPP, B. N., ANDERSON, T. F. & SANDBERG, P. A. 1986. Brachiopods as indicators of original isotopic composition in some Paleozoic limestones. *Geological Society of America Bulletin*, **56**, 715–727.

POSAMENTIER, H. W., ALLEN, G. P., JAMES, D. P. & TESSON, M. 1992. Forced regressions in a sequence stratigraphic framework. Concepts, examples and exploration significance. *American Association of Petroleum Geologists Bulletin*, **76**, 1687–1709.

REEDER, R. J. 1991. An overview of zoning in carbonate minerals. *In:* BARKER, C. E. & KOPP, O. C. (eds) *Luminesence Microscopy and Spectroscopy: Qualitative and Quantitative Applications*. Society of Economic Paleontologists and Mineralogists, Short Course Notes, 77–82.

—— & GRAMS, J. C. 1987. Sector zoning in calcite cement crystals: Implications for trace element distribution coefficients in carbonates. *Geochimica et Cosmochimica Acta*, **51**, 187–194.

ROEDDER, E. 1984. *Fluid inclusions*. Reviews in Mineralogy, **12**, Mineralogical Society of America.

SALLER, A. H. & MOORE, C. H. 1989. Meteoric diagenesis, marine diagenesis, and microporosity in Pleistocene and Oligocene limestones, Enewetok Atoll, Marshall Islands. *Sedimentary Geology*, **63**, 253–272.

SCHLAGER, W. 1991. Depositional bias and environmental change – important factors in sequence stratigraphy. *Sedimentary Geology*, **70**, 109–130.

—— 1993. Accommodation and supply – a dual control on stratigraphic sequences. *Sedimentary Geology*, **86**, 111–136.

SMALLEY, P. C., STIJFHORN, D. E., RAHEIM, A., JOHANSEN, H. & DICKSON, J. A. D. 1989. The laser microprobe and its application to the study of C and O isotopes in calcite and aragonite. *Sedimentary Geology*, **9**, 257–273.

STAFLEU, J., EVERTS, A. J. W. & KENTER, J. A. M. 1994. Seismic models of a prograding carbonate platform: Vercors, SE France. *Marine and Petroleum Geology*, **11**, 514–527.

STOESSEL, R. K., WARD, W. C., FORD, B. H. & SCHUFFERT, J. D. 1989. Water chemistry and CaCO₃ dissolution in the saline part of an open-flow mixing zone, coastal Yucatan Peninsula, Mexico. *Geological Society of America Bulletin*, **101**, 159–169.

VAN WAGONER, J. C., POSAMENTIER, H. W., MITCHUM, R. M., VAIL, P. R., SARG, J. F., LOUTIT, T. S. & HARDENBOL, J. 1988. An overview of the fundamentals of sequence stratigraphy and key definitions. *In:* WILGUS, C. K., HASTINGS, B. S., KENDALL, C. G. St. C., POSAMENTIER, H. W., ROSS, C. A. & VAN WAGONER, J. C. (eds) *Sea-Level Changes: An Integrated Approach*. Society of Economic Paleontologists and Mineralogists, Special Publication, **42**, 39–46.

WARD, W. B. & REEDER, R. J. 1993. The use of growth microfabrics and transmission electron microscopy in understanding replacement processes in carbonates. *In:* REZAK, R. & LAVOIE, D. (eds) *Carbonate Microfabrics: Frontiers in Sedimentary Geology*. Springer-Verlag, 253–264.

WEISSERT, H., McKENZIE, J. A. & CHANNEL, J. E. T. 1985. Natural variations in the carbon cycle during the Cretaceous. *In:* SUNDQUIST, E. T. & BROECKER, W. S. (eds) *The Carbon Cycle and Atmospheric CO₂: Natural Variations Archean to Present*. AGM Series, **32**, 531–547.

WEIMER, R. J. 1992. Developments in sequence stratigraphy: foreland and cratonic basins. *American Association of Petroleum Geologists Bulletin*, **76**, 965–982.

Compaction as a primary control on the architecture and development of depositional sequences: conceptual framework, applications and implications

DAVE HUNT,[1] TIM ALLSOP[2] & RICHARD E. SWARBRICK[2]

[1]*Department of Earth Sciences, The University of Manchester, Oxford Road, Manchester M13 9PL, UK*

[2]*Department of Geological Sciences, The University of Durham, South Road, Durham DH1 3LE, UK*

Abstract: A conceptual model is developed integrating the compaction process into a sequence stratigraphic framework, and incorporating an understanding of the ways that carbonate platforms respond to sea-level changes. The application of this model to a range of well-constrained examples allows examination of the compaction process within a high resolution temporal framework. This approach helps to gain a better understanding of the compaction process in the shallow subsurface. Conversely, the recognition of unconformities 'enhanced' by compaction-induced differential subsidence illustrates the dynamic and interactive role played by compaction during sequence development. It is this aspect of the compaction process, as a control of accommodation development, facies patterns and ultimately sequence architecture, that is the focus of interest here. Examples of compactionally 'enhanced' unconformities show compaction to be a dynamic process that can act as a primary control of sequence architecture and development. It is clear that compaction is a much underestimated process in extant sequence stratigraphic models.

The position, geometry and stacking patterns of depositional sequences and their component systems tracts (e.g. Fig. 1) is thought to be controlled by the interplay between tectonic subsidence, eustatic sea-level changes and sediment supply. These variables are often cited as 'primary' controls on sequence development, along with factors such as shelf physiography and environmental changes (e.g. Van Wagoner *et al.* 1990; Schlager 1993; Hunt & Tucker 1993; Gawthorpe *et al.* 1994; Helland-Hansen & Gjelberg 1994). Accordingly, other factors are considered to have only a secondary effect on sequence development. They are not thought to directly control stacking patterns or sequence architecture (e.g. Vail *et al.* 1991) because the rate or influence of 'secondary' processes is considered constant over the timescale(s) of sequence development (Jervey 1988). Until now, compaction has been considered to play only a 'secondary' role in sequence development. The role of compaction-induced subsidence as a control on sequence development and architecture is examined and evaluated here.

Geological background

Accommodation development

Within the concepts of sequence stratigraphy, the interplay between rates of sediment supply and accommodation development is thought to be fundamental in controlling the position and stacking patterns of stratal units (Jervey 1988; Schlager 1993). Accommodation is the space 'available for potential sediment accumulation', some of which may be inherited (Jervey 1988). Antecedent accommodation may be filled by deposition or destroyed by a relative base level fall, whereas 'new' accommodation space is only added by relative base level rise. The addition or destruction of accommodation is 'a function of both sea-level fluctuation (e.g. eustacy) and subsidence' (Jervey 1988).

Accommodation space is developed by basin-floor subsidence and/or a eustatic sea-level rise. Basin-floor subsidence creates accommodation, and is driven by local tectonics, lithospheric cooling, sediment loading and compaction. However, the relative roles played by these mechanisms depends on the basin setting, and are often difficult to quantify. In idealized models, rates of subsidence are assumed constant over the time-interval of sequence development, so that fluctuations of eustatic sea-level are proposed as the dominant control on stratigraphy (e.g. Vail 1987; Jervey 1988; Posamentier *et al.* 1988). Although the applicability of this assumption has been brought into question for active basins (e.g. Gawthorpe *et*

From Howell, J. A. & Aitken, J. F. (eds), *High Resolution Sequence Stratigraphy: Innovations and Applications*, Geological Society Special Publication No. 104, pp. 321–345.

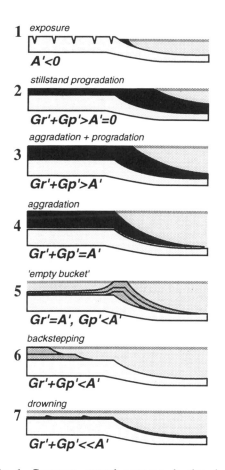

Fig. 1. Common stratal patterns developed on carbonate platforms with accretionary margins, reflecting the interplay between the rate of carbonate growth and production potential (G′) and the rate of change of accommodation (A′). Gr′, growth at the platform margin; Gp′, growth rates in the platform interior. The different geometries approximately correspond to the lowstand (1), highstand (2–3) and transgressive systems tracts (4–7) of sequence stratigraphy (modified after Schlager 1993).

al. 1994), the role of compaction-induced subsidence as a control on accommodation and sequence development is rarely considered.

Compaction-induced subsidence

Compaction-driven subsidence can contribute up to a fifth of the total subsidence in sedimentary basins (Steckler & Watts 1978; Watts 1981; Bond & Kominz 1984), yet as a control on stratigraphy and deformation mechanism, compaction is a much underestimated or even neglected process (Suppe 1985; Gay 1989). Many aspects of the compaction process are still relatively poorly understood. In particular, it is often difficult to constrain the rate at which compaction-driven subsidence takes place and its timing, or to quantify the contribution to total subsidence. Hence, compaction is commonly perceived as a rather undynamic 'background' process, in which the rate gradually decreases over a time interval comparable to the process of thermal subsidence (e.g. 10^{6-7}years). In this context subsidence could be assumed to be constant over the time interval of third (or higher) order sequence development.

It is perhaps for these reasons that compaction is considered as a 'secondary' control on accommodation and sequence development (Jervey 1988). However, the high resolution framework provided by sequence stratigraphy, especially in carbonates, allows significant insights to be made into the compaction process in the shallow subsurface. These insights have important implications for the timing of compaction-induced accommodation space in the framework of sequence stratigraphy. Relief generated by compaction-induced differential subsidence can directly control facies distributions and ultimately sequence development and architecture.

Differential compaction

To gain a better understanding of compaction-induced subsidence as a control on stratigraphy, attention is directed towards areas over which differential compaction has occurred (e.g. Fig. 2a). Where compaction is uniform, then there is little chance of being able to distinguish the contribution of compaction-driven subsidence relative to those of tectonic subsidence, or a eustatic sea-level rise (Fig. 2b). Furthermore, because of the problems of differentiating the effects of tectonism and compaction-induced subsidence, we focus on areas undergoing regional thermal or flexural subsidence with few active faults.

Compaction-induced differential subsidence rotates strata and generates relief when rates of compaction exceed those of deposition (Fig. 3). When differential compaction occurs in the shallow subsurface, the induced subsidence strongly deforms and rotates the depositional surface (Figs 2 and 3). The resulting relief can influence facies distributions (Gay 1989) (Fig. 3), stratal patterns and ultimately sequence architecture and development.

Compaction-induced differential subsidence takes place principally over areas of lateral

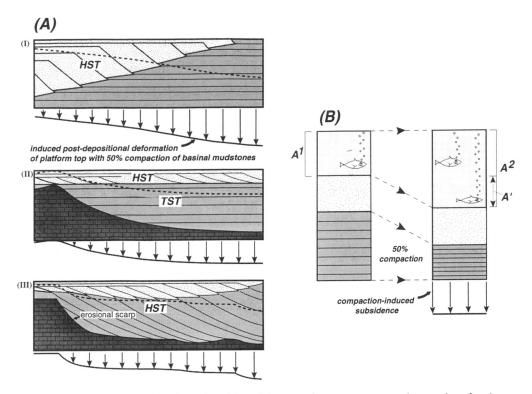

Fig. 2. (A) Some of the scenarios in which differential compaction can occur over the margins of carbonate platforms and buildups. (I) Depositional, through climbing progradation, (II) buried antecedent relief, and (III) combination of antecedent relief and depositional thickening. Following rapid progradation, differential compaction of wedge-shaped basinal mudstones results in the basinwards rotation and subsidence of toplap strata (as shown by the subsidence profiles below each sketch). The resulting architecture and position of the subsided platform top is shown by a dotted line on each diagram. Only the mudstones are shown to compact here, other strata are assumed incompactable. High energy foreslope deposits are stippled, pre-existing platform strata are blocked and periplatform mudstones have no ornament. (B) In contrast, where compaction is uniform, the contribution of compaction-induced subsidence to accommodation development is difficult to separate from the effects of eustacy and tectonic subsidence. A^1, initial accommodation, A^2, accommodation developed resulting from the addition of compaction-induced subsidence (A'). See text for further discussion.

facies change and/or the antecedent relief of buried relatively incompactable bodies (Fig. 2). Studies of differential compaction in siliciclastic systems have concentrated on the differences in compactability between channel sands and overbank muds (Heritier *et al.* 1979). In channel systems, the high compactability of overbank muds can control the stacking patterns of channel sands. This can be important in understanding the distribution of channel sand and overbank deposits in coal-bearing strata (e.g. Fielding 1986), and the connectivity of hydrocarbon-bearing sand bodies in both basin-floor fans (e.g. Timbrell 1993) and fluvial systems (e.g. Anderson 1991). The burial of antecedent topography is also of importance in siliciclastic systems (see Gay 1989). In contrast, interest in

carbonate systems has focused on the stratigraphy over buried carbonate buildups (e.g. Labute & Gretener 1969; Anderson & Franseen 1991; Hunt & Allsop 1993; Hunt *et al.* 1995).

Why study carbonate platforms?

From a historical perspective, studies of differential compaction have tended to focus on siliciclastic systems. However, we contend that the potential for initiating and understanding compaction-induced differential subsidence, and specifically its influence on stratigraphy, is probably greatest over the margins of carbonate platforms and buildups. In particular, the scale of change allows outcrop analysis as well as subsurface (seismic) observations. The following

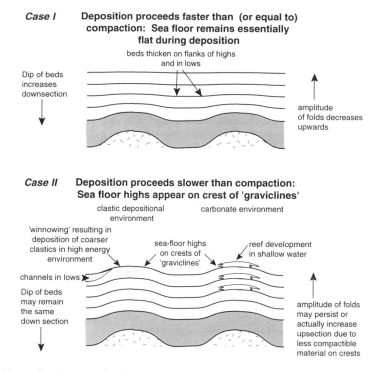

Fig. 3. Diagrams illustrating the interplay between rates of deposition and compaction-induced differential subsidence, and showing the possible influence of compaction-induced relief on patterns of siliciclastic and carbonate marine deposition (from Gay 1989, with minor modification).

factors are important pre-requisites for initiating differential compaction, and are relatively common on the margins of carbonate platforms: deposition of wedge-shaped stratal units; rapid lateral facies changes; juxtaposition of strata with markedly different compaction potentials; high depositional porosities of carbonate muds; and burial of incompactable antecedent topography, such as reefs or buildups (Fig. 2).

Other factors intrinsic to carbonate depositional systems are also important, including: abundance of fauna available for biostratigraphic control; abundance of geopetal fabrics (markers of the palaeohorizontal); the large scale of outcrops compared to the depositional system.

These factors allow a better constraint of the timing of compaction, a quantification of compaction-induced rotation, and therefore amount of compaction, as well as observation of the effects of compaction-induced differential subsidence on the depositional system as a whole.

In summary, the margins of carbonate platforms and buildups are excellent places to study the effects of differential compaction, and in

doing so, gain insights into the compaction process in general.

Carbonates and compaction

Compaction studies of limestones are divisible into five main types. (1) Largely petrographic (e.g. Meyers 1980; Meyers & Hill 1983; Shinn *et al.* 1983); (2) experimental (e.g. Bhattacharyya & Friedman 1979; Shinn & Robbin 1983); (3) down-hole porosity evaluation (Schmoker & Halley 1982; Scholle & Halley 1985; Audet 1995); and, more rarely, through (4) seismic study (Anderson & Franseen 1991); or (5) study of facies and large-scale stratal relationships in the field or subsurface (e.g. Pray 1960, 1965; Beach & Schumacher 1982; Doglioni & Goldhammer 1988). This study utilizes a combination of large-scale stratal relationships, detailed modelling and petrography. It has the advantage that observation of stratal relationships (as for example seen on seismic sections) can be used in a qualitative way to assess the role of compaction as a control of sequence development and architecture before detailed modelling is at-

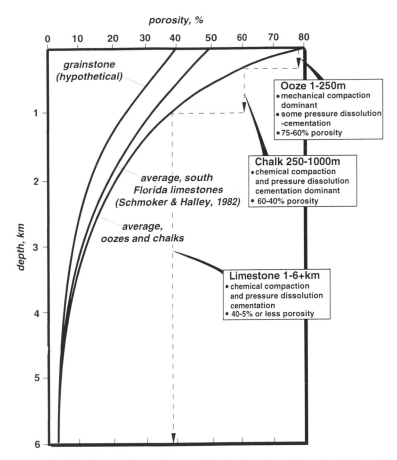

porosity, %

0 10 20 30 40 50 60 70 80

grainstone (hypothetical)

Ooze 1-250m
• mechanical compaction dominant
• some pressure dissolution -cementation
• 75-60% porosity

average, south Florida limestones (Schmoker & Halley, 1982)

Chalk 250-1000m
• chemical compaction and pressure dissolution cementation dominant
• 60-40% porosity

average, oozes and chalks

Limestone 1-6+km
• chemical compaction and pressure dissolution cementation
• 40-5% or less porosity

depth, km

Fig. 4. Idealized porosity–depth curves for carbonates. Note the large loss of porosity in fine-grained carbonates (not subject to early cementation) within the first few hundred metres of burial (redrawn from Choquette & James 1990).

tempted. Here we discuss the criteria for recognition of the effects on deposition.

Burial, compaction and porosity evolution

In the shallow subsurface (< 250 m), the compaction history of carbonate strata is highly dependent on initial depositional environment and texture, and original grain mineralogy and early cementation history (Bathurst 1975; see Choquette & James 1990 for review). At the depositional site, carbonate strata display a wide range of porosities, typically between 40 and 80% (Fig. 4). Lime-mudstones tend to have high porosities and low permeabilities, whereas in grainstones the converse is true; the initial range of porosities is largely a reflection of depositional texture.

Two 'idealized' porosity-depth curves for lime-mudstones and grainstones are shown in Fig. 4. They have a contrasting form, with the lime-mudstones showing a dramatic reduction of porosity within the first few hundred metres of burial. The loss of porosity in grainstones is more gradual. The relatively high rate of compaction for lime-mudstones in the shallow subsurface (compared to grain-supported rocks) is attributed to the effects of mechanical compaction rather than cementation occluding pore spaces.

Lime-mudstones can undergo large amounts of mechanical compaction within the first few hundred metres of burial (200–300 m). In this range, lime-mudstones are mechanically akin to soils, and behave in a manner predicted by soil mechanics (Audet 1995). These conclusions are

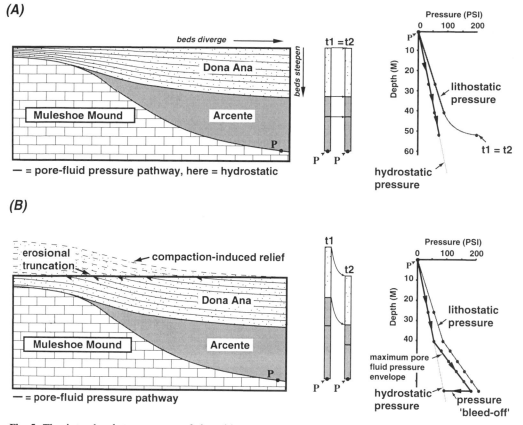

Fig. 5. The interplay between rates of deposition, compaction and pore fluid pressure development. Here, differential compaction occurred over relatively incompactable antecedent relief, such as a carbonate buildup (blocked ornament) whose relief was filled by lime-mudstones (shaded). (**A**) When rates of deposition are matched by those of compaction, then pore fluids remain hydrostatic during deposition. No additional compaction occurs once deposition ceases. (**B**) In contrast, when depositional rates exceed rates of pore-fluid loss, the loaded strata become overpressured, inhibiting compaction. In this case, compaction continues following deposition and, where differential, post-depositional compaction-induced relief is generated. If this topography is eroded an angular unconformity is developed. If not, then the induced relief will influence subsequent patterns of sedimentation (shown by dotted lines here). (Example based on Muleshoe Mound, Sacramento Mountains, USA, as illustrated in Fig. 14. The pressure profiles for point 'P' are based on station 11, Muleshoe Mound (see Fig. 14), and have been calculated with respect to sea-level which is assumed to have been constant).

largely based on down-hole data, but are also supported by experimental results. Compression experiments on modern carbonate mud-rich lithologies show that they can compact to half of their original thickness within 100 m of burial, with an accompanying porosity loss of 50–60% (Shinn & Robbin 1983). Such large amounts of compaction have the potential to develop significant accommodation space over short time intervals (e.g. 10^5–10^6 years).

Conceptual framework

Compaction is a load-driven process that takes place in response to the addition of overburden

that is comprised of two components: the load of overlying sediments, and the height of the water column above (if any).

Pore fluid pressure and equilibrium compaction

The rate and timing of compaction is linked to the rate of overburden addition or removal. In the shallow subsurface (e.g. $<250\pm50$ m), mechanical rather than chemical compaction predominates (Fig. 4). Mechanical compaction involves dewatering, and the rate at which this occurs is principally controlled by permeability.

As dewatering occurs, pore space is reduced and overburden stresses increase at grain contacts, inducing grain reorganization and compaction. When dewatering is complete, a state of compaction equilibrium is achieved, and pore fluid pressure is hydrostatic. In contrast, when dewatering is incomplete, part of the overburden stress is transferred to the pore fluids, which are overpressured.

Since fluids are generally considered to be incompressible, overpressured pore fluids support a part of the overburden, and this retards compaction. Because the effective stress is equal to the overburden (lithostatic) pressure minus the pore-fluid pressure; increases of pore pressure decrease the effective stress on grain contacts. Thus, overpressured strata are relatively 'undercompacted', and their porosity is no longer predictable as a simple function of depth. Rocks in a state of compaction disequilibrium therefore have a 'stored' potential to compact, and compaction will only occur as pore-fluids dissipate.

At the shallow depths considered here, overpressure is most likely to occur in low permeability strata subject to rapid addition of overburden (Audet & McConnell 1992). Potentially, rapid addition of overburden can produce relatively large amounts of overpressure (e.g. by rapid deposition of overlying grain-rich strata) (Fig. 5). However, such overpressuring is likely to be a temporary phenomenon, since even low permeabilities allow pore-fluids to bleed off in response to the induced pore-fluid pressure potential. The rate of bleed off is related in Darcy's equation to permeability and the pressure potential. As dissipation occurs, overburden stresses are transferred back onto grain contacts, and this induces compaction. As the pressure potential decreases with time, the rate of fluid flow, and hence the compaction rate decreases, and should stop as hydrostatic pressure and a state of equilibrium compaction is attained. However, the electrostatic and frictional forces between grains may halt fluid flow somewhat earlier (Hilbrecht 1989).

Thus, the development of pore-fluid overpressure and a state of disequilibrium compaction can be viewed as a two-fold process.

(1) During times of pore-fluid pressure build-up, rates of compaction are retarded. The overpressured strata develop a 'stored' capacity to compact.

(2) During the ensuing period of dissipation, compaction rates proceed at a relatively increased rate. The greater pore fluid pressure gradients drive off fluids and hence compact at a greater rate.

The record of pore fluid overpressure

As long as pore-fluid pressures fail to reach the tensile or cohesive strength of the overpressured sediments, there will be no direct record of the pressure build-up in these strata (the rock strength is here taken to approximate to lithostatic pressure, the value of which can be estimated from assumed bulk densities). Catastrophic failure and geologically instantaneous dewatering will occur if pore-fluid pressures reach lithostatic pressure.

In the examples considered here, there is no evidence that catastrophic failure and rapid dewatering occurred. Accordingly, pore-fluid pressures are interpreted to have fluctuated between the hydrostatic and lithostatic pressure gradients as shown in Fig. 5b. In these circumstances, the only tangible record of pore-fluid overpressure (and a delay in the attainment of compaction equilibrium) is derived from the relationships developed in overlying strata (i.e. compare Fig. 5a and b). Furthermore, this evidence is only likely to be easily deciphered when the subsequent compaction is differential and rapid, in comparison with rates of deposition.

The relative rates of compaction and deposition

The qualitative relationships between the relative rates of deposition and compaction-induced differential subsidence are relatively well established, as is the influence of compaction-induced topography on sedimentation patterns (e.g. Gay 1989; Fig. 3). However, the underlying mechanisms that control the dynamic interplay between differential compaction-induced subsidence and sequence development have not been previously explored. Here the interplay between depositional loading and compaction is examined, whilst sea-level is considered as constant.

A state of equilibrium compaction is maintained when rates of compaction match those of deposition. Where compaction is differential, there will be a tendency to fill or subdue existing topography and maintain a flat depositional surface during these times (e.g. Fig. 5a). All the compaction induced subsidence takes place during deposition, and no further compaction will occur if there is a pause in sedimentation. In the example shown in Fig. 5a, strata were deposited on a horizontal depositional surface undergoing differential compaction-induced subsidence. The resulting stratal unit (stippled) has a wedge shape reflecting that of the underlying compacting unit (shaded), as does the internal thickening of individual beds. The beds themselves form a divergent fan, with older strata dipping more steeply than younger ones. To achieve this situation, pore-fluid pressures must have remained close to hydrostatic during deposition (Fig. 5a).

In contrast, if compaction is in a disequilibrium state during deposition, it will continue following deposition, for example during a depositional hiatus (Fig. 5b). Where the resulting induced subsidence is differential, post-depositional relief is developed (dotted). In this instance, pore fluids will have been temporarily overpressured during deposition, and will attempt to return to hydrostatic pressure during the ensuing hiatus (Fig. 5b). During this time interval, rates of compaction-induced subsidence far exceed depositional rates. In the example shown in Fig. 5b, the resulting stratal patterns are quite different to those in Fig. 5a, yet the initial geometry of the compacting unit is the same. In this 'end-member' scenario, it is assumed that no fluids are lost from the compacting unit (shaded) during deposition of the grain-rich upper unit (stippled). This acts to inhibit compaction. Consequently, the rapidly deposited grain-rich unit has a sheet-like external form with parallel internal bedding relationships. Following deposition, relief is developed as the wedge-shaped unit compacts, rotating and steepening overlying strata (Fig. 5b).

Relief developed in this way can strongly influence patterns of sedimentation if it is preserved. Alternatively, if the compaction-induced relief is eroded, an angular unconformity is developed (e.g. Fig. 5b). Examples of these types of 'compactionally-enhanced' unconformities are discussed below.

Sea-level changes, pore fluids and compaction

Changes of relative sea-level can also directly alter the relative timing, rates and even the amounts of compaction. The magnitude of such changes is related to the amplitude, rate and direction of the fluctuations, as well as the prior condition of the strata in terms of equilibrium compaction.

Falls of relative sea-level may promote or augment the development of overpressure in low permeability sediments. This mechanism has been invoked to explain slope failure and widespread redeposition of pelagic sediment across Europe in the late Cretaceous (Hilbrecht 1989).

The amount of overpressure induced is related directly to the amplitude and indirectly to the rate of sea-level fall (Hilbrecht 1989). This process is especially important in sediments already in a state of compaction disequilibrium, since the augmentation of overpressure prolongs the time over which a state of disequilibrium compaction is maintained. In addition, the greater induced pore fluid pressure gradients will increase the rate of pore-fluid pressure dissipation and hence may increase compaction rates. However, the expulsion of fluids resulting from a sea-level fall will induce little additional compaction if pore-spaces remain fluid-filled. This is because a part of the overburden, represented by the weight of the water column, is reduced as sea-level falls.

The effect of sea-level rises is rather different and more complex. In sediments in perfect communication with the static water column above, changes in the height of the water column do not affect compaction since there is no change in the effective stress (the effective stress = lithostatic pressure − pore fluid pressure). Thus no compaction should occur if rates of fluid flow can match the change in the water column height, so maintaining normal pore fluid pressure. In contrast, when rates of fluid loss from low permeability units are exceeded by rates of loading (e.g. induced by rapid sea-level rise) then compaction may be induced. Thus, in terms of the relative timing and amounts of compaction, rapid sea-level rises can have much the same effect as that of rapid loading through deposition. This effect has been found to be important when modelling the subsidence history of early Permian mixed carbonate–siliciclastic depositional sequences in the Delaware Basin, Texas and New Mexico, USA (J. F. Sarg pers. comm. 1995).

Carbonate production and sea-level

Sedimentation on carbonate platforms is sensitive to changes of relative sea-level and environment that can cause marked changes of carbonate production rate (e.g. Schlager 1993; Hunt & Tucker 1993; Handford & Loucks 1993).

In general terms, highstands of sea-level are associated with progradation of accretionary platform margins (Fig. 6). On the platform, rates of carbonate production are high, reflecting the wide area of shallow water where sediment is overproduced and shed basinward, resulting in basin-filling and ultimately progradation. In contrast, times of falling and lowstands of sea-level are times of reduced carbonate production.

1 Highstand

Rimmed shelf with accretionary slope apron pattern of progradation.

2 Forced regressive

At the end of the third slope wedge sea-level is at its lowest point and the greatest area of the platform is exposed. The sequence boundary thus passes above slope and basin-floor sediments formed during forced regression. Sedimentation patterns reflect inherited topography.

3 Lowstand

As relative sea-level begins to rise the area available for sedimentation increases.

4 Transgressive

Most of the shelf drowns as sedimentation is outpaced by relative sea-level rise, and a condensed section develops across the shelf. Shelf margin is scalloped due to frequent collapse.

5 Highstand

Normal shelf sedimentation resumes as the rate of relative sea-level rise decreases. Facies on the shelf reflect inherited topography from the lowstand (eg. karst) and transgression (eg. build-ups). Shallow shelf-sediments bypass the slope to the basin-floor which aggrades, decreasing relief and correspondingly slope angles (highstand shedding).

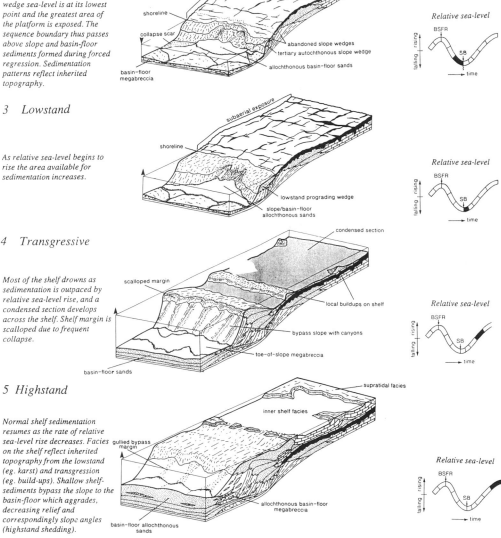

Fig. 6. Sequential block diagrams showing the evolution of a carbonate platform during a cycle of relative sea-level. Note the location of the lowstand wedge below and basinwards of the antecedent highstand shelfbreak, and the parallel and locally downlapping lower boundary of the transgressive systems tract onto the sequence boundary on the shelf (from Hunt & Tucker 1995).

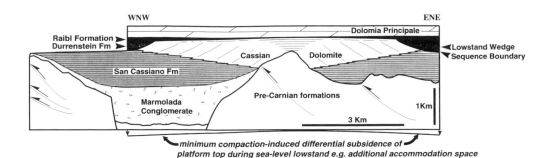

Fig. 7. Cross-section through the Triassic Sella platform, Dolomite Mountains, Italy. Here the Cassian platform prograded during a highstand of sea-level, with a basinwards climbing toe-of-slope. Platform top highstand strata are essentially parallel, showing that little differential subsidence occurred during highstand times. During the following lowstand of sea-level, when the Durrenstein lowstand wedge was deposited, compaction-induced differential subsidence of basin floor strata took place, arching the platform top into a gentle anticline. This created platform top relief over which the succeeding Raibl Formation thins (redrawn with minor modifications from Doglioni & Goldhammer 1988).

As sea-level falls, accommodation space is destroyed and the available area for shallow water carbonate production is restricted (Figs 1 and 6); the shallow-water 'carbonate factory' is limited to a relatively narrow zone along the antecedent highstand slope. Concurrently, the platform is subaerially exposed and chemically weathered. In humid climates karstic topography ·can be developed at these times (Purdy 1974a, b), and can act as a fundamental control on facies distributions when the platform is later flooded (Purdy 1974a, b; Purdy & Bertram 1993).

Rising sea-level is associated with the flooding and resumption of carbonate sedimentation on the shelf (Fig. 6). The response of carbonate platforms to rising sea-level is, however, complex. Stacking patterns can vary temporally and spatially, and may include backstepping, aggradation or drowning (Schlager 1993; Hunt & Tucker 1993). This reflects differences in sediment production across a carbonate platform, and the capacity of platform margins to aggrade and keep-up with rises of sea-level (Figs 1 and 6). The inherited physiography of the platform resulting from subaerial exposure and any environmental changes associated with transgression can also play an important role at these times (e.g. Purdy 1974a, b; Neumann & Macintyre 1985). Suffice to say that rapid (glacio-eustatic) rises in sea-level and/or environmental changes (e.g. Schlager 1981; Hallock & Schlager 1986; Blanchon & Shaw 1995) may temporarily drown carbonate platforms. During these times a non-depositional hiatus is developed.

Carbonate sequence stratigraphy and compactionally 'enhanced' unconformities

Changes of relative sea-level can directly control the timing and rates of both compaction and carbonate production. Because of the relationship between sediment production and depositional loading, sea-level changes can also indirectly control the sites, timing and rates of compaction that result from variations in depositional loading (e.g. Fig. 6).

The integration of data on compaction, with sequence stratigraphic studies relating carbonate production to sea-level changes, suggests that during the lowstand and transgressive systems tracts the effects of (1) decreased sedimentation rates, and (2) increased compaction rates (following a period of rapid loading and pore-fluid pressure build-up) can be combined. It is at these times that unconformities 'enhanced' by differential compaction are developed on carbonate platforms (see below). This is a process likely to be most important following rapid deposition (e.g. highstands in carbonates) when a state of compaction disequilibrium may be temporarily developed.

Differential compaction and sequence architecture

A depositional surface rotated by compaction-driven differential subsidence can form a compactionally 'enhanced' unconformity, onlapped by the subsequent lowstand and/or transgressive

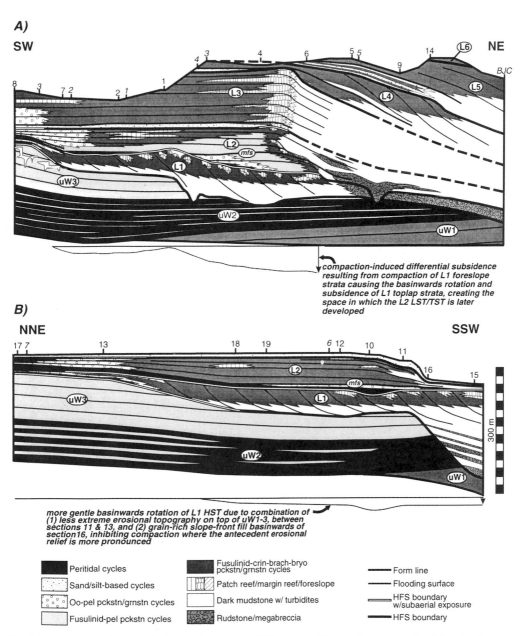

Fig. 8. Two cross-sections of (**A**) the northwestern and (**B**) a part of the southeastern walls of Apache Canyon, northwest Texas, USA. The Permian strata in this canyon clearly demonstrate the control of compaction-induced subsidence in modifying highstand platform architecture (L1), and the effects of this compaction-induced relief on the distribution and size of the overlying lowstand/transgressive systems tracts (L2). The maximum flooding surface (mfs) separates the lowstand/transgressive and highstand systems tract of sequence L2 (see text for further discussion). Also of significance is the compaction-modified basinwards descending geometries of the L4 and L5 sequences in (**A**). The toplap strata of these sequences dip basinwards where they have prograded over the antecedent aggradational margin of sequences L2–3, a relationship also evident in seismic sections from this area. HFS, high frequency sequence; uW1, uW2 and uW3, Hueco group–Wolfcampian; L1–5, Leonardian. These diagrams kindly supplied by W. M. Fitchen (reproduced with minor modifications from Fitchen 1995).

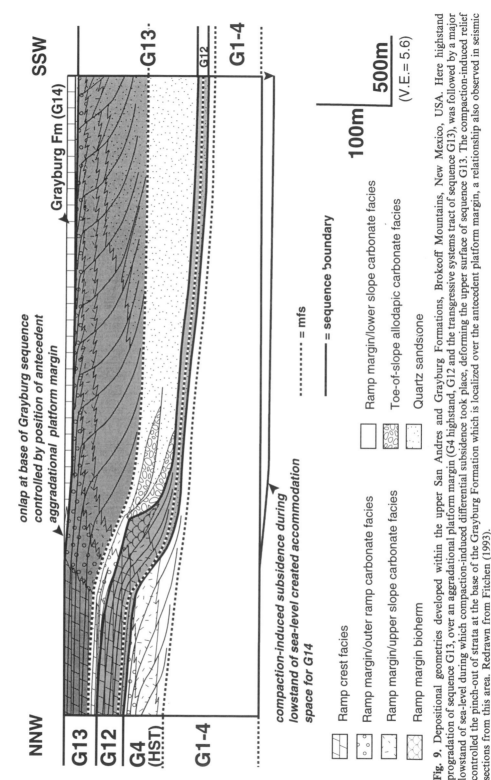

Fig. 9. Depositional geometries developed within the upper San Andres and Grayburg Formations, Brokeoff Mountains, New Mexico, USA. Here highstand progradation of sequence G13, over an aggradational platform margin (G4 highstand, G12 and the transgressive systems tract of sequence G13), was followed by a major lowstand of sea-level during which compaction-induced differential subsidence took place, deforming the upper surface of sequence G13. The compaction-induced relief controlled the pinch-out of strata at the base of the Grayburg Formation which is localized over the antecedent platform margin, a relationship also observed in seismic sections from this area. Redrawn from Fitchen (1993).

systems tracts. The compaction-induced relief and accommodation developed reflects the lithology, thickness, wedge shape and porosity decrease of the compacting unit. As a result of subsidence and rotation, the succeeding systems tracts inherit a quite different platform architecture to that remaining if a state of compaction equilibrium had been maintained during times of highstand deposition. Essentially, the effects are two-fold: the compaction-induced subsidence creates 'additional' accommodation space, and the platform physiography is changed as toplap strata are rotated. These effects exert a strong control on the position, extent and geometry of the ensuing lowstand and/or transgressive systems tracts.

Compactionally 'enhanced' unconformities: examples and control on sequence architecture and development

The examples discussed below demonstrate a variety of compactionally 'enhanced' unconformities in terms of scale, frequency and preservation. They illustrate the dynamic nature of the compaction process in the shallow subsurface, and the importance of this process over a wide range of scales. Furthermore, these examples demonstrate that the contribution of compaction-induced subsidence to total subsidence is not constant during sequence development; the effects of compaction-induced subsidence are localized at hiatal surfaces, especially sequence boundaries.

Antecedent compaction-induced relief and sedimentation

The Triassic Sella platform (Fig. 7) and Permian of both the Sierra Diablo and Guadalupe Mountains (Figs 8 and 9) illustrate the effects of compaction-induced subsidence at the platform-scale during third and fourth order lowstands, respectively. Subsequent examples illustrate more localized and smaller-scale examples of compaction-enhanced unconformities.

On the Sella Platform, compaction-induced differential subsidence took place during deposition of the Durrenstein lowstand wedge, and arched the platform into a gentle antiform over which the succeeding wedge of the Raibl Formation thins (Doglioni & Goldhammer 1988) (Fig. 7). If it were not for the compaction-induced differential subsidence, the Raibl Formation would have a sheet-like geometry and uniform thickness; it would not thicken towards the margins of the platform.

The effects of basinwards dipping topset surfaces on subsequent deposition is exemplified by lower Permian sequences exposed in Apache Canyon, Texas, USA as described by Fitchen (1995) (Fig. 8). Strong compaction-induced differential subsidence over antecedent erosional topography (developed in sequences uW1-3), caused basinwards rotation and subsidence of the L1 platform top during a lowstand of sea-level (Fig. 8a). The compaction-induced relief developed across the rotated L1 topset strata is onlapped and filled by a relatively large and thick aggrading to backstepping lowstand/transgressive systems tract of sequence L2 (Fig. 8a) (Fitchen 1995).

In direct contrast, on the opposite side of Apache Canyon, where the effects of compaction-induced subsidence are less pronounced, the upper surface of sequence L1 dips only gently basinwards (Fig. 8b). Here, the lower gradient developed across the L1 platform topsets would appear to have resulted in a more rapid transgression, and hence deposition of a more extensive, thinner sheet-like L2 transgressive systems tract (Fitchen 1995) (Fig. 8b). In Apache Canyon, it is clear that the development, extent and size of the L2 lowstand/transgressive systems tract is directly controlled by the antecedent relief developed across the L1 sequences by differential compaction-induced subsidence.

A similar relationship is observed at the fourth order sequence boundary that separates the upper San Andres and Grayburg Formations in the Brokeoff Mountains, New Mexico, USA (Fig. 9). Here, highstand strata of sequence G13 prograded basinwards over an antecedent, aggradational platform margin (Fitchen 1993). During the ensuing lowstand of sea-level, post-depositional compaction-induced relief was created along the upper surface of the G13 sequence (Fig. 9). This relief subsequently controlled the thickness and pinch-out of strata at the base of sequence G14 (the Grayburg Formation); a relationship that is also resolved on seismic sections from this area (Hunt et al. 1995). If it were not for the compaction-induced subsidence following G13 progradation, the pinch-out and thinning at the base of the Grayburg sequence (G14) would have occurred in a more basinward location.

At a smaller scale, compaction-'enhanced' sequence boundaries (5-6?th order) are well developed over a shelf-margin buildup on the Pennsylvanian Pedernal shelf, New Mexico (Goldstein et al. 1991; Goldstein pers. comm. 1993) (Fig. 10). In this example, Units 7, 8 and 9 are shoaling-up carbonate cycles deposited on

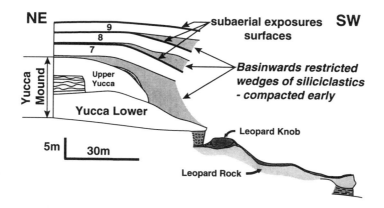

Fig. 10. High frequency mixed carbonate–siliciclastic sequences developed across the Pedernal shelf margin, New Mexico, USA. In this example, differential compaction over the platform margin Yucca Mound took place during lowstands of sea-level following deposition of shoaling-up carbonate highstand sediments. The carbonates were originally deposited as relatively flat-lying strata. The differential compaction-induced subsidence controlled the position of the landwards pinch-out of siliciclastic sediments. Redrawn from a diagram kindly supplied by R. Goldstein (Goldstein 1991).

relatively flat surfaces over basin-restricted wedges of siliciclastic sediment. During lowstands of sea-level, whilst the platform top was exposed, the siliciclastic strata compacted, rotating the carbonate units basinwards. This additional compaction-induced accommodation acted to control the pinchout of subsequent siliciclastic wedges.

Remaining in the Pennsylvanian, an example of the more local influence of compaction-induced subsidence on lowstand sedimentation is illustrated in Fig. 11. This example, exposed along the Goosenecks of the San Juan River, Paradox Basin, Utah, USA, illustrates how the course of incised valleys, cut during a lowstand of sea-level, was deflected around a compaction-induced high developed over an antecedent Desmoinesian buildup (Fig. 11. G. Gianniny pers. comm. 1993; Gianniny 1995).

Finally, relief created by compaction-induced subsidence at fourth order flooding surfaces has been recognized to control the development of subsequent parasequences and high frequency sequences in the late Dinantian–early Namurian of west Cumbria, UK (Thurlow 1994) (Fig. 12).

Angular unconformities

In contrast to the examples discussed above, compaction-induced relief developed at sequence boundaries is not always preserved to influence subsequent patterns of sedimentation. Instead, the induced relief can be eroded, developing angular unconformities (Figs 5, 13 and 14).

A cross-section through the mixed Permian carbonate–siliciclastic platform exposed in Last Chance Canyon, Guadalupe Mountains, New Mexico, is illustrated in Fig. 13. Here, overall progradation of Cycles 6–12 followed a period dominated by aggradation (Cycles 4–5). The effects of compaction-induced differential subsidence are most apparent along the upper boundary of the uSA4 sequence (M. Sonnenfeld pers. comm. 1995) (Fig. 13). During a lowstand of sea-level, compaction-induced subsidence developed relief that was subsequently eroded.

Erosional relief generated through differential compaction formed an angular unconformity (Fig. 13) and, of particular interest, a synformal structure located immediately basinwards of the aggradational margin of cycles 4 and 5 (Fig. 13a). This synformal structure is developed between the antecedent aggradational platform margin, and a relatively incompactable lower slope buildup (Fig. 13). Much of the dip observed in this section may well have been induced by the compaction of lower slope and basin-floor strata. Compaction also took place during hiatuses between the component cycles. Individual cycle boundaries converge landwards and older cycles dip more steeply than their successors (Cycles 6–11) (Fig. 13). However, the most significant compaction-induced subsidence occurred during a major lowstand of sea-level and basinal facies shift, between the upper San Andres and Grayburg Formations (Fig. 13). There is an interesting comparison to be made

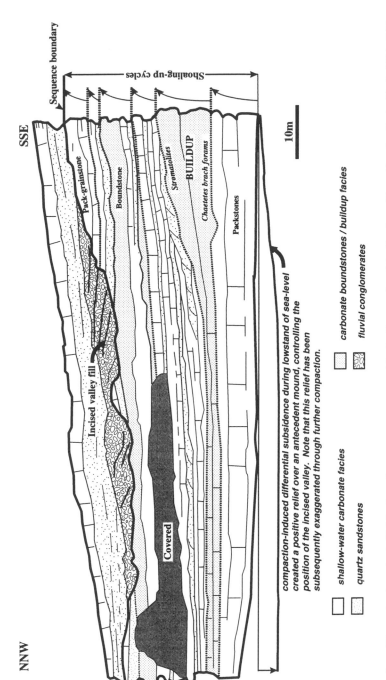

Fig. 11. Line drawing of Desmoinesian mixed siliciclastic–carbonate strata as exposed in the Goosenecks of the San Juan River, Paradox Basin, Utah, USA. Differential compaction over an antecedent buried carbonate buildup occurred during a major lowstand of sea-level. This developed relief controlled the position of incised valleys. Redrawn from an original line-drawing kindly supplied by G. Gianniny (from Gianniny 1995).

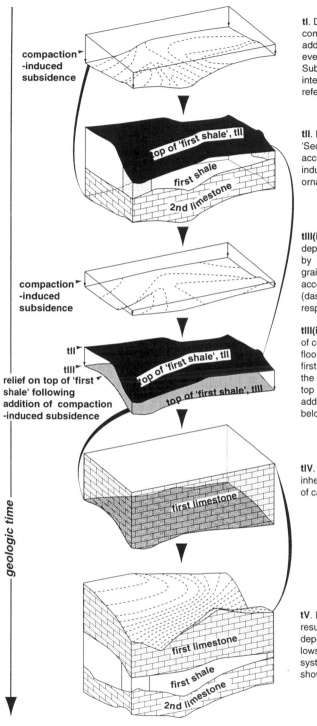

tI. Differential subsidence resulting from compaction of underlying clastics creates additional accommodation space prior to flooding event at the base of the 'Second Limestone'. Subsidence contours (dashed lines) at metre intervals shown with respect to a horizontal reference frame.

tII. Depositional package represented by the 'Second Limestone' and 'First Shale' fills accommodation space created by compaction-induced subsidence in underlying strata. Stippled ornament denotes location of major sandbodies.

tIII(i).During depositional hiatus following deposition of the 'first shale' subsidence induced by differential compaction of underlying fine-grained siliciclastics creates additional accommodation space. Subsidence contours (dashed lines) at metre intervals, here shown with respect to a horizontal reference frame.

tIII(ii). The diagram on the left illustrates the effect of compaction-induced subsidence on the sea-floor topography developed across the top of the first shale prior to the flooding event at the base of the 'First Limestone'. It shows the topography on top of the 'first shale' before (tII) and following the addition of compaction induced subsidence (tIII, below).

tIV. Deposition of the 'First Limestone' filled the inherited sea-floor topography, forming a blanket of carbonate.

tV. Lowstand-exposure of the carbonate shelf and resultant compaction of the underlying clastic deposits leads to the exploitation of topographic-lows (more compacted) areas by incised valley systems. Topographic contours (dashed lines) shown at metre intervals.

Fig. 12. Development of depositional architecture in response to changes of relative sea-level and the punctuated contribution of compaction-induced subsidence in a mixed siliciclastic-carbonate succession, late Dinantian–early Namurian, west Cumbria, UK. Diagrams kindly drawn by A. Thurlow (from Thurlow 1994).

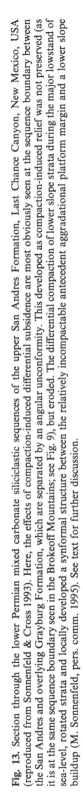

Cycles	Stratal Geometry	Progradation: Aggradation ratio	Offlap Angle
3-4	stratigraphic rise	+281:1	+0.2°
4-5	stratigraphic rise	+17:1	+3.4°
6-11a	stratigraphic rise	+209:1	+0.3°
11a-12b	stratigraphic fall	-47:1	-1.2°
12b-14	stratigraphic rise	≈+6.5:1	≈+9°

fusulinid grainstone/packstone

cherty brachiopod-fusulinid wackestone

● ≈20 m. bathymetric tie points within fusulinid tract

○ ≈20 m. bathymetric tie points within other facies tracts

Fig. 13. Section through the lower Permian mixed carbonate siliciclastic sequences of the upper San Andres Formation, Last Chance Canyon, New Mexico, USA (reproduced from Sonnenfeld & Cross 1993). Here, the effects of compaction-induced differential subsidence are most obviously seen at the sequence boundary between the San Andres and overlying Grayburg Formation, which are separated by an angular unconformity. This developed as compaction-induced relief was not preserved (as it is at the same sequence boundary seen in the Brokeoff Mountains; see Fig. 9), but eroded. The differential compaction of lower slope strata during the major lowstand of sea-level, rotated strata and locally developed a synformal structure between the relatively incompactable antecedent aggradational platform margin and a lower slope buildup (M. Sonnenfeld, pers. comm. 1995). See text for further discussion.

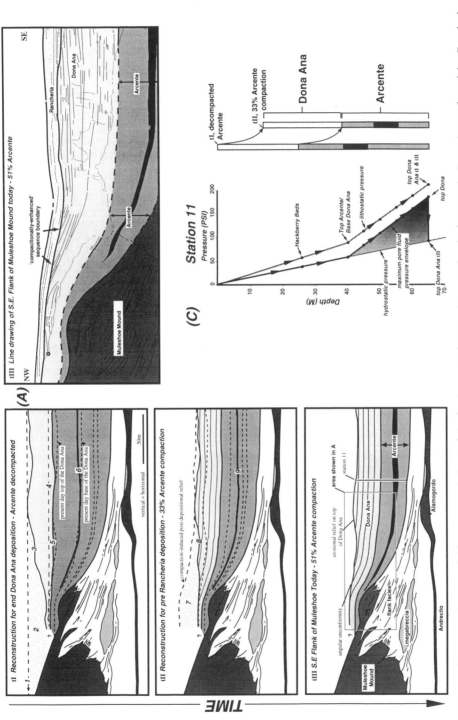

Fig. 14. Progressive compaction, pore-fluid pressure evolution and the development of an angular unconformity between the Dona Ana and overlying Rancheria on the flanks of Muleshoe Mound, Sacramento Mountains, USA. (**A**) Line drawing of part of the southern flank of Muleshoe Mound (location of line drawing shown on bottom diagram of part B). (**B**) Progressive compaction of the Arcente following deposition of the Dona Ana as shown for three time 'slices' representing: (tI) the time immediately following Dona Ana deposition, but prior to Arcente compaction, (tII) following compaction of the Arcente and rotation and erosion of the Dona Ana (prior to deposition of the Rancheria) and, (tIII) the relationships as seen today. (C) Shows the pore-fluid pressure pathway (with reference to sea-level, considered static, here) experienced at the base of the Arcente, located at station 11 during Dona Ana deposition, and the hiatus that followed (station 11 located in tIII of part B).

with another example of the same sequence boundary in the Brokeoff Mountains (Fig. 9), where compaction-induced relief was preserved to influence subsequent sedimentation patterns in the Grayburg Sequence. This illustrates the fact that there is lateral variability in the preservation of compactionally-'enhanced' unconformities.

A prominent angular unconformity formed by erosion of compaction-induced post-depositional relief is also developed on the margins of the Mississippian Muleshoe Mound in the Sacramento Mountains of New Mexico, USA (Figs 5 and 14). In this case, compaction-induced topography of up to 20 m was developed during sequence boundary development following deposition of the Dona Ana (Fig. 14B, tI-II). However, this compaction-induced relief was eroded prior to deposition of the overlying Rancheria sequence. Here, the amount and the timing of compaction have been accurately constrained through study of large-scale stratal relationships (e.g. as in comparison to Fig. 5), petrography and, most importantly, through study of numerous geopetal fabrics in combination with detailed numerical modelling (Hunt & Allsop 1993; Hunt *et al.* 1995). Pore-fluid pressure pathways have also been calculated using estimates of fluid and rock densities from comparable moden carbonate environments (Fig. 14C).

Discussion

Characteristic and diagnostic stratal patterns

In the platform-scale examples discussed above, compaction-induced relief was developed during lowstands of sea-level. In many cases (e.g. Figs 7–10), the induced relief was preserved and inherited by the succeeding sequence, where it acted to control the distribution, facies and pinchouts of subsequent systems tracts in much the same way that antecedent karstic topography can (e.g. Purdy, 1974*a*, *b*) (Figs 7–12).

The characteristic and diagnostic stratal patterns developed in these platform scale examples (Figs 7–10) is the onlap of basinwards dipping and rotated highstand topsets/toplap strata by the succeeding lowstand and/or transgressive systems tracts (e.g. Fig. 15). Where the platform margin is deflected downwards and basinwards by compaction-induced differential subsidence, the lowstand wedge can be developed over, and landwards, of the antecedent subsided highstand depositional shelf-slope break; it laps-out

against the rotated highstand shelf (e.g. Figs 15 and 16).

These stratal relationships are quite different from those classically portrayed for carbonate platforms (e.g. Figs 1 and 6). Normally the lowstand systems tract is located basinwards and below the antecedent highstand shelfbreak, not above and landwards of it (Fig. 15). It is also unusual for the transgressive systems tract to onlap the shelf-top, which is normally a sub-horizontal surface; the transgressive systems tract is often a sheet-like stratal unit over the highstand platform top, with a lower parallel relationship onto the sequence boundary (Figs 1 and 6). Locally, buildups developed during the transgressive systems tract also downlap onto the sequence boundary.

Clearly, onlap of rotated and subsided highstand toplap strata by the lowstand or transgressive systems tracts (e.g. Figs 7–10 & 15) is a key criterion and stratal relationship that can be used to identify compactionally 'enhanced' unconformities, where the induced relief is preserved. In contrast, where compaction-induced relief is eroded, the compaction-enhanced stratal boundaries are marked by angular unconformities (Figs 13 and 14).

Compaction-induced platform rotation as a primary control of sequence development and architecture

The subsidence induced by differential compaction can rotate and steepen toplap strata, whilst foreslope angles are concurrently reduced (Fig. 15). This rotation significantly alters platform architecture. For example, the reduction of gradients may act to stabilize the foreslope. However, this effect could be offset by shear stresses induced across outer parts of the platform by the basinward rotation of toplap strata. Such additional stresses may contribute towards the generation of megabreccias during times of falling sea-level and lowstands, as could the build-up of pore-fluid pressure along or below low permeability strata. This is because elevated pore-fluid pressures reduce shear strength (e.g. see Hilbrecht 1989).

The compaction-induced rotation and downwards deflection of highstand shelf strata can develop a post-depositional stratal geometry similar to that of a gradually downshifting shelf-margin, deposited during times of relative sea-level fall (e.g. see Figs 8, 15 and 16). This kind of post-depositional stratal pattern is particularly common where progradation occurs over an antecedent aggradational margin, as

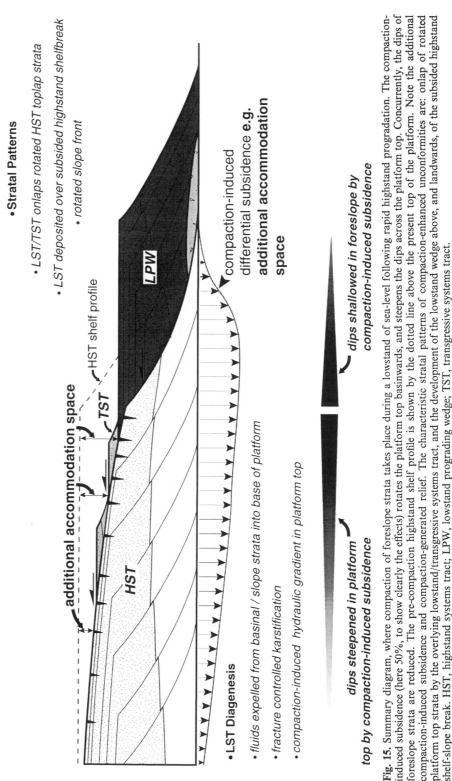

Fig. 15. Summary diagram, where compaction of foreslope strata takes place during a lowstand of sea-level following rapid highstand progradation. The compaction-induced subsidence (here 50%, to show clearly the effects) rotates the platform top basinwards, and steepens the dips across the platform top. Concurrently, the dips of foreslope strata are reduced. The pre-compaction highstand shelf profile is shown by the dotted line above the present top of the platform. Note the additional compaction-induced subsidence and compaction-generated relief. The characteristic stratal patterns of compaction-enhanced unconformities are: onlap of rotated platform top strata by the overlying lowstand/transgressive systems tract, and the development of the lowstand wedge above, and landwards, of the subsided highstand shelf-slope break. HST, highstand systems tract; LPW, lowstand prograding wedge; TST, transgressive systems tract.

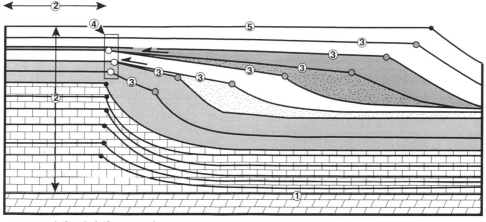

● = aggradational platform margin

◎ = subsided and downwards deflected depositional shelfbreak

○ = compaction-induced secondary break of slope developed over antecedent platform margin

Fig. 16. Schematic diagram illustrating common stratal patterns observed in seismic sections when platform progradation occurs over an antecedent aggradational margin (blocked) (e.g. compare to Fig. 8A), and the criteria necessary to distinguish the effects of compaction-induced differential subsidence from those of more regional tectonic or thermal subsidence. Note the pinch-out and landwards thinning of stratal units (numbered 3) is controlled by the position of the antecedent aggradational platform margin. 1, Regionally widespread reflector that is ideally subhorizontal or a gently dipping reflector from which the contribution of tectonically or thermally-induced subsidence can be evaluated. 2, Zone of subhorizontal and parallel lagoonal strata. 3, Basinwards rotated lagoonal/backreef strata, originally deposited along subhorizontal surfaces. Note that older strata are more strongly rotated and dip most steeply. 4, 'Hinge' or 'kink' zone, around which initially horizontal lagoonal strata dip basinwards. This zone is likely to be extensively fractured. 5, Horizontal lagoonal strata marking the end of syn-depositional compaction as a control of sequence architecture and development.

schematically shown in Fig. 16 (and compare to Fig. 8a, sequences L4–5). In seismic sections there is great potential to confuse these geometries developed by post-depositional rotation of highstand strata, and deposition during a gradual base-level fall (Hunt *et al.* 1995).

A combination of shelf-foreslope subsidence and rotation can lead to production of a more significant lowstand wedge, as there is a broader area available for shallow water lowstand sedimentation. Furthermore, the lowstand wedge may be located over and/or landwards of the preceding subsided highstand depositional shelf-slope break (e.g. Fig. 15). Compaction-induced subsidence results in the earlier transgression of the subsided platform-top. The basinwards tilting of toplap strata may also result in the more gradual flooding of the platform top. This can lead to the development of a larger, thicker transgressive systems tract, in a more landward location than would be predicted (in comparison to the antecedent highstand shelfbreak), as is

exemplified by Permian strata of the Diablo and Brokeoff Mountains illustrated in Figs 8 and 9.

If, in contrast, there had been no delay in the contribution of compaction to the development of accommodation during highstand times, then the additional space developed by compaction would have resulted in: (1) formation of thicker highstand topset strata which consequently (2) would have diverged basinwards, and their development in turn would (3) have acted to limit the lateral progradation of the highstand systems tract. This is because there would have been less excess sediment shed basinwards to enable basin-filling and hence lateral progradation. Thus, in summary, as a direct result of a delay in the contribution of compact-induced subsidence to accommodation development, the effects on the thickness and distribution of stratal packages are that (i) the highstand has a greater lateral extent, but has thinner topsets than expected, whereas (ii) the succeeding lowstand/transgressive systems tracts tend to be thicker and have a greater, and more landwards extent than would be otherwise expected (in comparison with the position of the antecedent subsided highstand shelfbreak).

Diagenesis

In diagenetic terms, the most important effects of compaction-induced differential subsidence are developed in close association with the 'compactional-hinge', around which strata are rotated. Here, the topography developed across the platform top not only exerts a strong control on patterns of early meteoric diagenesis, but also on subsequent burial diagenesis.

Differential compaction-induced subsidence deforms overlying strata (e.g. Figs 2, 15 and 16). Whilst the compacting strata inducing the subsidence are generally not subject to early cementation, the upper deforming surface of the platform is frequently subject to early lithification through the effects of marine and meteoric diagenesis. As the platform top is deformed into antiformal structures, its upper surface is stretched, and this can lead to the development of extensional fractures that can exert a profound influence on the development of early meteoric diagenesis and karst in particular. In Apache Canyon for example, karstic features are only locally developed along the L2 sequence boundary directly over the compaction-induced anticline developed in sequence L1 (W. M. Fitchen pers. comm. 1994) (e.g. Fig. 8a). A secondary feature of compactionally-enhanced unconformities is the generation of relief. This will result in the development of a hydraulic gradient within the exposed platform, and it is well known that this can exert a strong control on patterns of meteoric diagenesis (Goldstein 1988).

Finally, it is important to stress that the generation of extensional fracturing through differential compaction around compactional 'hinges' is also likely to be important during subsequent burial diagenesis. Furthermore, the strata in proximity to the compactional 'hinge' are likely to be subject to significant burial diagenesis due to the up-dip pinchout of stratigraphic units in its proximity (e.g. Fig. 16), concentrating fluid flow into this area.

Application to siliciclastic systems

The effects of rapid loading and a delay in the attainment of a state of compaction equilibrium have been discussed for carbonate systems. However, the response of carbonate and siliciclastic systems to changes of sea-level is rather different (Handford & Loucks 1993; Hunt & Tucker 1993; Schlager 1993). Sedimentation rates tend to be greatest during highstands of sea-level on and around carbonate shelves. In contrast, in siliciclastic systems sedimentation rates are typically greatest during times of falling and lowstands of sea-level (Posamentier *et al.* 1988; Van Wagoner *et al.* 1990). These are the times when pore-fluid overpressure due to rapid depositional loading is most significant in siliciclastic systems. Consequently, the contribution of compaction-induced subsidence to accommodation development is likely to be most important during the transgressive systems tract in siliciclastic systems following a major lowstand of sea-level. Compaction-induced subsidence generated at these times will exaggerate the effects of the ensuing relative sea-level rise.

Conclusions

Compaction is a dynamic process, the importance of which as a control of sequence architecture and development has been previously underestimated in sequence stratigraphy. Compaction-induced differential subsidence can substantially modify platform architecture, and in turn the distribution, extent, size and stacking patterns of succeeding stratal packages, be they systems tracts or sequences. Secondary platform geometries induced by differential compaction can also play an important role in controlling patterns of diagenesis, starting in the early meteoric realm and continuing through and during burial diagenesis.

Compaction-'enhanced' unconformities are developed during depositional hiatuses following the development of a state of compaction disequilibrium; on carbonate platforms they are best developed during sea-level lowstands following times of rapid highstand progradation. Compaction-induced relief developed at these times can be a primary control on sequence architecture and development. If it is eroded, then angular unconformities are formed. These effects are most profound when rapid platform progradation occurs over an antecedent aggradational and/or erosional platform margin. The role of compaction-induced differential subsidence should be carefully considered when interpreting the often complex stratal patterns of sequences developed on and over the margins of carbonate platforms and buildups.

This work has been supported by NERC fellowship (GT5/91/GS/7) to DH, and by an Amoco studentship to TA. Their contribution is gratefully acknowledged.

We must extend our thanks especially to Bill Fitchen, and also Kent Kirkby, Gary Gianniny, Bob Goldstein, Andy Thurlow, Mark Sonnenfeld and Rick Sarg for discussion of specific examples, and in many cases for providing line drawings of the examples presented. The comments of J. Kennard and an anonymous reviewer improved an earlier abbreviated version of this paper, and their contribution is much appreciated. Finally, DH must thank colleagues Toni Simo and Lloyd Pray at Wisconsin-Madison and Bob Ward, Gavin McAulay, Lesley McMurray and Rob Gawthorpe at Manchester for their discussion and input to the ideas presented here.

References

ANDERSON, N. L. & FRANSEEN, E. K. 1991. Differential compaction of Winnepegosis reefs: a seismic perspective. *Geophysics*, **56**, 142–147.

ANDERSON, S. 1991. *Differential compaction in alluvial sediments*. PhD Thesis, University College of Wales, Cardiff.

AUDET, D. M. 1995. Modelling of porosity evolution and mechanical compaction of calcareous sediments. *Sedimentology*, **42**, 355–373.

——— & McCONNELL, J. D. C. 1992. Forward modelling of porosity and pore pressure evolution in sedimentary basins. *Basin Research*, **4**, 147–162.

BATHURST, R. G. C. 1975. *Carbonate Sediments and their Diagenesis*. Elsevier Scientific Publications, Amsterdam, Developments in Sedimentology, **12**.

BEACH, D. K. & SCHUMACHER, A. 1982. Stanley field, North Dakota – a new model for exploration play. *In:* CHRISTOPHER, J. E. & KALDI, J. (eds) *4th International Williston Symposium*. Saskatchewan Geological Society, Special Publication, **6**, 235–243.

BHATTACHARYYA, A. & FRIEDMAN, G. M. 1979. Experimental compaction of ooids and lime mud and its implication for lithification during burial. *Journal of Sedimentary Petrology*, **49**, 1279–1286.

BLANCHON, P. & SHAW, J. 1995. Reef drowning during the last deglaciation: evidence for catastrophic sea-level rise and ice-sheet collapse. *Geology*, **23**, 4–9.

BOND, G. C. & KOMINZ, M. A. 1984. Construction of tectonic subsidence curves for early Paleozoic miogeocline, southern Canadian Rocky Mountains: implications for subsidence mechanisms, age of breakup, and crustal thinning. *Geological Society of America Bulletin*, **95**, 155–173.

CHOQUETTE, P. & JAMES, N. P. 1990. Limestones – the burial diagenetic environment. *In:* McILREATH, A. & MORROW, D. W. (eds) *Diagenesis*. Geoscience Canada, reprint series, **4**, 75–111.

DOGLIONI, C. & GOLDHAMMER, R. K. 1988. Compaction-induced subsidence in the margins of a carbonate platform. *Basin Research*, **1**, 237–246.

FIELDING, C. R. 1986. Fluvial channel and overbank deposits from the Westphalian of the Durham Coalfield, N.E. England. *Sedimentology*, **33**, 119–140.

FITCHEN, W. M. 1993. Sequence stratigraphic framework of the upper San Andres Formation and equivalent basinal strata in the Brokeoff Mountains, Otero County, New Mexico. *New Mexico Geological Society Guidebook, 44th Field Conference, Carlsbad Region, New Mexico and West Texas*, 185–193.

——— 1995. *Early Permian Sequence Stratigraphy of the Diablo Platform, Sierra Diablo, West Texas: March 10–11th, 1995.* Unpublished Fieldguide, University of Texas, Austin.

GAWTHORPE, R. L., FRASER, A. J. & COLLIER, R. E. Ll. 1994. Sequence stratigraphy in active extensional basins. Implications for the interpretation of ancient basin-fills. *Marine and Petroleum Geology*, **11**, 642–658.

GAY, S. P. 1989. Gravitational compaction, a neglected process in structural and stratigraphic studies: new evidence from mid-continent, USA. *American Association of Petroleum Geologists Bulletin*, **73**, 641–657.

GIANNINY, G. 1995. *Facies and sequence stratigraphic evolution of the mixed carbonate–siliciclastic strata, lower Desmoinesian, southwest Paradox Basin, Utah*. PhD Thesis, University of Wisconsin-Madison, USA.

GOLDSTEIN, R. H. 1988. Paleosols of late Pennsylvanian cyclic strata, New Mexico. *Sedimentology*, **35**, 777–804.

———, HALLEY, R. B. & SCHOLLE, P. A. 1991. Upper Paleozoic bioherms in the northern Sacramento Mountains (Part II). *In:* SCHOLLE, P. A. & GOLDSTEIN, R. H. (eds) *Classic Upper Paleozoic Reefs and Bioherms of West Texas and New Mexico: Guidebook*. American Association of Petroleum Geologists, Classic Carbonates Course, 147–246.

HALLOCK, P. & SCHLAGER, W. 1986. Nutrient excess and the demise of coral reefs and carbonate platforms. *Palaios*, **1**, 389–398.

HANDFORD, C. R. & LOUCKS, R. G. 1993. Carbonate depositional sequences and systems tracts – responses of carbonate platforms to relative sea-level changes. *In:* LOUCKS, R. G. & SARG, J. F. (eds) *Carbonate Sequence Stratigraphy – Recent Developments and Applications*. American Association of Petroleum Geologists, Memoir, **57**, 435–474.

HELLAND-HANSEN, W. & GJELBERG, J. G. 1994. Conceptual basis and variability in sequence stratigraphy: a different perspective. *Sedimentary Geology*, **92**, 31–52.

HERITIER, F. E., LOSSEL, P. & WATHNE, E. 1979. Frigg field: large submarine fan trap in lower Eocene rocks, North Sea, Viking Graben. *American Association of Petroleum Geologists Bulletin*, **63**, 1999–2020.

HILBRECHT, H. 1989. Redeposition of late Cretaceous pelagic sediments controlled by sea-level fluctuations. *Geology*, **17**, 1072–1075.

HUNT, D. & ALLSOP, T. 1993. Syndepositional differential compaction and the development of 'enhanced' unconformities at major stratal surfaces: examples from Mississippian strata of the Sacramento Mountains, Otero County, New Mexico, U.S.A. (abstract). *In:* BURCHETTE, T. &

HARWOOD, G. (eds) *Abstracts for Carbonate Petroleum Reservoirs: Models for Exploration and Development*, Geological Society, London, March 17–18th, 1993.

—— & TUCKER, M. E. 1993. The sequence stratigraphy of carbonate shelves with an example from the mid-Cretaceous of S.E. France. *In:* POSAMENTIER, H. W., SUMMERHAYES, C. P., HAQ, B. U. & ALLEN, G. P. (eds) *Sequence Stratigraphy and Facies Associations.* International Association of Sedimentologists, Special Publication, **18**, 307–342.

—— & —— 1995. Stranded parasequences and the forced regressive wedge systems tract: deposition during base-level fall-reply. *Sedimentary Geology*, **95**, 147–160.

——, ALLSOP, T. & SWARBRICK, R. 1995. Compaction as a primary control of sequence development on carbonate platforms (abstract). *Official Program, 1995 Annual Convention, American Association of Petroleum Geologists*, Houston, USA, March 5–8th 1995.

JERVEY, M. T. 1988. Quantitative geological modelling of siliciclastic rock sequences and their seismic expression. *In:* WILGUS, C. K., HASTINGS, B. S., KENDALL, C. G. St. C., POSAMENTIER, H. W., ROSS, C. A. & VAN WAGONER, J. C. (eds) *Sea-Level Changes: An Integrated Approach.* Society of Economic Paleontologists and Mineralogists, Special Publication, **42**, 47–69.

LABUTE, G. J. & GRETENER, P. E. 1969. Differential compacion around a Leduc reef – Wizard Lake Area, Alberta. *Bulletin of Canadian Petroleum Geology*, **17**, 304–325.

MEYERS, W. J. 1980. Compaction in Mississippian skeletal limestones, New Mexico. *Journal of Sedimentary Petrology*, **50**, 457–474.

MEYERS, W. J. & HILL, B. E. 1983. Quantitative studies of compaction in Mississippian skeletal limestones, New Mexico. *Journal of Sedimentary Petrology*, **53**, 231–242.

NEUMANN, A. C. & MACINTYRE, I. G. 1985. Response to sea-level rise; keep-up, catch-up or give-up. *Proceedings of the 5th Coral Reef Congress, Tahiti*, **3**, 763–786.

POSAMENTIER, H. W., JERVEY, M. T. & VAIL, P. R. 1988. Eustatic controls on clastic deposition I. Conceptual framework. *In:* WILGUS, C. K., HASTINGS, B. S., KENDALL, C. G. St. C., POSAMENTIER, H. W., ROSS, C. A. & VAN WAGONER, J. C. (eds) *Sea-Level Changes: An Integrated Approach.* Society of Economic Paleontologists and Mineralogists, Special Publication, **42**, 108–142.

PRAY, L. C. 1960. Compaction in calcilutites (abstract). *Geological Society of America Bulletin*, **71**, 1946.

—— 1965. Clastic limestone dikes and marine cementation, Mississippian bioherms, New Mexico (abstract). *Permian Basin Section, Society of Economic Paleontologists and Mineralogists, Annual meeting, Program with Abstracts*, 21–22.

PURDY, E. G. 1974a. Reef configurations: causes and effects. *In:* LAPORTE, L. F. (ed.) *Reefs In Time And*

Space. Society of Economic Paleontologists and Mineralogists, Special Publication, **18**, 9–76.

—— 1974b. Karst-determined facies patterns in British Honduras: Holocene carbonate sedimentation model. *American Association of Petroleum Geologists Bulletin*, **58**, 825–855.

—— & BERTRAM, G. T. 1993. *Carbonate Concepts from the Maldives, Indian Ocean.* American Association of Petroleum Geologists, Studies In Geology, **34**.

SCHLAGER, W. 1981. The paradox of drowned reefs and carbonate platforms. *Geological Society of America Bulletin*, **92**, 197–211.

—— 1993. Accommodation and supply – a dual control on stratigraphic sequences. *Sedimentary Geology*, **86**, 111–136.

SCHMOKER, J. W. & HALLEY, R. B. 1982. Carbonate porosity versus depth: a predictable relation for south Florida. *American Association of Petroleum Geologists Bulletin*, **66**, 2561–2570.

SCHOLLE, P. A. & HALLEY, R. B. 1985. Burial diagenesis: out of sight, out of mind! *In:* SCHEIDERMANN, N. & HARRIS, P. M. (eds) *Carbonate Cements.* Society of Economic Paleontologists and Mineralogists, Special Publication, **36**, 309–334.

SHINN, E. A. & ROBBIN, D. M. 1983. Mechanical and chemical compaction in fine-grained shallow-water limestones. *Journal of Sedimentary Petrology*, **53**, 595–618.

——, ——, LIDZ, B. H. & HUDSON, J. H. 1983. Influence of deposition and early diagenesis on porosity and chemical compaction in two Paleozoic buildups: Mississippian and Permian age rocks in the Sacramento Mountains, New Mexico. *In:* HARRIS, P. M. (ed.) *Carbonate Buildups: a Core Workshop.* Society of Economic Paleontologists and Mineralogists, Core Workshop, **4**, 182–222.

SONNENFELD, M. D. & CROSS, T. A. 1993. Volumetric partitioning and facies differentiation within the Permian upper San Andres Formation of Last Chance Canyon, Guadalupe Mountains, New Mexico. *In:* LOUCKS, R. G. & SARG, J. F. (eds) *Carbonate Sequence Stratigraphy – Recent Developments and Applications.* American Association of Petroleum Geologists, Memoir, **57**, 435–474.

STECKLER, M. S. & WATTS, A. B. 1978. Subsidence of the Atlantic-type continental margin off New York. *Earth and Planetary Science Letters*, **41**, 1–13.

SUPPE, J. 1985. *Principles of Structural Geology.* Prentice-Hall, New Jersey.

THURLOW, A. D. 1994. Modelling the relative roles of overpressure, differential compaction and sequence stratigraphy in thickness variations within a mixed carbonate clastic environment: the upper Limestone Series of west Cumbria, U.K. (abstract). *In:* JONES, G, LI. (ed.) *European Dinantian Environments II: Developments in Lower Carboniferous Geology*, University College Dublin, September 6th–8th, 1994.

TIMBRELL, G. 1993. Sandstone architecture of the Balder Formation depositional system, UK

Quadrant 9 and adjacent areas. *In:* PARKER, J. R. (ed.) *Petroleum Geology of Northwest Europe: Proceedings of the 4th Conference*. The Geological Society, London. 107–121.

VAIL, P. R. 1987. Seismic stratigraphy interpretation using sequence stratigraphy. Part 1: seismic stratigraphy interpretation procedure. *In:* BALLY, A. W. (ed.) *Atlas of Seismic Stratigraphy, Volume 1*. American Association of Petroleum Geologists, Studies in Geology, **27**, 1–10.

———, AUDEMARD, F., BOWMAN, S. A., EINSER, P. N. & PEREZ-CRUZ, C. 1991. The stratigraphic signatures of tectonics, eustacy and sedimentology – an overview. *In:* EINSELE, G., RICKEN, W. & SEILACHER, A. (eds) *Cycles and Events in Stratigraphy*. Springer-Verlag, Berlin, 617–659.

VAN WAGONER, J. C., MITCHUM, R. M., CAMPION, K. M. & RAHMANIAN, V. D. 1990. *Siliciclastic Sequence Stratigraphy in Well Logs, Cores, and Outcrops*. American Association of Petroleum Geologists, Methods in Exploration, **7**.

WATTS, A. B. 1981. *The U.S. Atlantic continental margin: subsidence history, crustal structure and thermal evolution*. American Association of Petroleum Geologists, Education Course Notes Series, **19**, 2–75.

High frequency sequence stratigraphy of a siliciclastic influenced carbonate platform, lower Moscovian, Amdrup Land, North Greenland

LARS STEMMERIK

Grønlands Geologiske Undersøgelse, Øster Voldgade 10, DK-1350 København K, Denmark

Abstract The Moscovian succession in southern Amdrup Land, North Greenland consists of seven, 40–135 m thick, third-order depositional sequences of mixed carbonates, siliciclastics and evaporites. The sequences have sheet-like or wedge-shape geometries and are composed of stacked high frequency sequences with a duration of *c.* 100 000 years. The Kap Jungersen 2 Sequence delineates an up to 135 m thick carbonate platform composed of 10 high frequency sequences. Each individual high frequency sequence shows great lateral facies and thickness variations. In the platform area they are dominated by bioclastic grainstones, *Palaeoaplysina*–phylloid algae build-ups and bryozoan build-ups. Deposition took place mainly during sea-level highstands; the transgressive systems tracts are generally thin and shaley, and sea-level lowstand led to subaerial exposure and karstification.

The Moscovian successions of North Greenland, Bjørnøya and Spitsbergen are mainly composed of cyclic shelf carbonates and siliciclastics with minor evaporites (Steel & Worsley 1984; Stemmerik & Håkansson 1989; Stemmerik & Worsley 1989). Carbonate build-ups are relatively rare and have been described so far only from Amdrup Land in eastern North Greenland (Fig. 1). In this region, the Moscovian succession is more than 600 m thick and is composed of seven third-order depositional sequences of mixed siliciclastics, carbonates and evaporites (Stemmerik 1994). Most third-order sequences are tabular within the outcrop area and composed of several high frequency sequences of mixed siliciclastics and carbonates with abundant levels of tabular or lenticular carbonate build-ups. Larger build-ups are restricted to the lower part of the succession where they stack to form a wedge-shaped carbonate platform sequence of early Moscovian age, the Kap Jungersen 2 sequence (Fig. 2). This lower Moscovian carbonate platform has more than 50 m of relief and is comparable in size to seismic anomalies interpreted to represent carbonate build-ups in the offshore areas of the Norwegian Barents Shelf (e.g. Gerard & Buhrig 1990). The relief created during carbonate platform growth was filled with evaporites during deposition of the succeeding depositional sequence.

This paper discusses the high frequency sequence stratigraphy of the Kap Jungersen 2 Sequence in order to understand the controls on internal architecture, facies distribution and depositional evolution of the platform. The platform is exposed in a 350 m high and several kilometres long, east–west trending coastal cliff. Outcrops thus allow detailed two dimensional studies of internal architecture and facies distribution of the platform. This is important when considering these sediments as a potential analogue for seismically-defined mounds in the Barents Sea.

Carbonate platform growth apparently took place during several high frequency, most likely glacio-eustatic driven sea-level cycles within a third-order highstand. The overlying evaporite-dominated sequence was deposited during basin restriction and lowering of relative sea-level.

Geological setting and stratigraphy

The marginal parts of North Greenland, together with the Barents Sea, Spitsbergen and Bjørnøya formed an extensive mosaic of interconnected intracratonic basins during the late Palaeozoic. Post-Caledonian sedimentation started during the latest Devonian to earliest Carboniferous, and thick successions of continental siliciclastics were deposited in localized half-grabens. Renewed rifting took place in the late Bashkirian and thick successions of marginal marine conglomerates and evaporites were deposited in narrow half-grabens on Spitsbergen and Bjørnøya (Steel & Worsley 1984). The marine transgression reached North Greenland during the Moscovian and thick successions of shallow marine sediments were deposited in rapidly subsiding basins along the margins of North Greenland (Fig. 1).

The Moscovian succession in northern Holm Land and Amdrup Land in easternmost North

From Howell, J. A. & Aitken, J. F. (eds), *High Resolution Sequence Stratigraphy: Innovations and Applications,* Geological Society Special Publication No. 104, pp. 347–365.

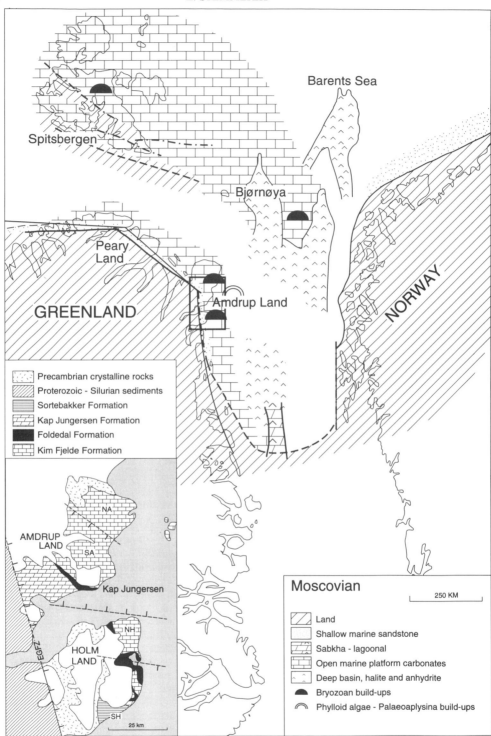

Fig. 1. Moscovian palaeogeographic reconstruction of the western Barents Sea area with names mentioned in the text. Inset map shows position of Kap Jungersen. NA, Northern Amdrup Land block; SA, Southern Amdrup Land block; NH, Northern Holm Land block; SH, Southern Holm Land block; EGFZ: East Greenland Fault Zone.

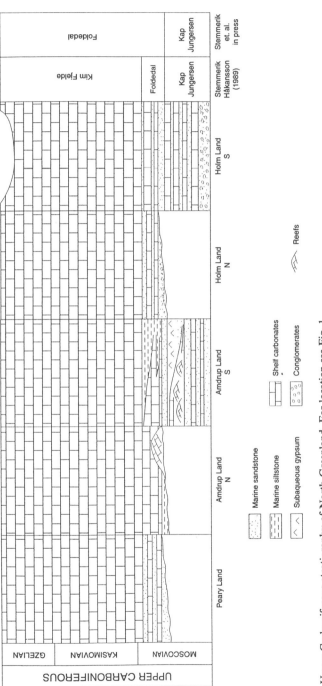

Fig. 2. Upper Carboniferous stratigraphy of North Greenland. For location see Fig. 1.

Fig. 3. Kap Jungersen Sequence 2 to 7 in the western part of the study area. Note the wedge shape of Sequence 2 and the onlap of gypsum (white) onto the platform margin. Cliff is *c.* 400 m high.

Greenland directly overlies Caledonian-affected basement and is conformably overlain by younger Carboniferous sediments (Håkansson *et al.* 1981). Deposition took place on small isolated fault blocks bounded to the west by the East Greenland Fault Zone (Fig. 1). This fault zone separated the post-Caledonian sedimentary basins to the east from the stable Greenland craton (Håkansson & Stemmerik 1989). Furthermore, a series of less important NW–SE trending faults influenced sedimentation during the Moscovian (Fig. 1; Stemmerik & Håkansson 1989, 1991).

The southern Amdrup Land and the southern Holm Land blocks were transgressed during the early Moscovian and not until later in the Moscovian did the marine transgression reach northern Holm Land and northern Amdrup Land so that a laterally continuous carbonate platform became established (Stemmerik & Håkansson 1989, 1991). The section at Kap Jungersen in southern Amdrup Land includes the Kap Jungersen, Foldedal and Kim Fjelde Formations (Stemmerik & Håkansson 1989). The lower Moscovian Kap Jungersen Formation forms a 300–400 m thick fining-upwards succession of mixed siliciclastics and carbonates that at the base is dominated by coarse-grained siliciclastics and at the top by shallow shelf carbonates and evaporites. Deposition was

confined to the southern Holm Land and southern Amdrup Land blocks. It differs significantly between the two fault blocks. On the southern Holm Land block, cyclic interbedded shelf carbonates and shallow water siliciclastics form laterally continuous cycles that can be traced for several kilometres (Stemmerik & Håkansson 1989). Similar cycles are developed in the lower part of the formation on the southern Amdrup Land block. There the upper part of the formation is characterized by development of the laterally restricted Kap Jungersen Sequence 2 platform (Fig. 2) which pass westwards into lagoonal evaporites and fluvial red beds. The relief created during deposition of this platform was filled by evaporites, and the top surface of the formation is flat within the outcrop area. The base of the Foldedal Formation records renewed, tectonic induced supply of siliciclastic material to the region. The depositional area expanded greatly during the upper Moscovian, and shallow shelf sediments were deposited in Holm Land, Amdrup Land and northern Peary Land (Fig. 1). The cyclic shelf sedimentation expanded to cover both southern and northern Holm Land, while on the southern Amdrup Land block, laterally restricted carbonate platforms continued to form during the late Moscovian. This led

Fig. 4. Stacked *Palaeoaplysina*-phylloid algae build-ups from the upper part of Kap Jungersen Sequence 2. Note dark (red) incipient soil horizons on top of build-ups (arrows). G, gypsum. Upper build-up is 8-10 m thick.

to the development of a western carbonate dominated succession and an eastern shale dominated succession. Sedimentation changed during latest Moscovian times, and the uppermost Moscovian to Gzelian succession consists of stacked shallow water carbonates. These rocks were previously included in the Kim Fjelde Formation (Stemmerik & Håkansson 1989, 1991) but later fieldwork indicates that they rather should be included in the Foldedal Formation (Fig. 2; Stemmerik & Elvebakk 1994; Stemmerik *et al.* 1995). The studied carbonate platform forms the upper part of the Kap Jungersen Formation and is dated as upper lower Moscovian based on fusulinids (Dunbar *et al.* 1962; Stemmerik & Håkansson 1989).

The Moscovian succession in Amdrup Land belongs to the upper lower Moscovian *Profusulinella* zone and the upper Moscovian *Wedekindellina* zone (Dunbar *et al.* 1962; Nilsson 1994). These two zones roughly correspond to the three youngest fusulinid zones in the standard zonation (Nilsson 1994). The entire Moscovian spans 8 Ma (Harland *et al.* 1990), and within the present biostratigraphic framework the Kap Jungersen succession spans *c.* 6 Ma (Fig. 2). It is divided into seven third-order depositional sequences giving an average duration of *c.* 1 Ma

for each sequence (Fig. 2; Stemmerik 1994). The sequences range in thickness from 40–135 m and have sheet-like or wedge-shaped geometries. Each sequence is a composite of several high frequency (HF) sequences.

The basal Moscovian sequence, Kap Jungersen 1 Sequence, is composed of cyclic interbedded shallow marine carbonates, sandstones and shales. Individual beds are laterally persistent within the outcrop area, and the sequence has a sheet-like geometry. The overlying Kap Jungersen 2 Sequence records a marked change in depositional conditions, possibly related to tilting and increased rates of subsidence in the area. In the western part of the outcrop area it is dominated by cyclic deposits of shallow marine carbonates, including various types of carbonate build-up, with minor evaporites and siliciclastics (Fig. 3). These sediments are stacked to form a 135 m thick carbonate platform with up to 50 m of depositional relief. The sequence has a wedge-shaped geometry and is relatively thin towards the east (Fig. 3). In the platform area, the upper sequence boundary is a marked subaerial exposure surface. The boundary between the Kap Jungersen 2 Sequence and the overlying Kap Jungersen 3 Sequence is an onlap surface along the platform margin (Fig. 3).

Kap Jungersen 2 Sequence

Depositional environments

The Kap Jungersen 2 Sequence is dominated by shallow marine carbonates deposited under environmetally stressed conditions. The volumetrically most important carbonate facies are associated with *Palaeoaplysina*–phylloid algae build-ups, bryozoan build-ups, bioclastic grainstone shoals and protected, environmentally stressed lagoons. Shallow marine shales and sandstones and shallow subtidal to supratidal gypsum are also present.

Palaeoaplysina–phylloid algae build-ups. Build-ups dominated by *Palaeoaplysina* and phylloid algae are usually less than 10 m thick. They vary from small isolated lenticular bodies, less than 10 m wide to large tabular to domal bodies which can be traced for hundreds of metres. The larger build-ups are usually a composite of several small build-ups. The top surface of the build-ups is undulating and shows evidence of subaerial exposure. Small build-ups are capped by thin, less than 50 cm red-coloured horizons interpreted as incipient soils (Fig. 4). The larger build-up complexes are in most cases capped by thicker soils that connects to up to 40 m deep karst fissures (Fig. 5).

The build-ups vary from boundstones of tightly packed *Palaeoaplysina* plates and phylloid algae cementstones to packstones and wackestones with isolated plates of phylloid algae and *Palaeoaplysina*. They include a relatively rich fauna of brachiopods, algae, corals and crinoids. Steeply dipping flank deposits are mainly found around build-ups with a core of boundstone or cementstone. The build-ups are completely dolomitized and have considerable macroporosity (Stemmerik & Elvebakk 1994).

Palaeoaplysina and phylloid algae build-ups were common in the Arctic region during the mid-Carboniferous to early Permian (e.g. Beauchamp 1993; Stemmerik *et al.* 1994), and were most likely deposited below wave base in shallow subtidal environments.

Bryozoan build-ups are restricted to the uppermost part of the platform. The build-ups form isolated lenticular bodies, up to 20 m high and 100 m wide, and are surrounded by steeply dipping flank deposits (Stemmerik 1989). The build-ups are divided into a lower unit composed of small stromatolites and crinoid wackestone. Brachiopods, nautiloids, trilobites and bivalves are also present in this unit. The upper part of the build-ups is composed of bryozoan cementstone. The build-ups are completely dolomitized.

Bryozoan-crinoid dominated build-ups are common in the Arctic region from the mid-Carboniferous to the late Permian (e.g. Scholle *et al.* 1991; Beauchamp 1993; Stemmerik 1993).

Fig. 5. Large *Palaeoaplysina*-phylloid algae build-up surrounded by prograding clinoforms of grainstone and cut by 30–40 m deep vertical karst fissures (arrows). B, build-up.

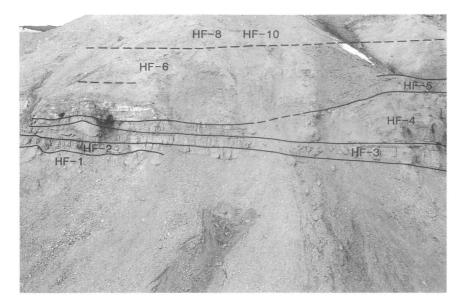

Fig. 6. Western part of the Kap Jungersen 2 carbonate platform showing thick beds of gypsum (white) at the base of HF Sequence 5, that onlaps the HF Sequence 4 build-up.

Fig. 7. Downlapping biogenic grainstone (HST) on bioturbated wackestone (TST).

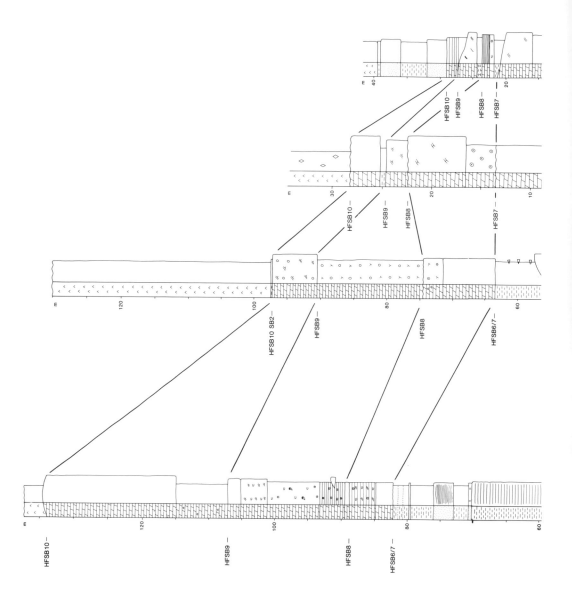

The build-ups at Kap Jungersen show a marked vertical facies zonation. Initial deposition apparently took place in protected environments judging from the abundance of carbonate mud, while deposition of the upper cement-rich part most likely occurred in more well agitated environments.

Grainstone shoals.
The volumetrically most important facies within the platform succession is cross-bedded bioclastic grainstones. This facies is composed of low diversity grainstones where the most important biogenic components are algae, encrusting foraminifers and bivalves. Micritized grains are common and ooids are rare. Fragments of normal marine fossils like bryozoans, brachiopods and crinoids are missing or very rare.

The lack of a normal marine fauna points towards environmentally stressed depositional conditions, and the absence of ooids points towards moderately agitated environments. The grainstones form strongly progradational units of up to 20 m high, internally cross-bedded

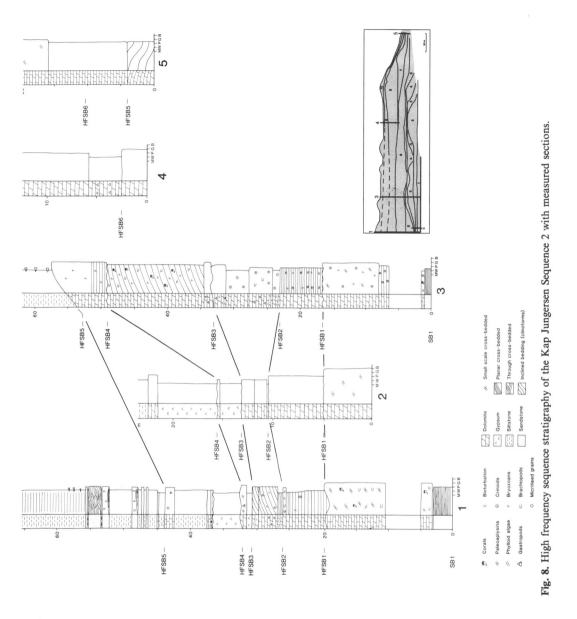

Fig. 8. High frequency sequence stratigraphy of the Kap Jungersen Sequence 2 with measured sections.

clinoforms (Fig. 5). Evidence of subaerial exposure is seen both on the top of the shoals and along clinoform surfaces.

Lagoonal/restricted shelf carbonates. Dolomitized biogenic packstones and wackestones with a low diversity fauna of foraminifers, algae and crinoids, and with abundant micritized grains are locally abundant in the western part of the platform. The fauna resembles that of the grainstones pointing towards environmentally stressed conditions. The abundance of carbonate mud suggests deposition in protected environments below wave base, in accordance with the dominance of this facies on the inner part of the platform.

The biogenic wackestones and packstones often form the basal part of upwards shoaling inner platform cycles capped by biogenic grainstones. The lagoonal proportion of these cycles becomes thicker towards the west, where small isolated lenticular mounds are associated with the packstones and wackestones.

Fig. 9. High frequency sequences at section 3 in Fig. 8. Note the erosinal upper surface of HF Sequence 1 (HFSB1). For lithological details see Fig. 8.

Shallow shelf siliciclastics. In the western part of the platform, the middle part of the succession is composed of a thick succession of poorly exposed, bioturbated, organic-lean siltstone with thin sandstone and carbonate layers (Fig. 3). The siltstones are intensely bioturbated and most likely were deposited in a protected shallow marine environment below wave base. Thin erosive and bioturbated sandstone layers interbedded with the siltstones most likely represent storm deposits. Thicker, bioturbated or cross-bedded fine-grained sandstone was deposited in the shallow marine environments.

Evaporites. Massive to bedded gypsum is common in the lower part of the western platform. The beds are 0.5–8 m thick and can be traced laterally for several hundred metres (Fig. 6). Most gypsum is strongly recrystallized and shows little evidence of primary structures. However, outlines of bottom nucleated gypsum crystals are locally present. This points towards deposition in a shallow subaqueous environment (cf. Warren 1982).

High frequency sequence stratigraphy

High frequency sequence stratigraphic analysis of the Kap Jungersen 2 Sequence is based on tracing sequence stratigraphic key surfaces in the field and from helicopter using photos. The high frequency sequence boundaries are in most cases subaerial exposure surfaces with incipient soils or more well developed karstic features. They often show considerable depositional relief and are onlapped by the transgressive deposits of the succeeding high frequency sequence. The maximum flooding interval is often distinguished as an intensely bioturbated facies with numerous *Thallasinoides* burrows on which the highstand deposits downlap (Fig. 7). The studied outcrop is *c.* 500 m long and the maximum thickness of the sequence is 135 m. The sequence thins towards the east and in the easternmost exposures, the sequence is less than 50 m thick. The internal facies distribution is based on measurements of five sedimentological sections through various parts of the sequence from the platform interior in the west to the platform margin in the east (Fig. 8).

HF Sequence 1 is the first high frequency cycle that contains carbonate build-ups. It is tabular in form and consists of a mixture of siliciclastics and carbonates and has a total thickness of *c.* 20 m. The lowstand systems tract consists of less than 3 m of poorly exposed cross-bedded, fine-grained sandstone. It is overlain by up to 7 m of poorly exposed biogenic wackestones and packstones with a basal thin shale layer. This succession is interpreted as the transgressive systems tract and the early highstand systems tract. The highstand systems tract is dominated by a tabular complex of *Palaeoaplysina*-dominated, mud-supported carbonate build-ups. The build-ups are 8–10 m thick and consist of massive to crudely bedded biogenic packstone with up to 1 m long plates of *Palaeoaplysina*. The build-ups show no vertical facies trends and were apparently deposited by accretion in a protected, shallow subtidal environment.

The top surface of the sequence, HFSB-1 is an erosional surface with more than 1 m of relief (Fig. 9).

HF Sequence 2 is mound-shaped with a maximum thickness of 30 m (Figs 8 and 9). In the western part the HF sequence is dominated by 6–7 m of well bedded biogenic packstones and wackestones. These sediments are fining-upwards and in the upper part shale interbeds and bioturbation by *Zoophycus* are common. In the eastern part of the study area the sequence is composed of thick massive to crudely bedded *Palaeoaplysina*-phylloid algae build-up, up to 30 m thick.

The stacking pattern in the western part of the area indicates that the sediments were deposited during rising sea-level and so form part of the transgressive systems tract. In this area sedimentation was not able to keep pace with the sea-level rise and the transgressive deposits are overlain directly by late highstand deposits in the form of a thin, 50 cm thick grainstone bed. Thick highstand deposits are confined to the localized build-ups towards the east and consist of aggradational to slightly progradational units of *Palaeoaplysina*-dominated facies.

The HF sequence boundary terminating HF Sequence 2 is a subaerial exposure surface with well developed red soil. It has more than 20 m of relief due to differential sedimentation and deposition of the lenticular build-up.

HF Sequence 3. The depositional relief created during deposition of HF Sequence 2 controlled the architecture and facies distribution of HF Sequence 3. The topographic highest part of the old build-up remained exposed during deposition of HF Sequence 3 and formed a palaeohigh around which lenticular units of biogenic grainstones were deposited (Figs 8 and 10). Main deposition in HF Sequence 3 took place 100 m west of the old build-up where a new lenticular *Palaeoaplysina*–phylloid algae build-up occurred. This build-up has a steep eastward facing slope. Towards the west it becomes gradually thinner and passes laterally into a succession of biogenic packstones and grainstones (Fig. 8).

The transgressive systems tract is a 50–100 cm thick unit composed of bioturbated clay-rich carbonate mudstone and wackestone with abundant *Zoophycos* burrows in the low lying areas west and east of the HF Sequence 2 mound. The boundary between the transgressive systems tract and the highstand systems tract is a downlap surface in most areas (Fig. 7). The highstand systems tract is composed of clinoforms of biogenic grainstones and *Palaeoaplysina*–phylloid algae build-ups. The clinoform dominated units are strongly progradational and most likely represent late highstand deposition in accordance with their position around the pre-depositional high. Progradation was both to the east and the west away from the palaeohigh. The build-ups are aggradational to slightly progradational.

The upper HF sequence boundary is a subaerial exposure surface with evidence of soil formation and karst fissures in the elevated areas. It defines two isolated highs. The western high has steeply dipping margins and the eastern high has a steeply dipping eastern margin and a more gently dipping western margin (Fig. 8).

HF Sequence 4 is 1.5–20 m thick and has a very irregular base due to the mounded topography of the underlying HF sequence. It is thin in the western part of the study area, 1.5–4 m. It is 5–10 m thick above the most elevated parts of the previous HF Sequence and attains its maximum thickness between these highs (Fig. 8).

The transgressive systems tract is restricted to the western part of the platform where 1–3 m of massive to bedded gypsum were deposited in an isolated shallow lagoon behind the carbonate build-ups. The gypsum onlaps the build-ups and disappears immediately west of the main build-ups (Fig. 6). There the TST consists of a less than 1 m thick carbonate build-up.

The highstand systems tract is composed of strongly progradational units of biogenic grainstones with internally cross-bedded clinoforms. The two systems tracts are separated by a downlap surface. The upper HF sequence boundary is a subaerial exposure surface with evidence of soil formation and karstification. It defines one large isolated carbonate platform with a steeply dipping seaward margin to the east and a more gently dipping landward slope to the west (Fig. 8).

HF Sequence 5 is from 5 m to more than 30 m thick. In the back platform areas to the west it is dominated by evaporites and siliciclastics. On

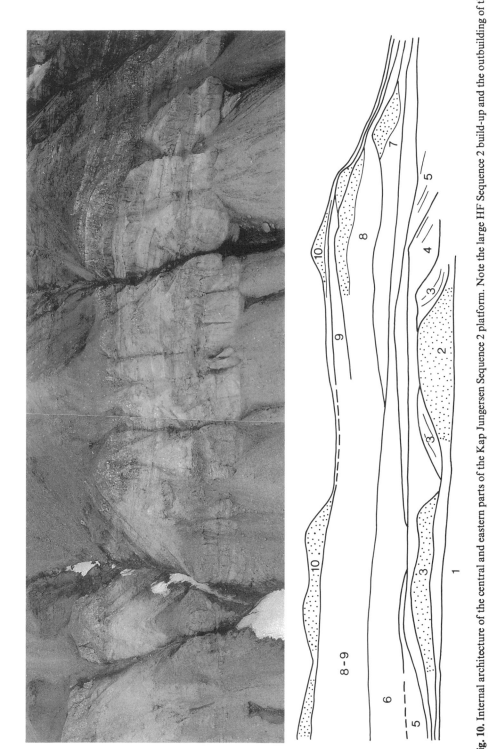

Fig. 10. Internal architecture of the central and eastern parts of the Kap Jungersen Sequence 2 platform. Note the large HF Sequence 2 build-up and the outbuilding of the platform during several stages.

the platform and towards the east this HF sequence is composed mainly of carbonates.

The transgressive systems tract is best developed in the back platform area where it consists of up to 7 m of massive to bedded gypsum overlain by a 1–2 m thick biogenic wackestone. The gypsum onlaps the platform topography and disappears in the area between sections 2 and 3 in Fig. 8.

The highstand systems tract is composed of biogenic grainstones on the carbonate platform and along the seaward margin of the platform. It forms a strongly progradational unit with more than 20 m high clinoforms (Fig. 10). In the back platform areas the highstand systems tract is composed mainly of shallow marine to lagoonal siltstones. Correlation to the carbonate platform is difficult due to lack of prominent surfaces and the position of the sequence boundary is based on facies analysis in this area. The highstand systems tract consists of c. 5 m of bioturbated siltstone overlain by 2 m of biogenic packstone. The packstones most likely formed as a result of shedding during late highstand and the HF sequence boundary is therefore placed above this horizon (Fig. 8). The sequence boundary is a mature karst surface in the carbonate platform area with more than 30 m deep karst fissures (Fig. 10).

HF Sequence 6 is almost 40 m thick in the back platform area. It thins to 10–15 m above the platform top (Fig. 8). This thickness variation is primarily due to variations in the underlying relief. In the palaeotopographic low behind the carbonate platform, fine-grained siliciclastics were trapped during sea-level lowstand and the lowstand systems tract is composed of 10–15 m of interbedded bioturbated siltstone and sandstone and cross-bedded shallow marine sandstone (Fig. 8). Lowstand deposits are confined to the low lying areas to the west and disappear between section 2 and 3 in Fig. 8.

The transgressive systems tract is composed of bioturbated siltstone in the back platform area. These siltstones onlap the western margin of the carbonate platform and overlies HF sequence boundary 5 in the western part of the platform (see section 3 in Fig. 8). The shales disappear in the eastern, elevated areas of the platform and the HF sequence boundary is directly overlain by biogenic packstones and wackestones.

The highstand systems tract is composed of shales with minor sandstone in the western part of the platform. In the eastern part of the platform biogenic packstones are dominant. They migrated westwards during sea-level highstand and overlie the transgressive shales in the

central part of the platform (between section 3 and 4 in Fig. 8). East of the platform resedimented packstones and grainstones were deposited during sea-level highstand.

The top surface of HF Sequence 6 defines a relatively even platform with a gently dipping seaward slope. The HF sequence boundary is a subaerial exposure surface marked by intense weathering of the siltstones in the west. In the eastern part of the platform it is an incipient karst surface.

HF Sequence 7 is limited to the easternmost part of the platform and the platform margin. It is wedge shaped, 5–30 m thick with a maximum thickness east of the former platform break (Figs 8 and 11). The transgressive systems tract is composed of bedded biogenic carbonates with abundant chert. They onlap the eastern platform margin and wedge out towards the west.

The highstand systems tract is fairly thin above the platform. On the platform slope, the highstand systems tract is composed of a 20–25 m thick, lenticular phylloid algae build-up. The build-up has an erosive base and downlaps in an eastwards direction on the transgressive deposits (Fig. 10).

HF Sequence 8 forms a thin, 5–10 m sheet of open marine packstones across the platform. The transgressive systems tract is composed of a thin onlapping shale unit along the platform margin. On the platform the transgressive deposits are mainly composed of bioturbated packstones. The highstand systems tract is composed of westward migrating cross-bedded packstones in the western part of the platform. Along the platform margin, the highstand systems tract is composed of a thin lenticular build-up and farther eastwards the highstand systems tract is composed of an upward thickening unit of resedimented carbonates (Fig. 8).

The top surface of HF Sequence 8 is a subaerial exposure surface with incipient soil formation in the platform areas.

HF Sequence 9 is 5–20 m thick. It resembles HF Sequence 8 and is composed mainly of biogenic wackestones and packstones in the platform area. Onlapping transgressive shales are restricted to the seaward margin of the platform. They are overlain by a phylloid algae build-up. Along the platform margin and across the platform the transgressive systems tract is composed of biogenic wackestones. The highstand systems tract is composed of a lenticular build-up along the platform margin, and in the

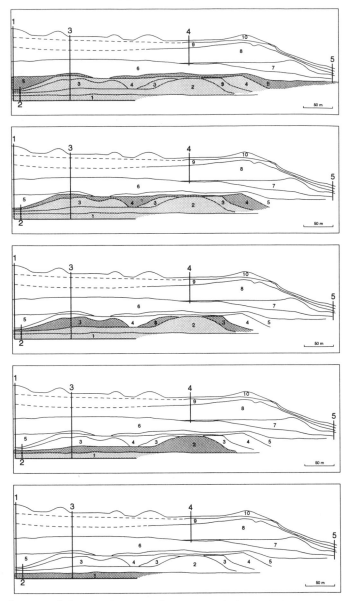

Fig. 11. Depositional stages of the Kap Jungersen Sequence 2 platform. Each stage corresponds to a high frequency sequence.

inner part of the platform it is composed of cross-bedded packstones. The top surface of HF Sequence 9 is a subaerial exposure surface with incipient soil formation in the platform areas and on the slope.

HF Sequence 10 is from less than 1 m to more than 25 m thick. It is composed of a series of isolated lenticular bryozoan-dominated build-

ups. Each build-up consists of a transgressive unit with stromatolites overlain by mud-rich wackestones and terminated by cementstones (Stemmerik 1989). The build-ups thus form a transgressive–regressive unit where the cement-dominated part is regarded as highstand deposits together with the surrounding flank deposits. The areas between the build-ups are character-ized by non-deposition. The HF sequence is there composed of thin, less than 1 m of siltstone

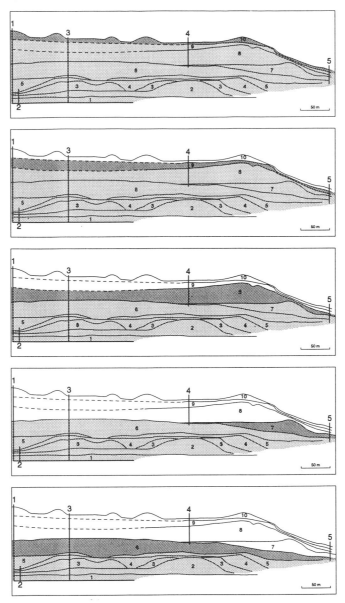

Depositional evolution

The sequence stratigraphic analysis indicates that the Kap Jungersen 2 Sequence is composed of 10 high frequency depositional sequences (Fig. 8). The Kap Jungersen 2 Sequence spans c. 1 Ma, and each high frequency sequence represents on average c. 100 000 years.

The carbonate platform was initiated by deposition of a series of lenticular *Palaeoaplysi-*

(Fig. 8). The top surface of this HF sequence corresponds to the top surface of the Kap Jungersen 2 Sequence. It is highly irregular with a relief of 50–60 m and below this surface intense diagenetic alteration of the platform carbonates took place. Along the platform margin, this surface is onlapped by subaqueously deposited gypsum that eventually reached the platform top and buried the carbonates (Fig. 3).

na–phylloid algae build-ups which merged to form a tabular build-up dominated highstand systems tract of HF Sequence 1. The build-ups became subaerially exposed and during the following rise in relative sea-level the firm surface of these build-ups acted as foundation for new build-ups. Carbonate production and build-up growth were not able to keep up with the rise in relative sea-level in most areas. Continuous build-up growth was restricted to a localized area towards the east where an isolated build-up with more than 30 m of relief formed (Fig. 11). This build-up is the foundation of the platform, in this incipient stage the platform was less than 100 m wide.

During the following HF Sequence 3 rise in relative sea-level, a new carbonate build-up formed in the protected and environmentally stressed setting west of the previous build-up (Fig. 11). During sea-level highstand, parts of the old build-up became flooded and large volumes of carbonate grains were produced in shallow, medium to high energy environments. These grains were transported down the slopes and form thick prograding units of clinoforms on both sides. The lack of accommodation space above the old build-up indicates that relative sea-level was lower than during HF Sequence 2. At the end of HF Sequence 3 time the platform was composed of two, 150–200 m wide highs, separated by a 25 m deep and 30 m wide low (Fig. 11). The changes in platform morphology were mainly due to build-up growth in the protected areas behind the old build-up. The seaward margin moved only slightly eastwards during this stage.

During the rise in relative sea-level accompanying the accumulation of HF Sequence 4, gypsum was deposited west of the platform. This indicates that the platform formed a laterally continuous barrier with isolated lagoons behind it at that time. Carbonate production took place in shallow marine environments above the platform highs and accommodation space was created both atop the highs and around their margins. The platform prograded 50 m eastwards during deposition of HF Sequence 4 and the two highs merged to form one large, more than 350 m wide and 50 m high platform (Fig. 11).

The platform also acted as a barrier during the following rise in relative sea-level and gypsum was deposited in isolated lagoons west of the platform. Evaporite deposition ceased as the platform became flooded and the lagoons became fully marine. During sea-level highstand carbonate production mainly took place along the eastern, seaward margin of the platform and a new prograding wedge of grainstones led to eastwards expansion of the platform. More limited carbonate production and accumulation took place on the inner part of the platform. There was a marked change in depositional conditions behind the platform during deposition of HF Sequence 5. Large amounts of fine-grained siliciclastic material were trapped in the lagoons behind the platform and bioturbated siltstones were deposited directly on top of the transgressive evaporites.

The succeeding fall in relative sea-level exposed the carbonate platform and carbonate production stopped. Siliciclastic material continued to be deposited in the lagoons west of the platform. These lowstand deposits are composed of shallow marine sandstones and siltstones (Fig. 11). The boundary between the lowstand systems tract and the transgressive systems tract is placed at the base of a thick, uniform unit of bioturbated siltstones in the lagoonal section (Section 1 in Fig. 8). This siltstone unit onlaps the western margin of the platform and during sea-level rise the lagoonal area gradually expanded eastward. During maximum flooding, the western part of the platform was drowned and carbonate production became limited to the easternmost part of the platform (Fig. 11). There the carbonates form an aggradational to slightly progradational wedge of packstones and grainstones with maximum accumulation of sediments east of the former platform break. During sea-level highstand, carbonates also migrated westwards and the area of siliciclastic sedimentation gradually became smaller. The increased supply of siliciclastic material led to gradual infill of the lagoons and by the end of HF Sequence 6 the former lagoons were included in the platform (Fig. 11). The following fall in sea-level led to subaerial exposure of the entire platform including the former lagoons to the west.

During deposition of HF Sequence 7 only the eastern part of the platform was flooded (Fig. 11). The transgressive systems tract is composed of bedded biogenic packstones and grainstones that wedge out towards the west. The highstand systems tract is dominated by a lenticular phylloid algae build-up situated east of the former platform margin. Behind the build-up bedded biogenic carbonates were deposited in more protected environments. The platform margin migrated *c*. 100 m eastwards during deposition of HF Sequence 7.

The following sea-level rise pushed carbonate production back on the shelf and the lower, transgressive to early highstand part of the sequence is composed of well bedded aggradational units of biogenic packstones. These carbonates onlap the lenticular build-up formed during the previous highstand (Fig. 11). Main accumulation of sediments took place along the edge of the platform and during the ?late highstand, carbonate build-ups started to grow in a position c. 50 m landwards of the previous platform edge. In the off-platform areas, the trangressive systems tract is composed of a thin shale and the highstand systems tract is composed of an upward thickening unit of resedimented carbonates. The facies patterns seen in HF Sequence 8 were repeated in HF Sequence 9. Main sediment accumulation took place behind the build-ups on the outer part of the platform where cross-bedded grainstones were deposited. The carbonate platform was terminated by deposition of a series of small bryozoan build-ups. Apparently, most of the platform was drowned during the HF Sequence 10 rise in sea-level and only in localized areas was carbonate sedimentation maintained.

The main control on platform sedimentation seems to be high frequency, most likely glacio-eustatic driven sea-level cycles. Deposition took place mainly during transgression and sea-level highstand, while the platform in most cases was subaerially exposed during sea-level lowstand. The platform is dominated by subtidal facies with very little evidence of intertidal and supratidal sedimentation. Similar patterns have been described from carbonate successions elsewhere in the region (Stemmerik et al. 1994) and seem to characterize carbonate cycles in glacial (icehouse) periods (Wright 1992). This, together with the shape of many of the HF sequences, suggests that most sea-level rises and falls were very rapid. This is particularly well documented in HF Sequence 2 and 3 where the platform only locally was able to keep up with rising sea-level and where carbonate production was unable to fill adjacent lows during falling sea-level (Fig. 11).

Discussion

During recent years it has become increasingly evident that depositional sequences are controlled not only by changes in relative sea-level but also factors like climate and sediment supply (e.g. Schlager 1991, 1993). The shape of the Kap Jungersen 2 Sequence and the internal architecture and facies distribution of the platform seem to be controlled both by climate, long-term differential subsidence due to tectonic tilting and high frequency changes in sea-level. Supply of siliciclastic sediments generally was low and main accumulation took place in the back

platform areas and the poorly exposed basinal areas.

The Kap Jungersen 2 Sequence contains a fauna dominated by algae, foraminifers, bryozoans and crinoids; this is typical for the tropical chloroforam fauna of Beauchamp (1994) and suggests deposition on a tropical shelf. The abundance of evaporites suggests that the climate generally was arid. However, the occurrence of more humid climatic conditions is indicated by the well developed karstic surfaces defining most sequence boundaries that were developed during sea-level lowstands. These rapid climatic changes support the inferrred glacio-eustaic origin of the high frequency sea-level cycles related to glaciation in the southern hemisphere (cf. Veevers & Powell 1987).

The wedge shaped geometry of the Kap Jungersen sequence differs from the pattern seen on the Holm Land block, 10 km to the south and in more distant areas like eastern Peary Land and Bjørnøya. In these areas, the usual pattern seen in the Moscovian is sheet-like sequences composed of stacked, laterally persistent high frequency cycles/sequences. This suggests deposition in areas with slow subsidence rates where sedimentation was able to fill tectonically created accommodation space during each cycle (cf. Goldhammer et al. 1990). In contrast the Kap Jungersen 2 Sequence was deposited in an area with high tectonic subsidence rates, where sedimentation only locally was able to fill the tectonically created accommodation space. Tectonic tilting of southern Amdrup Land apparently took place during the Moscovian. The Moscovian succession thins from more than 600 m to the east to less than 100 m in the westernmost outcrops along the East Greenland Fault Zone.

While tectonic tilting apparently controlled the shape and location of the Kap Jungersen 2 platform, the combined effect of tectonic tilting and high frequency sea-level fluctuations controlled the internal facies distribution and architecture. In areas with low subsidence rates these sea-level cycles led to deposition of laterally persistent cycles of shelf carbonates and siliciclastics without any platform or reef development (Steel & Worsley 1984; Stemmerik & Worsley 1989).

It is thus concluded that the Kap Jungersen 2 carbonate platform forms a wedge-shaped third-order depositional sequence with a complex internal architecture related to high frequency fluctuations in sea-level. Internal platform architecture is controlled mainly by the high frequency changes in sea-level, while the overall shape and location of the platform seem to be controlled by a combination of tectonic subsidence rates and sea-level. Deposition took place during a third-order highstand of sealevel, and the platform margin is onlapped by transgressive evaporites and siliciclastics of the succeeding Kap Jungersen 3 Sequence.

Sequence stratigraphic analyses based on tracing of significant surfaces in the field allows the division of the Kap Jungersen 2 Sequence into 10 HF sequences with a complex two dimensional geometry. These sequences are mainly composed of transgressive and highstand deposits, while lowstands are indicated by periods of non-deposition and karstification. However, lowstand deposits have been recognized in HF Sequence 5.

Published with permission of the Geological Survey of Greenland.

References

BEAUCHAMP, B. 1993. Carboniferous and Permian reefs of Sverdrup Basin, Canadian Arctic: an aid to Barents Sea exploration. *In:* VORREN, T. O. *ET AL.* (eds) *Arctic Geology and Petroleum Potential.* Norwegian Petroleum Society (NPF), Special Publication No. 2, Elsevier, Amsterdam, 217–242.

—— 1994. Permian cooling in the Canadian Arctic. *In:* KLEIN, G. D. (ed.) *Pangea: Paleoclimate, Tectonics and Sedimentation during accretion, zenith and break-up of a supercontinent.* Geological Society of America, Special Paper, **288**, 229–246.

DUNBAR, C. P., TROELSEN, J., ROSS, C., ROSS, J. P. & NORFORD, B. 1962. Faunas and correlation of the Late Paleozoic rocks of northeast Greenland. *Meddelelser om Grønland*, **167**, 4.

GERARD, J. & BUHRIG, C. 1990. Seismic facies of the Permian section of the Barents Shelf: analysis and interpretation. *Marine and Petroleum Geology*, **7**, 234–252.

GOLDHAMMER, R. K., DUNN, P. A. & HARDIE, L. A. 1990. Depositional cycles, composite sea-level changes, cycle stacking patterns and the hierarchy of stratigraphic forcing: examples from Alpine Triassic platform carbonates. *Geological Society of America Bulletin*, **102**, 535–562.

HÅKANSSON, E. & STEMMERIK, L. 1989. Wandel Sea basin – A new synthesis of the late Paleozoic to Tertiary accumulation in North Greenland. *Geology*, **17**, 683–686.

——, HEINBERG, C. & STEMMERIK, L. 1981. The Wandel Sea Basin from Holm Land to Lockwood Ø, eastern North Greenland. *Rapport Grønlands Geologiske Undersøgelse*, **106**, 47–63.

HARLAND, W. B., ARMSTRONG, R. L., COX, A. V., CRAIG, L. E., SMITH, A. G. & SMITH, D. G. 1990. *A geologic time scale 1989.* Cambridge University Press. Cambridge.

NILSSON, I. 1994. Upper Palaeozoic fusulinid assemblages, Wandel Sea Basin, North Greenland. *Rapport Grønlands Geologiske Undersøgelse*, **161**, 45–71.

SCHLAGER, W. 1991. Depositional bias and environmental change – important factors in sequence stratigraphy. *In:* BIDDLE, K. T. & SCHLAGER, W. (eds) *The Record of Sea-Level Fluctuations.* Sedimentary Geology, **70**, 109–130.

—— 1993. Accommodation and supply – a dual control on stratigraphic sequences. *In:* CLOETINGH, S., SASSI, W., HARVATH, F. & PUIGDEFABREGAS, C. (eds) *Basin Analysis and Dynamics of Sedimentary Basin Evolution.* Sedimentary Geology, **86**, 111–136.

SCHOLLE, P. A., STEMMERIK, L. & ULMER, D. S. 1991. Diagenetic history and hydrocarbon potential of Upper Permian carbonate buildups, Wegener Halvø area, Jameson Land basin, East Greenland. *American Association of Petroleum Geologists Bulletin*, **75**, 701–725.

STEEL, R. J. & WORSLEY, D. 1984. Svalbard's Post-Caledonian strata: An atlas of sedimentational patterns and palaeogeographic evolution. *In:* SPENCER, A. M. (ed.) *Petroleum Geology of the North European Margin.* Graham & Trotman, Norwegian Petroleum Society, London, 109–135.

STEMMERIK, L. 1989. Crinoid-bryozoan reef mounds, Upper Carboniferous, Amdrup Land, eastern North Greenland. *In:* GELDSETZER, H. H. J., JAMES, N. P. & TEBBUTT, G. E. (eds) *Reefs, Canada and adjacent areas.* Canadian Society of Petroleum Geology, Memoir, **13**, 690–693.

—— 1993. Moscovian bryozoan-dominated build-ups, northern Amdrup Land, eastern North Greenland. *In:* VORREN, T. O., BERGSAGER, E., DAHL-STAMNES, Ø. A., HOLTER, E., JOHANSEN, B., LIE, E. & LUND, T. B. (eds) *Arctic geology and petroleum potential.* Norwegian Petroleum Society, Special Publication, **2**, Elsevier, Amsterdam 99–106.

—— 1994. Sequence stratigraphy of a mixed carbonate, siliciclastic and evaporite succession, Upper Carboniferous, North Greenland. *In:* JOHNSON, S. D. (ed.) *High Resolution sequence stratigraphy: Innovations and applications.* Abstract Volume, University of Liverpool, 159–160.

—— & ELVEBAKK, G. 1994. A newly discovered mid-Carboniferous–?early Permian reef complex in the Wandel Sea Basin, eastern North Greenland. *Rapport Grønlands Geologiske Undersøgelse*, **161**, 39–44.

—— & HÅKANSSON, E. 1989. Stratigraphy and depositional history of the Upper Palaeozoic and Triassic sediments in the Wandel Sea Basin, central and eastern North Greenland. *Rapport Grønlands Geologiske Undersøgelse*, **143**, 21–45.

—— & —— 1991. Carboniferous and Permian history of the Wandel Sea Basin, North Greenland. *Grønlands Geologiske Undersøgelse Bulletin*, **160**, 141–151.

—— & WORSLEY, D. 1989. Late Palaeozoic sequence correlations, North Greenland, Svalbard and the Barents Shelf. *In:* COLLINSON, J. D. (ed.) *Correlation in Hydrocarbon Exploration.* Graham & Trotman, Norwegian Petroleum Society, London, 99–111.

———, LARSON, P. A., LARSSEN, G. B., MORK, A. & SIMONSEN, B. T. 1994. Depositional evolution of Lower Permian *Palaeoaplysina* build-ups, Kapp Duner Formation, Bjørnøya, Arctic Norway. *Sedimentary Geology*, **92**, 161–174.

———, HÅKANSSON, E., MADSEN, L., NILSSON, I., PIASECKI, S., PINARD, S. & RASMUSSEN, J. A. 1995. Stratigraphy and depositional evolution of the Upper Palaeozoic sedimentary succession in eastern Peary Land, North Greenland. *Rapport Grønlands Geologiske Undersøgelse*, in press.

VEEVERS, J. J. & POWELL, C. M. C. A. 1987. Late Paleozoic glacial episodes in Gondwana Land reflected in transgressive–regressive depositional sequences in Euramerica. *Geological Society of America Bulletin*, **98**, 475–487.

WARREN, J. K. 1982. The hydrological setting, occurrence and significance of laminated selenite and other gypsum fabrics in Late Quaternary salt lakes in Southern Australia. *Sedimentology*, **29**, 609–637.

WRIGHT, V. P. 1992. Speculations on the controls on cyclic peritidal carbonates: ice-house versus greenhouse eustatic controls. *Sedimentary Geology*, **76**, 1–5.

Index